Dingyü Xue

Differential Equation Solutions with MATLAB®

Also of Interest

Fractional-Order Control Systems, Fundamentals and Numerical Implementations
Dingyü Xue, 2017
ISBN 978-3-11-049999-5, e-ISBN (PDF) 978-3-11-049797-7,
e-ISBN (EPUB) 978-3-11-049719-9

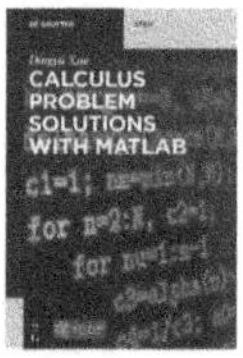

Calculus Problem Solutions with MATLAB®
Dingyü Xue, 2020
ISBN 978-3-11-066362-4, e-ISBN (PDF) 978-3-11-066697-7,
e-ISBN (EPUB) 978-3-11-066375-4

MATLAB® Programming, Mathematical Problem Solutions
Dingyü Xue, 2020
ISBN 978-3-11-066356-3, e-ISBN (PDF) 978-3-11-066695-3,
e-ISBN (EPUB) 978-3-11-066370-9

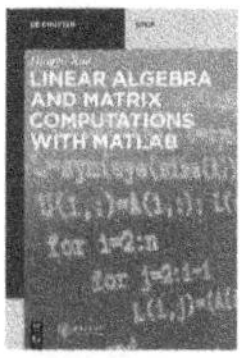

Linear Algebra and Matrix Computations with MATLAB®
Dingyü Xue, 2020
ISBN 978-3-11-066363-1, e-ISBN (PDF) 978-3-11-066699-1,
e-ISBN (EPUB) 978-3-11-066371-6

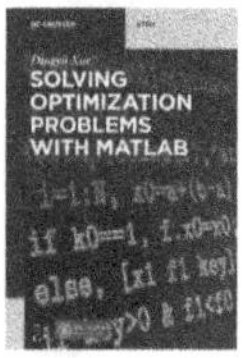

Solving Optimization Problems with MATLAB®
Dingyü Xue, 2020
ISBN 978-3-11-066364-8, e-ISBN (PDF) 978-3-11-066701-1,
e-ISBN (EPUB) 978-3-11-066369-3

Dingyü Xue

Differential Equation Solutions with MATLAB®

DE GRUYTER

清華大學出版社
TSINGHUA UNIVERSITY PRESS

Author
Prof. Dingyü Xue
School of Information Science and Engineering
Northeastern University
Wenhua Road 3rd Street
110819 Shenyang
China
xuedingyu@mail.neu.edu.cn

ISBN 978-3-11-067524-5
e-ISBN (PDF) 978-3-11-067525-2
e-ISBN (EPUB) 978-3-11-067531-3

Library of Congress Control Number: 2020931439

Bibliographic information published by the Deutsche Nationalbibliothek
The Deutsche Nationalbibliothek lists this publication in the Deutsche Nationalbibliografie; detailed bibliographic data are available on the Internet at http://dnb.dnb.de.

Cover image: Dingyü Xue
Typesetting: VTeX UAB, Lithuania
Printing and binding: CPI books GmbH, Leck

www.degruyter.com

Preface

Scientific computing is commonly and inevitably encountered in course learning, scientific research and engineering practice for each scientific and engineering student and researcher. For the students and researchers in the disciplines which are not pure mathematics, it is usually not a wise thing to learn thoroughly low-level details of related mathematical problems, and also it is not a simple thing to find solutions of complicated problems by hand. It is an effective way to tackle scientific problems, with high efficiency and in accurate and creative manner, with the most advanced computer tools. This method is especially useful in satisfying the needs for those in the area of science and engineering.

The author had made some effort towards this goal by addressing directly the solution methods for various branches in mathematics in a single book. Such a book, entitled "MATLAB based solutions to advanced applied mathematics", was published first in 2004 by Tsinghua University Press. Several new editions were published afterwards: in 2015, the second edition in English by CRC Press, and in 2018, the fourth edition in Chinese were published. Based on the latest Chinese edition, a brand new MOOC project was released in 2018,[1] and received significant attention. The number of registered students was around 14 000 in the first round of the MOOC course, and reached tens of thousands in later rounds. The textbook has been cited tens of thousands times by journal papers, books, and degree theses.

The author has over 30 years of extensive experience of using MATLAB in scientific research and education. Significant amount of materials and first-hand knowledge has been accumulated, which cannot be covered in a single book. A series entitled "Professor Xue Dingyü's Lecture Hall" of such works are scheduled with Tsinghua University Press, and the English editions are included in the DG STEM series with De Gruyter. These books are intended to provide systematic, extensive and deep explorations in scientific computing skills with the use of MATLAB and related tools. The author wants to express his sincere gratitude to his supervisor, Professor Derek Atherton of Sussex University, who first brought him into the paradise of MATLAB.

The MATLAB series is not a simple revision of the existing books. With decades of experience and material accumulation, the idea of "revisiting" is adopted in authoring the series, in contrast to other mathematics and other MATLAB-rich books. The viewpoint of an engineering professor is established and the focus in on solving various applied mathematical problems with tools. Many innovative skills and general-purpose solvers are provided to solve problems with MATLAB, which is not possible by any other existing solvers, so as to better illustrate the applications of computer tools in solving mathematical problems in every mathematics branch. It also helps the readers broaden their viewpoints in solving scientific computing, and even find

1 MOOC address: https://www.icourse163.org/learn/NEU-1002660001

https://doi.org/10.1515/9783110675252-201

innovative solutions by themselves to scientific computing which cannot be solved by any other existing methods.

The first title in the MATLAB series, “MATLAB Programming”, can be used as an entry-level textbook or reference book to MATLAB programming, so as to establish a solid foundation and deep understanding for the application of MATLAB in scientific computing. Each subsequent volume tries to cover a branch or topic in mathematical courses. Bearing in mind the “computational thinking” in authoring the series, deep understanding and explorations are made for each mathematics branch involved. These MATLAB books are suitable for the readers who have already learnt the related mathematical courses, and revisit the courses to learn how to solve the problems by using computer tools. It can also be used as a companion in synchronizing the learning of related mathematics courses, and viewing the course from a different angle, so that the readers may expand their knowledge in learning the related courses, so as to better learn, understand and practice the materials in the courses.

This book is the fifth one in the MATLAB series and fully devoted to the solutions of various differential equations, with extensive use of MATLAB. The analytical solutions of ordinary differential equations are studied first, followed by numerical solutions to initial value problems of various ordinary differential equations, including conventional and special equations, delay differential equations, and fractional differential equations. Some property analysis tasks are also covered in this book, and block diagram-based patterns to various differential equations are addressed. Discussions are also made to the numerical solution approaches to boundary value problems and partial differential equations.

At the time the books are published, the author wishes to express his sincere gratitude to his wife, Professor Yang Jun. Her love and selfless care over the decades provide the author immense power, which supports the authors’ academic research, teaching, and writing.

September 2019 Xue Dingyü

Contents

1 An introduction to differential equations

Equations are equalities containing unknown variables. Equations are often classified into algebraic and differential equations. In Volumes III and IV in the series, algebraic equations and their solutions were fully addressed. Algebraic equations are used to describe static relationships among the unknowns, while differential equations are used to describe dynamic relationships. Differential equations are the mathematical foundations of dynamic systems.

In Volume II, foundations and computations of functions and calculus were fully covered. Ordinary differential equations are used to describe relationships among unknowns, their derivatives and independent variables such as time t. Besides, relationships among the present and past values of the unknowns are also involved. Generally speaking, the description and solutions of differential equations are much more complicated than those of their algebraic counterparts.

In Section 1.1, examples are studied for the modeling of electric, mechanical and social systems. A method is described of how to establish differential equation models from phenomena. In Section 1.2, a brief history of differential equations is presented. In Section 1.3, the outline and brief introduction to the materials in this book are proposed.

1.1 Introduction to differential equation modeling

In scientific research, including social sciences, and even in daily life, a phenomenon or a system can be described by differential equations. In this section several simple examples are given to demonstrate the modeling of circuit and mechanical systems, and some examples of the models in social sciences are presented.

1.1.1 Modeling of an electric circuit

In electric circuit theory, it is known from Ohm's law that the relationship between the current and voltage of a resistor is static, $u(t) = Ri(t)$. That is, the voltage $u(t)$ can be computed from the current value of the current $i(t)$. Therefore for a resistive network, the system can be modeled by algebraic equations. The current in any branch and the voltage across any component can be modeled by static equations.

If two other types of electric components – capacitors and inductors – are involved, differential equations must be employed to describe the relationship between the instantaneous voltage and current.

https://doi.org/10.1515/9783110675252-001

(1) If there is a capacitor C in a circuit, the following differential equation is needed to describe the instantaneous voltage $u(t)$ and current $i(t)$:

$$i(t) = R\frac{\mathrm{d}u(t)}{\mathrm{d}t}. \tag{1.1.1}$$

(2) If there is an inductor element L in a circuit, the following differential equation can be used to describe the relationship between its instantaneous voltage $u(t)$ and current $i(t)$:

$$u(t) = C\frac{\mathrm{d}i(t)}{\mathrm{d}t}. \tag{1.1.2}$$

An example is given next to demonstrate the differential equation modeling procedures of a simple electric circuit.

Example 1.1. Consider the simple circuit in Figure 1.1. Establish the mathematical model of the circuit.

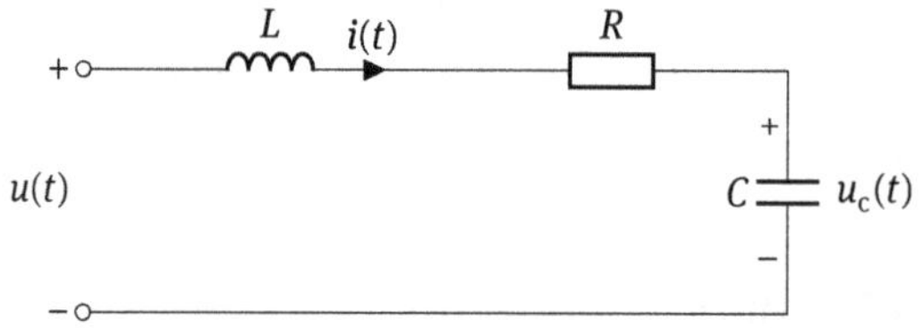

Figure 1.1: Serially connected resistor, capacitor and inductor.

Solutions. In the circuit, the signals of common interest are the supplied voltage $u(t)$ and the voltage $u_c(t)$ across the capacitor. The former is referred to as the input signal, while the latter, the output signal. The current $i(t)$ in the loop is the intermediate signal. It is known from the fundamental circuit theory that the voltage across the inductor L and the loop current $i(t)$ satisfy the following differential equation:

$$u_l(t) = L\frac{\mathrm{d}i(t)}{\mathrm{d}t}. \tag{1.1.3}$$

Therefore, from the well-known Kirchhoff's law, the equation between the input and output signals can be written as

$$u(t) = Ri(t) + L\frac{\mathrm{d}i(t)}{\mathrm{d}t} + u_c(t). \tag{1.1.4}$$

Besides, it is known from circuit theory that the voltage $u_c(t)$ across the capacitor and the loop current $i(t)$ satisfy the following differential equation:

$$i(t) = C\frac{\mathrm{d}u_c(t)}{\mathrm{d}t}. \tag{1.1.5}$$

Substituting (1.1.5) into (1.1.4), it is found that the entire differential equation for the circuit can be written as

$$u(t) = RC\frac{\mathrm{d}u_c(t)}{\mathrm{d}t} + LC\frac{\mathrm{d}^2u_c(t)}{\mathrm{d}t^2} + u_c(t). \tag{1.1.6}$$

Besides, if $u_c(t)$ and $i(t)$ are selected as the states of the circuit, the following state space model can be written; more state space models will be studied later in this book:

$$\begin{cases} \dfrac{\mathrm{d}u_c(t)}{\mathrm{d}t} = \dfrac{1}{C}i(t), \\ \dfrac{\mathrm{d}i(t)}{\mathrm{d}t} = \dfrac{1}{L}u(t) - \dfrac{R}{L}i(t) + \dfrac{1}{L}u_c(t). \end{cases} \tag{1.1.7}$$

It can be seen that differential equation models can be set up easily for simple circuits, with the essential knowledge of calculus. If there are more loops, high-order differential equations can be established. If there are nonlinear elements such as diodes and transistors, nonlinear differential equations for the circuits may be created. It can be seen that differential equations are the ubiquitous modeling tool in describing dynamic behaviors in electric circuits.

The commonly used modeling technique introduced here is for lumped parameter circuits. In real circuits, if there exists leakage resistance or other factor, ordinary differential equations are not adequate in modeling the circuit. Partial differential equations must be introduced to model distributed parameter circuits.

1.1.2 Modeling in mechanical systems

The Newton's second law in mechanics, which is usually learnt in high school, is mathematically represented as

$$F(t) = Ma(t) \tag{1.1.8}$$

where the mass M is a constant. For better describe the dynamic relationship among the variables, the external force and acceleration are both represented as functions of time t. They can be regarded as the instantaneous external force $F(t)$ and acceleration $a(t)$. It seems that they satisfy an algebraic relation.

In practice, the acceleration of the mass cannot be measured directly. The measurable variable is the instantaneous position $\hat{x}(t)$ of the mass. The instantaneous displacement can be defined as $x(t) = \hat{x}(t) - x(t_0)$. What kind of relationship is there between the displacement and the external force? Since acceleration is the second-order derivative of the displacement, it can be seen from Newton's second law that the following differential equation can be established:

$$F(t) = Mx''(t). \tag{1.1.9}$$

A simple modeling example will be given next to show how to write differential equation models for a mechanical system.

Example 1.2. Consider a simple mechanical system shown in Figure 1.2. If an external force $F(t)$ is applied, what is the mathematical model between the displacement $x(t)$ and the external force $F(t)$?

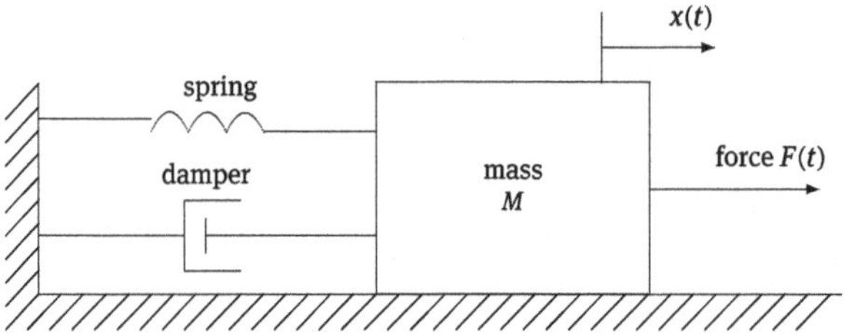

Figure 1.2: Illustration of a mechanical system.

Solutions. It is known from Newton's second law that the composite force is required, where we assume that the external force is in the positive direction. It is known from Hooke's law that the resisting force in the spring is proportional to the displacement, denoted as $-Kx(t)$; the other resisting force is from the damper, which is proportional to the speed of the mass, denoted as $-R\,\mathrm{d}x(t)/\mathrm{d}t$. Therefore it is not hard to establish the following differential equation for the mechanical system from Newton's second law:

$$F(t) - Kx(t) - R\frac{\mathrm{d}x(t)}{\mathrm{d}t} = M\frac{\mathrm{d}^2x(t)}{\mathrm{d}t^2}. \tag{1.1.10}$$

1.1.3 Models in social systems

In natural science engineering, differential equation models are everywhere, since all continuous dynamic behaviors must be described by differential equations. Apart from that, even in the fields such as social science, differential equations are often employed in describing phenomena and laws. Here several differential equation models in social systems are illustrated.

Example 1.3 (Population model). Malthus's population model was proposed by a British cleric and economist Thomas Robert Malthus (1766–1834), while he was analyzing the population in the Great Britain. If the population at time t can be expressed by $x(t)$, and the natural growth rate is a constant r, the Malthus population model can be expressed by a differential equation:

$$x'(t) = rx(t), \quad \text{with known } x(0) = x_0.$$

The analytical solution to the differential equation is $x(t) = x_0 e^{r(t-t_0)}$, which implies that the population is increasing as an exponential function. If t is large, the population may tend to infinity. This is not feasible, since when the total population reaches a certain size, the factors such a space and resources may restrict its growth such that it may not increase infinitely.

A Belgian mathematician Pierre-François Verhulst constructed a function $r(t)$, later known as the logistic function (also known as a sigmoid function, or S-shaped function), to replace the constant in Malthus model, and the population model becomes the following time-varying differential equation:

$$x'(t) = r(t)x(t), \quad \text{with known } x(0) = x_0.$$

Example 1.4. Richardson's arms race model[10] was proposed by a British mathematician and physicist Lewis Fry Richardson (1881–1953).[59] There are three basic premises in Richardson's arms race model. We assume that there are two nations X and Y. If one nation finds that the other is spending a huge amount of wealth on purchasing weapons, it may spend more money on purchasing weapons as well. Of course, the money spent may create economic and social burden. More money spent on purchasing weapons may inhibit future increases in spending; furthermore, there are grievances and ambitions relating to both cultures and national leaders that either encourage or discourage changes in military spending. Therefore Richardson proposed the arms race model using differential equations:

$$\begin{cases} x'(t) = ay(t) - mx(t) + g, \\ y'(t) = bx(t) - ny(t) + h. \end{cases} \tag{1.1.11}$$

Example 1.5 (Lotka–Volterra's predator–prey equations). Assume that in an enclosed region, there are two kinds of animals – foxes and rabbits. At the beginning there are certain numbers of them. If the number of foxes increases, more rabbits will be eaten and the number of rabbits may decrease. If the number of rabbits decreases, the number of foxes may also decrease since more foxes will starve to death, which in turn will increase the number of rabbits. This kind of phenomenon happens in a repeated manner. Assuming that at time t, the numbers of rabbits and foxes are respectively expressed as $x(t)$ and $y(t)$, the following differential equations can be used to describe the dynamic changes in the quantities of the two animals:

$$\begin{cases} x'(t) = \alpha x(t) - \beta x(t)y(t), \\ y'(t) = \delta x(t)y(t) - \gamma y(t) \end{cases}$$

where α, β, δ, and γ are positive real numbers. The model was first proposed by an American mathematician Alfred James Lotka (1880–1949), and a similar model was proposed later by an Italian mathematician Vito Volterra (1860–1940). This model is commonly used in the fields such as economics.

Example 1.6 (A model in epidemiology). An epidemic model was proposed by a British epidemiologist William Ogilvy Kermack (1898–1970) and a British mathematician Anderson Gray McKendrick (1876–1943) in 1927.[39] Later the variable names were substituted into the following form:[36]

$$\begin{cases} S'(t) = -\beta I(t)S(t), & S(0) = S_0, \\ I'(t) = \beta I(t)S(t) - \gamma I(t), & I(0) = I_0, \\ R'(t) = \gamma I(t), & R(0) = R_0. \end{cases}$$

Therefore the model is also called as the SIR model, where $S(t)$, $I(t)$ and $R(t)$ are respectively the numbers of susceptible (meaning, healthy persons), infective, and removed (removed by immunity, isolation or death) persons, with $S(t) + I(t) + R(t) = N(t)$, and $N(t)$ is the total population. For convenience of handling the problem, the numbers are normalized such that $S(t) + I(t) + R(t) = 1$. In the model parameters, β is the contact rate, which is the number of contacts by an infective per unit time, also known as infection rate; γ is the removal rate, indicating that the number of removed is proportional to the number of infected.

If the latent period is τ days, then the SIR model can be rewritten as the following delay differential–algebraic equations:[36]

$$\begin{cases} S'(t) = -\beta I(t)S(t), \\ I'(t) = \beta I(t)S(t) - \beta I(t-\tau)S(t-\tau), \\ S(t) + I(t) + R(t) = 1. \end{cases}$$

It can be seen that even though one is carrying research in social sciences, if differential equations are mastered as a tool in research, a new viewpoint may still be introduced and quantitative results may be obtained.

1.2 A brief history of differential equations

A British scientist Sir Isaac Newton (1643–1727) studied first differential equations. The power series method was adopted. For instance, in his work "*Methodus Fluxionum et Serierum Infinitarum*" published in 1671, Newton studied the differential equation[32]

$$y'(x) = 1 - 3x + y(x) + x^2 + xy(x). \tag{1.2.1}$$

With the power series method, $y(x)$ is expressed as a power series whose coefficients are unknown. It can be substituted into the equation, and letting the coefficients on both sides of the equation of equal power of x be the same, algebraic equation can be solved, and the solution he found was

$$y(x) = x - xx + \frac{1}{3}x^3 - \frac{1}{6}x^6 + \frac{1}{30}x^5 - \frac{1}{45}x^6$$

where xx was the notation used in Newton's works, and it should be x^2.

Of course, according to the standard in contemporary mathematics, this is far from a perfect solution. In later chapters, computers can be used in finding the analytical solution of the differential equation.

A German mathematician Gottfried Wilhelm Leibniz (1646–1716) solved in 1693 a differential equation from the field of geometry[32]

$$y'(x) = -\frac{y(x)}{\sqrt{a^2 - y(x)}}. \tag{1.2.2}$$

He introduced the integral method to find the solution of the differential equation, namely

$$x = \int_y^a \frac{\sqrt{a^2 - y^2}}{y} \mathrm{d}y = -\sqrt{a^2 - y^2} - a \ln \frac{a - \sqrt{a^2 - y^2}}{y}.$$

Leibniz also proposed the solution methods for differential equations with separable variables, which can be used to solve a large category of differential equations.

In fact, before Newton and Leibniz invented the calculus theory, an Italian physicist and astronomer Galileo Galilei (1564–1642) studied some problems which commonly required the basis of differential equations. Since calculus was not established at his time, he mostly used traditional geometric and algebraic tools.

A French mathematician Alexis Claude Clairaut (1713–1765) was interested in a particular field of application problems. He was the first to study implicit differential equations:[32]

$$y(x) - xy'(x) + f(y'(x)) = 0. \tag{1.2.3}$$

Swiss mathematicians, the Bernoulli brothers, namely Jacob Bernoulli (also known as James or Jacques, 1654–1705) and John Bernoulli (also Jean, 1667–1748) studied differential problems, and proposed "analytical" methods for a class of nonlinear differential equations, later named as Bernoulli equations.

In 1768, a Swiss mathematician Leonhard Euler (1707–1783) published his famous three volumes of works "*Institutionum calculi integralis* (Introduction to integral calculus), where many creative ideas are presented, including the well-known integrating factor method and total differential equation, and also the very first numerical method, later known as Euler's method.

Earlier research on differential equations concentrated on analytical solutions and properties of the equations. Since there was no support of tools such as computers, the numerical solution methods did not receive adequate attention.

A German mathematician Carl David Tolmé Runge (1856–1927) introduced more interpolation points to Euler's method, in order to increase the accuracy, and proposed in 1895 some solution algorithms, and extended the methods to the solutions of differential equation sets. In 1900, a German mathematician Karl Heun (1859–1929) revised Runge algorithm, by introducing more general methods based on Gaussian integral formulas. In 1901, a German mathematician Martin Wilhelm Kutta (1867–1944) extended the algorithms proposed by Runge and Heun, to a generalized high-order form, and proposed the fifth order algorithm. A French scholar A Huťa published in 1956 the sixth-order algorithm.[13] In 1963, a New Zealand mathematician John Charles Butcher (1933–) formulated more a general Runge–Kutta table, known as Butcher's tableau.[12] In a research report by NASA in 1968, Erwin Felhberg proposed the 7th and 8th order Runge–Kutta method formulas.[24] An Austrian mathematician Ernst Hairer (1949–) proposed the 10th order Runge–Kutta algorithm in 1978.[30] All these algorithms are uniformly named as Runge–Kutta methods.

The Runge–Kutta methods can be classified as one-step numerical methods. Apart from one-step methods, there are a family of linear multistep algorithms, such as Adams algorithms, named after a British mathematician and astronomer John Couch Adams (1819–1892). Compared with Runge–Kutta methods, Adams methods have longer history.[5] If the first few points of the solutions are known, Adams methods can be used to recursively find the numerical solutions of differential equations. There is an inevitable question for multistep methods. That is, how can the information of the first few points be found. If this question cannot be answered successfully, multistep algorithms cannot be used effectively.

The tools and computer mathematical languages are now all equipped with powerful differential equation solvers. Numerical and even analytical solutions of differential equations can be found easily with such tools. The eventual target of this book is to instruct the readers how to solve various differential equations with MATLAB.

1.3 Outline and main topics in the book

In this volume, we concentrate on how to use MATLAB to solve various differential equations directly.

In Chapter 1, simple differential equation modeling problems are illustrated for some practical applications.

In Chapter 2, analytical solutions for some differential equations are studied. Some fundamental knowledge on analytical solution methods are presented first, followed by the use of MATLAB in solving linear differential equations with constant coefficients. Then an introduction is made on the solutions of matrix differential equations and some special nonlinear differential equations.

From Chapter 3 onward, numerical solution methods are presented for various differential equations. In Chapter 3, and several subsequent chapters, initial value problems in ordinary differential equations are studied systematically. Some conventional numerical algorithms and their MATLAB implementations for first-order explicit differential equations are presented first. Validations of the numerical solutions to the differential equations are also made.

In Chapter 4, the task is to convert differential equations in different forms into the directly solvable first-order explicit ones. Single high-order differential equations and differential equation sets are considered. If the conversion to the standard form is successful, the methods in Chapter 3 can be called directly for solving differential equations.

In Chapter 5, some special forms of differential equations unsuitable for the previously discussed methods are explored, including stiff differential equations, implicit differential equations, and differential–algebraic equations. Also, numerical solution methods for switched differential equations and linear stochastic differential equations are discussed.

In Chapter 6, the general forms of delay differential equations are introduced, and then simple delay differential equations, equations with variable delays, and neutral-type delay differential equations are all studied.

In Chapter 7, properties and behaviors of differential equations are analyzed. Properties such as stability and periodicity of differential equations are studied systematically. Behaviors such as limit cycles, chaos, and bifurcations are studied. The concepts such as equilibrium points and linearization are also presented.

In Chapter 8, solutions of fractional-order differential equations are provided. Fundamental definitions and properties of fractional calculus are introduced first, followed by the introduction of linear and nonlinear fractional-order differential equations. In particular, high precision algorithms in numerical solution of fractional-order differential equations are proposed.

In Chapter 9, a block diagram-based solution approach for various differential equations is discussed, with the powerful tool Simulink. An introduction to Simulink is presented. Then the Simulink modeling and solution methods are demonstrated through examples of ordinary differential equations, differential–algebraic equations, switched differential equations, stochastic differential equations, and delay differential equations. General purpose modeling and simulation methods for various fractional-order differential equations are also discussed.

In Chapter 10, boundary value problems of ordinary differential equations are presented. The basic ideas of shooting methods are proposed first. Then a powerful MATLAB solver and its applications in boundary value problems are demonstrated for various problems. Finally, optimization-based boundary value problem solvers are proposed to tackle problems not solvable by the conventional methods.

Finally, in Chapter 11, numerical solutions to partial differential equations are introduced. The algorithm and MATLAB implementation of diffusion equations are discussed, and then with the use of MATLAB Partial Differential Equation Toolbox, partial differential equations in several special forms are discussed.

The focus of this book is to show how to solve various differential equations. If analytical solutions can be found, the solutions can be adopted, and the numerical solutions are implied. In many cases where analytical solutions are not available, numerical solutions should be found. The philosophy of this book is to instruct the readers on how to utilize the powerful facilities provided in MATLAB to study differential equations in a reliable manner.

1.4 Exercises

1.1 In the circuit shown in Figure 1.1, if the resistor is replaced by a serially connected resistor R and another capacitor C_1, write again the differential equation. If it is replaced by R and C_1 connected in parallel, how to rewrite the differential equation?

1.2 In the Malthus population model, if the world population at year t_0 is 6 billion, and it is known that the natural growth rate is 2 %, find the population after 10 years. What will be the population after 100 years?

1.3 Find the first 8 terms of the Newton's equation in (1.2.1), with the power series method.

1.4 If you have already acquired a certain knowledge on differential equations, solve the Clairaut's implicit differential equation in (1.2.3), if function $f(c) = 5(c^3 - c)/2$ is known.

2 Analytical solutions of ordinary differential equations

Differential equation modeling of physical phenomena was demonstrated in the previous section. If there exists a differential equation model for a certain dynamical system, or for a physical phenomenon, the next step to carry out is to solve the differential equation, so as to find the system response. Compared with the solutions of algebraic equations, differential equations are much more complicated. Normally, more skills and tactics in dedicated methods are needed when solving differential equations. The aim of the book is to extensively use computer tools to explore solutions of various differential equations.

In Section 2.1, major analytical methods are explored for the first-order differential equations. The low-level methods are introduced when addressing various types of first-order differential equations. The analytical solutions for some differential equations can be found in this way. In Section 2.2, analytical solutions of the second-order linear differential equations are explored, and various skills are needed to apply the low-level methods. Some commonly used special functions are introduced. In Section 2.3, Laplace transform-based methods are discussed for finding analytical solutions of differential equations. With Laplace transform tools, the linear differential equations with constant coefficients can be mapped into algebraic equations, and direct solution methods are explored when finding solutions of linear differential equations with nonzero initial values. In Section 2.4, Symbolic Math Toolbox is introduced to find solutions directly of some ordinary differential equations. With such a tool, complicated ordinary and time-varying differential equations can be solved directly. In Section 2.5, solution methods are explored for linear differential equations, state space equations, and Sylvester matrix equations. In Section 2.6, attempts are made to solve directly certain particular nonlinear differential equations. It is worth mentioning that only a few nonlinear differential equations have analytical solutions. For most nonlinear differential equations, analytical solutions do not exist. Numerical solutions will be discussed in later chapters.

Manual formulation methods are discussed in the first few sections in this chapter, which may be useful background knowledge for the readers to understand the low-level solution process of differential equations. If the readers are only interested in how to solve differential equations with computers, the materials here can be skipped, and one can start reading the materials from Section 2.4.

2.1 Analytical solutions of first-order differential equations

In this book, solution methods for the first-order differential equations are presented first. Several special cases are discussed, including simple equations, as well as homo-

https://doi.org/10.1515/9783110675252-002

geneous and inhomogeneous nonlinear differential equations. Solutions of nonlinear differential equations with separable variables are finally discussed.

2.1.1 Differential equation solvable by simple integrals

There are various formats for the first-order differential equations. Therefore the solutions of the differential equations become extremely complicated, since a tremendous amount of skills and tactics should be mastered before one is able to solve differential equations. For some differential equations, even if one is an expert who has mastered a significant number of methods and tactics, he/she may still not able to find the analytical solutions. In this section, some simple first-order differential equations are discussed, and solution patterns are explored.

Definition 2.1. The simplest form of a class of differential equations is

$$\frac{\mathrm{d}y(x)}{\mathrm{d}x} = f(x). \tag{2.1.1}$$

It can also be simply denoted as $y'(x) = f(x)$. The analytical solution of this differential equation is, in fact, the indefinite integral of $f(x)$, namely

$$y(x) = \int f(x)\mathrm{d}x + C. \tag{2.1.2}$$

The solution to such a differential equation is, in fact, the first-order indefinite integral of the given function. If the indefinite integral has an analytical expression, then there is an analytical solution for the differential equation. Otherwise, there is no analytical solution for the differential equation.

Example 2.1. Solve the following first-order differential equation:

$$\frac{\mathrm{d}}{\mathrm{d}x}y(x) = \frac{\cos x}{x^2 + 4x + 3} - \frac{\sin x(2x + 4)}{(x^2 + 4x + 3)^2}.$$

Solutions. Comparing the mathematical form given here with that in the definition, it is immediately found that the right-hand side is $f(x)$. Therefore, direct integration can be performed to find the primitive function which is the analytical solution of the differential equation with the following statements:

```
>> syms x
   f(x)=cos(x)/(x^2+4*x+3)-sin(x)*(2*x+4)/(x^2+4*x+3)^2;
   Y=simplify(int(f)) % evaluate integral and simplify the result
```

and the analytical solution can be found as $\sin x/(x^2 + 4x + 3)$. In fact, the actual solution is not the only primitive function of the indefinite integral, an arbitrary constant

C should be added to the result. Several different values of C are selected, and the family of differential equation solutions are shown in Figure 2.1. It can be seen that the solutions of the differential equation are simple translations of a certain solution.

```
>> fplot([Y,Y+1,Y+2,Y+3],[0,10])
```

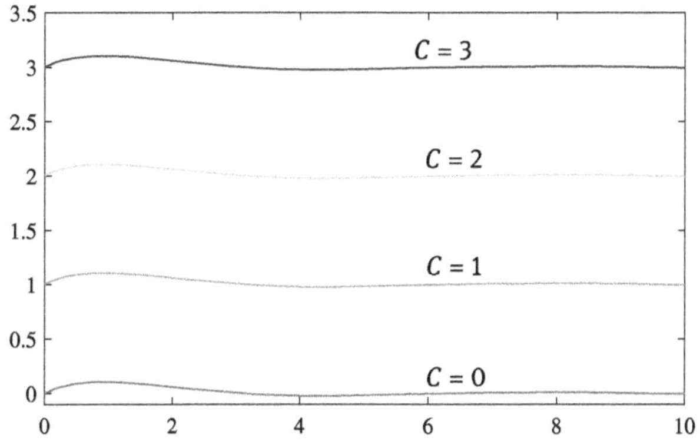

Figure 2.1: Solutions of the differential equation.

2.1.2 Homogeneous differential equations

The aforementioned differential equations directly solvable by integration are not genuine differential equations. Normally, a differential equation may also contain the function $y(x)$ itself. Such differential equations cannot be solved directly by merely using direct indefinite integral evaluation methods. In this section, simple first-order homogeneous linear differential equations are introduced, as well as their general solution methods.

Definition 2.2. The mathematical form of the first-order homogeneous linear differential equation is

$$\frac{\mathrm{d}y(x)}{\mathrm{d}x} + f(x)y(x) = 0. \tag{2.1.3}$$

With simple conversion, it is not hard to find the analytical solution of the differential equation through the following procedure:

$$\frac{\mathrm{d}y(x)}{y(x)} = -f(x)\mathrm{d}x. \tag{2.1.4}$$

It can then be found that

$$\ln y(x) = -\int f(x)\mathrm{d}x. \tag{2.1.5}$$

Finally, the analytical solution of the differential equation is

$$y(x) = Ce^{-\int f(x)\mathrm{d}x}. \tag{2.1.6}$$

For the first-order differential equations, there exists an undetermined coefficient C. If the exact value of the function $y(x)$ at a certain time is given, an algebraic equation solution technique is used to determine uniquely the value of C. For high-order differential equations, there may exist more undetermined coefficients. If some values of the function $y(x)$ and its first few derivatives are known, analytical solutions of such equations may be uniquely determined.

Example 2.2. Solve the differential equation $y'(x) = -xy(x)$.

Solutions. For this type of differential equation, with separable variables, the equation can be converted into the form of $y'(x)/y = -x$. Computing integrals of the left- and right-hand side of the equation

```
>> syms x y(x); L=int(diff(y)/y), R=int(-x)
```

the solution of the differential equation can be found as

$$\ln y(x) = \frac{-x^2}{2} + C_1 \Rightarrow y(x) = Ce^{-x^2/2}.$$

2.1.3 Inhomogeneous linear differential equations

If a term x is added to the right-hand side function $g(x)$, the original homogeneous differential equation is converted into an inhomogeneous one. In this section, an idea is explored when finding analytical solutions of the first-order inhomogeneous linear differential equation.

Definition 2.3. The mathematical form of a first-order inhomogeneous linear differential equation is

$$\frac{\mathrm{d}y(x)}{\mathrm{d}x} + f(x)y(x) = g(x). \tag{2.1.7}$$

It may be complicated with manual formulation methods. Some special tactics should be introduced. For instance, one can try to find an auxiliary function $\phi(x)$ such that

$$\phi(x)\left[\frac{\mathrm{d}y(x)}{\mathrm{d}x} + f(x)y(x)\right] = \frac{\mathrm{d}}{\mathrm{d}x}(\phi(x)y(x)). \tag{2.1.8}$$

Therefore the original differential equation can be converted into

$$\frac{\mathrm{d}}{\mathrm{d}x}[\phi(x)y(x)] = \phi(x)g(x). \tag{2.1.9}$$

Evaluating definite integrals of both sides of the equation over the integration range $[x_0, x]$, it is found that

$$\phi(x)y(x) - \phi(x_0)y(x_0) = \int_{x_0}^{x} \phi(t)g(t)\mathrm{d}t. \tag{2.1.10}$$

Finally, the analytical solution of the equation can be found as

$$y(x) = \frac{\phi(x_0)}{\phi(x)}y(x_0) + \frac{1}{\phi(x)}\int_{x_0}^{x} \phi(t)g(t)\mathrm{d}t. \tag{2.1.11}$$

It can be seen that the first key step in solving this equation is to find an auxiliary function $\phi(x)$. The method of finding an auxiliary function depends upon the expression of function $f(x)$. The following integral is needed in solving the problem:

$$\phi(x) = \exp\left(\int f(x)\mathrm{d}x\right). \tag{2.1.12}$$

Example 2.3. Find the analytical solution of the following first-order inhomogeneous differential equation:[38]

$$\frac{\mathrm{d}y(t)}{\mathrm{d}t} + \mathrm{e}^{\lambda t}y(t) = k\mathrm{e}^{\lambda t}, \quad y(0) = y_0.$$

Solutions. It can be seen from the standard form in Definition 2.3 that $f(t) = \mathrm{e}^{\lambda t}$, $g(t) = k\mathrm{e}^{\lambda t}$, and $x_0 = 0$. Therefore, the following MATLAB statements can be used to find directly the analytical solution of the differential equation:

```
>> syms t x k lam y0
   f(t)=exp(lam*t); g(t)=k*exp(lam*t); phi=exp(int(f))
   y=phi(0)*y0/phi+1/phi*int(phi(x)*g(x),x,0,t)
```

The analytical solution and auxiliary function are respectively

$$y(t) = k + (y_0 - k)\,\mathrm{e}^{(-\mathrm{e}^{\lambda t}-1)/\lambda} \text{ and } \phi(t) = \mathrm{e}^{\mathrm{e}^{\lambda t}/\lambda}.$$

Of course, the analytical solution algorithm is rather complicated to implement. No further comments and explanation are made about this method. Later, a more general and direct solution method with MATLAB will be illustrated to solve the problems directly.

2.1.4 Nonlinear differential equations with separable variables

Definition 2.4. A first-order nonlinear differential equation with separable variables can be written as

$$\frac{\mathrm{d}y(x)}{\mathrm{d}x} = g(y(x))f(x), \quad y(x_0) = y_0. \tag{2.1.13}$$

Theorem 2.1. *The analytical solution for the nonlinear differential equation with separable variables in Definition 2.4 is*

$$\int_{y(x_0)}^{y(x)} \frac{d\tau}{g(\tau)} = \int_{x_0}^{x} f(t)dt. \tag{2.1.14}$$

If the expressions of the two integrals in (2.1.14) can be obtained, the analytical solution of the original solution can be found. In fact, there is no analytical solution for most nonlinear differential equations, even those with separable variables may have no analytical solutions.

Example 2.4. Solve the following first-order nonlinear differential equation:

$$\frac{dy(t)}{dt} + 8y(t) + y^2(t) = -15, \quad y(0) = 0.$$

Solutions. It can be seen from the nonlinear differential equation that $g(y) = 15 + 8y + y^2, f(x) = 1$. The following statements can be used to solve the equation directly:

```
>> syms x t u y
   g(y)=15+8*y+y^2; f(t)=-sym(1);
   G=simplify(int(1/g(u),u,0,y)), F=int(f,0,t)
   eq=G==F, y=solve(eq,y)
```

where the left-hand side of (2.1.14), i. e., G, is $\operatorname{arctanh}(4) - \operatorname{arctanh}(y+4)$, while on the right-hand side, the solution is $-t$. Therefore an implicit equation can be constructed as $\operatorname{arctanh}(4) - \operatorname{arctanh}(y+4) = -t$. Solving the implicit equation, the solution to the nonlinear differential equation can be found as $y(t) = \tanh(t + \operatorname{arctanh}(4)) - 4$.

It can be seen by computing the integral of G that, if the manual method is used, $1/g(y)$ can be expressed in a partial fraction form. The result obtained is then different from the $\operatorname{arctanh}(\cdot)$ function obtained here. If similar results are expected, the format of G should be rewritten. For instance, use `rewrite()` function to modify its default display format.

```
>> G=rewrite(G,'exp'), F=int(f,0,t)
   eq=G==F, y=simplify(solve(eq,y)) %solve implicit equation
```

The implicit equation obtained is then $\ln(y+3)/2 - \ln(y+5)/2 - \ln 3/2 + \ln 5/2 = -t$. The analytical solution to the nonlinear differential equation is then written as $y(t) = 6/(5e^{2t} - 3) - 3$. The result obtained here is equivalent to that obtained earlier.

It can be seen that for an ordinary linear differential equation, the manual solution method is complicated and tedious. If integrals are involved in the solution process, the equations are usually not solvable. Special functions should be invented to describe the analytical solutions of the differential equations. The solutions of nonlinear differential equations are even more complicated, and most nonlinear differential equations do not have analytical solutions. In exploring analytical solutions of certain types of differential equations, mathematicians invented various special functions. Despite this, there is still a significant number of nonlinear differential equations which do not have analytical solutions. Therefore form Chapter 3 onward, we will concentrate on exploring differential equations with numerical methods using MATLAB.

2.2 Special functions and second-order differential equations

Compared with first-order differential equations, the solution process of second-order differential equations is much more difficult, if not impossible. In this section, the mathematical models of some commonly used second-order linear differential equations are explored, and “analytical solutions” are found, with the dedicated special functions.

Definition 2.5. The general form of a second-order linear differential equation is

$$a(x)\frac{\mathrm{d}^2y(x)}{\mathrm{d}x^2}+p(x)\frac{\mathrm{d}y(x)}{\mathrm{d}x}+q(x)y(x)=f(x) \tag{2.2.1}$$

where the coefficient functions $a(x)$, $p(x)$, $q(x)$ and $f(x)$ are treated as given.

With the variations of the coefficient functions, the solution methods for second-order linear differential equations are also different. In [58], more than 1 000 different coefficient combinations are discussed. Most of the cases are under the assumption that $f(x)=0$, that is, homogeneous equations are considered. For the linear differential equations with no analytical solutions, various special functions are invented by mathematicians to describe certain “analytical solutions” of the equations.

For a better understanding of certain analytical solutions, it is necessary to introduce first some commonly used special functions. They include gamma functions, hypergeometric functions, various Bessel functions, Legendre functions, and Airy functions. Some of the special functions are related to certain second-order linear differential equations. In this section, a simple introduction to these equations and special functions is presented.

2.2.1 Gamma function

Definition 2.6. The gamma function is defined by the following infinite integral:

$$\Gamma(z)=\int_0^\infty \mathrm{e}^{-t}t^{z-1}\mathrm{d}t. \tag{2.2.2}$$

Theorem 2.2. *An important property of the gamma function is*

$$\Gamma(z+1) = z\Gamma(z). \tag{2.2.3}$$

Theorem 2.3. *As a special case, for a nonnegative integer z, the factorial formula can be derived directly from* (2.2.3)

$$\Gamma(z+1) = z\Gamma(z) = z(z-1)\Gamma(z-1)\cdots = z!. \tag{2.2.4}$$

The gamma function can be regarded as an interpolation of the factorial, or it can be understood as an extension of the factorial to the noninteger z domain. If z is a negative integer, $\Gamma(z+1)$ explodes to $\pm\infty$. Function **y**=`gamma(`**x**`)` in MATLAB can be used to evaluate the gamma function directly. If **x** is a vector, the result **y** is also a vector. Besides, **x** can be a matrix or any other data type, and the result **y** is of the same data type as **x**. Note that function `gamma()` here can only handle real input arguments.

Example 2.5. Draw the gamma function curve in the interval $x \in (-5, 5)$.

Solutions. The following statements can be used to draw directly the gamma function curve, as shown in Figure 2.2.

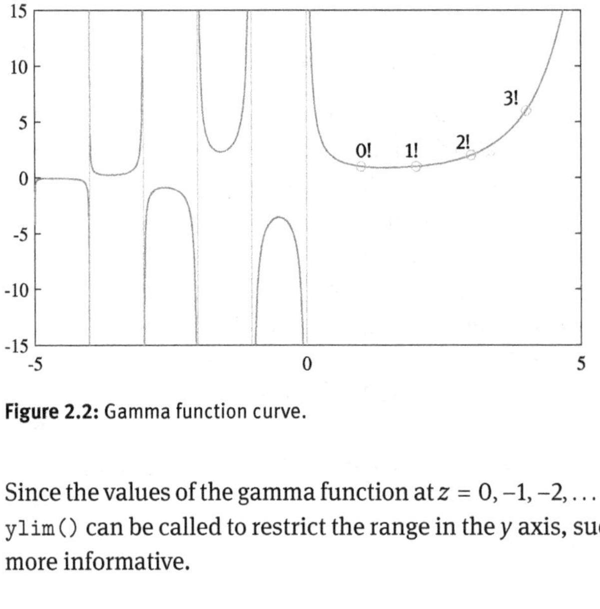

Figure 2.2: Gamma function curve.

Since the values of the gamma function at $z = 0, -1, -2, \ldots$ explode to infinity, function `ylim()` can be called to restrict the range in the y axis, such that the curve obtained is more informative.

```
>> a=-5:0.002:5; plot(a,gamma(a)), ylim([-15,15])
   hold on, v=[1:4]; plot(v,gamma(v),'o')
```

For some particular values of z, the gamma function values are

$$\Gamma\left(\frac{1}{2}\right) = \sqrt{\pi}, \quad \Gamma\left(\frac{3}{2}\right) = \frac{\sqrt{\pi}}{2}, \quad \Gamma\left(\frac{5}{2}\right) = \frac{3\sqrt{\pi}}{4}, \quad \Gamma\left(\frac{7}{2}\right) = \frac{15\sqrt{\pi}}{8}, \ldots \tag{2.2.5}$$

Example 2.6. Show (2.2.5) with MATLAB.

Solutions. The gamma function in (2.2.5) can be shown with the following MATLAB statements:

```
>> syms t z; Gam=int(exp(-t)*t^(z-1),t,0,inf);
   I1=subs(Gam,z,sym(1/2)), I2=subs(Gam,z,sym(3/2))
   I3=subs(Gam,z,sym(5/2)), I4=subs(Gam,z,sym(7/2))
```

Theorem 2.4. *Generally, if z is not an integer, the following formula is satisfied:*

$$\Gamma\left(z+\frac{1}{m}\right)=(2\pi)^{(m-1)/2}m^{1/2-mz}\,\Gamma(mz). \tag{2.2.6}$$

2.2.2 Hypergeometric functions

Hypergeometric functions are commonly used special functions, and they comprise also the foundation for some other special functions. In this section, definitions of hypergeometric functions are presented, and then computations of hypergeometric functions are addressed.

Definition 2.7. The general form of a hypergeometric function is[1]

$$\begin{aligned}&{}_p\mathrm{F}_q(a_1,\ldots,a_p;b_1,\ldots,b_q;z)\\&\quad=\frac{\Gamma(b_1)\cdots\Gamma(b_q)}{\Gamma(a_1)\cdots\Gamma(a_p)}\sum_{k=0}^{\infty}\frac{\Gamma(a_1+k)\cdots\Gamma(a_p+k)}{\Gamma(b_1+k)\cdots\Gamma(b_q+k)}\frac{z^k}{k!}\end{aligned} \tag{2.2.7}$$

where b_i cannot be nonpositive integers. If $p\leqslant q$, the function is convergent for all z; If $p=q+1$, the function is convergent when $|z|<1$; If $p>q+1$, the function is divergent for all z.

Function `hypergeom()` provided in the Symbolic Math Toolbox in MATLAB can be used to compute the hypergeometric function

$${}_p\mathrm{F}_q(a_1,\ldots,a_p;b_1,\ldots,b_q;z),$$

with the syntax

`y=hypergeom([a1,...,ap],[b1,...,bq],z)`

For instance, the hypergeometric function ${}_1\mathrm{F}_1(a,b;z)$ can be evaluated with `y=hypergeom(a,b,z)`, while function ${}_2\mathrm{F}_1(a,b;c;z)$ can be evaluated directly with `y=hypergeom([a,b],c,z)`.

Example 2.7. Draw the curve of ${}_2\mathrm{F}_1\,(1.5,-1.5;1/2;(1-\cos x)/2)$, the Gaussian hypergeometric function.

Solutions. The hypergeometric function ${}_2F_1(1.5, -1.5; 1/2; (1 - \cos x)/2)$ can be simplified as $\cos 1.5x$. If function `hypergeom()` is used, the curve for the hypergeometric function can be drawn, as shown in Figure 2.3. It can be seen that the curve coincides completely with that of $\cos 1.5x$.

```
>> syms x, y=hypergeom([1.5,-1.5],0.5,0.5*(1-cos(x)));
   fplot([cos(1.5*x),y],[-pi,pi])
```

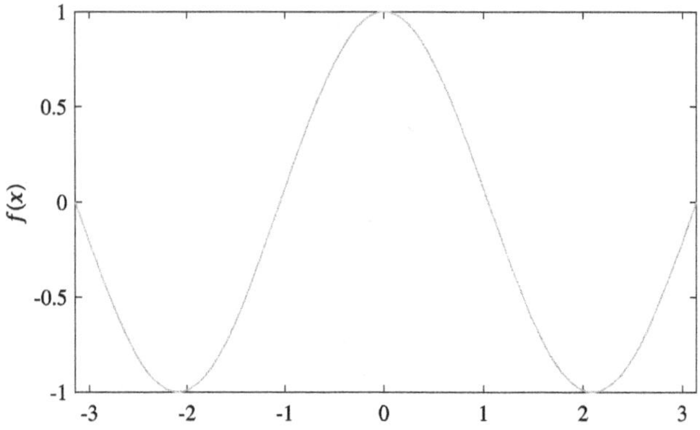

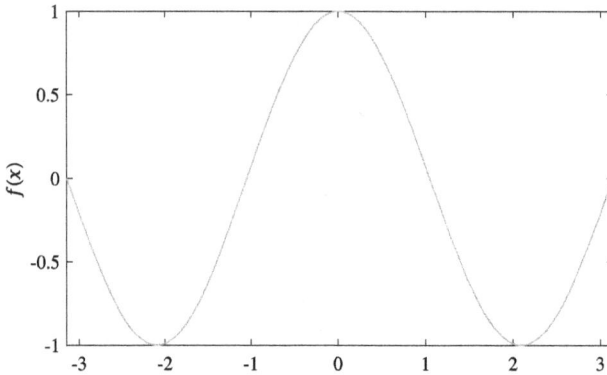

Figure 2.3: The hypergeometric function curve in Example 2.7.

Theorem 2.5. *As a special case, the hypergeometric function* ${}_2F_1(a, b; c; z)$ *is the solution of the following differential equation:*

$$z(1-z)\frac{d^2y(z)}{dz^2} + [c - (a+b+1)]\frac{dy(z)}{dz} - aby(z) = 0 \tag{2.2.8}$$

where we deliberately denote the independent variable by z*, indicating that it can also be a complex variable.*

2.2.3 Bessel differential equations

If the coefficients in the second-order differential equation in (2.2.1) have different forms, different forms of special differential equations can be constructed. For instance, the commonly used Bessel differential equations can be formulated. In this section, the mathematical form of Bessel differential equations is presented, and then various Bessel functions and their computations are summarized.

Definition 2.8. If the general form of a second-order homogeneous differential equation can be written as

$$x^2\frac{d^2y(x)}{dx^2} + x\frac{dy(x)}{dx} + (x^2 - \nu^2)y(x) = 0, \tag{2.2.9}$$

the function is referred to as a Bessel differential equation.

Theorem 2.6. *The general "analytical solution" of a νth-order Bessel differential equation can be written as*

$$y(x) = C_1 J_\nu(x) + C_2 J_{-\nu}(x) \tag{2.2.10}$$

where C_1 and C_2 are arbitrary constants, $J_\nu(x)$ is the νth order Bessel function of the first kind, defined as

$$J_\nu(t) = \sum_{m=0}^{\infty} (-1)^m \frac{t^{\nu+2m}}{2^{\nu+2m} m!\,\Gamma(\nu+m+1)} \tag{2.2.11}$$

where $\Gamma(\cdot)$ is the gamma function.

In other words, since genuine analytical solutions to the original differential equations do not exist, relevant special functions were invented by mathematicians. It is the same as in the case where the indefinite integral of e^{-t^2} does not exist, where mathematicians invented the special function $\operatorname{erf}(\cdot)$ to study its analytical properties.

Theorem 2.7. *If $\nu = n$ is a positive integer, the Bessel function of the first kind has the following properties:*

$$J_\nu(x) = (-1)^\nu J_{-\nu}(x), \tag{2.2.12}$$

$$\frac{J_\nu(x)}{dx} = \frac{\nu}{x} J_\nu(x) - J_{\nu+1}(x), \tag{2.2.13}$$

$$\int_0^x t^\nu J_{\nu-1}(t)\,dt = x^\nu J_\nu(x). \tag{2.2.14}$$

Theorem 2.8. *The νth order Bessel function of the first kind is a special case of the hypergeometric function, namely*

$$J_\nu(x) = \frac{(x/2)^\nu}{\Gamma(\nu+1)}\, {}_0F_1\left(\nu+1; -\frac{x^2}{4}\right). \tag{2.2.15}$$

Bessel functions are named after a German astronomer and mathematician Friedrich Wilhelm Bessel (1784–1846). Bessel functions were first proposed by a Swiss mathematician Daniel Bernoulli (1700–1782), and further extended by Bessel. Apart from Bessel functions of the first kind defined earlier, Bessel functions of the second and third kind are also used.

Definition 2.9. The νth order Bessel function of the second kind is defined as

$$Y_\nu(t) = \frac{J_\nu(t)\cos \nu t - J_{-\nu}(t)}{\sin \nu t}. \tag{2.2.16}$$

Bessel function of the second kind is also known as Neumann function, named after a German mathematician Carl Gottfried Neumann (1832–1925).

Definition 2.10. Bessel functions of the third kind are defined as

$$H_\nu^{(1)}(x) = J_\nu(x) + jY_\nu(x), \quad H_\nu^{(2)}(x) = J_\nu(x) - jY_\nu(x). \tag{2.2.17}$$

Functions `besselj()`, `bessely()` and `besselh()` are provided in MATLAB to compute respectively Bessel functions of the first, second, and third kind. The syntaxes of the functions are

```
y=besselj(ν,x);  y=bessely(ν,x);  y=besselh(ν,k,x)
```

where $k = 1$ and $k = 2$ correspond to the two cases of Bessel functions of the third kind.

Example 2.8. Draw Bessel functions of the first kind of orders $\nu = 0, 1, -1, 2$.

Solutions. Several orders are selected like these, and the following statements can be used to draw directly the Bessel function curves, as shown in Figure 2.4. It can be seen that $J_1(x) = -J_{-1}(x)$.

```
>> x=-5:0.1:5;
   y1=besselj(0,x); y2=besselj(1,x); y3=besselj(-1,x);
   y4=besselj(2,x); plot(x,y1,x,y2,x,y3,x,y4)
```

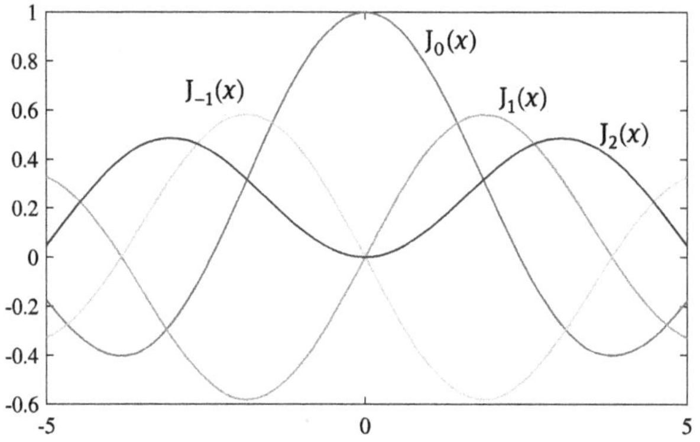

Figure 2.4: Bessel function curves of different orders.

2.2.4 Legendre differential equations and functions

Legendre functions satisfy Legendre differential equations. The definitions of Legendre differential equations and functions are given next, followed by their computation methods. Legendre differential equations and functions are named after a French mathematician Adrien-Marie Legendre (1752–1833).

Definition 2.11. The mathematical form of Legendre differential equations is

$$(1-z^2)\frac{\mathrm{d}^2y(z)}{\mathrm{d}z^2}-2z\frac{\mathrm{d}y(z)}{\mathrm{d}z}+\left[\lambda(1+\lambda)-\frac{\mu^2}{1-z^2}\right]y(z)=0 \tag{2.2.18}$$

where the complex quantities λ and μ are referred to respectively as the degrees and orders of the Legendre function.

Definition 2.12. The mathematical form of the Legendre function can also be depicted by the hypergeometric function

$$\mathrm{P}_\lambda^\mu(z)=\frac{1}{\Gamma(1-\lambda)}\left(\frac{1+z}{1-z}\right)^{\mu/2}{}_2\mathrm{F}_1\left(-\lambda,\lambda+1;1-\mu;\frac{1-z}{2}\right) \tag{2.2.19}$$

with the range of convergence being $|1-z|<2$, where $\mu=0,1,2,\dots,\lambda$.

Example 2.9. Draw the Legendre function curve for $\mu=3$.

Solutions. If $\mu=3$, the following MATLAB commands can be used to draw the four different Legendre functions, as shown in Figure 2.5. Of course, the readers may consider the following commands to draw different Legendre functions:

```
>> x=-1:0.04:1; Y=legendre(3,x); plot(x,Y)
```

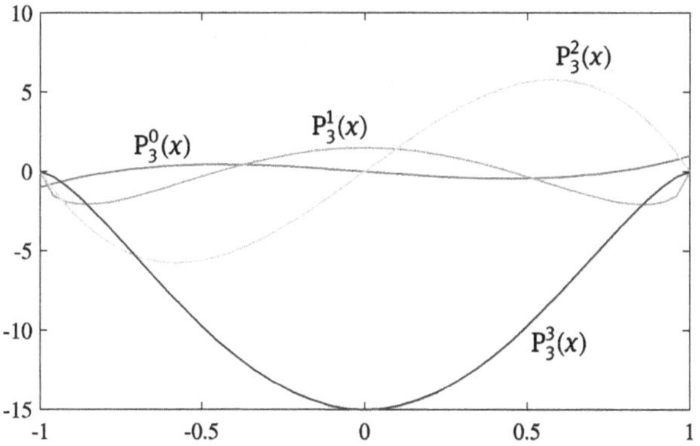

Figure 2.5: Legendre function curves for different orders.

2.2.5 Airy functions

Airy functions are a class of special functions named after a British mathematician and astronomer George Biddell Airy (1801–1892). These functions are defined as the analytical solutions of Airy differential equations. In this section the general form of Airy differential equations is presented first, followed by the computation and sketching of Airy functions.

Definition 2.13. Airy differential equations are simple and defined as

$$\frac{\mathrm{d}^2 y(z)}{\mathrm{d}z^2} - zy(z) = 0 \tag{2.2.20}$$

where the independent variable z can be a complex quantity.

Definition 2.14. Airy functions of the first and second kind are defined as

$$\mathrm{Ai}(z) = \frac{1}{\pi}\int_0^\infty \cos\left(\frac{t^3}{3} + zt\right)\mathrm{d}t, \tag{2.2.21}$$

$$\mathrm{Bi}(z) = \frac{1}{\pi}\int_0^\infty \left[\exp\left(-\frac{t^3}{3} + zt\right) + \sin\left(\frac{t^3}{3} + zt\right)\right]\mathrm{d}t. \tag{2.2.22}$$

Function `z=airy(k,z)` is provided in MATLAB to evaluate Airy function values of complex matrices, where $k = 0$ and 1 compute respectively $\mathrm{Ai}(\boldsymbol{z})$ and its first-order derivative, while $k = 2, 3$ compute respectively $\mathrm{Bi}(\boldsymbol{z})$ and its first-order derivative. If one is expecting only $\mathrm{Ai}(\boldsymbol{z})$ function, the argument k can be neglected. Of course, if the input argument is a real matrix, the output is the real matrix of the same size.

Example 2.10. It is known that Airy function is a complex one. Draw the contours of the real part of the Airy function in the complex plane.

Solutions. The complex mesh grids matrices $\boldsymbol{z} = \boldsymbol{x} + \mathrm{j}\boldsymbol{y}$ can be generated first. The Airy function matrix can be obtained. In order to increase the resolution of the complex matrix, the thresholds of ±4 can be used, and the contours obtained are shown in Figure 2.6.

```
>> [x,y]=meshgrid(-4:0.1:4); z=x+y*1i; y1=airy(z);
   y1=real(y1); ii=find(y1>=4); y1(ii)=4;
   ii=find(y1<=-4); y1(ii)=-4; contourf(x,y,y1,80)
```

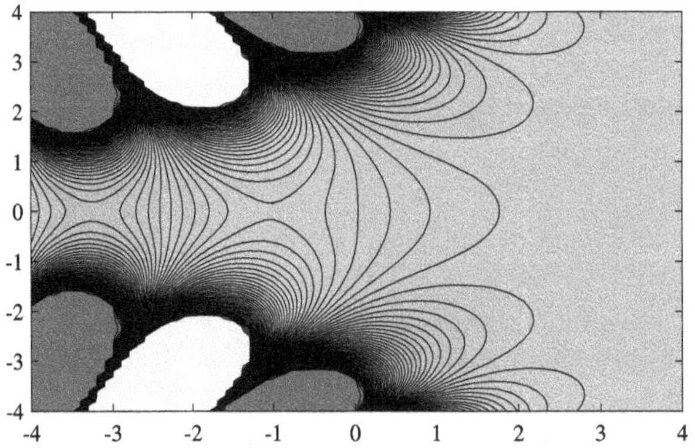

Figure 2.6: Contours of the real part of Airy function.

It was pointed out earlier that, with different forms of the coefficients, different forms of second-order linear differential equations can be constructed, such that manual solutions of the differential equations become extremely difficult. Second-order nonlinear differential equations or third- and higher order linear or nonlinear differential equations are even more complicated to handle. The methods studied here are only applicable to extremely simple differential equations.

Dedicated tools for analytically solving differential equations will be illustrated later, such that second- or even higher order differential equations can be solved directly. If there are no analytical solutions, numerical solutions will be studied next for differential equations.

2.3 Solutions of linear differential equations with constant coefficients

High-order nonlinear differential equations with variable coefficients are hard to be solved analytically, since almost all such equations do not have analytical solutions. In this section, we concentrate on discussing the solutions of linear differential equations with constant coefficients. The mathematical form of such differential equations is given first, and Laplace transforms and Laplace transform-based solution methods for linear differential equations with constant coefficients are presented. Finally, analytical solutions for differential equations with nonzero initial values are formulated.

2.3.1 Mathematical modeling of linear constant-coefficient differential equations

The mathematical form of linear differential equations with constant coefficients is given here, and in the subsequent sections, different methods will be used to explore the solutions of such differential equations.

Definition 2.15. Linear differential equations with constant coefficients are mathematically described as

$$\begin{aligned}\frac{\mathrm{d}^n y(t)}{\mathrm{d}t^n} &+ a_1\frac{\mathrm{d}^{n-1} y(t)}{\mathrm{d}t^{n-1}} + a_2\frac{\mathrm{d}^{n-2} y(t)}{\mathrm{d}t^{n-2}} + \cdots + a_{n-1}\frac{\mathrm{d}y(t)}{\mathrm{d}t} + a_n y(t)\\ &= b_1\frac{\mathrm{d}^m u(t)}{\mathrm{d}t^m} + b_2\frac{\mathrm{d}^{m-1} u(t)}{\mathrm{d}t^{m-1}} + \cdots + b_m\frac{\mathrm{d}u(t)}{\mathrm{d}t} + b_{m+1}u(t)\end{aligned} \tag{2.3.1}$$

where a_i and b_i are constants, and n is the order of the differential equation.

If the differential equation is regarded as a dynamical system model, signal $y(t)$ is the output of the system, while $u(t)$ is the input signal of the system. Differential equations are used to describe a dynamical relationship between the input and output signals.

2.3.2 Laplace transform-based solutions

In many books on linear differential equations, it is suggested to convert linear differential equations into algebraic equations to find their characteristic roots and then construct the analytical solutions. Unfortunately, very few explain explicitly why algebraic equations should be solved first. Here Laplace transform-based solution is presented.

Definition 2.16. Laplace transform of a time domain function $f(t)$ is defined as

$$\mathscr{L}[f(t)] = \int_0^{\infty} f(t)\mathrm{e}^{-st}\mathrm{d}t = F(s) \tag{2.3.2}$$

where $\mathscr{L}[f(t)]$ is the short-hand notation of Laplace transform.

Definition 2.17. If Laplace transform expression $F(s)$ of a time domain function is known, the following formula can be used to invert its inverse Laplace transform

$$f(t) = \mathscr{L}^{-1}[F(s)] = \frac{1}{2\pi\mathrm{j}}\int_{\sigma-\mathrm{j}\infty}^{\sigma+\mathrm{j}\infty} F(s)\mathrm{e}^{st}\mathrm{d}s \tag{2.3.3}$$

where σ is larger than the real part of any singularity of $F(s)$.

Theorem 2.9. *The differentiation property of Laplace transform, being the most important property when solving differential equations, reads:*

$$\mathscr{L}\left[\frac{\mathrm{d}f(t)}{\mathrm{d}t}\right] = sF(s) - f(0^+). \tag{2.3.4}$$

More generally, Laplace transform of the nth order derivative can be evaluated from the following equation:

$$\mathscr{L}\left[\frac{\mathrm{d}^n f(t)}{\mathrm{d}t^n}\right] = s^n F(s) - s^{n-1}f(0^+) - s^{n-2}f'(0^+) - \cdots - f^{(n-1)}(0^+). \tag{2.3.5}$$

Assuming that the initial values of $f(t)$ and its derivatives are all zero, (2.3.5) *can be simplified as*

$$\mathscr{L}\left[\frac{\mathrm{d}^n f(t)}{\mathrm{d}t^n}\right] = s^n F(s). \tag{2.3.6}$$

With the property (2.3.6), it is known that for zero initial value problems, $\mathscr{L}[\mathrm{d}^m y(t)/\mathrm{d}t^m] = s^m \mathscr{L}[y(t)]$, the following polynomial equation can be derived

$$s^n + a_1 s^{n-1} + a_2 s^{n-2} + \cdots + a_{n-1}s + a_n = 0 \tag{2.3.7}$$

Theorem 2.10. *Assuming that the characteristic roots s_i of the algebraic equation can be found, and they are distinct, the general form of the analytical solution of the original differential equation can be constructed as*

$$y(t) = C_1 e^{r_1 t} + C_2 e^{r_2 t} + \cdots + C_n e^{r_n t} + y(t) \tag{2.3.8}$$

where C_i are undetermined coefficients and $y(t)$ is a particular solution under $u(t)$ signal input.

Similarly, analytical solutions can also be formulated if some of the s_i are repeated roots.

Theorem 2.11. *If there is an m-tuple repeated root r_i, the corresponding terms can be constructed for the general solution:*

$$(C_1 + C_2 t + \cdots + C_m t^{m-1}) e^{-r_i t}. \tag{2.3.9}$$

It can be seen from the two theorems that, if the roots of the characteristic equation can be found, the general form of the analytical solution of a differential equation can be manually composed.

It can be seen from the algebraic equation in (2.3.7) that, according to the well-known Abel–Ruffini theorem, all the roots of the characteristic equations of degree 4 or lower can be found analytically, which implies that the low-order differential equations have analytical solutions. High-order differential equations may also have quasi-analytical solutions, since high precision solutions to the polynomial equations can be found such that quasi-analytical solutions of high order linear differential equations with constant coefficients can be composed.

Example 2.11. Solve the following homogeneous linear differential equation with constant coefficients:

$$y^{(4)}(t) + 10y'''(t) + 35y''(t) + 50y'(t) + 24y(t) = 0.$$

Solutions. If the initial values of the output signal $y(t)$ and its derivatives are all zero, it can be found according to the Laplace transform property that

$$\begin{aligned} s^4 Y(s) + 10s^3 Y(s) + 35s^2 Y(s) + 50sY(s) + 24Y(s) \\ = (s^4 + 10s^3 + 35s^2 + 50s + 24)Y(s) = 0. \end{aligned}$$

Solving the characteristic equation $s^4 + 10s^3 + 35s^2 + 50s + 24 = 0$, it is found that the roots are $s_1 = -1$, $s_2 = -2$, $s_3 = -3$, and $s_4 = -4$. Therefore, the general solution of the differential equation can be written as

$$y(t) = C_1 e^{-t} + C_2 e^{-2t} + C_3 e^{-3t} + C_4 e^{-4t}.$$

Example 2.12. Find the general solution of the following homogeneous differential equation:

$$y^{(5)}(t) + 12y^{(4)}(t) + 57y'''(t) + 134y''(t) + 156y'(t) + 72y(t) = 0$$

under the assumption that the initial values of $y(t)$ and its derivatives are all zero.

Solutions. The characteristic equation $s^5 + 12s^4 + 57s^3 + 134s^2 + 156s + 72 = 0$ can be manually created. It can be expressed and solved with the following MATLAB commands:

```
>> syms s;                                  % declare symbolic variables
   F=s^5+12*s^4+57*s^3+134*s^2+156*s+72;   % describe the polynomial
   r=solve(F)                               % solve the polynomial equation
```

It is the found that the characteristic roots are $s_1 = -2$, $s_2 = -2$, $s_3 = -2$, $s_4 = -3$, and $s_5 = -3$. There are repeated roots, and the general solution to the differential equation can be manually constructed as

$$y(t) = (C_1 + C_2 + C_3t^2)e^{-2t} + (C_4 + C_5t)e^{-3t}.$$

2.3.3 Solutions of inhomogeneous differential equations

In the previous section, the differential equations were assumed to be homogeneous. A linear combination of $y(t)$ and its derivatives can be found to compose the analytical solutions. If the differential equations are inhomogeneous, Laplace transform can be taken of both sides and then Laplace transform of the output signal can be found. For this expression, inverse Laplace transform can be taken to find analytical solutions of the differential equations.

If computer tools are not used, partial fraction expansion should be employed such that Laplace transform can be expressed in simple form, from which the analytical solution of the corresponding differential equation can be composed. Equipped with the powerful tools such as MATLAB, the partial fraction expansion manipulation can be bypassed, function `ilaplace()` can be called directly to find the analytical solution of the differential equations. Examples are given next to show solution procedures.

Example 2.13. Consider again the differential equation in Example 2.12. If it is given as the following inhomogeneous one, find the general solution of the differential equation:

$$y^{(5)}(t) + 12y^{(4)}(t) + 57y'''(t) + 134y''(t) + 156y'(t) + 72y(t) = e^{-t}\sin t.$$

Solutions. The following statements can be used in finding the Laplace transform and for computing the Laplace transform expression of the output signal, such that the analytical solution of the differential equation can be constructed. The input and output signals can be obtained, as shown in Figure 2.7. In this particular example, since the output signal is too small, 20 times of $y(t)$ is drawn instead.

```
>> syms s t; u(t)=exp(-t)*sin(t);
   F=s^5+12*s^4+57*s^3+134*s^2+156*s+72;
   U=laplace(u); Y=U/F; y=ilaplace(Y), y=simplify(y)
   fplot([u,20*y],[0,10])  % draw the curves of the input and output signals
```

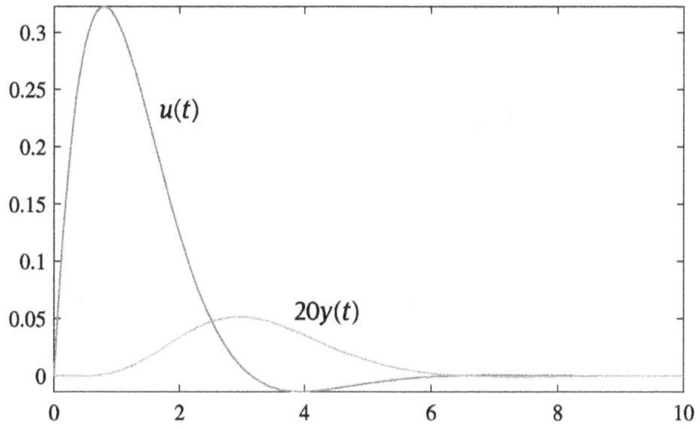

Figure 2.7: The curves of the input and output signals.

From the above commands, the analytical solution can be found as

$$y(t) = \frac{3}{4}e^{-2t} - \frac{19}{25}e^{-3t} - \frac{1}{2}te^{-2t} - \frac{1}{5}te^{-3t} + \frac{1}{100}e^{-t}(\cos t - 7\sin t) + \frac{1}{4}t^2e^{-2t}.$$

Collecting the like terms manually, the analytical solution can further be simplified as

$$y(t) = \frac{1}{4}(3 - 2t + t^2)e^{-2t} - \frac{1}{25}(18 + 5t)e^{-3t} + \frac{1}{100}e^{-t}(\cos t - 7\sin t).$$

2.3.4 Solutions of differential equations with nonzero initial values

If the initial values of the differential equations are not zero, the expression in (2.3.6) does not hold. Initial conditions should be taken into account. From (2.3.5), the differential equation can be converted into an algebraic equation. With inverse Laplace transform, the analytical solution of the original differential equation can be found. Examples will be given next to show solution procedures.

Example 2.14. Assuming that the input signal is $u(t) = \mathrm{e}^{-5t}\cos(2t+1)+5$, and the initial values are $y(0)=3$, $y'(0)=2$, $y''(0)=y'''(0)=y^{(4)}(0)=0$, find the analytical solution of the following differential equation:

$$y^{(4)}(t)+10y'''(t)+35y''(t)+50y'(t)+24y(t)=u(t).$$

Solutions. For simplicity, the following statements can be used to write directly the Laplace transform on the left, where part of the results (`eqn`) represents the coefficients of $Y(s)$, the other part (`eqn1`) is the initial value related expression. In this case, the expression $Y(s)$ can be found, for which function `ilaplace()` can be called to find the analytical solution of the differential equation, and the output can also be drawn as in Figure 2.8.

```
>> syms t s; y0=3; y1=2; u=exp(-5*t)*cos(2*t+1)+5;
   eqn=s^5+12*s^4+57*s^3+134*s^2+156*s+72;
   eqn1=(s^4*y0+s^3*y1)+12*(s^3*y0+s^2*y1)+...
          57*(s^2*y0+s*y1)+134*(s*y0+y1)+156*y0;
   Y(s)=(laplace(u)+eqn1)/eqn; y=ilaplace(Y);
   y=simplify(y), fplot(y,[0,10]) %simplify and draw the solution
```

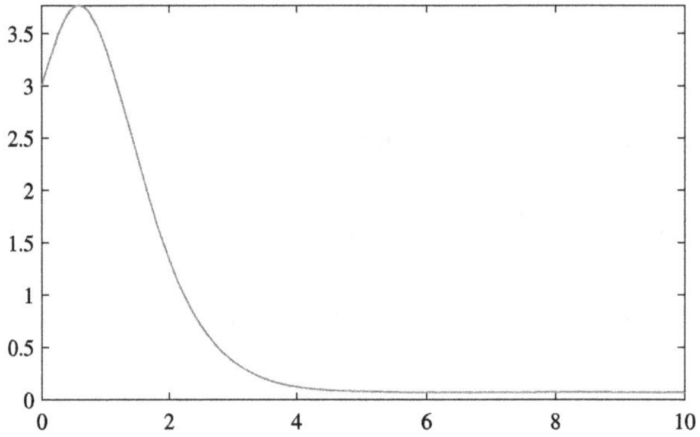

Figure 2.8: Output of the differential equation.

It can be found that the analytical solution obtained is

$$\begin{aligned} y(t) = {} & \mathrm{e}^{-2t}\left(\frac{1642\cos 1}{2197}-\frac{1372\sin 1}{2197}+\frac{4203}{8}\right)-\mathrm{e}^{-3t}\left(\frac{3\cos 1}{4}-\frac{5\sin 1}{8}+\frac{4702}{9}\right)\\ & -t\mathrm{e}^{-3t}\left(\frac{\cos 1}{4}-\frac{\sin 1}{4}+\frac{475}{3}\right)-t\mathrm{e}^{-2t}\left(\frac{83\cos 1}{169}-\frac{64\sin 1}{169}+\frac{1425}{4}\right) \end{aligned}$$

$$+\mathrm{e}^{-5t}\left(\cos 2t-\frac{\sin 2t\,(9\cos 1+46\sin 1)}{46\cos 1-9\sin 1}\right)\left(\frac{23\cos 1}{8\,788}-\frac{9\sin 1}{17\,576}\right)$$
$$+\frac{t^2\mathrm{e}^{-2t}}{52}\left(\frac{3\cos 1}{13}-\frac{2\sin 1}{13}+\frac{451}{2}\right)+\frac{5}{72}.$$

Example 2.15. Solve the following high-order linear differential equation set with constant coefficients:

$$\begin{cases} x''(t)-x(t)+y(t)+z(t)=0,\\ x(t)+y''(t)-y(t)+z(t)=0,\\ x(t)+y(t)+z''(t)-z(t)=0 \end{cases}$$

where $x(0)=1$, $y(0)=z(0)=x'(0)=y'(0)=z'(0)=0$.

Solutions. Consider the initial values given. If x'' in the equations is replaced by $s^2X(s)-sx(0)=s^2X(s)-s$, x, y and z are replaced by $X(s)$, $Y(s)$ and $Z(s)$, respectively, with y'' and z'' replaced by $s^2Y(s)$ and $s^2Z(s)$, respectively, the following algebraic equations can be formulated:

$$\begin{cases} s^2X(s)-s-X(s)+Y(s)+Z(s)=0,\\ X(s)+s^2Y(s)-Y(s)+Z(s)=0,\\ X(s)+Y(s)+s^2Z(s)-Z(s)=0. \end{cases}$$

With the following statements, the algebraic equations can then be solved.

```
>> syms s X Y Z
   eqns=[s^2*X-s-X+Y+Z==0, X+s^2*Y-Y+Z==0, X+Y+s^2*Z-Z==0];
   sol=solve(eqns,[X,Y,Z]); sol.X, sol.Y, sol.Z
```

and the following Laplace expressions can be found:

$$X(s)=\frac{s^3}{(s^2+1)(s^2-2)},\quad Y(s)=Z(s)=-\frac{s}{(s^2+1)(s^2-2)}.$$

Taking inverse Laplace transform of the solutions

```
>> x=ilaplace(sol.X), y=ilaplace(sol.Y)
   z=ilaplace(sol.Z); fplot([x,y],[0,0.5])
```

the analytical solution of the original differential equation can be found as

$$x(t)=\frac{2}{3}\cosh\sqrt{2}t+\frac{1}{3}\cos t,\quad y(t)=z(t)=\frac{1}{3}\cos t-\frac{1}{3}\cosh\sqrt{2}t,$$

from which the signals $x(t)$ and $y(t)$ can be drawn as shown in Figure 2.9.

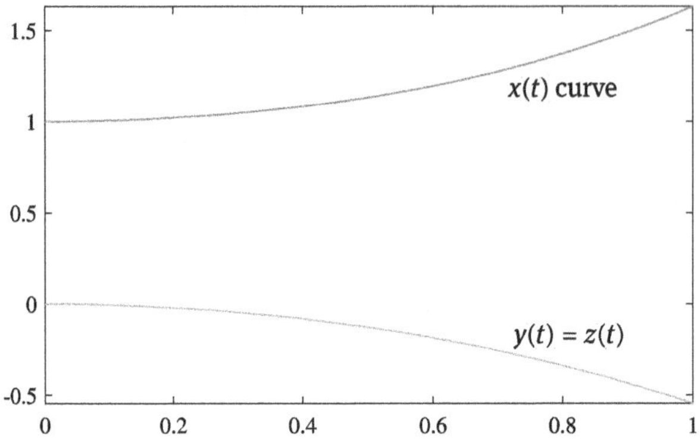

Figure 2.9: The solution of the differential equations.

2.4 Analytical solutions of ordinary differential equations

In this section, the ordinary differential equation solver provided in MATLAB Symbolic Math Toolbox is presented to directly find the analytical solutions of linear differential equations with constant coefficients.

Low-level differential equation solving methods were explored in the previous sections, and it can be seen that the methods have apparent limitations. They are not suitable for handling complicated problems. Also different differential equations may need different solution methods, which is not a good way of handling problems in practice. A unified method is needed to find the analytical solutions in a formal way. In this section, the dedicated solver in the Symbolic Math Toolbox is presented, aiming at finding analytical solutions of ordinary differential equations.

2.4.1 Analytical solutions of simple differential equations

Function `dsolve()` is provided in MATLAB Symbolic Math Toolbox. It can be used in finding analytical solutions of ordinary differential equations. With such a tool, different complicated differential equations can be handled directly. If a certain differential equation is shown to have no analytical solutions, numerical methods should be used instead to solve it.

The syntaxes of `dsolve()` function are

```
sols=dsolve(f1,f2,...,fm), %default independent variable t
y=dsolve(f1,f2,...,fm,varlist), %indicate the independent variable
[x,y,...]=dsolve(f1,f2,...,fm,varlist), %indicate return variables list
```

In the earlier versions of MATLAB, f_i could be expressed by both symbolic expressions and strings, to describe the equations and known conditions. In the current and subsequent versions, less support is provided for string expressions. The focus is on symbolic expression description and solutions.

Example 2.16. Solve the differential equation in (1.2.1) with computer tools.

Solutions. With the powerful computer tools, the problem which could not be fully studied by Newton, can be solved easily and in a simple and unified way

```
>> syms x y(x)      % declare symbolic variable and function
   y1=dsolve(diff(y)==1-3*x+y+x^2+x*y)
```

The analytical solution found is

$$y_1(x) = Ce^{x(x+2)/2} - x + 3\sqrt{2\pi}\,\mathrm{erf}\left(\sqrt{2}\left(\frac{x}{2}+\frac{1}{2}\right)\right)\mathrm{e}^{(x+1)^2/2} + 4$$

where C is an arbitrary constant, and erf(·) is a special function. Note that in describing the equation, the equality sign should be expressed as ==, otherwise there may be error messages. In fact, the solution obtained by Newton is only a particular solution when $y(0) = 0$. Taylor series expansion can also be written for the solution obtained above.

```
>> y2=dsolve(diff(y)==1-3*x+y+x^2+x*y,'y(0)==0');
   y2=expand(taylor(y2,'Order',10))
```

It can be seen that the last 6 terms are identical to the Newton's result:

$$y_2(x) \approx -\frac{x^9}{9072} - \frac{13x^8}{5040} + \frac{x^7}{630} - \frac{x^6}{45} + \frac{x^5}{30} - \frac{x^4}{6} + \frac{x^3}{3} - x^2 + x.$$

Example 2.17. Solve the differential equation in (1.2.2).

Solutions. The following MATLAB commands can be used to try to solve the original differential equation. Unfortunately, it is prompted that there is no analytical solution.

```
>> syms x y(x); syms a positive
   y1=dsolve(diff(y)==-y/sqrt(a^2-y^2)) % directly solve differential equation
```

Why is there no analytical solution found, if Leibniz found one? According to Leibniz's solution, y was used as the independent variable, and the explicit solution of x was found. It is not possible to find the analytical solution in the form of $y = f(x)$. According to Leibniz's method, if the fact that $\mathrm{d}y/\mathrm{d}x = 1/(\mathrm{d}x/\mathrm{d}y)$ is used, the following statements can be employed to solve the differential equation:

```
>> syms y x(y);
   x1=dsolve(1/diff(x)==-y/sqrt(a^2-y^2)) % analytical solution
```

from which the analytical solution is found as

$$x_1(y) = C - a\ln\left(\sqrt{\frac{a^2}{y^2}-1} - a\sqrt{\frac{1}{y^2}}\right) - \sqrt{a^2-y^2}.$$

It appears that Leibniz obtained only a particular solution.

Example 2.18. Solve the first-order differential equation in Example 2.4 directly by using the unified solver.

Solutions. If function `dsolve()` is called, the symbolic variable x and symbolic function $y(x)$ should be declared first. A symbolic expression can be used to describe the differential equation directly. Then the solver can be called to solve it. Note that when describing the differential equation, since the right side is 0, `==0` can be omitted from the symbolic expression. The following statements can be written to describe and solve the differential equation:

```
>> syms x y(x)
   eqn=diff(y)+8*y+y^2==-15; y=dsolve(eqn, y(0)==0)
   y=simplify(y)   % solve and simplify the result
```

It is found that the analytical solution is $y(x) = -(15e^{2x}-15)/(5e^{2x}-3)$. In fact, although the solution obtained here is apparently different from that obtained in Example 2.4, they are equivalent.

In the earlier versions, strings could be used to describe the differential equations and initial values. For instance, the following statements could be used and identical results were found:

```
>> y=dsolve('Dy+8*y+y^2=-15','y(0)=0'); y=simplify(y)
```

In a string description, the short-hand notation `D3y` can be used to describe the third-order derivative of y. In this book, string descriptions of differential equations are not recommended.

Example 2.19. Solve the following linear differential equation set:

$$\begin{cases} y_1'(t) = y_2(t), \\ y_2'(t) = -\lambda y_1(t) \end{cases}$$

where the initial values are $y_1(0) = a$, $y_2(0) = b$, and $\lambda > 0$.

Solutions. A positive symbolic variable λ should be declared first. The following commands can be written to describe the differential equations and given initial values, and then the solution can be found:

```
>> syms t a b y1(t) y2(t); syms lam positive
   [y1,y2]=dsolve(diff(y1)==y2,diff(y2)==-lam*y1,...
                  y1(0)==a,y2(0)==b) % differential equation solution
```

The analytical solution of the differential equation set obtained is

$$y_1(t) = a\cos\sqrt{\lambda}\,t + \frac{b}{\sqrt{\lambda}}\sin\sqrt{\lambda}\,t, \quad y_2(t) = b\cos\sqrt{\lambda}\,t - a\sqrt{\lambda}\sin\sqrt{\lambda}\,t.$$

2.4.2 Analytical solutions of high-order linear differential equations with constant coefficients

The function `dsolve()` discussed earlier can be applied directly in solving high-order linear differential equations with constant coefficients, or any other form of complicated differential equations. The differential equations and given conditions should be expressed directly so that the function can be called to find the analytical solutions. In the earlier versions, strings could be used to describe the differential equations, while in a future version, the string description might not be supported.

To concisely describe the differential equation and given conditions, intermediate variables can be defined to record the derivatives of $y(t)$. The definition of the intermediate variables will be demonstrated through examples.

Example 2.20. Solve again the differential equation in Example 2.14 with the unified solver and compare the results.

Solutions. Since function $y(x)$ and the initial values of several derivatives are needed, some intermediate variables should be defined to describe the derivatives of $y(x)$. For instance, variable `d3y` can be used to describe the intermediate function $y'''(x)$. This kind of notation is recommended such that the description of the differential equations become more concise and easy to understand. The following commands can be used to directly solve the differential equation. The result obtained is exactly the same as that obtained in Example 2.14. It can be seen that the statements used here are more concise and easy to validate. The final plotting command yields the same curve as that shown in Figure 2.8.

```
>> syms t y(t); u=exp(-5*t)*cos(2*t+1)+5;
   d1y=diff(y); d2y=diff(y,2); d3y=diff(y,3); d4y=diff(y,4);
   y=dsolve(diff(d4y)+12*d4y+57*d3y+134*d2y+156*d1y+72*y==u,...
            y(0)==3,d1y(0)==2,d2y(0)==0,d3y(0)==0,d4y(0)==0);
   y=simplify(y), fplot(y,[0,10]) % solution and its plot
```

Example 2.21. Assuming that the input signal is $u(t) = \mathrm{e}^{-5t}\cos(2t+1)+5$, find the general solution of the following differential equation:

$$y^{(4)}(t)+10y'''(t)+35y''(t)+50y'(t)+24y(t)=5u''(t)+4u'(t)+2u(t).$$

Solutions. Since no initial values of $y(t)$ are involved in this example, intermediate variables may be declared as in the case of Example 2.20, or they may be ignored.

```
>> syms t y(t); u(t)=exp(-5*t)*cos(2*t+1)+5;
   d1y=diff(y); d2y=diff(y,2); d3y=diff(y,3); d4y=diff(y,4);
   y=dsolve(d4y+10*d3y+35*d2y+50*d1y+24*y==...
            5*diff(u,t,2)+4*diff(u,t)+2*u);
   y=simplify(y) %solve the differential equation and simplify the solution
```

With the above statements, the general solution of the differential equation can be found as

$$\begin{aligned}y(t)=&\frac{5}{12}-\frac{343}{520}\mathrm{e}^{-5t}\cos(2t+1)-\frac{547}{520}\mathrm{e}^{-5t}\sin(2t+1)\\&+C_1\mathrm{e}^{-4t}+C_2\mathrm{e}^{-3t}+C_3\mathrm{e}^{-2t}+C_4\mathrm{e}^{-t}\end{aligned}$$

where C_i are arbitrary constants. If the initial or boundary values are known, they can be used to automatically solve the algebraic equations so as to find the values of C_i. The idea is the same as that discussed in calculus courses.

To validate the general solution found for the differential equation, it can be substituted back into the differential equation, and sometimes function `simplify()` can be called to evaluate the error. If the error is zero, the solution is validated. For this example, it can be seen that the error is zero, which means that the general solution satisfies the original differential equation, no matter what are the values of C_i.

```
>> err=diff(y,4)+10*diff(y,3)+35*diff(y,2)+50*diff(y)+24*y-...
      (5*diff(u,t,2)+4*diff(u,t)+2*u)   %solution validation
```

Example 2.22. In the differential equations studied so far, only real poles appear. The Symbolic Math Toolbox applies to the cases where there are also complex poles. Assume that a differential equation is given by

$$y^{(5)}(t)+5y^{(4)}(t)+12y'''(t)+16y''(t)+12y'(t)+4y(t)=u'(t)+3u(t).$$

Assuming that the input signal is sinusoidal, $u(t)=\sin t$, and $y(0)=y'(0)=y''(0)=y'''(0)=y^{(4)}(0)=0$, find the analytical solution of the differential equation.

Solutions. With the following commands, the analytical solution of the original differential equation can be found. Some intermediate variables should be defined first, and the equation can be expressed and solved. The result is shown graphically in Figure 2.10.

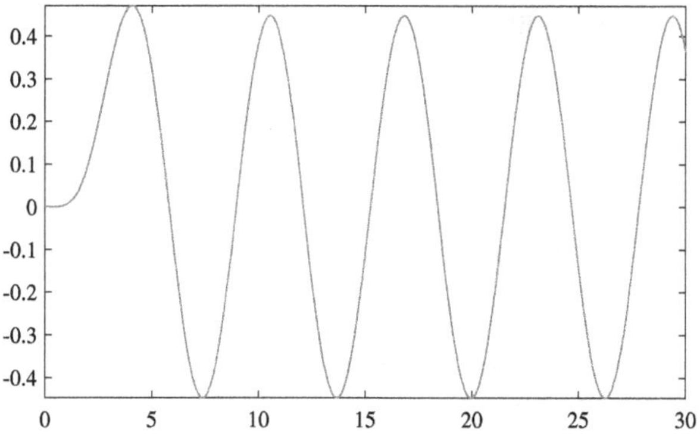

Figure 2.10: The nearly sustained oscillation curve.

```
>> syms t y(t); u(t)=sin(t);
   d1y=diff(y); d2y=diff(d1y); d3y=diff(d2y); d4y=diff(d3y);
   y=dsolve(diff(d4y)+5*d4y+12*d3y+16*d2y+12*d1y+4*y==...
                diff(u)+3*u,...
             y(0)==0,d1y(0)==0,d2y(0)==0,d3y(0)==0,d4y(0)==0);
   y=simplify(y), fplot(y,[0,30])  %solve equation and draw plot
```

The mathematical form of the analytical solution is

$$y(t) = \mathrm{e}^{-t} - \frac{1}{5}\cos t - \frac{2}{5}\sin t - \frac{4}{5}\mathrm{e}^{-t}\cos t + \frac{11}{10}\mathrm{e}^{-t}\sin t - \frac{1}{2}t\mathrm{e}^{-t}\cos t,$$

or it can be modified manually as

$$y(t) = \mathrm{e}^{-t} - \frac{1}{5}\cos t - \frac{2}{5}\sin t - \left(\frac{4}{5} + \frac{t}{2}\right)\mathrm{e}^{-t}\cos t + \frac{11}{10}\mathrm{e}^{-t}\sin t.$$

It can be seen from the curve that when t is relatively large, the curve is almost sustained oscillation. This is because the first two terms are sine and cosine functions which are sustained oscillating ones, while the other terms will vanish when t is large.

Example 2.23. Solve again the linear differential equation set in Example 2.15.

Solutions. The following commands can be used directly to solve the differential equation set and the results are identical to those obtained in Example 2.15:

```
>> syms t x(t) y(t) z(t)
   d1x=diff(x); d1y=diff(y); d1z=diff(z);
   [x,y,z]=dsolve(diff(x,2)-x+y+z==0,x+diff(y,2)-y+z==0,...
                  x+y+diff(z,2)-z==0, x(0)==1, d1x(0)==0,...
                  y(0)==0, d1y(0)==0,z(0)==0, d1z(0)==0);
   x=simplify(x), y=simplify(y), z=simplify(z)
```

Example 2.24. Solve the following linear differential equation set

$$\begin{cases} x''(t) + 2x'(t) = x(t) + 2y(t) - \mathrm{e}^{-t}, \\ y'(t) = 4x(t) + 3y(t) + 4\mathrm{e}^{-t}. \end{cases}$$

Solutions. The linear differential equation can also be solved directly with `dsolve()` function. The following MATLAB commands can be used to solve the differential equation:

```
>> syms t x(t) y(t) %declare symbolic variable and function
   [x,y]=dsolve(diff(x,2)+2*diff(x)==x+2*y-exp(-t),...
                 diff(y)==4*x+3*y+4*exp(-t)) %solve differential equation
```

The result obtained is

$$\begin{cases} x(t) = -6t\mathrm{e}^{-t} + C_1\mathrm{e}^{-t} + C_2\mathrm{e}^{(1+\sqrt{6})t} + C_3\mathrm{e}^{-(-1+\sqrt{6})t}, \\ y(t) = 6t\mathrm{e}^{-t} - C_1\mathrm{e}^{-t} + 2(2+\sqrt{6})C_2\mathrm{e}^{(1+\sqrt{6})t} + 2(2-\sqrt{6})C_3\mathrm{e}^{-(-1+\sqrt{6})t} + \mathrm{e}^{-t}/2. \end{cases}$$

2.4.3 Analytical solutions of linear time-varying differential equations

The direct solution method was demonstrated for linear differential equations with constant coefficients. If the coefficients are functions of the independent variable, and the variable is regarded as time, the differential equation is known as a time-varying one. Time-varying differential equations can also be explored with the solver `dsolve()`. Several examples are given next to demonstrate direct solutions of time-varying differential equations.

Example 2.25. Solve the following second-order time-varying differential equation:

$$y''(x) + ay'(x) + (bx + c)y(x) = 0.$$

Solutions. The differential equation can be expressed directly with a symbolic expression and solved with the following statements:

```
>> syms x a b c y(x)
   y=dsolve(diff(y,2)+a*diff(y)+(b*x+c)*y==0)
```

The analytical solution of the differential equation can be obtained as follows, where Airy special functions are involved:

$$y(x) = \frac{C_1}{\sqrt{\mathrm{e}^{ax}}}\mathrm{Ai}\left(-\frac{-a^2+4c+4bx}{4b(1/b)^{1/3}}\right) + \frac{C_2}{\sqrt{\mathrm{e}^{ax}}}\mathrm{Bi}\left(-\frac{-a^2+4c+4bx}{4b(1/b)^{1/3}}\right).$$

Note that for this particular problem, the following string expression format is not suitable:

```
>> y=dsolve('D2y+a*Dy+(b*x+c)*y')
```

This is because the notation D2y, by default, describes the second-order derivative of y with respect to t, not x. Therefore, the above command yields the wrong result. In order to solve it correctly, an 'x' option should be appended to the function call. Since in the future versions string descriptions might not be supported, this format is not recommended in the book.

Example 2.26. Find the analytical solution to the time-varying differential equation

$$(2x+3)^3y'''(x)+3(2x+3)y'(x)-6y(x)=0.$$

Solutions. For this particular time-varying differential equation, the solver dsolve() can be used directly to solve the equation

```
>> syms x y(x)
   y=dsolve((2*x+3)^3*diff(y,3)+3*(2*x+3)*diff(y)-6*y==0);
   y=simplify(y) % differential equation solution and simplification
```

The analytical solution obtained is as follows:

$$y(x)=C_1\left(x+\frac{3}{2}\right)-2C_2\sqrt{x+\frac{3}{2}}+C_3\left(2x+6\right)\sqrt{x+\frac{3}{2}}.$$

When it is substituted back into the original equation, the error is zero.

Example 2.27. Solve the following time-varying differential equation:

$$x^2(2x-1)y'''(x)-(4x-3)xy''(x)-2xy'(x)+2y(x)=0.$$

Solutions. If the independent variable x is regarded as time, the coefficients are all functions of time. Such differential equations can be regarded as time-varying differential equations. For this particular problem, since no initial values are provided, there is no need to define intermediate variables. The differential equation can be expressed directly, then with function dsolve() the solution of the time-varying differential equation can be found.

```
>> syms x y(x)
   y=dsolve(x^2*(2*x-1)*diff(y,3)+...
         (4*x-3)*x*diff(y,2)-2*x*diff(y)+2*y==0);
   simplify(y)                    % solution and simplification
   simplify(x^2*(2*x-1)*diff(y,3)+(4*x-3)*x*diff(y,2)...
         -2*x*diff(y)+2*y) % solution validation
```

The analytical solution obtained is as follows. When it is substituted back to the original equation, the error is zero. In fact, since C_i are free variables, $2C_1-C_3$ can be combined into a new free constant C_0 such that the result can further be simplified into

$$y(x) = -\frac{1}{16x}(2C_1 - C_3 + 8C_3x + 32C_2x^2 + 8C_3x^2 \ln x).$$

2.4.4 Solutions of time-varying differential equation sets

Linear time-varying differential equation sets can also be handled directly with the solver `dsolve()` in the Symbolic Math Toolbox. It is fine if the analytical solutions can be found, and it is also normal that the analytical solutions do not exist. In the following chapters, numerical solutions of various differential equations will be explored. The analytical solution is demonstrated through the following example.

Example 2.28. Find the analytical solutions for the following linear time-varying differential equation:

$$\begin{cases} \mathrm{d}^2y(x)/\mathrm{d}x^2 + 2y(x) + 4z(x) = \mathrm{e}^x, \\ \mathrm{d}^2z(x)/\mathrm{d}x^2 - y(x) - 3z(x) = -x. \end{cases}$$

Solutions. Before the solution process, the variables y and z should be declared as functions of x. The following statements can be used to solve the linear differential equation set:

```
>> syms x y(x) z(x)
   [y1,z1]=dsolve(diff(y,2)+2*y+4*z==exp(x),...diff(z,2)-y-3*z==-x)
   y1=simplify(y1),
   z1=simplify(z1) % differential equation solution and simplification
   simplify(diff(y1,x,2)+2*y1+4*z1-exp(x))
   simplify(diff(z1,x,2)-y1-3*z1+x) % solution validation
```

The analytical solution obtained is as follows:

$$\begin{cases} y_1(x) = \mathrm{e}^x - 2x - 4C_1\cos x - 4C_2\sin x - \dfrac{\sqrt{2}C_3\mathrm{e}^{\sqrt{2}x}}{2} + \dfrac{\sqrt{2}C_4\mathrm{e}^{-\sqrt{2}x}}{2}, \\ z_1(x) = x - \dfrac{\mathrm{e}^x}{2} + C_1\cos x + C_2\sin x + \dfrac{\sqrt{2}}{2}C_3\mathrm{e}^{\sqrt{2}x} - \dfrac{\sqrt{2}}{2}C_4\mathrm{e}^{-\sqrt{2}x}. \end{cases}$$

It can be validated that the general solution satisfies the original equations.

2.4.5 Solutions of boundary value problems

Function dsolve() can also be used to solve boundary value problems directly. Examples will be shown next to demonstrate how to solve boundary value problems with computers.

Example 2.29. Consider the differential equation in Example 2.20. Assume that the boundary values are known as $y(0) = 1/2$, $y'(0) = 1$, $y(2\pi) = 0$ and $y'(2\pi) = 0$. Solve the differential equation.

Solutions. Let us review the solution process in Section 2.3, where Laplace transform was involved. In Laplace transforms, only initial values are needed. While in this example, the values of $y(x)$ at different time instances are expected. Therefore such a problem cannot be handled with Laplace transform method. It is not necessary to worry about these trivial things, if the solver dsolve() is used. Therefore the unified framework of differential equation solution procedures introduced here can be adopted.

```
>> syms t y(t); u(t)=exp(-5*t)*cos(2*t+1)+5;
   d1y=diff(y); d2y=diff(y,2); d3y=diff(y,3); d4y=diff(y,4);
   y=dsolve(d4y+10*d3y+35*d2y+50*d1y+24*y==...
                  5*diff(u,t,2)+4*diff(u,t)+2*u,...
            y(0)==1,d1y(0)==0,y(2*pi)==0,d1y(2*pi)==0);
   y=simplify(y), fplot(y,[0,10]) %solution and graphical display
```

The analytical solution of the equation can be found, however, it is too complicated to display the result here. The solution curve is shown in Figure 2.11.

In fact, function dsolve() can also be used to directly solve the problems of complicated initial value problems. If the conditions are mutually independent, and the

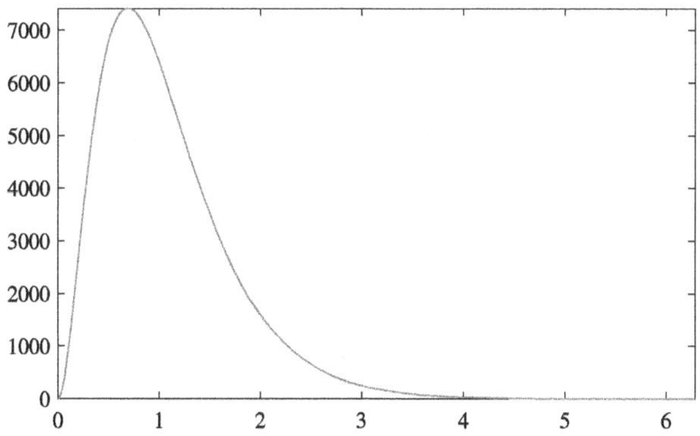

Figure 2.11: Solution of the differential equation.

number of conditions is the same as that of undetermined coefficients, the function can be called to uniquely find the analytical solution of the original differential equations.

Example 2.30. Consider again the differential equation in Example 2.20, and assume that the given conditions are $y(0) = 1/2$, $y'(\pi) = 0$ and $y''(2\pi) = y'(2\pi) = 0$. Solve again the differential equation.

Solutions. Let us review the Laplace transform method studied in Example 2.3. Since only initial values can be accepted in Laplace transform method, while in this example, the values of $y(x)$ at different time instances are needed, Laplace transform cannot be used in solving this kind of problem, while the solver `dsolve()` can be called directly.

```
>> syms t y(t); u(t)=exp(-5*t)*cos(2*t+1)+5;
   d1y=diff(y); d2y=diff(y,2); d3y=diff(y,3); d4y=diff(y,4);
   y=dsolve(d4y+10*d3y+35*d2y+50*d1y+24*y==...
                     5*diff(u,t,2)+4*diff(u,t)+2*u,...
               y(0)==1/2,d1y(pi)==0,d2y(2*pi)==0,d1y(2*pi)==0);
   y=simplify(y), fplot(y,[0,2*pi])
```

The corresponding solution curve is shown in Figure 2.12. Even though there are many zero equations, the analytical solution is still far too complicated to display.

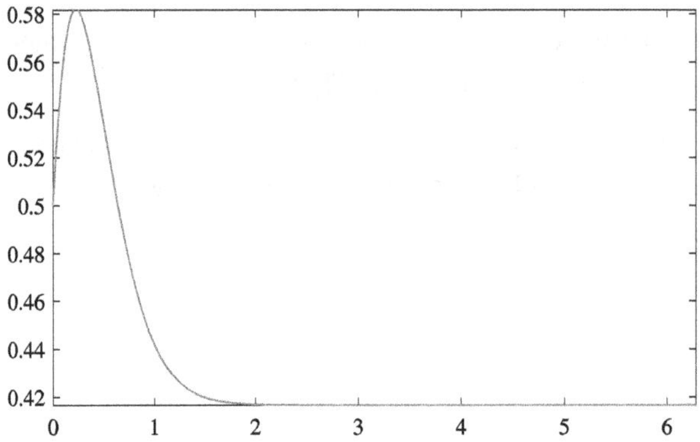

Figure 2.12: The solution curve.

2.5 Solutions of linear matrix differential equations

High-order linear differential equations with constant coefficients can sometimes be expressed in the form of state space equations. In this section, the closed-form solution of the linear state space equations is presented, followed by the direct solution method. Also matrix differential equations such as Sylvester equations will also be studied in this section.

2.5.1 Analytical solutions of linear state space equations

Linear state space models are the most widely used linear differential equations in many fields. They provide an alternative description format for linear differential equations with constant coefficients. For instance, in control science, linear state space models are used to describe linear dynamical systems. In this section, the mathematical form of a state space model is presented, and analytical solution methods are illustrated.

Definition 2.18. A linear state space model can be expressed as

$$\begin{cases} \boldsymbol{x}'(t) = \boldsymbol{A}\boldsymbol{x}(t) + \boldsymbol{B}\boldsymbol{u}(t), \\ \boldsymbol{y}(t) = \boldsymbol{C}\boldsymbol{x}(t) + \boldsymbol{D}\boldsymbol{u}(t) \end{cases} \tag{2.5.1}$$

where $\boldsymbol{A}, \boldsymbol{B}, \boldsymbol{C}, \boldsymbol{D}$ are constant matrices. Vector $\boldsymbol{x}(t)$ is referred to as a state variable vector. The initial state vector is known $\boldsymbol{x}(t_0) = \boldsymbol{x}_0$. Signals $\boldsymbol{u}(t)$ and $\boldsymbol{y}(t)$ are regarded respectively as the input and output of the system.

The first equation in (2.5.1) can be rewritten as

$$\boldsymbol{x}'(t) - \boldsymbol{A}\boldsymbol{x}(t) = \boldsymbol{B}\boldsymbol{u}(t). \tag{2.5.2}$$

Multiplying both sides by $\mathrm{e}^{-\boldsymbol{A}t}$, it is found that

$$\mathrm{e}^{-\boldsymbol{A}t}\boldsymbol{x}'(t) - \mathrm{e}^{-\boldsymbol{A}t}\boldsymbol{A}\boldsymbol{x}(t) = \mathrm{e}^{-\boldsymbol{A}t}\boldsymbol{B}\boldsymbol{u}(t). \tag{2.5.3}$$

Since $\boldsymbol{A}\mathrm{e}^{-\boldsymbol{A}t} = \mathrm{e}^{-\boldsymbol{A}t}\boldsymbol{A}$, the left side happens to be the derivative formula

$$\frac{\mathrm{d}}{\mathrm{d}t}[\mathrm{e}^{-\boldsymbol{A}t}\boldsymbol{x}(t)] = \mathrm{e}^{-\boldsymbol{A}t}\boldsymbol{B}\boldsymbol{u}(t). \tag{2.5.4}$$

Computing integrals of both sides with respect to t, the following formula can be derived directly:

$$\mathrm{e}^{-\boldsymbol{A}t}\boldsymbol{x}(t) - \mathrm{e}^{-\boldsymbol{A}t_0}\boldsymbol{x}(t_0) = \int_{t_0}^{t} \mathrm{e}^{-\boldsymbol{A}\tau}\boldsymbol{B}\boldsymbol{u}(\tau)\mathrm{d}\tau. \tag{2.5.5}$$

From this formula, the analytical solutions to linear state space equations can be established.

Theorem 2.12. *The analytical solution of the state space equation in Definition 2.18 can be written as*

$$\boldsymbol{x}(t) = \mathrm{e}^{\boldsymbol{A}(t-t_0)}\boldsymbol{x}(t_0) + \int_{t_0}^{t} \mathrm{e}^{\boldsymbol{A}(t-\tau)}\boldsymbol{B}\boldsymbol{u}(\tau)\,\mathrm{d}\tau \tag{2.5.6}$$

where $\mathrm{e}^{\boldsymbol{A}(t-t_0)}$ *is also known as the state transition matrix, denoted as* $\boldsymbol{\Phi}(t,t_0) = \mathrm{e}^{\boldsymbol{A}(t-t_0)}$*, which can be used to describe state transition from time* t_0 *to time* t*.*

For the given input signal $\boldsymbol{u}(t)$, matrix integral can be computed to find the analytical solutions of the differential equations. It can be seen from the formula that the exponential function of matrices is involved and integrals are evaluated. These problems can easily be handled with the facilities provided in the Symbolic Math Toolbox. With the toolbox, (2.5.6) can be evaluated directly with the following command:

```
X=expm(A*(t-t0))*x(t0)+int(expm(A*(t-τ))*B*subs(u,t,τ),τ,t0,t)
```

Definition 2.19. If in Definition 2.18, the external input signal $\boldsymbol{u}(t) \equiv \boldsymbol{0}$, the system will be driven solely by the initial value $\boldsymbol{x}(t_0)$, and the model is changed to

$$\boldsymbol{x}'(t) = \boldsymbol{A}\boldsymbol{x}(t), \quad \text{with given } \boldsymbol{x}(t_0) = \boldsymbol{x}_0. \tag{2.5.7}$$

Example 2.31. Assume that the input signal is $u(t) = 2 + 2\mathrm{e}^{-3t}\sin 2t$, and the matrices in the state space model are given as follows:

$$\boldsymbol{A} = \begin{bmatrix} -19 & -16 & -16 & -19 \\ 21 & 16 & 17 & 19 \\ 20 & 17 & 16 & 20 \\ -20 & -16 & -16 & -19 \end{bmatrix}, \quad \boldsymbol{B} = \begin{bmatrix} 1 \\ 0 \\ 1 \\ 2 \end{bmatrix}, \quad \boldsymbol{C}^{\mathrm{T}} = \begin{bmatrix} 2 \\ 1 \\ 0 \\ 0 \end{bmatrix}, \quad \boldsymbol{D} = 0.$$

Find the analytical solution of the differential equation if the initial state vector is known as $\boldsymbol{x} = [0,1,1,2]^{\mathrm{T}}$.

Solutions. The matrices can be entered into MATLAB workspace first, and then according to (2.5.6), the analytical solutions of the equation can be found as

```
>> syms t tau; u(t)=2+2*exp(-3*t)*sin(2*t); %input signal
   A=[-19,-16,-16,-19; 21,16,17,19; 20,17,16,20; -20,-16,-16,-19];
   B=[1; 0; 1; 2]; C=[2 1 0 0]; x0=[0; 1; 1; 2];
   y=C*(expm(A*t)*x0+int(expm(A*(t-tau))*B*u(tau),tau,0,t));
   simplify(y)  %direct implementation of (2.5.6)
```

The mathematical form of the analytical solution is

$$y(t) = \frac{119}{8}\mathrm{e}^{-t} + 57\mathrm{e}^{-3t} + \frac{127t}{4}\mathrm{e}^{-t} + 4t^2\mathrm{e}^{-t} - \frac{135}{8}\mathrm{e}^{-3t}\cos 2t + \frac{77}{4}\mathrm{e}^{-3t}\sin 2t - 54.$$

Manual like-term collection yields the simpler form

$$y(t) = \left(\frac{119}{8} + \frac{127t}{4} + 4t^2\right)\mathrm{e}^{-t} + 57\mathrm{e}^{-3t} - \frac{135}{8}\mathrm{e}^{-3t}\cos 2t + \frac{77}{4}\mathrm{e}^{-3t}\sin 2t - 54.$$

2.5.2 Direct solutions of state space models

Function `dsolve()` can be used to solve the state space equations directly. The difficulties of using such a function appear in constructing a symbolic function matrix $\boldsymbol{X}(t)$. A MATLAB `any_matrix()` function is written to define immediately any symbolic function matrix. The listing of the function is as follows. The syntaxes of the function will be demonstrated later through examples.

```
function A=any_matrix(nn,sA,varargin) % generate arbitrary matrix function
v=varargin; n=nn(1);
if length(nn)==1, m=n; else, m=nn(2); end
s=''; k=length(v); K=0; if n==1 || m==1, K=1; end
if k>0, s='('; for i=1:k, s=[s ',' char(v{i})]; end
   s(2)=[]; s=[s ')'];
end
for i=1:n, for j=1:m % loop structure to handle each element individually
   if K==0, str=[sA int2str(i),int2str(j) s];
   else, str=[sA int2str(i*j) s]; end
   eval(['syms ' str]); eval(['A(i,j)=' str ';']);
end, end
```

Example 2.32. Input the following two symbolic function matrices:

$$\boldsymbol{A} = \begin{bmatrix} a_{11}(x,y) & a_{12}(x,y) & a_{13}(x,y) & a_{14}(x,y) \\ a_{21}(x,y) & a_{22}(x,y) & a_{23}(x,y) & a_{24}(x,y) \\ a_{31}(x,y) & a_{32}(x,y) & a_{33}(x,y) & a_{34}(x,y) \\ a_{41}(x,y) & a_{42}(x,y) & a_{43}(x,y) & a_{44}(x,y) \end{bmatrix}, \quad \boldsymbol{B} = \begin{bmatrix} f_{11}(t) & f_{12}(t) \\ f_{21}(t) & f_{22}(t) \\ f_{31}(t) & f_{32}(t) \\ f_{41}(t) & f_{42}(t) \end{bmatrix}.$$

Solutions. The following commands can be used directly to generate the two symbolic function matrices, and the results are the same, as expected:

```
>> syms x y t
   A=any_matrix(4,'a',x,y), B=any_matrix([4,2],'f',t)
```

Example 2.33. Solve again the state space equation in Example 2.31.

Solutions. With function `any_matrix()`, a symbolic function expression can be used to describe the original differential equation, and then the equation can be solved. The result obtained is the same as that obtained previously.

```
>> syms t; u(t)=2+2*exp(-3*t)*sin(2*t);
   A=[-19,-16,-16,-19; 21,16,17,19; 20,17,16,20; -20,-16,-16,-19];
   B=[1; 0; 1; 2]; C=[2 1 0 0]; x0=[0; 1; 1; 2];
   x(t)=any_matrix([4,1],'x',t);
   y=dsolve(diff(x)==A*y+B*u,x(0)==x0);
   y=C*[y.x1; y.x2; y.x3; y.x4]
```

2.5.3 Solution of Sylvester differential equation

Sylvester differential equation is a special form of matrix differential equations. The definition of the equation is presented first in the section, and then three solution methods are explored.

Definition 2.20. Matrix Sylvester differential equation is expressed as[43]

$$\boldsymbol{X}'(t) = \boldsymbol{A}\boldsymbol{X}(t) + \boldsymbol{X}(t)\boldsymbol{B}, \quad \boldsymbol{X}(0) = \boldsymbol{X}_0 \tag{2.5.8}$$

where $\boldsymbol{A} \in \mathscr{R}^{n\times n}$, $\boldsymbol{B} \in \mathscr{R}^{m\times m}$, and $\boldsymbol{X}, \boldsymbol{X}_0 \in \mathscr{R}^{n\times m}$.

Theorem 2.13. *The analytical solution of the matrix Sylvester differential equation*[43] *is* $\boldsymbol{X}(t) = \mathrm{e}^{\boldsymbol{A}t}\boldsymbol{X}_0\mathrm{e}^{\boldsymbol{B}t}$.

Example 2.34. Solve the following matrix differential equation:

$$\boldsymbol{X}'(t) = \begin{bmatrix} -1 & -2 & 0 & -1 \\ -1 & -3 & -1 & -2 \\ -1 & 1 & -2 & 0 \\ 1 & 2 & 1 & 1 \end{bmatrix} \boldsymbol{X}(t) + \boldsymbol{X}(t) \begin{bmatrix} -2 & 1 \\ 0 & -2 \end{bmatrix}, \quad \boldsymbol{X}(0) = \begin{bmatrix} 0 & -1 \\ 1 & 1 \\ 1 & 0 \\ 0 & 1 \end{bmatrix}.$$

Solutions. Input the relevant matrices into MATLAB workspace first, then Theorem 2.13 can be used to solve the differential equation directly.

```
>> A=[-1,-2,0,-1; -1,-3,-1,-2; -1,1,-2,0; 1,2,1,1];
   B=[-2,1; 0,-2]; X0=[0,-1; 1,1; 1,0; 0,1];
   A=sym(A); B=sym(B); X0=sym(X0); syms t
   X=simplify(expm(A*t)*X0*expm(B*t))          % solve differential equation
   simplify(diff(X)-A*X-X*B), subs(X,t,0)-X0  % validation
```

The solution obtained is as follows:

$$\boldsymbol{X}(t) = \begin{bmatrix} (t^2/2-2)\mathrm{e}^{-3t} + 2\mathrm{e}^{-4t} & (2t+1)\mathrm{e}^{-4t} + (-2+t^2+t^3-4t)\mathrm{e}^{-3t} \\ 2\mathrm{e}^{-4t} - (t+1)\mathrm{e}^{-3t} & (2t+1)\mathrm{e}^{-4t} - (t^2+3t)\mathrm{e}^{-3t} \\ -\mathrm{e}^{-3t}(t^2-2)/2 & -t\mathrm{e}^{-3t}(t^2+2t-6)/2 \\ (2+t)\mathrm{e}^{-3t} - 2\mathrm{e}^{-4t} & (2+t^2+4t)\mathrm{e}^{-3t} - (2t+1)\mathrm{e}^{-4t} \end{bmatrix}$$

Substituting the solution back to the original equation, it can be seen that the equation and initial values are satisfied, since the error matrices are both zero.

Function dsolve() can be tried to solve the Sylvester differential equation directly. The following example is used to demonstrate the solution process.

Example 2.35. Use dsolve() function to solve again the Sylvester differential equation in Example 2.34.

Solutions. As in the case of earlier examples, declare $\boldsymbol{X}$ as a symbolic function matrix. Then the following commands can be employed to solve the differential equation. The results obtained are exactly the same as in Example 2.34.

```
>> A=[-1,-2,0,-1; -1,-3,-1,-2; -1,1,-2,0; 1,2,1,1];
   B=[-2,1; 0,-2]; X0=[0,-1; 1,1; 1,0; 0,1];
   A=sym(A); B=sym(B); X0=sym(X0); syms t
   X(t)=any_matrix([4,2],'x',t);            % construct matrix function
   x=dsolve(diff(X)==A*X+X*B, X(0)==X0) % solve the differential equation
   X=[x.x11, x.x12; x.x21 x.x22; x.x31 x.x32; x.x41 x.x42]
```

2.5.4 Kronecker product-based solutions of Sylvester differential equations

Kronecker product studied in Volume III can be introduced to convert the matrix Sylvester equation into the standard vector differential equation. With Kronecker product, (2.5.8) can be converted into

$$\boldsymbol{x}'(t) = (\boldsymbol{I}_m \otimes \boldsymbol{A} + \boldsymbol{B}^{\mathrm{T}} \otimes \boldsymbol{I}_n)\boldsymbol{x}(t) \tag{2.5.9}$$

where $\boldsymbol{x}(t) = \mathrm{vec}(\boldsymbol{X}(t))$ is the column vector expanded column-wise from matrix $\boldsymbol{X}(t)$. The original matrix equation can be converted into the standard form of first-order explicit differential equations. With function dsolve(), the equation can be solved directly through function expm(). After the solution process, function reshape() can be used to convert the vector back to the expected matrix.

Example 2.36. Use Kronecker product to solve again Sylvester differential equation in Example 2.34.

Solutions. With Kronecker product, the coefficient matrix can be computed, from which the solution of the original Sylvester differential equation can be found, which is the same as that in Example 2.34.

```
>> A=[-1,-2,0,-1; -1,-3,-1,-2; -1,1,-2,0; 1,2,1,1];
   B=[-2,1; 0,-2]; X0=[0,-1; 1,1; 1,0; 0,1];
   A0=kron(eye(2),A)+kron(B',eye(4)); syms t
   x=expm(A0*t)*X0(:); x1=reshape(x,4,2) % solve the equation
```

If the solver is employed instead, the following commands can be used and the solutions obtained are exactly the same:

```
>> x(t)=any_matrix([8,1],'x',t);
   y=dsolve(diff(x)==A0*x, x(0)==X0(:)); % an alternative method
   x2=[y.x1 y.x5; y.x2 y.x6; y.x3 y.x7; y.x4 y.x8]
```

2.6 Analytical solutions to special nonlinear differential equations

Some nonlinear differential equations can also be tackled with the solver `dsolve()`. The description format of the differential equations is the same as that demonstrated earlier. If the differential equation is expressed, the solver can be called directly to try to find the solution. Several examples will be demonstrated later, and an example is given to show the case where analytical solutions are not available.

2.6.1 Solvable nonlinear differential equations

The is no other method which can be used to judge whether a nonlinear differential equation is solvable or not. The only way is to express the differential equation in the standard way as shown earlier, and call the solver to try to solve it. The solvability of differential equations can only be probed in this way. Examples are shown later to show the solution methods of nonlinear differential equations.

Example 2.37. Some of the low-order differential equations studied earlier are nonlinear, for instance, in Example 2.4, the first-order nonlinear differential equation was studied. Solve the differential equation again with the solver:

$$\frac{\mathrm{d}y(t)}{\mathrm{d}t} + 8y(t) + y^2(t) = -15, \quad y(0) = 0.$$

Solutions. The standard method as follows can be used to describe the differential equation, and then call the solver to find the analytical solution:

$$Y = -(15\mathrm{e}^{2t} - 15)/(5\mathrm{e}^{2t} - 3).$$

The result obtained is equivalent to that obtained earlier.

```
>> syms t y(t)
   Y=dsolve(diff(y)+8*y+y^2==-15, y(0)==0);
   simplify(Y)  % solution and simplification
```

It can be seen that there is no need to apply any tactics. What the user needs to do is to express the differential equation in a formal way, and then call the solver to find the solution. The analytical solution of the differential equation, if possible, can be obtained directly.

Example 2.38. Find the analytical solution of the first-order nonlinear differential equation $x'(t) = x(t)(1 - x^2(t))$.

Solutions. The differential equation can be tried directly with `dsolve()`.

```
>> syms t x(t); x=dsolve(diff(x)==x*(1-x^2)) % direct solution
```

The analytical solution can be found as

$$x(t) = \sqrt{-1/\left(\mathrm{e}^{C-2t} - 1\right)}.$$

Besides, the constants ±1 and 0 are also solutions.

Example 2.39. Solve the following second-order differential equation:[58]

$$\left(\frac{\mathrm{d}^2y(x)}{\mathrm{d}x^2}\right)^2 + (2ay(x) + bx + c)\frac{\mathrm{d}^2y(x)}{\mathrm{d}x^2} - a\left(\frac{\mathrm{d}y(x)}{\mathrm{d}x}\right)^2 - b\frac{\mathrm{d}y(x)}{\mathrm{d}x} + k = 0.$$

Solutions. This nonlinear differential equation seems to be too complicated, and it may be very hard to solve it manually for non-professionals. MATLAB can be used to describe the equation in a formal way, and the following commands can be applied in solving it:

```
>> syms a b c k x y(x)
   d1y=diff(y); d2y=diff(y,2);
   y=dsolve(d2y^2+(2*a*y+b*x+c)*d2y-a*d1y^2-b*d1y+k==0)
```

The analytical solution obtained is as follows. It can be seen that both branches obtained are valid:

$$y(x) = \begin{cases} \dfrac{C_1x^2}{2} + C_2x - \dfrac{1}{2C_1a}(C_1^2 + cC_1 - aC_2^2 - bC_2 + k), \\ C_3\mathrm{e}^{-\sqrt{a}\,x} - \dfrac{1}{2a}(c + bx) + \dfrac{1}{16C_3a^3}\mathrm{e}^{\sqrt{-a}\,x}(b^2 + 4ak). \end{cases}$$

Example 2.40. Consider now the following third-order nonlinear differential equation and find the analytical solution in the interval $t \in (0.2, \pi)$:[58]

$$x^5y'''(x) = 2(xy'(x) - 2y(x)), \quad y(1) = 1, \quad y'(1) = 0.5, \quad y''(1) = -1.$$

Draw the curve of the solution.

Solutions. Compared with first- and second-order differential equations, there are rare cases where a third-order differential equation has analytical solutions. In fact, the formal solution method discussed earlier can be tried. Here for the given differential equation and initial values, intermediate variables can be defined, and the differential equation can be solved directly, with the solution curve shown in Figure 2.13.

```
>> syms x y(x)
   d1y=diff(y); d2y=diff(d1y); d3y=diff(d2y);
   y=dsolve(x^5*d3y==2*(x*d1y-2*y),...
            y(1)==1, d1y(1)==0.5, d2y(1)==-1)
   fplot(y,[0.2,pi])  % solve differential equation and draw the curve
```

The analytical solution obtained is

$$y(x) = x^2 - \frac{3\sqrt{2}e^{\sqrt{2}}}{8}x^2e^{-\sqrt{2}/x} + \frac{3\sqrt{2}e^{-\sqrt{2}}}{8}x^2e^{\sqrt{2}/x}.$$

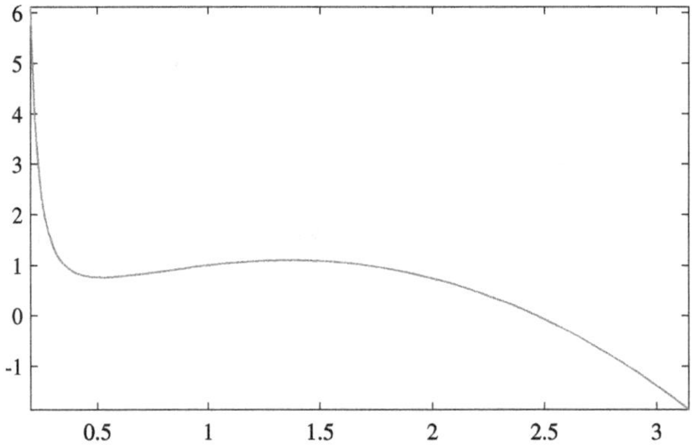

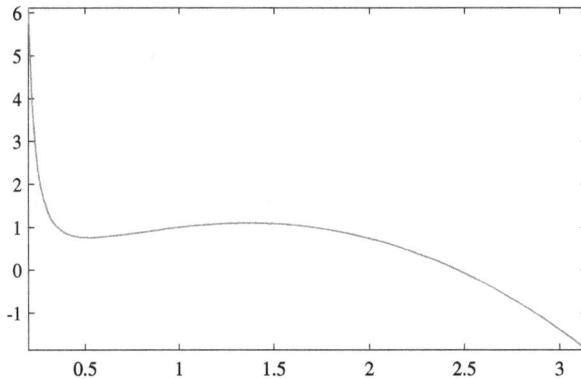

Figure 2.13: The curve of the solution.

2.6.2 Nonlinear differential equations where analytical solutions are not available

It can be seen that the differential equation solver `dsolve()` is rather powerful. This may give the readers an illusion that it can be applied directly to any differential equation. The solver can be tried, of course, but for many cases, especially for nonlinear differential equations, no analytical solutions are available. Several examples are presented in this section.

Example 2.41. Solve the nonlinear differential equation $x'(t) = x(t)(1 - x^2(t)) + 1$.

Solutions. In fact, this equation is that of Example 2.38, with a slight modification, namely, we have added 1 to the right-hand side of the equation. The following commands can be tried, yet the solution process is unsuccessful, since the warning message “Warning: Unable to find explicit solution. Returning implicit solution instead” is displayed, meaning that there is no analytical solution to the equation.

```
>> syms t x(t);
   x=dsolve(diff(x)==x*(1-x^2)+1) % equation has no solution
```

Example 2.42. A Dutch physicist Balthasar van der Pol (1889–1959) and his colleagues reported an oscillating circuit in 1927.[69] The corresponding differential equation is often referred to as van der Pol equation, with the mathematical form of

$$\frac{\mathrm{d}^2 y(t)}{\mathrm{d}t^2}+\mu[y^2(t)-1]\frac{\mathrm{d}y(t)}{\mathrm{d}t}+y(t)=0. \tag{2.6.1}$$

Use `dsolve()` to solve this differential equation and see what happens.

Solutions. For many differential equations studied so far, it seems that they can be studied with the `dsolve()` solver. It is natural to try to solve the given nonlinear differential equation with such a solver.

If the following commands are tried for van der Pol equation, the warning message “Unable to find explicit solution” is obtained, indicating that the solution process is unsuccessful.

```
>> syms t y(t) mu;
   y=dsolve(diff(y,2)+mu*(y^2-1)*diff(y)+y==0) % direct solution
```

It can be seen that analytical solutions of many nonlinear differential equations cannot be found with the solver `dsolve()`. Therefore numerical solutions are the only way available to study nonlinear differential equations. In the subsequent chapters, we will concentrate on numerical solution analysis of various nonlinear differential equations.

2.7 Exercises

2.1 Solve the following first-order differential equations:[58]

(1) $y'(x)=\dfrac{\sqrt{x(x+1)}}{\sqrt{x}+\sqrt{1+x}}$,

(2) $y'(t)=\left[\dfrac{t-\sin 2t}{t^2+\cos 2t}\right]y(t)$, $y(0)=4$,

(3) $y'(t)+\dfrac{3}{(1+t^2/4)^{3/2}}y(t)=\dfrac{2t}{(1+t^2/4)^2}$, $y(0)=8$.

2.2 Validate the following properties for hypergeometric functions:[71]

(1) ${}_1\mathrm{F}_1(a;a;x)=\mathrm{e}^x$,

(2) $_2F_1(a,1;1;z) = \dfrac{1}{(1-z)^a}$,

(3) $_2F_1(1,1;2;z) = \dfrac{1}{z}\ln(z+1)$,

(4) $_2F_1(1/2,1;3/2;z^2) = \dfrac{1}{2z}\ln\dfrac{z+1}{1-z}$.

(5) $_2F_1(1/2,1/2;3/2;z^2) = \dfrac{1}{z}\arcsin z$.

2.3 Solve the following differential equations with Laplace transform method:

$$\begin{cases} x''(t) + y''(t) + x(t) + y(t) = 0, \\ 2x''(t) - y''(t) - x(t) + y(t) = \sin t \end{cases}$$

where $x(0) = 2$, $y(0) = 1$, and $x'(0) = y'(0) = -1$.

2.4 Find the general solution to the following linear differential equation:

$$\begin{aligned} y^{(5)} + 13y^{(4)} + 64y'''(t) + 152y''(t) + 176y'(t) + 80y(t) \\ = \mathrm{e}^{-2t}\left[\sin\left(2t + \frac{\pi}{3}\right) + \cos 3t\right]. \end{aligned}$$

Assuming that the given conditions for the above equation are $y(0) = 1$, $y(1) = 3$, $y(\pi) = 2$, $y'(0) = 1$, and $y'(1) = 2$, find the analytical solution satisfying all the conditions and draw the curve of the solution.

2.5 Find the general solutions for the following differential equations:

(1) $\begin{cases} x''(t) - 2y''(t) + y'(t) + x(t) - 3y(t) = 0, \\ 4y''(t) - 2x''(t) - x'(t) - 2x(t) + 5y(t) = 0, \end{cases}$

(2) $\begin{cases} 2x''(t) + 2x'(t) - x(t) + 3y''(t) + y'(t) + y(t) = 0, \\ x''(t) + 4x'(t) - x(t) + 3y''(t) + 2y'(t) - y(t) = 0. \end{cases}$

2.6 Find the general solution of the following differential equation with variable coefficients:

$$x^4y^{(4)}(x) + 14x^3y'''(x) + 55x^2y''(x) + 65xy'(x) + 16y(x) = 0.$$

If the following conditions are known, $y(0) = y(\pi) = 1$, $y'(0) = y'(\pi) = 1$, solve the differential equation again and draw the curve of the solution.

2.7 Find the general solution of the following differential equation, and also find the analytical solution satisfying the conditions $x(0) = 1$, $x(\pi) = 2$ and $y(0) = 0$:

$$\begin{cases} x''(t) + 5x'(t) + 4x(t) + 3y(t) = \mathrm{e}^{-6t}\sin 4t, \\ 2y'(t) + y(t) + 4x'(t) + 6x(t) = \mathrm{e}^{-6t}\cos 4t. \end{cases}$$

2.8 Find the analytical solutions of the following time-varying linear differential equations:

(1) Legendre equation, $(1-t^2)x''(t) - 2tx'(t) + n(n+1)x(t) = 0$,
(2) Bessel equation, $t^2x''(t) + tx'(t) + (t^2 - n^2)x(t) = 0$.
Find the solution when $n = 2$, $x(0) = 1$ and $x'(0) = 0$, and draw the curve of the solution.

2.9 Find the general solution of the differential equation

$$y''(x) - (2 - 1/x)\,y'(x) + (1 - 1/x)\,y(x) = x^2 \mathrm{e}^{-5x},$$

and also find the solution satisfying the boundary conditions $y(1) = \pi$ and $y(\pi) = 1$.

2.10 Solve the following boundary value problem:

$$y''(x) + xy'(x) + y(x) = \cos x, \quad y(0) = 0, \quad y(2) = 1.$$

Draw the curve of the solution and see whether the boundary conditions are satisfied.

2.11 Find the general solutions for the following differential equations:
(1) $x''(t) + 2tx'(t) + t^2x(t) = t + 1$,
(2) $y'(x) + 2xy(x) = x\mathrm{e}^{-x^2}$,
(3) $y'''(t) + 3y''(t) + 3y'(t) + y(t) = \mathrm{e}^{-t} \sin t$.

2.12 Use MATLAB solver to directly tackle the differential equation in Exercise 2.3 and validate the solution.

2.13 Find the general solution of the following first-order differential equation:

$$\frac{\mathrm{d}y(x)}{\mathrm{d}x} - \frac{2x}{x^2+4}\,y(x) = (4 + x^2)^3 \mathrm{e}^{-2x}.$$

If it is known that $y(0) = 1$, solve the equation again.

2.14 Find the analytical solutions of the following nonlinear differential equations:
(1) $y'(x) = y^4(x) \cos x + y(x) \tan x$,
(2) $xy^2(x)y'(x) = x^2 + y^2(x)$,
(3) $xy'(x) + 2y(x) + x^5y^3(x)\mathrm{e}^x = 0$.

2.15 Find the solutions of the following differential equations and draw the (x, y) trajectory:

$$\begin{cases} (2x''(t) - x'(t) + 9x(t)) - (y''(t) + y'(t) + 3y(t)) = 0, \\ (2x''(t) + x'(t) + 7x(t)) - (y''(t) - y'(t) + 5y(t)) = 0. \end{cases}$$

The initial values are $x(0) = x'(0) = 1$, $y(0) = y'(0) = 0$.

2.16 Solve the following time-varying linear differential equations:
(1) $(x^2 - 2x + 3)y'''(x) - (x^2 + 1)y''(x) + 2xy'(x) - 2y(x) = 0$,
(2) $x^2 \ln xy''(x) - xy'(x) + y(x) = 0$,

(3) $(e^t + 1)y''(t) - 2y'(t) - e^t y(t) = 0$.

2.17 Solve the following nonlinear differential equations:

(1) $(y''(t))^2 = \alpha(ty'(t) - y(t)) + \beta y'(t) + \gamma$,

(2) $y''(t) = (\alpha e^{\beta y} y(t) + \beta)y'(t)$,

(3) $t^5 y'''(t) = a(ty'(t) - 2y(t))$,

(4) $y'(x) = \dfrac{y(x)(2x^2 - xy(x) + y^2(x))}{2x - y(x)}$.

2.18 Find the analytical solutions of the following nonlinear differential equation:

$$x^2 y(x) y''(x) - \left[2x^2 (y'(t))^2 + axy(x)y'(x) + ay^2(x)\right] = 0$$

2.19 Solve the following linear differential equation:

$$\boldsymbol{x}'(t) = \begin{bmatrix} -3 & -2 & 0 & -2 \\ 3 & 2 & 0 & 3 \\ 2 & 2 & -1 & 2 \\ -2 & -2 & 0 & -3 \end{bmatrix} \boldsymbol{x}(t) + \begin{bmatrix} 1 & 0 \\ 3 & 1 \\ 2 & 1 \\ 1 & 0 \end{bmatrix} \boldsymbol{u}(t)$$

where $\boldsymbol{x}(0) = [0, 1, 2, 1]^{\mathrm{T}}$, $u_1(t) = \sin t$, and $u_2(t) = e^{-t} + \cos t$.

3 Initial value problems

Analytical solution methods were studied in Chapter 2. It was indicated that there are many differential equations, especially nonlinear, where analytical solutions are not available. Numerical methods should be employed to study these differential equations. From this chapter on, numerical solutions for various differential equations are presented.

In Section 3.1, the mathematical model of a first-order explicit differential equation and its initial value problem is presented. In Section 3.2, fixed-step Runge–Kutta methods and multistep algorithms are introduced, with low-level MATLAB implementations. Examples are used to assess the accuracy of the algorithms, through comparative studies.

In real applications, fixed-step methods are not actually utilized, since the accuracy and efficiency are not satisfactory. Adaptive variable-step methods with precision monitoring facilities are usually adopted. In Section 3.3, the variable-step solver provided in MATLAB is demonstrated directly. The solution procedures are proposed for ordinary differential equations. Examples are used to demonstrate the numerical solution process. In Section 3.4, a very important step in numerical solution process – the validation step – is presented and demonstrated through examples. With such a step, the solutions obtained may be ensured to be valid. High precision algorithms and fixed-step display of simulation results are also presented.

In the first part of this chapter, some ideas and low-level implementations of simple numerical algorithms are presented. It may be helpful for the reader to understand numerical solutions and algorithms. If the readers are only interested in how to use MATLAB to directly find numerical solutions of initial value problems, it is suggested to read this chapter from Section 3.3.

3.1 Initial value descriptions for first-order explicit differential equations

Many algorithms for solving differential equations are constructed for the solution of initial value problems. Normally the algorithms are designed for solving differential equations described using first-order explicit forms. This kind of model is described first in this section, and then some theoretical results on existence and uniqueness of solutions are proposed.

3.1.1 Mathematical forms of initial value problems

In this section, the first-order explicit differential equation format is presented, which can be regarded as the fundamental basis for numerical algorithms presented later.

https://doi.org/10.1515/9783110675252-003

Definition 3.1. The mathematical model of a first-order explicit differential equation is given by

$$\boldsymbol{x}'(t) = \boldsymbol{f}(t, \boldsymbol{x}(t)) \tag{3.1.1}$$

where the vector $\boldsymbol{x}(t) = [x_1(t), x_2(t), \dots, x_n(t)]^{\mathrm{T}}$ is known as the state vector, and the function vector $\boldsymbol{f}(\cdot) = [f_1(\cdot), f_2(\cdot), \dots, f_n(\cdot)]^{\mathrm{T}}$ is composed by any nonlinear functions.

Definition 3.2. If the initial state vector $\boldsymbol{x}_0(t_0) = [x_1(t_0), x_2(t_0), \dots, x_n(t_0)]^{\mathrm{T}}$ is known, and the first-order explicit differential equation in Definition 3.1 is to be solved, the problem is regarded as an initial value problem.

Numerical solutions of initial value problems aim at finding the solutions $\boldsymbol{x}(t)$ of the differential equations over a given interval $t \in [t_0, t_{\mathrm{n}}]$ in a numerical format. The quantity t_{n} is referred to as the terminal time. The numerical solution reliably finds how the state vector evolves from the given initial state vector $\boldsymbol{x}_0$ in the predefined interval $t \in [t_0, t_{\mathrm{n}}]$.

It can be seen from the initial value problem model that if the differential equations are provided in the first-order explicit form, numerical methods can be used to find the solutions of the equations. On the contrary to the analytical method, numerical methods do not significantly change when the format of differential equations changes. It will be shown in later chapters that even if the equations are not provided in the first-order explicit form, conversions can be made such that numerical solvers can still be used to solve the equations.

3.1.2 Existence and uniqueness of solutions

Two important theorems are presented in the differential equation theory to describe the existence and uniqueness of the solutions. The two theorems are listed below,[8] with some necessary illustrations.

Theorem 3.1 (Existence theorem). *Assuming that $f(t, x(t))$ is continuous in the rectangular region $a < t < b$ and $c < x(t) < d$, and the initial value $(t_0, x(t_0))$ is also defined in this region, there exists an $\epsilon > 0$, such that in the interval $t_0 - \epsilon < t < t_0 + \epsilon$, the function $x(t)$ satisfies* (3.1.1).

In simple terms, if $f(t, x(t))$ is a continuous function, there exists solutions to the initial value problems. However, this does not imply that if the function is not continuous, there is no solutions.

Theorem 3.2 (Uniqueness theorem). *Assume that $f(t, x(t))$ is continuous over the rectangular region $a < t < b$ and $c < x(t) < d$, and the initial value $(t_0, x(t_0))$ is also defined in the region. If $\partial f/\partial x$ is also a continuous function, there exists an $\epsilon > 0$, such that in* *the time interval* $t_0 - \epsilon < t < t_0 + \epsilon$, *if the equation has two solutions* $x_1(t)$ *and* $x_2(t)$, *one has* $x_1(t) = x_2(t)$. *In other words, the solution is unique.*

3.2 Implementation of fixed-step numerical algorithms

Classical numerical algorithms for differential equations are originated from fixed-step algorithms. That is, if at initial time $t = t_0$, the initial state $\boldsymbol{x}(t_0)$ is given, the numerical solution $\boldsymbol{x}(t_0+h)$ at time t_0+h can be found. If this point is used as an initial point, the solution $\boldsymbol{x}(t_0 + 2h)$ can be found in the same way. With such an evolution process, the numerical solutions within the interval of interest can be found.

In this section, some commonly used first-step algorithms are introduced. For instance, the oldest Euler's method, second- and fourth-order Runge–Kutta algorithms, Runge–Kutta algorithms of any order, and so on. Also, the linear multistep algorithms, such as Adams–Bashforth and Adams–Mouton algorithms, are presented. Examples are used to compare the algorithms with respect to error propagation and computational efficiency. A solid foundation can be established for the readers to better understand numerical solutions to be illustrated later.

3.2.1 Euler's method

For the initial value problem of ordinary differential equations, Euler's method is the most straightforward. Euler's method is named after a Swiss mathematician Leonhard Euler (1707–1783), who proposed this algorithm in his book "*Institutionum calculi integralis*" in 1768. It is the oldest numerical algorithm for differential equations.

Although the algorithm appears to be very simple, it is helpful in understanding other complicated algorithms to be presented later. Therefore, Euler's method is introduced, and its MATLAB implementation is completed.

Assume that at time t_0, the initial state vector $\boldsymbol{x}(t_0)$ is known. A very small step-size h can be selected, and from the definition of derivatives in (3.1.1),

$$\lim_{h\to 0} \frac{\boldsymbol{x}(t_0+h) - \boldsymbol{x}(t_0)}{t_0 + h - t_0} = \boldsymbol{f}(t, \boldsymbol{x}(t)). \tag{3.2.1}$$

It can be seen that in mathematics there is no step-size h satisfying $h \to 0$. A relatively small step-size h can be chosen instead, and the limit sign in (3.2.1) can be removed, such that the $\boldsymbol{x}(t)$ at time $t_0 + h$ can be approximately expressed as

$$\hat{\boldsymbol{x}}(t_0 + h) \approx \boldsymbol{x}(t_0) + h\boldsymbol{f}(t_0, \boldsymbol{x}(t_0)). \tag{3.2.2}$$

Strictly speaking, such an approximation may yield errors. Therefore the state vector at time $t_0 + h$ can be expressed as

$$\boldsymbol{x}(t_0 + h) = \hat{\boldsymbol{x}}(t_0 + h) + \boldsymbol{R}_0 = \boldsymbol{x}(t_0) + h\boldsymbol{f}(t, \boldsymbol{x}(t_0)) + \boldsymbol{R}_0. \tag{3.2.3}$$

If Taylor series expansion is used to approximate $\boldsymbol{x}(t_0+h)$, then

$$\boldsymbol{x}(t_0+h) = \boldsymbol{x}(t_0) + h\boldsymbol{x}'(t_0) + \frac{h^2}{2}\boldsymbol{x}''(t_0) + o(h^3). \tag{3.2.4}$$

It can be seen by comparing the two expressions that the cumulative error is $h^2\boldsymbol{x}''(t_0)/2$. In mathematics, it is said that the algorithm is of accuracy $o(h)$. In practical algorithm presentation, the "$\approx$" sign in (3.2.2) can be dropped, and the numerical solution is denoted directly as $\boldsymbol{x}_1$. Besides $\boldsymbol{x}(t_0+kh)$ is simply denoted as $\boldsymbol{x}_k$.

Theorem 3.3. *Assuming that the state vector $\boldsymbol{x}_k$ at time t_k is known, the numerical solution at time t_k+h can be obtained by Euler's method as*

$$\boldsymbol{x}_{k+1} = \boldsymbol{x}_k + h\boldsymbol{f}(t, \boldsymbol{x}_k). \tag{3.2.5}$$

An iterative method can be used to evaluate the numerical solution of the differential equation in the interval $t \in [0, T]$, which means that the numerical solutions at time instances $t_0+h, t_0+2h, \ldots$ can be found.

Based on the Euler's method, the following MATLAB function can be written to find numerical solutions of first-order explicit differential equations:

```
function [t,x]=ode_euler(f,vec,x0)
if length(vec)==3, h=vec(2); t=[vec(1):h:vec(3)]';
else, h=vec(2)-vec(1); t=vec(:); end
n=length(t); x0=x0(:); x=[x0'; zeros(n-1,length(x0))];
for i=1:n-1
   x1=f(t(i),x0); x1=x0+h*x1(:); %iteratively solve the equation for one step
   x(i+1,:)=x1'; x0=x1;          %store and update the solutions
end
```

Note that the action of `(:)` in the program is to force the vector into a column vector. This is the fault tolerance facility in the program. If the vector of the function is erroneously written as a row vector, by careless action, the program can still be executed normally.

The syntax of the function is `[`$\boldsymbol{x}$`,`$\boldsymbol{t}$`]=ode_euler(`f`,vec,`$\boldsymbol{x}_0$`)`, where the differential equation can be described by the function handle f. Later, examples are used to show how to express differential equations by function handles. The argument $\boldsymbol{x}_0$ stores the initial state vector. There are two definitions in the argument `vec`: one is to specify the actual fixed-step time vector $\boldsymbol{t}$, the other is to supply the parameters in a vector as `[`t_0`,`h`,`$t_{\rm n}$`]`, where h is the step-size while $(t_0, t_{\rm n})$ is the time interval of interest. The returned argument $\boldsymbol{t}$ is a time vector, while $\boldsymbol{x}$ is a matrix composed of n columns, each corresponding to the numerical expression of a state variable, and n is the number of states. Examples are given next to demonstrate the use of Euler's method, and also its behaviors.

Example 3.1. Consider the second-order differential equation studied in Example 2.19. Let $\lambda = 2$, $a = 1$, and $b = 0$. For simplicity, the differential equation is given again, and $y_i(t)$ is rewritten as $x_i(t)$:

$$\begin{cases} x_1'(t) = x_2(t), \\ x_2'(t) = -2x_1(t), \quad x_1(0) = 1, \quad x_2(0) = 0. \end{cases}$$

First find the analytical solution of the differential equation. Besides, select different step-sizes and find the numerical solution of the differential equation with Euler's method. Assess the impact of the step-size on the computational accuracy.

Solutions. If the method in Example 2.19 is used, the following commands can be employed to find the analytical solution of the differential equation:

```
>> syms t x1(t) x2(t);
   [x10,x20]=dsolve(diff(x1)==x2,diff(x2)==-2*x1,...
             x1(0)==1,x2(0)==0)    % find analytical solution
```

It is found that the analytical solution is $x_1(t) = \cos\sqrt{2}t$ and $x_2(t) = -\sqrt{2}\sin\sqrt{2}t$. Later this solution is used to assess the accuracy of the algorithms.

If numerical algorithms are applied, the equation must be expressed in MATLAB first. If the first-order explicit differential equation is given, a MATLAB or anonymous function should be written to compute the derivative $\boldsymbol{x}'(t)$ from the known t and $\boldsymbol{x}(t)$. The standard anonymous function for this problem is as follows:

```
>> f=@(t,x)[x(2); -2*x(1)]; % describe equation with anonymous function
```

Now selecting different step-sizes $h = 0.001, 0.01, 0.02$, and computing the numerical solutions, it is found that the maximum errors are respectively $e_1 = 0.0086$, $e_2 = 0.0874$, and $e_3 = 0.1789$. Comparisons of the numerical solutions with the analytical solution are shown in Figure 3.1. It is evident that when the step-size is large, there exist large discrepancies in the numerical solutions. It can be concluded from the example that Euler's method is far from satisfactory.

```
>> t0=0; tn=5; sq=sqrt(2); fplot([x10,x20],[0,5])
   H=[0.001,0.01,0.02,0.05,0.1,0.2]; E=[];
   for h=H % different step-sizes can be tried for numerical solutions
      [t1,x1]=ode_euler(f,[t0,h,tn],[1;0]);
      sol=[cos(sq*t1), -sq*sin(sq*t1)]; % analytical solution
      e1=norm(sol-x1,inf); E=[E,e1]; line(t1,x1)
   end
```

In the displayed result, the thick curves are composed of several curves, and they are almost the same. They represent the analytical solution and the numerical solution

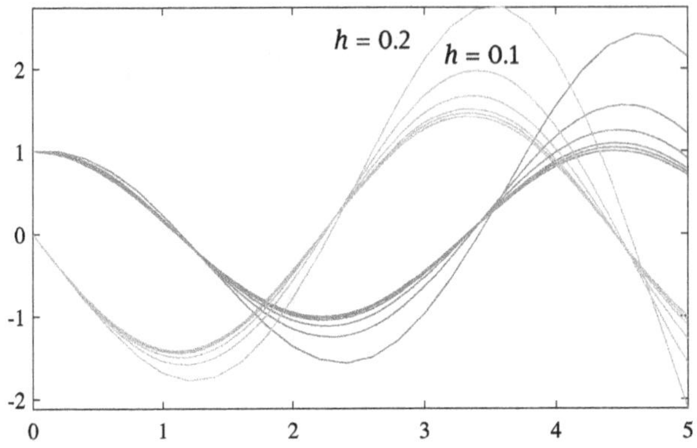

Figure 3.1: Comparisons for different step-sizes.

at $h = 0.001$. The other curves correspond to the cases when $h = 0.01$ and $h = 0.02$. It can be seen that the errors are relatively large. When $h = 0.001$, the difference may even not be witnessed from the curves, since the maximum error is only $e = 0.0086$. If h is increased, the error tends to increase significantly, and the curve deviates from the theoretical one. Even if $h = 0.001$, the error is still considered too large under the double-precision standard. Better algorithms are needed for such problems.

It has been commented that there are two ways to describe the differential equations. The anonymous function method was demonstrated above. The other method is to write a genuine MATLAB function. For instance, the function below can be written

```
function dx=c3mtest(t,x)
dx=[x(2); -2*x(1)]; % use MATLAB function for the differential equation
```

In this case, the following commands can be used to solve the first-order explicit differential equation, and the result is identical to that given above.

```
>> h=0.001; [t1,x1]=ode_euler(@c3mtest,[t0,h,tn],[1;0]);
```

3.2.2 Second-order Runge–Kutta algorithm

Since the accuracy of the Euler's method described earlier is too low, better algorithms with higher accuracy are expected. Runge–Kutta methods are a family of widely used numerical methods. Runge–Kutta methods are named after two German mathematicians Carl David Tolmé Runge (1856–1927) and Martin Wilhelm Kutta (1867–1944), where Runge proposed the method in 1895, and Kutta proposed a series of computation formulas of orders less than 4.

Euler's method discussed earlier is also a special case of these formulas, known as the first-order Runge–Kutta method. Here, another simple Runge–Kutta method – second-order Runge–Kutta algorithm – is illustrated.

Theorem 3.4. *Assuming that the state vector $\boldsymbol{x}_k$ at time t_k is known, the numerical solution of the differential equation at $t_k + h$, using second-order Runge–Kutta algorithm, can be written as*

$$\boldsymbol{x}_{k+1} = \boldsymbol{x}_k + \frac{h}{2}(\boldsymbol{k}_1 + \boldsymbol{k}_2) \tag{3.2.6}$$

where the intermediate variables are

$$\begin{cases} \boldsymbol{k}_1 = \boldsymbol{f}(t_k, \boldsymbol{x}_k), \\ \boldsymbol{k}_2 = \boldsymbol{f}(t_{k+1}, \boldsymbol{x}_k + h\boldsymbol{k}_1). \end{cases} \tag{3.2.7}$$

It can be shown that the cumulative error is $o(h^2)$.

What is the difference in the notations $o(h)$ and $o(h^2)$? A simple illustration is given below. If the step-size is selected as $h = 0.01$, then $o(h)$ means that the cumulative error is about the level of 0.01, while $o(h^2)$ yields a cumulation error at level 0.01^2, i. e., 0.0001. It is obvious that for relatively small step-sizes h, the $o(h^2)$ algorithms are evidently superior to the $o(h)$ algorithms. For the same step-size h, one should select $o(h^p)$ algorithms with larger values of p.

Using the framework of the function designed for Euler's method, the following MATLAB function can be written for the second-order fixed-step Runge–Kutta algorithm, and the syntax is exactly the same as for `ode_euler()` function:

```
function [t,x]=ode_rk2(f,vec,x0)
if length(vec)==3, h=vec(2); t=[vec(1):h:vec(3)]';
else, h=vec(2)-vec(1); t=vec(:); end
n=length(t); x0=x0(:); x=[x0'; zeros(n-1,length(x0))];
for i=1:n-1
   k1=f(t(i),x0); k2=f(t(i+1),x0+h*k1);
   x1=x0+h*(k1(:)+k2(:))/2; x(i+1,:)=x1'; x0=x1;
end
```

Example 3.2. Solve again the differential equation in Example 3.1 using the second-order Runge–Kutta method, and assess the accuracy.

Solutions. The following commands can be used to solve the differential equation directly. Again different step-sizes can be tried, and the maximum errors for the step-sizes are respectively $e_1 = 4.0246\times10^{-6}$, $e_2 = 4.0299\times10^{-4}$, and $e_3 = 0.0016$, which are significantly smaller than those for Euler's method. A larger step-size of $h = 0.1$ can be selected, and the maximum error is $e_4 = 0.0407$. Comparisons of the numerical solutions are shown in Figure 3.2. It can be seen that apart from the case $h = 0.1$, the other results can hardly be distinguished from the plot. Even if a larger step-size of $h = 0.2$ is selected, the result obtained is still much better than that with Euler's method.

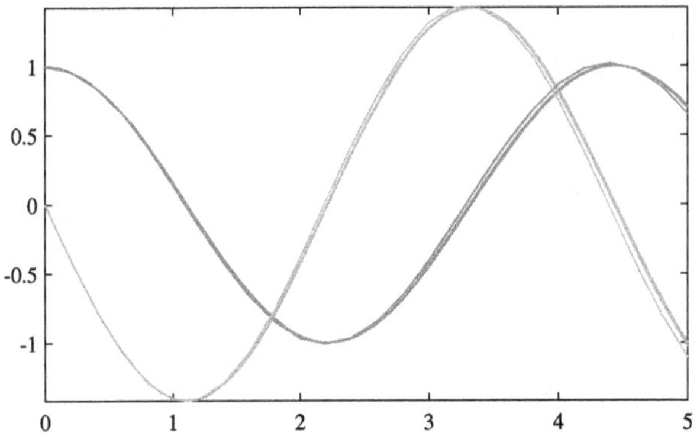

Figure 3.2: Comparisons for different step-sizes.

```
>> f=@(t,x)[x(2); -2*x(1)]; t0=0; tn=5; sq=sqrt(2);
   H=[0.001,0.01,0.02,0.05,0.1,0.2]; E=[];
   for h=H  % try different step-sizes to solve the equation
      [t1,x1]=ode_rk2(f,[t0,h,tn],[1;0]);
      sol=[cos(sq*t1), -sq*sin(sq*t1)]; % analytical solution
      e1=norm(sol-x1,inf); E=[E,e1]; line(t1,x1)
   end
```

It can be seen that, compared with Euler's method, the errors are significantly reduced for the same step-size. For relatively large step-sizes, the error may still be rather large.

3.2.3 Fourth-order Runge–Kutta algorithm

The fourth-order fixed-step Runge–Kutta algorithm is often taught in numerical analysis and system simulation courses. It is regarded as an effective algorithm. The structure of the algorithm is simple and suitable for computer implementation.

Theorem 3.5. *With the fourth-order Runge–Kutta algorithm, the states can be recursively computed from*

$$\boldsymbol{x}_{k+1} = \boldsymbol{x}_k + \frac{h}{6}(\boldsymbol{k}_1 + 2\boldsymbol{k}_2 + 2\boldsymbol{k}_3 + \boldsymbol{k}_4) \tag{3.2.8}$$

where the four intermediate vectors can be computed from

$$\begin{cases} \boldsymbol{k}_1 = \boldsymbol{f}(t_k, \boldsymbol{x}_k), \\ \boldsymbol{k}_2 = \boldsymbol{f}\,(t_k + h/2, \boldsymbol{x}_k + h\boldsymbol{k}_1/2)\,, \\ \boldsymbol{k}_3 = \boldsymbol{f}\,(t_k + h/2, \boldsymbol{x}_k + h\boldsymbol{k}_2/2)\,, \\ \boldsymbol{k}_4 = \boldsymbol{f}(t_k + h, \boldsymbol{x}_k + h\boldsymbol{k}_3) \end{cases} \tag{3.2.9}$$

where h is the step-size, which can be a constant.

It can be shown that the accuracy of the algorithm is supposed to be $o(h^4)$. We can still use $h = 0.01$ as an example. If the fourth-order Runge–Kutta algorithm is used, the error may reach the level of 0.01^4, i. e., 10^{-8}. Therefore this algorithm's accuracy is far higher than for Euler's method and second-order Runge–Kutta algorithm.

A recursive method can still be used to compute from the initial state vector to find the states at time instances $t_0 + h, t_0 + 2h, \ldots$, such that the numerical solutions in the interval $t \in [t_0, t_n]$ can be found.

With the above mathematical formula, the MATLAB solver can be written for the fourth-order Runge–Kutta algorithm:

```
function [t,x]=ode_rk4(f,vec,x0)
if length(vec)==3, h=vec(2); t=[vec(1):h:vec(3)]';
else, h=vec(2)-vec(1); t=vec(:); end
n=length(t); x0=x0(:); x=[x0'; zeros(n-1,length(x0))];
for i=1:n-1                    % use loop structure to solve the differential equation
   k1=f(t(i),x0); k2=f(t(i)+h/2,x0+h*k1/2);
   k3=f(t(i)+h/2,x0+h*k2/2); k4=f(t(i+1),x0+h*k3);
   x1=k1+2*k2+2*k3+k4; x1=x0+x1(:)*h/6;
   x(i+1,:)=x1'; x0=x1; % update the initial values and store the solutions
end
```

Example 3.3. Use the fourth-order Runge–Kutta algorithm to solve again the problem in Example 3.1, and assess the accuracy.

Solutions. With the fourth-order Runge–Kutta algorithm, and selecting the step-sizes $h = 0.001$, $h = 0.01$, $h = 0.1$, and an even larger one $h = 0.2$, the maximum errors obtained are respectively $e_1 = 4.0712 \times 10^{-13}$, $e_2 = 4.0307 \times 10^{-9}$, $e_3 = 4.0756 \times 10^{-5}$, and $e_4 = 0.00065$. It can be seen that with the fourth-order Runge–Kutta algorithm, the accuracy is significantly increased. Even though the step-size is selected as $h = 0.2$, the algorithm is still applicable, and the error is smaller than for Euler's method, if $h = 0.001$ is used. It can be seen that the efficiency of the algorithm is much superior to those of Euler's and second-order Runge–Kutta algorithms.

```
>> f=@(t,x)[x(2); -2*x(1)]; t0=0; tn=5; sq=sqrt(2);
   H=[0.001,0.01,0.02,0.05,0.1,0.2]; E=[];
   for h=H % try different step-sizes for the differential equation
      [t1,x1]=ode_rk4(f,[t0,h,tn],[1;0]);
      sol=[cos(sq*t1), -sq*sin(sq*t1)]; % analytical solution
      e1=norm(sol-x1,inf); E=[E,e1];      % find maximum error
   end
```

3.2.4 Gill's algorithm

Gill's algorithm is a revised form of the typical fourth-order Runge–Kutta algorithm. The reference points and weighting coefficients in the algorithm are modified, and the algorithm is given below.

Theorem 3.6. *Gill's algorithm is a recursive way to compute the states as*

$$\boldsymbol{x}_{k+1} = \boldsymbol{x}_k + \frac{h}{6}\left[\boldsymbol{k}_1 + (2-\sqrt{2})\boldsymbol{k}_2 + (2+\sqrt{2})\boldsymbol{k}_3 + \boldsymbol{k}_4\right] \tag{3.2.10}$$

where the four intermediate vectors can be computed from

$$\begin{cases} \boldsymbol{k}_1 = \boldsymbol{f}(t_k, \boldsymbol{x}_k), \\ \boldsymbol{k}_2 = \boldsymbol{f}(t_k + h/2, \boldsymbol{x}_k + h\boldsymbol{k}_1/2), \\ \boldsymbol{k}_3 = \boldsymbol{f}(t_k + h/2, \boldsymbol{x}_k + h(\sqrt{2}-1)\boldsymbol{k}_1/2 + (2-\sqrt{2})h\boldsymbol{k}_2/2), \\ \boldsymbol{k}_4 = \boldsymbol{f}(t_k + h, \boldsymbol{x}_k - \sqrt{2}h\boldsymbol{k}_2/2 + (2+\sqrt{2})h\boldsymbol{k}_3/2) \end{cases} \tag{3.2.11}$$

where h is the step-size, and the error level of Gill's algorithm is $o(h^4)$.

Based on the above algorithm, it is not difficult to write the following MATLAB implementation:

```
function [t,x]=ode_gill(f,vec,x0)
if length(vec)==3, h=vec(2); t=[vec(1):h:vec(3)]';
else, h=vec(2)-vec(1); t=vec(:); end
n=length(t); x0=x0(:); x=[x0'; zeros(n-1,length(x0))];
for i=1:n-1
   k1=f(t(i),x0); k2=f(t(i)+h/2,x0+k1*h/2);
   k3=f(t(i)+h/2,x0+(sqrt(2)-1)*h*k1/2+(2-sqrt(2))*h*k2/2);
   k4=f(t(i+1),x0-sqrt(2)*h*k2/2+(2+sqrt(2))*h*k3/2);
   x1=k1+(2-sqrt(2))*k2+(2+sqrt(2))*k3+k4;
   x1=x0+x1(:)*h/6; x(i+1,:)=x1'; x0=x1;
end
```

Example 3.4. Solve again the problem in Example 3.1 with Gill's algorithm and assess the accuracy.

Solutions. Similar to the above cases, the following statements can be tried for various step-sizes, and the maximum errors can be found:

```
>> f=@(t,x)[x(2); -2*x(1)]; t0=0; tn=5; sq=sqrt(2);
   H=[0.001,0.01,0.02,0.05,0.1,0.2]; E=[];
   for h=H
```

```
    [t1,x1]=ode_gill(f,[t0,h,tn],[1;0]);
    sol=[cos(sq*t1), -sq*sin(sq*t1)];
    e1=norm(sol-x1,inf); E=[E,e1];
end
```

3.2.5 The *m*th order Runge–Kutta algorithm

Several numerical algorithms were discussed above, where Euler's method can be regarded as a special case of Runge–Kutta methods, or as the first-order Runge–Kutta algorithm. The second- and fourth-order Runge–Kutta algorithms were also presented. It can be seen from the structures that the two algorithms are similar, and the accuracies are respectively $o(h^2)$ and $o(h^4)$. More generally, a class of the mth Runge–Kutta algorithms can be constructed.

Interpolation methods are used in the Runge–Kutta algorithms, and from the given values one computes the next function values through the definite integral

$$\boldsymbol{x}(b) = \boldsymbol{x}(a) + \int_a^b \boldsymbol{f}(t, \boldsymbol{x}(t))\mathrm{d}t \tag{3.2.12}$$

where $a = t_k$ and $b = t_{k+1}$. The step-size is $h = t_{k+1} - t_k$. Within the interval (t_k, t_{k+2}), some interpolation points are defined

$$t_{ij} = t_k + \alpha_j h, \quad j = 1, 2, \dots, m. \tag{3.2.13}$$

From the specifically selected interpolation points, the mth order Runge–Kutta algorithm is formulated.

Theorem 3.7. *The general formula for the mth order Runge–Kutta algorithm is*

$$\boldsymbol{x}_{k+1} = \boldsymbol{x}_k + \sum_{i=1}^{m} h\gamma_i \boldsymbol{k}_i \tag{3.2.14}$$

where the intermediate vectors can be evaluated from

$$\boldsymbol{k}_i = \boldsymbol{f}\left(t_k + \alpha_i h, \boldsymbol{x}_k + h\sum_{j=1}^{i-1} \beta_{ij}\boldsymbol{k}_j\right), \quad i = 1, 2, \dots, m. \tag{3.2.15}$$

It can be seen that the second- and fourth-order Runge–Kutta algorithms as well as Gill's algorithm are all special cases of the mth order Runge–Kutta algorithms. The cumulative error of the mth order Runge–Kutta algorithm is $o(h^m)$.

Theorem 3.8. *In Runge–Kutta methods, the relationship between the coefficients is described in Butcher tableau*[13]

$$\begin{array}{c|cccccc} \alpha_1=0 & & & & & \\ \alpha_2 & \beta_{21} & & & & \\ \alpha_3 & \beta_{31} & \beta_{32} & & & \\ \vdots & \vdots & \vdots & \ddots & & \\ \alpha_m & \beta_{m1} & \beta_{m2} & \cdots & \beta_{m,m-1} & \\ \hline & \gamma_1 & \gamma_2 & \cdots & \gamma_{m-1} & \gamma_m \end{array} \tag{3.2.16}$$

where the coefficients satisfy the following formulas:

$$\sum_{j=1}^{i-1}\beta_{ij}=\alpha_j,\quad j=2,3,\dots,m \tag{3.2.17}$$

and

$$\sum_{i=1}^{m-1}\gamma_i\alpha_i^{\lambda-1}=\frac{1}{\lambda},\quad \lambda=1,2,\dots,m-1, \tag{3.2.18}$$

$$\sum\gamma_i\beta_{ij}\alpha_j=\frac{1}{6},\quad \sum\gamma_i\alpha_i\beta_{ij}\alpha_j=\frac{1}{8},\dots \tag{3.2.19}$$

Butcher invented the tree-like symbols to denote combination forms of the coefficient terms. For instance, with a solid dot • and symbols ⁞ and Y, we can denote the terms for $\lambda=1$, $\lambda=2$, and $\lambda=3$ in (3.2.18).

Example 3.5. Find all the undetermined coefficients for the third-order Runge–Kutta algorithm, and write down the formulas.

Solutions. If one selects $m=3$, then six equations can be established as follows:

$$\begin{cases} \beta_{21}=\alpha_2,\quad \beta_{31}+\beta_{32}=\alpha_3,\\ \gamma_1+\gamma_2+\gamma_3=1,\quad \gamma_2\alpha_2+\gamma_3\alpha_3=\dfrac{1}{2},\quad \gamma_2\alpha_2^2+\gamma_3\alpha_3^2=\dfrac{1}{3},\\ \gamma_3\beta_{32}\alpha_2=\dfrac{1}{6}. \end{cases}$$

It can be seen that there are eight undetermined variables to define. One should then assign artificially two of them. For simplicity, one may assign the values α_i. For instance, select $\alpha_2=1/2$ and $\alpha_3=1$. With MATLAB symbolic computation, the other coefficients can be solved directly from the following algebraic equations:

```
>> syms b21 b31 b32 g1 g2 g3
   a2=1/2; a3=1;
   [b21 b31 b32 g1 g2 g3]=solve(b21==a2,b31+b32==a3,...
      g1+g2+g3==1,g2*a2+g3*a3==1/2, g2*a2^2+g3*a3^2==1/3,...
      g3*b32*a2==1/6)   % find Runge–Kutta algorithm coefficients
```

The solution obtained is

$$\beta_{21} = \frac{1}{2}, \quad \beta_{31} = -1, \quad \beta_{32} = 2, \quad \gamma_1 = \frac{1}{6}, \quad \gamma_2 = \frac{2}{3}, \quad \gamma_3 = \frac{1}{6}.$$

Based on this result, the third-order Runge–Kutta algorithm is formulated as

$$\boldsymbol{x}_{k+1} = \boldsymbol{x}_k + \frac{h}{6}(\boldsymbol{k}_1 + 4\boldsymbol{k}_2 + \boldsymbol{k}_3)$$

where

$$\begin{cases} \boldsymbol{k}_1 = \boldsymbol{f}(t_k, \boldsymbol{x}_k), \\ \boldsymbol{k}_2 = \boldsymbol{f}(t_k + h/2, \boldsymbol{x}_k + h\boldsymbol{k}_1/2), \\ \boldsymbol{k}_1 = \boldsymbol{f}(t_k + h, \boldsymbol{x}_k + h(-\boldsymbol{k}_1 + 2\boldsymbol{k}_2)). \end{cases}$$

Of course, α_2 and α_3 can be assigned to other values. For instance, letting $\alpha_2 = 1/2$ and $\alpha_3 = 3/4$, the corresponding algebraic equations can be solved again:

```
>> syms b21 b31 b32 g1 g2 g3
   a2=1/2; a3=3/4;
   [b21 b31 b32 g1 g2 g3]=solve(b21==a2,b31+b32==a3,...
      g1+g2+g3==1,g2*a2+g3*a3==1/2, g2*a2^2+g3*a3^2==1/3,...
      g3*b32*a2==1/6)
```

The solution found is

$$\beta_{21} = \frac{1}{2}, \quad \beta_{31} = 0, \quad \beta_{32} = \frac{3}{4}, \quad \gamma_1 = \frac{2}{9}, \quad \gamma_2 = \frac{1}{3}, \quad \gamma_3 = \frac{4}{9}.$$

So another third-order Runge–Kutta algorithm can be formulated as

$$\boldsymbol{x}_{k+1} = \boldsymbol{x}_k + \frac{h}{9}(2\boldsymbol{k}_1 + 3\boldsymbol{k}_2 + 4\boldsymbol{k}_3)$$

where

$$\begin{cases} \boldsymbol{k}_1 = \boldsymbol{f}(t_k, \boldsymbol{x}_k), \\ \boldsymbol{k}_2 = \boldsymbol{f}(t_k + h/2, \boldsymbol{x}_k + h\boldsymbol{k}_1/2), \\ \boldsymbol{k}_3 = \boldsymbol{f}(t_k + 2h/3, \boldsymbol{x}_k + 3h\boldsymbol{k}_2/4). \end{cases}$$

The Butcher tableaux for the two algorithms can be described by

$$\begin{array}{r|ccc} 0 & & & \\ 1/2 & 1/2 & & \\ 1 & -1 & 2 & \\ \hline & 1/6 & 2/3 & 1/6 \end{array} \qquad \begin{array}{r|ccc} 0 & & & \\ 1/2 & 1/2 & & \\ 2/3 & 0 & 3/4 & \\ \hline & 2/9 & 1/3 & 4/9 \end{array}$$

It may be difficult to derive formulas for the fourth- or even high-order Runge–Kutta algorithms. The interested readers may refer to [25, 65], where the fourth-order Runge–Kutta algorithm requires solving 11 algebraic equations, with 13 unknowns. Two of the unknowns should be assigned artificially. For instance, one may select α_2 and α_3 to find others.

3.2.6 Multistep algorithms and implementation

The Runge–Kutta methods discussed so far are one-step methods. That is, we compute $\boldsymbol{x}_{k+1}$ based on the given $\boldsymbol{x}_k$. Although in the solution process, several intermediate variables $\boldsymbol{k}_i$ are computed, they can be regarded as interpolations within a step-size h. In real applications, if some initial points $\boldsymbol{x}_0, \boldsymbol{x}_1, \dots$ are known, multistep algorithms can be introduced.

The main idea of a linear multistep method is to express the solution of the differential equation in the form of linear difference equations, and use a recursive method to find the approximate solutions of the differential equations. The typical mathematical formula for a linear multistep method is as follows:

$$\boldsymbol{x}_{k+m} = -a_m\boldsymbol{x}_{k+m-1} - \cdots - a_1\boldsymbol{x}_k + h\big[b_m\boldsymbol{f}(t_{n+m}, \boldsymbol{x}_{k+m}) + \cdots + b_1\boldsymbol{f}(t_k, \boldsymbol{x}_k)\big]. \tag{3.2.20}$$

The so-called linear multistep methods need to determine the weights a_i and b_i, $i = 1, 2, \dots, m$, and use a recursive method to find the numerical solutions of the differential equations. The commonly used ones include the fourth-order Adams–Bashforth and Adams–Mouton algorithms.

Theorem 3.9. *Selecting $a_m = 1$ and $a_i = 0$, $i = 1, 2, \dots, m-1$, and computing suitable values of b_i, the fourth-order Adams–Bashforth formula can be established to find numerical solutions of the differential equations:*

$$\boldsymbol{x}_{k+1} = \boldsymbol{x}_k + \frac{h}{24}\Big[55\boldsymbol{f}(t_k, \boldsymbol{x}_k) - 59\boldsymbol{f}(t_{k-1}, \boldsymbol{x}_{k-1}) + 37\boldsymbol{f}(t_{k-2}, \boldsymbol{x}_{k-2}) - 9\boldsymbol{f}(t_{k-3}, \boldsymbol{x}_{k-3})\Big] \tag{3.2.21}$$

where $k = 3, 4, \dots$ The accuracy of the algorithm is $o(h^4)$.[65]

Theorem 3.10. *The numerical solution with the fourth-order Adams–Mouton algorithm is*

$$\boldsymbol{x}_{k+1} = \boldsymbol{x}_k + \frac{h}{24}\Big[9\boldsymbol{f}(t_{k+1}, \boldsymbol{x}_{k+1}) + 19\boldsymbol{f}(t_k, \boldsymbol{x}_k) + 37\boldsymbol{f}(t_{k-1}, \boldsymbol{x}_{k-1}) - 9\boldsymbol{f}(t_{k-2}, \boldsymbol{x}_{k-2})\Big] \tag{3.2.22}$$

where $k = 2, 3, \dots$ The accuracy is also $o(h^4)$.[65]

Since in the Adams–Mouton algorithm, when computing x_{k+1}, the same variable appears also on the right-hand side of the formula, this may cause problems in actual computation. Sometimes a predictor–corrector method may be needed to implement the algorithm. This algorithm is not discussed further in this book. The implementation of the fourth-order Adams–Bashforth algorithm is studied instead.

It can be seen from the algorithm that, if x_3 (or in Adams–Mouton formula, x_2) can be provided, the algorithm can be initiated. In actual initial value problems, only the initial states x_0 are given, the other ones, such as x_3, are not known. Other methods should be used to determine the initial states. In a classical linear multistep method, Euler's method is used to determine x_1, and from x_0 and the newly found x_1, a low-accuracy trapezoidal method is used to approximate x_2, and three-step methods can be used to find x_3. It is obvious that there may exist relatively large errors at the beginning, which may impact all other subsequent computations. Therefore this algorithm is not a good choice. High precision algorithms should be introduced to reconstruct the initial points. For instance, the fourth-order Runge–Kutta algorithm can be embedded into the solver to compute x_1, x_2, and x_3:

```
function [t,x]=ode_adams(f,vec,x0)
if length(vec)==3, h=vec(2); t=[vec(1):h:vec(3)]';
else, h=vec(2)-vec(1); t=vec(:); end
[t0,x]=ode_rk4(f,[vec(1),h,vec(1)+3*h],x0)
n=length(t); x=[x; zeros(n-4,length(x0))];
for i=4:n-1  % find the numerical solutions with multistep algorithm
   k1=f(t(i),x(i,:)); k2=f(t(i-1),x(i-1,:));
   k3=f(t(i-2),x(i-2,:)); k4=f(t(i-3),x(i-3,:));
   x1=55*k1-59*k2+37*k3-9*k4; x1=x(i,:).'+h*x1(:)/24;
   x1=x1(:); x(i+1,:)=x1';
end
```

Example 3.6. Solve the differential equation in Example 3.1 with Adams–Bashforth method and assess the accuracy.

Solutions. In contrast to the statements in the previous examples, the following MATLAB commands can be used, and for different step-sizes, the numerical solution can be found, as well as the errors assessed.

```
>> f=@(t,x)[x(2); -2*x(1)]; t0=0; tn=5; sq=sqrt(2);
   H=[0.001,0.01,0.02,0.05,0.1,0.2]; E=[];
   for h=H  % try different step-sizes for the difference equation
      [t1,x1]=ode_adams(f,[t0,h,tn],[1;0]); % solve differential equation
      sol=[cos(sq*t1), -sq*sin(sq*t1)];      % analytical solution
```

```
    e1=norm(sol-x1,inf); E=[E,e1];          % maximum error
end
```

For the same differential equation, different solvers were tried in the previous examples. The accuracy information is collected as shown in Table 3.1 for different algorithms. It can be seen that the fourth-order Runge–Kutta and Gill's algorithms yield almost the same accuracy. Besides, the fourth-order Runge–Kutta algorithm (with $o(h^4)$) is evidently more accurate than the second-order Runge–Kutta algorithm (with $o(h^2)$) and Euler's method (with $o(h)$). Even though a large step-size such as $h = 0.2$ is used, rather high accuracy results can still be witnessed for the $o(h^4)$ algorithms. Although Adams–Bashford linear multistep algorithm claims to have $o(h^4)$ accuracy, its behavior is much worse than that of the fourth-order Runge–Kutta algorithm for this particular problem.

Table 3.1: Comparison of maximum errors for different algorithms.

algorithm	$h = 0.001$	$h = 0.01$	$h = 0.02$	$h = 0.05$	$h = 0.1$	$h = 0.2$
Euler's method	0.00856	0.08741	0.1789	0.4792	1.069	2.554
2nd order RK algorithm	4.025×10^{-6}	0.000403	0.00161	0.01013	0.040735	0.16457
4th order RK algorithm	4.071×10^{-13}	4.03×10^{-9}	6.46×10^{-8}	2.535×10^{-6}	4.076×10^{-5}	0.00065
Gill's algorithm	4.070×10^{-13}	4.03×10^{-9}	6.46×10^{-8}	2.535×10^{-6}	4.076×10^{-5}	0.00065
multistep	1.68×10^{-11}	1.68×10^{-7}	2.68×10^{-6}	0.0001	0.0016	0.0229

Compared with the one-step methods such as Runge–Kutta methods, the disadvantages are that other methods should be used to compute the first few points, so that the multistep algorithms can be started. The accuracy of the computed initial points may impact the total accuracy of the algorithm. In the implementation, the initial points are evaluated by the fourth-order Runge–Kutta algorithm. From another viewpoint, multistep methods introduced here are only applicable to fixed-step algorithms, otherwise the previous information cannot be adopted. Variable multistep methods may have even more complicated structures which are not easy to implement.[32] They are not discussed further in this book.

3.3 Variable-step numerical algorithms and implementations

All the algorithms studied so far were fixed-step ones. That is, we selected a constant step-size h and used them step-by-step to recursively find the numerical solutions of the differential equations. If the order of the algorithm is selected, the way of increase the accuracy is to reduce the step-size h. However, in real applications, the step-size h cannot be reduced infinitely. There are two reasons for that:

(1) **Slow computational speed.** For the same time of interest, reducing step-size implies increasing the total number of points in the interval, therefore the speed may be slow.
(2) **Increased cumulative error.** No matter how small the step-size is selected, there will be a roundoff error. If the step-size is reduced, the number of points will be increased, and the roundoff error may have more time to propagate, which implies a larger cumulative error. The roundoff, cumulative and total errors are illustrated in Figure 3.3.

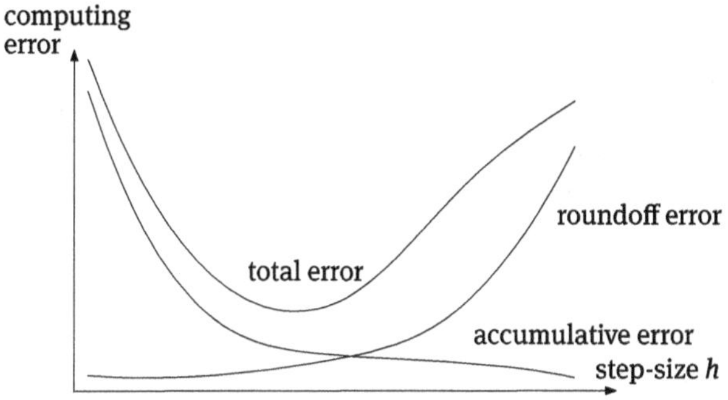

Figure 3.3: Illustration of the errors.

In this section, variable-step algorithms are mainly discussed. With the variable-step solver provided in MATLAB, the solutions of ordinary differential equations are demonstrated through examples. The syntaxes of the solver are addressed so as to better solve differential equations.

3.3.1 Measures to increase efficiency

Although fixed-step algorithms are mainly taught in the courses such as numerical analysis, almost nobody uses them in practice, since there are too many limitations. If one wants to increase the efficiency in solving differential equations, the following measures can be considered:

(1) **Selecting a suitable step-size.** As in the case of Euler's method, the step-size should be properly chosen. It should not be too small or too large.
(2) **Improving the accuracy of the algorithm.** Since the accuracy of Euler's method is too low, better algorithms should be selected. For instance, the fourth-order Runge–Kutta algorithm is a better choice.

(3) **Using variable-step mechanism.** The "suitable" step-size mentioned earlier is a vague concept. In fact, many variable-step algorithms are available, which allow changing the step-size adaptively. When the error is small, a larger step-size can be automatically chosen to increase the speed, while when the error is too large, a smaller step-size is adopted to ensure the accuracy. Variable-step algorithms are the top choice in solving differential equations.

3.3.2 An introduction to variable-step algorithms

The principles of some of the variable-step algorithms are illustrated in Figure 3.4. If the state vector $\boldsymbol{x}_k$ at time t_k is known, the initial step-size h can be used to compute the state $\tilde{\boldsymbol{x}}_{k+1}$ at time $t_k + h$. On the other hand, the step-size can be reduced by half such that the state vector at $\hat{\boldsymbol{x}}_{k+1}$ can be evaluated in two steps. If the error using the two step-sizes $\boldsymbol{\epsilon} = \|\hat{\boldsymbol{x}}_{k+1} - \tilde{\boldsymbol{x}}_{k+1}\|$ is smaller than the prespecified error tolerance, the step-size can be used, or increased; if the error is too large, the step-size should be reduced, and checked again. An adaptive variable-step algorithm monitors the error in the solution process, and adjusts the step-size whenever necessary, to ensure fast speed and high accuracy.

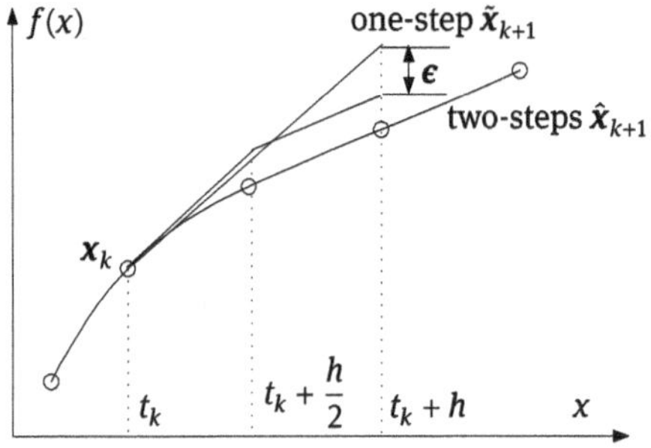

Figure 3.4: Illustration of a variable-step mechanism.

An alternative variable-step mechanism is that two methods can be used to compute $\boldsymbol{x}_{k+1}$ and $\hat{\boldsymbol{x}}_{k+1}$. If $\|\boldsymbol{x}_{k+1} - \hat{\boldsymbol{x}}_{k+1}\| \leqslant \varepsilon$, then the current step-size can be selected, or even increased to ensure fast speed; If the condition is not satisfied, a smaller step-size should be taken.

A fixed-step algorithm is similar to the case in open-loop control, where the step-size is selected once. The step-size is used throughout the solution process. The algorithm does not care about whether the error is acceptable or not. On the other hand, the variable-step algorithms behave more like closed-loop control, where the error is monitored throughout the solution process. If the error is large, the step-size can be adjusted automatically to ensure the accuracy of the solution. Meanwhile, since sometimes very large step-sizes are allowed in variable-step algorithms, the variable-step algorithms are more efficient in terms of speed.

3.3.3 The 4/5th order Runge–Kutta variable-step algorithm

Erwin Felhberg, a researcher working for NASA, improved the classical Runge–Kutta method,[25] where within each computation step, the $f_i(\cdot)$ function is evaluated 6 times, so as to ensure high precision and numerical stability. The efficiency of the algorithm is high and the accuracy can be controlled by the user. Therefore it is usually used in practical applications.

Theorem 3.11. *Assuming that the current step-size is h_k, 6 immediate variables $\boldsymbol{k}_i$ are evaluated as*

$$\boldsymbol{k}_i = \boldsymbol{f}\left(t_k + \alpha_i h_k, \boldsymbol{x}_k + \sum_{j=1}^{i-1} \beta_{ij}\boldsymbol{k}_j\right), \quad i = 1, 2, \dots, 6 \tag{3.3.1}$$

where t_k is the current time, and the coefficients α_i, β_{ij}, and γ_i are shown in Table 3.2. The coefficients α_i and β_{ij} are also referred to as Dormand–Prince pairs. The state variable in the next step can be computed from

$$\boldsymbol{x}_{k+1} = \boldsymbol{x}_k + \sum_{i=1}^{6} \gamma_i h_k \boldsymbol{k}_i. \tag{3.3.2}$$

Table 3.2: Butcher tableau for the 4/5th order RKF coefficients.

0						
1/4	1/4					
3/8	3/32	9/32				
12/13	1 932/2 197	−7 200/2 197	7 296/2 197			
1	439/216	−8	3 680/513	−845/4 104		
1/2	−8/27	2	−3 544/2 565	1 859/4 104	−11/40	
γ_i	16/135	0	6 656/12 825	28 561/56 430	−9/50	2/55
γ_i^*	25/216	0	1 408/2 565	2 197/4 104	−1/5	0

Of course, a fixed-step algorithm can be formulated like this. While in real applications, an alternative formula can be used to evaluate the solution with another method. The coefficients γ_i^* are used. The errors are defined as follows:

$$\boldsymbol{\epsilon}_k = \sum_{i=1}^{6} (\gamma_i - \gamma_i^*) h \boldsymbol{k}_i. \tag{3.3.3}$$

The value of the error can be used to update the step-size. The algorithm can be used to adaptively change the step-size, and ensure the accuracy is satisfied. In the algorithm, the result obtained with y_i is of $o(h^5)$ accuracy, while that used in monitoring the error, i. e., evaluating y_i^*, is of $o(h^4)$ accuracy. Therefore the algorithm is also known as the 4/5th order Runge–Kutta–Felhberg (RKF) algorithm.

3.3.4 The differential equation solver provided in MATLAB

It can be seen from the numerical examples studied earlier that the fixed-step algorithms are not really used in practical situations. Examples are given to illustrate differential equations with variable-step algorithms. Comparative studies are also made for existing MATLAB solvers.

Function `ode45()` is provided in MATLAB to solve first-order explicit differential equations. The variable-step 4/5th order Runge–Kutta–Felhberg algorithm is adopted. The syntaxes of the function are

```
[t,x]=ode45(Fun, [t0,tn],x0),                 %direct solution
[t,x]=ode45(Fun, [t0,tn],x0,options),         %with control options
[t,x]=ode45(Fun, [t0,tn],x0,options,p1,p2,...,pm)
```

in the latter, additional parameters $p_1, p_2, \ldots$ are allowed. The differential equation can be described by either anonymous or MATLAB function, with `Fun` as its function handle. Examples will be used later to demonstrate the use of function handles. The argument `tspan` is used to describe the solution interval. Normally `tspan=`$[t_0, t_n]$, which is used to describe the solution interval, and if only one value, t_n, is used, the default initial time is $t_0 = 0$, and t_n is the terminal time. Apart from these, the initial state vector $\boldsymbol{x}_0$ should be supplied.

Besides, in the function call, $t_n < t_0$ is allowed, meaning that backward solutions of differential equations are possible, where t_0 can be regarded as terminal time, and t_n the initial time. Vector $\boldsymbol{x}_0$ stores the terminal state value. The function can then be used in solving terminal value problems directly.

Apart from the solver `ode45()`, other similar solvers such as `ode113`, `ode15s()`, and `ode23()` are also provided. They share the same syntax, but the internal algorithms inside the functions are different.

The key step in differential equation solution is to write a MATLAB function to describe the first-order explicit differential equation. The interface of the function should be

```
function xd=funname(t,x),                   %without additional parameters
function xd=funname(t,x,p1,p2,...,pm),      %with additional parameters
```

where scalar t is the time variable, so that time-varying differential equations can also be handled. If the equation does not explicitly contain t, it should still be used to hold the space, otherwise errors may occur in the MATLAB solution process. The other input argument $\boldsymbol{x}$ is the state vector. The returned argument $\boldsymbol{x}_\mathrm{d}$ computes the derivative of the state vector.

In some application examples, there are cases where additional parameters should be assigned. These parameters can be passed to the model with variables $p_1, p_2, \ldots, p_m$, which should be exactly matched. Examples will be given later to demonstrate in detail the syntaxes of the solvers.

Example 3.7. Consider the well-known Lorenz model with the state space equation

$$\begin{cases} x_1'(t) = -\beta x_1(t) + x_2(t)x_3(t), \\ x_2'(t) = -\rho x_2(t) + \rho x_3(t), \\ x_3'(t) = -x_1(t)x_2(t) + \sigma x_2(t) - x_3(t) \end{cases}$$

where the coefficients are $\beta = 8/3$, $\rho = 10$, and $\sigma = 28$. The initial values are $x_1(0) = x_2(0) = 0$ and $x_3(0) = \epsilon$, where ϵ is a small quantity recognizable by the machine. For instance, let $\epsilon = 10^{-10}$. Solve the differential equation.

Solutions. Since the system of differential equations is nonlinear, there is no analytical solution. Numerical solution is the only choice in studying such a system. To solve this equation system, it must be converted into the standard form $\boldsymbol{x}'(t) = \boldsymbol{f}(t, \boldsymbol{x}(t))$. Constructing the state variable vector $\boldsymbol{x}(t) = [x_1(t), x_2(t), x_3(t)]^\mathrm{T}$, the original system of differential equations can be written in the following format:

$$\boldsymbol{x}'(t) = \boldsymbol{f}(t, \boldsymbol{x}(t)), \quad \text{where } \boldsymbol{f}(t, \boldsymbol{x}(t)) = \begin{bmatrix} -\beta x_1(t) + x_2(t)x_3(t) \\ -\rho x_2(t) + \rho x_3(t) \\ -x_1(t)x_2(t) + \sigma x_2(t) - x_3(t) \end{bmatrix}.$$

In the actual solution process, vector $\boldsymbol{f}(t, \boldsymbol{x}(t))$ should be expressed in MATLAB. For instance, an anonymous function can be written to declare the dynamic system model, with

```
>> f=@(t,x)[-8/3*x(1)+x(2)*x(3);
            -10*x(2)+10*x(3);
            -x(1)*x(2)+28*x(2)-x(3)];
```

Function `ode45()` can then be called to numerically solve the differential equation described by the anonymous function f. A graphical display of the solution can be found when the numerical solutions are obtained.

```
>> tn=100; x0=[0;0;1e-10];
   [t,x]=ode45(f,[0,tn],x0); plot(t,x) %find numerical
```

where t_n is the terminal time and $\boldsymbol{x}_0$ is the initial state. The relationship between the states versus time is obtained as shown in Figure 3.5.

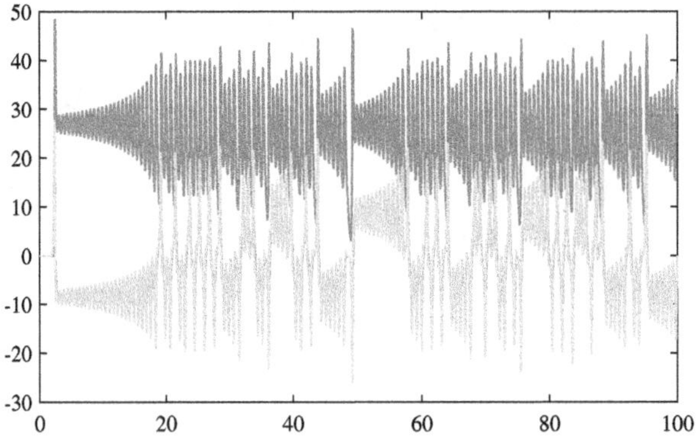

Figure 3.5: Time responses of the states in Lorenz equation.

If the state space trajectory of the three states is expected, the following statements can be used, and the result is shown in Figure 3.6.

```
>> plot3(x(:,1),x(:,2),x(:,3)); grid %state space trajectory
```

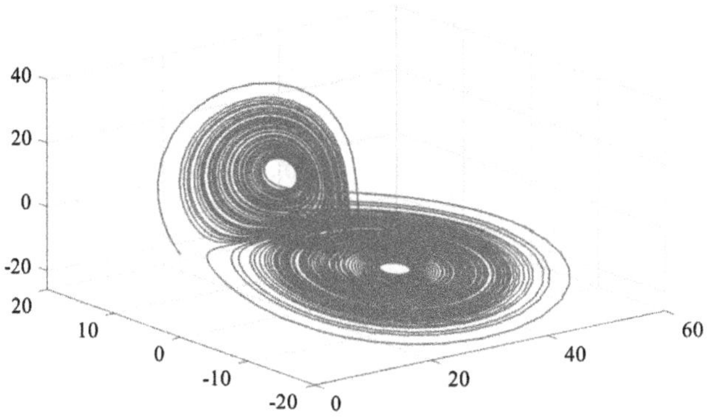

Figure 3.6: State space trajectory of Lorenz equation.

It can be seen that a complicated nonlinear differential equation with three states can be solved directly by a few simple statements. Besides, simple scientific visualization commands can be used to display various results. This is the reason why MATLAB is selected as the working language in the book.

In fact, the best way to observe the phase space trajectory is by using function `comet3()`, where animation is used to show the trajectory dynamically. The reader only needs to change the last command to `comet3(x(:, 1),x(:,2),x(:,3))` to show the animation.

Lorenz equation was proposed by an American meteorologist Edward Norton Lorenz (1919–2008) in 1963, which is the mathematical model describing atmospheric convection.[49] The order of the states is rearranged slightly from the original model, to better display the shape of the butterfly in the phase state trajectory. Lorenz initiated the research of chaos. More information on chaos will be given in Chapter 7.

Now let us summarize the procedures in numerical solutions of ordinary differential equations:

(1) **Express the standard form of differential equation**, which is $\boldsymbol{x}'(t) = \boldsymbol{f}(t, \boldsymbol{x}(t))$, with given $\boldsymbol{x}(t_0)$.
(2) **Use MATLAB to describe the differential equation.** Either anonymous or MATLAB function can be used to describe the standard form of the differential equation. Note that the input argument t must be used, even though the original equation does not explicitly contain t.
(3) **Solve the differential equation.** The solver `ode45()` or other related function can be called directly to solve the equation.
(4) **Validate the solution.** Solution validation is the most important step in numerical solutions. This topic will be further discussed in later sections.

With the standardized procedures in numerical solutions, examples will be shown later to illustrate the solution of various differential equations.

Example 3.8. Consider the Lotka–Volterra's predator–prey equations in Example 1.5, given again as follows. Solve the differential equation and draw the solutions in curves:

$$\begin{cases} x'(t) = 4x(t) - \alpha x(t)y(t), \\ y'(t) = \beta x(t)y(t) - 3y(t) \end{cases}$$

where $\alpha = 2$ and $\beta = 1$. The initial values are $x(0) = 2$, $y(0) = 3$.

Solutions. The functions to be solved for are $x(t)$ and $y(t)$, and in the standard form we need the $\boldsymbol{x}(t)$ vector. The state variables should be introduced. For instance, letting $x_1(t) = x(t)$ and $x_2(t) = y(t)$, the standard form of the differential equations can be written as

$$\boldsymbol{x}'(t) = \boldsymbol{f}(t, \boldsymbol{x}(t)), \quad \text{where } \boldsymbol{f}(t, \boldsymbol{x}(t)) = \begin{bmatrix} 4x_1(t) - 2x_1(t)x_2(t) \\ x_1(t)x_2(t) - 3x_2(t) \end{bmatrix}, \quad \boldsymbol{x}(0) = \begin{bmatrix} 2 \\ 3 \end{bmatrix}.$$

With the standard model, an anonymous function can be written to describe the differential equation, and the relation between $x(t)$ and $y(t)$ can be drawn, also known as the phase plane trajectory, as shown in 3.7. The elapsed time for the solution process is 0.00648 seconds, and the number of points computed is 153.

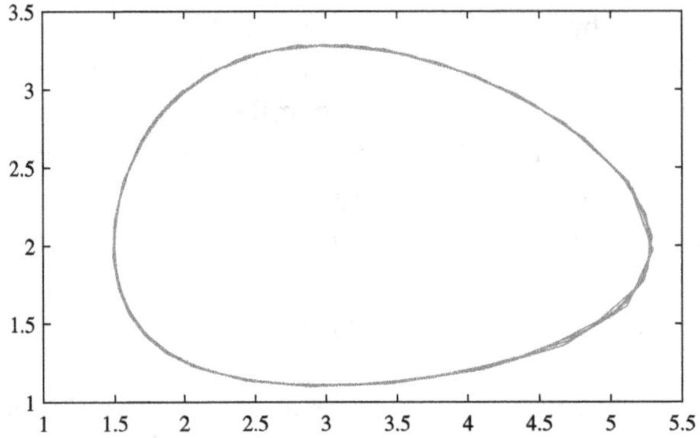

Figure 3.7: Phase plane trajectory (the curves are rather rough).

```
>> x0=[2; 3]; tn=10;
   f=@(t,x)[4*x(1)-2*x(1)*x(2); x(1)*x(2)-3*x(2)];
   tic, [t,x]=ode45(f,[0,tn],x0); toc %solve and measure time
   length(t), plot(x(:,1),x(:,2))     %draw phase plane trajectory
```

It can be seen from the example that, if the equation can be expressed by an anonymous or MATLAB function, the solver `ode45()` can be called to find the numerical solutions directly. It can be seen that writing a MATLAB function is the key step in the numerical solution process.

Now let us consider an even more complicated example and see how this example can be solved with MATLAB.

Example 3.9. Consider the three-body model.[13] The notation (t) is omitted from the following mathematical model:

$$\begin{cases} y_1' = y_4, \\ y_2' = y_5, \\ y_3' = y_6, \\ y_4' = 2y_5 + y_1 - \dfrac{\mu(y_1+\mu-1)}{\sqrt{(y_2^2+y_3^2+(y_1+\mu-1)^2)^3}} - \dfrac{(1-\mu)(y_1+\mu)}{\sqrt{(y_2^2+y_3^2+(y_1+\mu)^2)^3}}, \\ y_5' = -2y_4 + y_2 - \dfrac{\mu y_2}{\sqrt{(y_2^2+y_3^2+(y_1+\mu-1)^2)^3}} - \dfrac{(1-\mu)y_2}{\sqrt{(y_2^2+y_3^2+(y_1+\mu)^2)^3}}, \\ y_6' = -\dfrac{\mu y_3}{\sqrt{(y_2^2+y_3^2+(y_1+\mu-1)^2)^3}} - \dfrac{(1-\mu)y_3}{\sqrt{(y_2^2+y_3^2+(y_1+\mu)^2)^3}} \end{cases}$$

where the two larger bodies are the Earth and the Moon, while the smaller one is a satellite. The spatial coordinates of the satellite are expressed by y_1, y_2, and y_3, while y_4, y_5, and y_6 are the projections of the speed onto the axes. Assume that the initial state values are given by

$$\boldsymbol{y}_0 = [0.994, 0, 0, 0, -2.0015851063790825224, 0]^{\mathrm{T}},$$

and $\mu = 1/81.45$ is a constant. Draw the spatial trajectory of the satellite.

Solutions. It can be seen from the first-order explicit differential equations that anonymous functions can be used to express the equations. Then the solver `ode45()` can be called to solve the differential equations. It is not hard to see from the mathematical model that there are common terms,

$$\sqrt{(y_2^2 + y_3^2 + (y_1 + \mu - 1)^2)^3} \text{ and } \sqrt{(y_2^2 + y_3^2 + (y_1 + \mu)^2)^3},$$

which appear many times. It is better to compute them and use as pre-assigned variables so that they can be reused in other cases. Since anonymous functions do not support intermediate variables, MATLAB functions can be used instead to describe the differential equations:

```
function dy=threebody(t,y)
mu0=1/81.45;
D1=sqrt((y(2)^2+y(3)^2+(y(1)+mu0-1)^2)^3);
D2=sqrt((y(2)^2+y(3)^2+(y(1)+mu0)^2)^3);
dy=[y(4:6);
    2*y(5)+y(1)-mu0*(y(1)+mu0-1)/D1-(1-mu0)*(y(1)+mu0)/D2;
    -2*y(4)+y(2)-mu0*y(2)/D1-(1-mu0)*y(2)/D2;
    -mu0*y(3)/D1-(1-mu0)*y(3)/D2];
```

Selecting $t_0 = 0$ and $t_n = 40$, the following commands can be used to solve the differential equations, and the trajectory of the satellite can be obtained, as shown in Figure 3.8. Unfortunately, the result obtained is not correct. This brings us a new

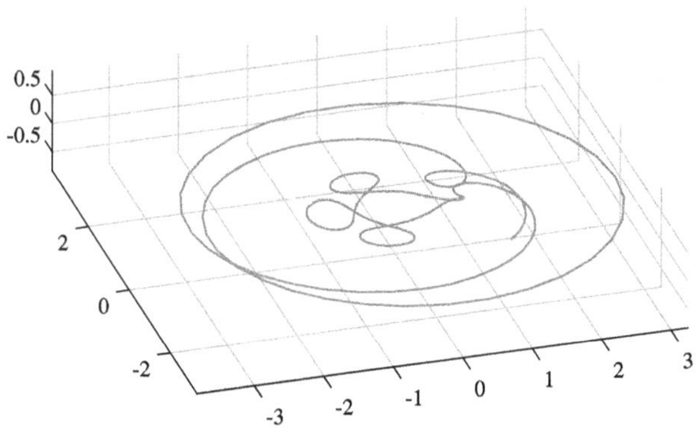

Figure 3.8: The phase space trajectory (erroneous results).

question: How to decide whether a solution is correct or not?

```
>> y0=[0.994,0,0,0,-2.001585106379082522 4,0]';
   [t,y]=ode45(@threebody,[0,40],y0); % solve the differential equations
   plot3(y(:,1),y(:,2),y(:,3))         % draw the phase space trajectory
```

3.3.5 Solutions of differential equations with additional parameters

In the solution process, the aim of introducing additional parameters is that, if a differential equation contains certain parameters, and different values of the parameters are to be tried, they can be used as additional parameters, such that there is no need for the user to modify the model file every time the parameters are changed. Consider the case in Example 3.7 where there are parameters β, ρ, and σ in the Lorenz equation. If they are selected as additional parameters, there is no need to change the MATLAB function for the Lorenz equation each time the parameters are changed.

Example 3.10. Write a MATLAB function containing additional parameters for the Lorenz equation in Example 3.7. Solve the equation again and compare the results. Select a new set of parameters $\beta = 2$, $\rho = 5$, and $\sigma = 20$. Solve numerically the Lorenz equation again.

Solutions. Selecting additional parameters β, ρ, and σ, the following MATLAB function can be written to describe the Lorenz equation:

```
function dx=lorenz1(t,x,b,r,s) % with additional parameters
dx=[-b*x(1)+x(2)*x(3);
    -r*x(2)+r*x(3); -x(1)*x(2)+s*x(2)-x(3)]; % the differential equations
```

Then the differential equations can be solved with the following commands. It can be seen that in the calling command, the variable names need not be the same as in the function. If the corresponding relationships are the same, the parameters can be passed to the variables correctly. In the `ode45()` function call, an empty matrix is used to indicate that the default control option is used.

```
>> b1=8/3; r1=10; s1=28; tn=100; % set additional parameters
   x0=[0;0;1e-10];               % set initial values
   [t,x]=ode45(@lorenz1,[0,tn],x0,[],b1,r1,s1);
   plot(t,x) % solve the differential equations and draw time response
```

If a three-dimensional phase space trajectory is needed, the following commands can be used. It can be seen that the results are exactly the same, as when an anonymous function was used.

```
>> plot3(x(:,1),x(:,2),x(:,3));   % draw phase space trajectory again
```

With the new MATLAB function to describe the Lorenz equation, the parameters β, ρ, and σ can be assigned to other values, and the user does not have to modify the model file `lorenz1.m`. For instance, selecting $\beta = 2$, $\rho = 5$, and $\sigma = 20$, the following commands can be used to find the numerical solutions directly. The results obtained are as shown in Figures 3.9 and 3.10.

```
>> tn=100; x0=[0;0;1e-10]; b2=2; r2=5; s2=20; % another set of data
   [t2,x2]=ode45(@lorenz1,[0,tn],x0,[],b2,r2,s2); % solve the equations
   plot(t2,x2), figure;                          % draw time response
   plot3(x2(:,1),x2(:,2),x2(:,3)); grid; % phase space trajectory
```

It can be seen from the results that if the parameters in the chaotic system are changed, the behavior of the system may also be changed. In this example, chaotic behaviors are no longer so apparent.

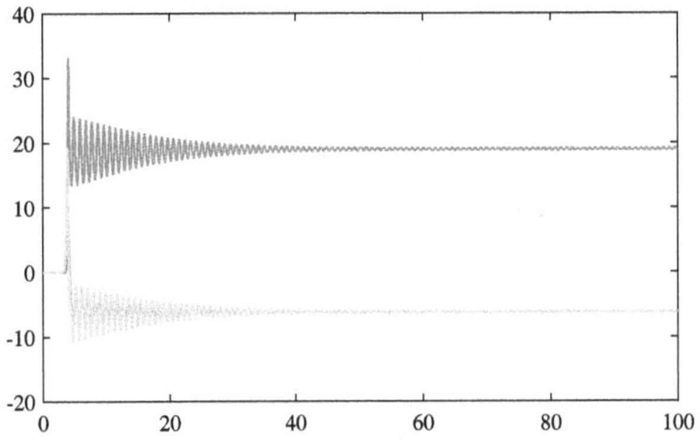

Figure 3.9: The state curves under the new parameters in Lorenz equation.

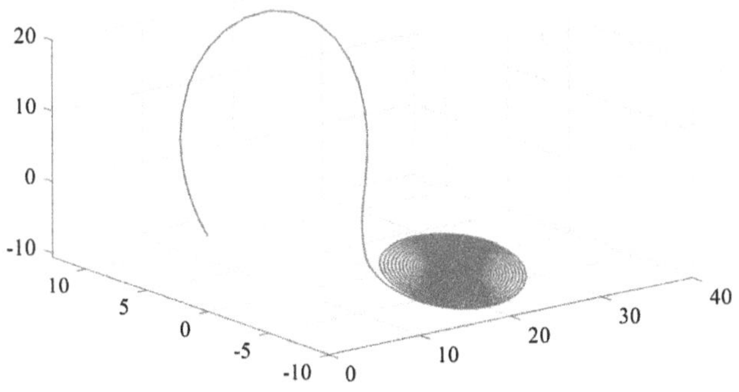

Figure 3.10: The phase space trajectory of Lorenz equation under new parameters.

If the differential equations are simple, anonymous functions can be used to describe them. In this case, there is no need to assign additional parameters, since the variables in MATLAB workspace can be extracted directly in anonymous functions. The following commands can then be used and identical results can be obtained:

```
>> b=8/3; r=10; s=28;             % use of anonymous function
   f=@(t,x)[-b*x(1)+x(2)*x(3); -r*x(2)+r*x(3);
            -x(1)*x(2)+s*x(2)-x(3)];
   [t,x]=ode45(f,[0,tn],x0); % solve the differential equations
   plot(t2,x2), figure;          % draw the time responses
   plot3(x2(:,1),x2(:,2),x2(:,3)); grid; % draw the phase space trajectory
```

3.3.6 Avoiding the use of additional parameters

In real applications, sometimes the use of additional parameters may not be very convenient, since in certain function calls, additional parameters are not allowed. For simple problems, anonymous functions can be used to express the differential equations, and the variables in MATLAB workspace can be extracted directly. There is no need to use additional parameters.

In other applications, the differential equations may be too complicated to describe. Intermediate variables are needed. Therefore, anonymous functions are not suitable to describe differential equations. MATLAB functions may become the only choice. In these cases, a method is demonstrated on how to avoid additional parameters.

An anonymous function can be designed as an interface to such a MATLAB function. Therefore the additional parameters in MATLAB workspace can be passed to the anonymous function, then to the MATLAB function through the interface. In this case, there is no need to use additional parameters. An example is given next to demonstrate the interface manipulations.

Example 3.11. Consider the differential equations model in Example 3.10, where additional parameters are introduced to describe the differential equations, and file `lorenz1.m` is written. Set up the differential equations solution mechanism such that no additional parameters are needed.

Solutions. In order to avoid additional parameters, the parameters should be specified first in MATLAB. Then an anonymous function can be written as an interface to `lorenz1.m` file. Therefore, the following commands can be used to solve again the differential equations. The results are identical to those obtained earlier.

```
>> b=8/3; r=10; s=28;               % input parameters to MATLAB workspace
   f=@(t,x)lorenz1(t,x,b,r,s) % write an interface with anonymous function
```

Therefore, the following commands can be used to solve again the differential equations. The results are identical to those obtained earlier.

```
>> tn=100; x0=[0;0;1e-10];
   [t,x]=ode45(f,[0,tn],x0); plot(t2,x2) %solve and draw
```

3.4 Validations of numerical solutions

It has been demonstrated that if a simulation algorithm or control parameters – such as the relative error tolerance – are not chosen correctly, unreliable or even wrong results can be obtained. Therefore the most important step in differential equation solution is to validate the results.

In real applications, most problems to be solved may not have analytical solutions. How to validate the solutions? The following methods can be tried to validate the results:

(1) A feasible way is to modify the control parameters. For instance, the acceptable error tolerance. The members `RelTol` and `AbsTol` in the control options can be set to smaller numbers, and one can observe the results to see whether consistent results can be found. If there are unacceptable differences, the error tolerances can be reduced and solutions checked again. The error tolerances in the control options can be set to the smallest possible values, such as `eps`, so as to find the most accurate and reliable results under the double precision framework.

(2) One can select different differential equation solution algorithms to apply cross-validation for the simulation results.

3.4.1 Validation of the computation results

Further adjustments to the control options can be made in the differential equation solution process. The control `options` can be modified with function `odeset()`. The variable `options` thus created is a structured variable. In Table 3.3, some of the commonly used members in the structured variable are listed. There are two ways to modify the members. One is to modify the options with `odeset()` function, the other is to directly change the members in the variable `options`. For instance, if one wants to set the relative error tolerance to 10^{-7}, the following two methods can be used:

```
options=odeset('RelTol',1e-7);
options=odeset; options.RelTol=1e-7;
```

Examples are given next to demonstrate the validation of numerical solutions of differential equations.

Table 3.3: Members in the control options of the solvers.

name	member explanation
`RelTol`	Relative error tolerance, with default value of 0.001 (i. e., 0.1 % of relative error). This value is too large in practice, and should be reduced in the solution
`AbsTol`	The absolute error tolerance, with default value of 10^{-6}. Again this value is too large and should be decreased
`MaxStep`	The maximum allowed step-size
`Mass`	The mass matrix in differential–algebraic equations, which should be assigned in describing differential–algebraic equations
`Jacobian`	Describes Jacobian matrix function $\partial \boldsymbol{f}/\partial \boldsymbol{x}$. It should be a function handle to describe Jacobian matrix, so as to speed up the simulation process
`Events`	Event response property, used to set event response function handle
`OutputFcn`	Call user-defined function in each successful computation

Example 3.12. It can be seen from Figure 3.7 that the curve is rather rough. It seems that there is something wrong in the precision setting. Select a more strict control option and see whether accurate results can be found.

Solutions. The relative error tolerance can be set to a very small value, for instance, 10^{-10}. The differential equations can be solved again, and the new phase plane trajectory can be obtained as shown in Figure 3.11. It can be seen that the curves are smooth. It implies that the precision is rather high. No better results may be found by further reducing the error tolerance.

```
>> x0=[2; 3]; tn=10;
   f=@(t,x)[4*x(1)-2*x(1)*x(2); x(1)*x(2)-3*x(2)];
   ff=odeset; ff.RelTol=1e-10; ff.AbsTol=1e-10;
```

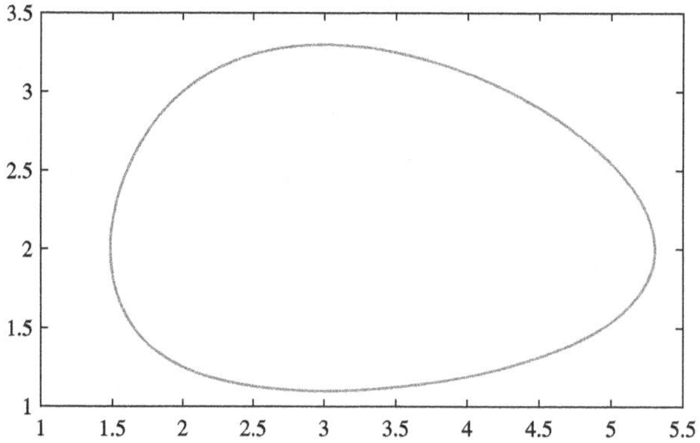

Figure 3.11: A smooth phase plane trajectory.

```
   tic, [t,x]=ode45(f,[0,tn],x0,ff); toc  % solve the equations
   length(t), plot(x(:,1),x(:,2))          % draw phase plane plot
```

Since the precision criterion is increased by 10^7 times, is it true that the computational load increased significantly? It can be seen that the elapsed time is 0.02007 seconds, and the number of points is increased to 3 593. It can be seen that the elapsed time has not increased significantly. It is worthwhile to use this method to find more accurate solutions.

If the error tolerances are set to `eps` (in fact, the minimum allowed error tolerance is 100`eps` for the solvers), the elapsed time is only increased to 0.0487 seconds, and the number of points is increased to 19 329.

```
>> ff.RelTol=eps; ff.AbsTol=eps;  % set error tolerances
   tic, [t,x]=ode45(f,[0,tn],x0,ff); toc, length(t)
```

Example 3.13. Solve again the problem in Example 3.9 with better accuracy.

Solutions. Since there is no analytical solution for the original problem, validation of the solution is not a simple thing. A reliable way is to set tough control options for solving the differential equations. It can be seen that the two members, `AbsTol` and `RelTol`, are the most important control options regarding the accuracy of the equations. If necessary, they can be set to the toughest `eps`. If under such tough control, accurate solutions can still not be found, it means that `ode45()` function is not suitable for solving this problem. Other alternative effective methods should be used to tackle it.

The two error tolerances can be set to the toughest `eps`, and the following commands can be used to find a numerical solution of the equation:

```
>> y0=[0.994,0,0,0,-2.001585106379082522 4,0]';
   ff=odeset; ff.AbsTol=eps; ff.RelTol=eps;
   tic, [t,y]=ode45(@threebody,[0,40],y0,ff); toc
   plot3(y(:,1),y(:,2),y(:,3)), length(t) % phase space trajectory
```

Running the above code, a warning message "Warning: `RelTol` has been increased to 2.22045×10^{-14}" indicates that the error tolerances are not allowed to be set to such small values. They have been set automatically to 2.22045×10^{-14}. This is the most accurate possible solution under the double precision data type. The new three-dimensional phase space trajectory obtained is as shown in Figure 3.12. It can be seen by closely observing the curves that the trajectory has the tendency to diverge at the point marked in the figure. If the terminal time is increased, divergent results can be found, and erroneous results may be obtained. It can also be seen that with `length()` function, it is found that the number of points computed is 83 485.

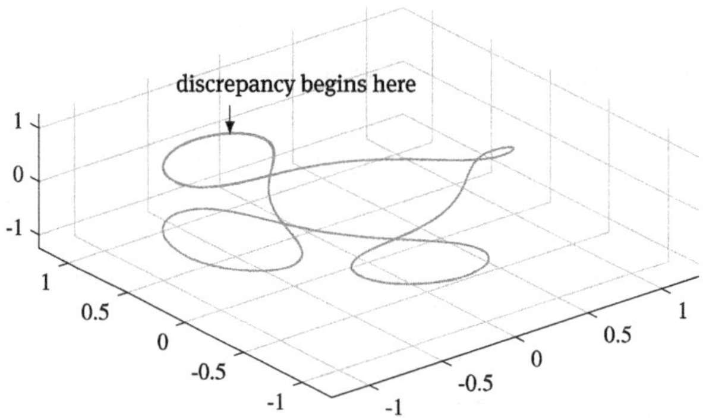

Figure 3.12: Accurate phase space trajectory.

3.4.2 Dynamic manipulation of intermediate results

In normal cases, the user may use an anonymous or MATLAB function to describe the differential equations, then call the solver `ode45()` to find the numerical solutions. If the user wants to display the intermediate results during the solution process, the member `OutputFcn` in the control options can be set to a user-defined function such that, when the solution at each point is successfully found, the MATLAB mechanism calls the user-defined function once automatically and the intermediate result can be handled.

Four existing functions are already provided in MATLAB to routinely handle the intermediate results. The four functions are:

`@odeplot` – dynamically draw the time response of the states;
`@odephas2` – draw the phase plane plot of the first two states;
`@odephas3` – draw the phase space plot of the first three states;
`@odeprint` – display the current time and states in digits.

An example is shown next to demonstrate how to handle intermediate results during the solution process.

Example 3.14. Consider the Lorenz equation in Example 3.7. Draw dynamically the phase space trajectory during the solution process.

Solutions. Compared with Example 3.7, no other changes should be made to the solution commands. The only thing to modify is the `OutputFcn` member of the control options. It can be set to `@odephas3`. To ensure the accuracy in the solutions, tough error tolerances can be set. The following statements can be used to solve the differential equations. During the solution process, the dynamic phase space trajectory can be drawn automatically.

```
>> f=@(t,x)[-8/3*x(1)+x(2)*x(3);
        -10*x(2)+10*x(3); -x(1)*x(2)+28*x(2)-x(3)];
   tn=100; x0=[0;0;1e-10];
   ff=odeset('RelTol',100*eps,'AbsTol',100*eps,...
             'OutputFcn',@odephas3); % modify several members
   [t,x]=ode45(f,[0,tn],x0,ff);         % show dynamic phase space plot
```

To better display the dynamic behaviors, in default options, exaggerated width in the dynamic trajectory is used in MATLAB. This is adjusted by the `'Marker'` option in the plotting facilities. If the user is not satisfied with that, he/she can write his/her own response function. Alternatively, the low-level command in file `odephas3.m` can be modified. More specifically, the following statement can be located with source code editing interface:

```
ud.line = plot3(y(1),y(2),y(3),'-o','Parent',TARGET_AXIS);
```

We can change the option `'-o'` into `'-'`. If necessary, other source files, such as `odeplot.m` and `odephas2.m`, can be modified accordingly.

3.4.3 More accurate solvers

It can be seen from the examples that, although very tough error tolerances are set in the control options, the simulation results may contain large errors. It seems that this is beyond the capabilities of the `ode45()` solvers. More accurate solvers may be needed.

An 8/7th order Runge–Kutta variable-step solver was developed by a Russian scholar Vasiliy Govorukhin, which is the solver `ode87()`.[29] The theoretical accuracy may reach $o(h^8)$. In each computation step, the model function is called 13 times. Therefore the efficiency may well exceed that of the `ode45()` solver. Higher accuracy solutions may be achieved. The syntaxes of the solver are exactly the same as those for the function `ode45()`. It should be noted that the error tolerances should not be assigned to too small values. Otherwise error message may be returned.

Example 3.15. Solve again the three-body problem in Example 3.9 again with the more accurate solver.

Solutions. The more accurate solver can be called again for the problem. It can be seen that the same divergence phenomenon may happen again, since the code is executed under the same double precision framework. Since the `ode45()` solver has already found the most accurate solution under the double precision framework, the difference is that the total number of points is significantly reduced to 2 511 – only 3 %

of the points with solver ode45(). It can be seen that the efficiency of this solver is much higher than that of ode45().

```
>> ff=odeset; ff.AbsTol=1e-15; ff.RelTol=1e-15;
   y0=[0.994,0,0,0,-2.001585106379082522 4,0]';
   [t,y]=ode87(@threebody,[0,40],y0,ff);   % solve equations
   plot3(y(:,1),y(:,2),y(:,3)), length(t) % phase space trajectory
```

It should be noted that since function ode87() is written and executed under the double precision framework, it may not yield a more accurate result than that obtained with the ode45() solver, if the error tolerances are already set to the toughest values. Besides, the two solvers can be executed to solve the same problem and the user can compare whether consistent results can be found. In this way, the solutions can be cross-validated.

3.4.4 Step-sizes and fixed-step display

In the ode45() solver provided in MATLAB and the third-party solver such as ode87(), variable-step algorithms are implemented. This is different from the functions such as ode_rk2() discussed earlier. Examples are used to show the benefits of variable-step algorithms. Also, we illustrate how to return fixed-step results using the variable-step solvers.

Example 3.16. Solve the differential equations in Example 3.8 again, and observe the changes in the step-sizes in the solution process.

Solutions. Since in the function call of ode45(), the time vector $\boldsymbol{t}$ can also be returned, differences (the latter term subtracted by the former term) of vector $\boldsymbol{t}$ can be taken, such that each step-size can be found. Differences can be obtained by calling function diff($\boldsymbol{t}$). Note that the length of the result is 1 less than that of $\boldsymbol{t}$ vector. The error tolerances are set to 100 times the machine precision eps, and the following commands can be used to solve the differential equations. The step-sizes during the solution process are shown in Figure 3.13. It can be seen that the step-sizes are changing in the period to ensure that the error tolerances are obeyed.

```
>> x0=[2; 3]; tn=10;
   f=@(t,x)[4*x(1)-2*x(1)*x(2); x(1)*x(2)-3*x(2)];
   ff=odeset; ff.RelTol=100*eps; ff.AbsTol=100*eps;
   [t,x]=ode45(f,[0,tn],x0,ff);                % solve the differential equations
   plot(t(1:end-1),diff(t)), min(diff(t)) % draw step-size curves
```

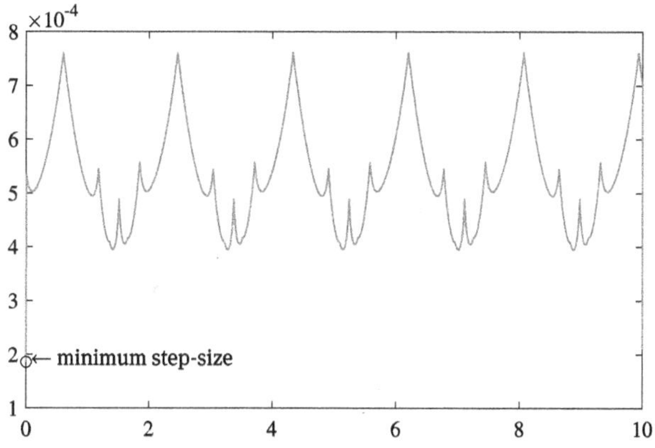

Figure 3.13: The step-size in the solution process.

It can be seen that at the initial stage, the smallest step-size was taken, and with `min()` function call, it is found that the minimum step-size at this point is 1.859×10^{-4}.

It has been shown that in the function call of the solver `ode45()`, the argument `tspan` can be assigned to a given time vector $\boldsymbol{t}$ and function `ode45()` can be called in variable-step format within each subinterval of vector $\boldsymbol{t}$. The exact expressions of the states at the instances in vector $\boldsymbol{t}$ can be found. Even if $\boldsymbol{t}$ is an equally spaced vector, it does not imply that a fixed-step algorithm is used in solving the differential equations.

Example 3.17. In Example 3.1, a differential equation is demonstrated. In the subsequent examples, many different fixed-step algorithms are compared. It is concluded that when h is relatively large, the errors become larger, and sometimes exceed the acceptable region. Use `ode45()` function to find the solutions at the points with step-size of $h = 0.2$.

Solutions. If the argument `tspan` is set to an equally spaced $\boldsymbol{t}$ vector, with increment of $h = 0.2$, and tough error tolerances are also selected, the solutions at the points in vector $\boldsymbol{t}$ can be found, and marked "o" in Figure 3.14. It can be seen that the points are exactly located on the analytical solution curves. The maximum error at the samples is 2.6423×10^{-14}, which is much more accurate than when using the fixed-step algorithms.

```
>> f=@(t,x)[x(2); -2*x(1)]; t0=0; tn=5; sq=sqrt(2);
   h=0.2; t=t0:h:tn;
   ff=odeset; ff.RelTol=100*eps; ff.AbsTol=100*eps;
   [t1,x1]=ode45(f,t,[1;0],ff); % find numerical solutions
   sol=[cos(sq*t1), -sq*sin(sq*t1)]; norm(sol-x1,inf)
   plot(t,sol,t1,x1,'o')             % compare solutions
```

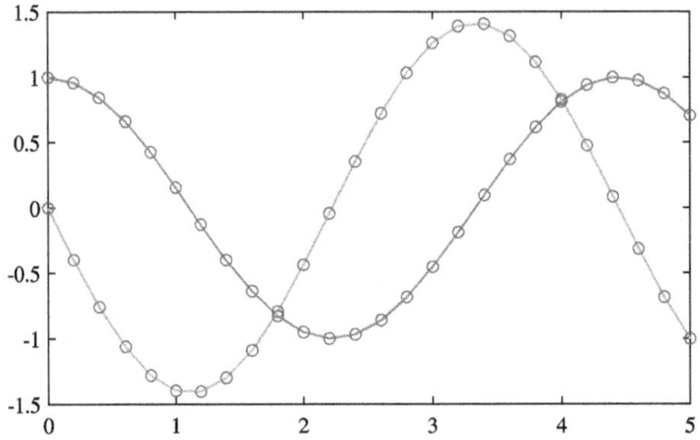

Figure 3.14: Time responses of the states at fixed-step points.

Note that also in the returned vectors an equally spaced vector $\boldsymbol{t}$ is found, yet the computation process applies a variable-step algorithm all the time. The results obtained are accurate and reliable.

Fixed-step displays of simulation results are useful and meaningful in some particular applications. An example is given next to show an application of the fixed-step display.

Example 3.18. Consider the second-order differential equation in Example 2.19. Find the solutions $x_2(t)$ to the differential equation for different values of λ. For convenience, the differential equation is given below:

$$\begin{cases} x_1'(t) = x_2(t), \\ x_2'(t) = -\lambda x_1(t), \quad x_1(0) = 1, \quad x_2(0) = 0. \end{cases}$$

Solutions. If function `ode45()` is called directly, for each value of λ, the numerical expression can be found. However, since a variable-step mechanism is used, the length of $\boldsymbol{t}$ and the solution vectors may be different. It may be hard to draw the responses in three-dimensional surfaces. If a fixed-step display format is adopted, the solutions at preselected $\boldsymbol{t}$ samples can be found, for each value of λ. Therefore the results can be stored in a matrix, and the surface of $x_2(t)$ can be drawn, as shown in Figure 3.15.

```
>> t=0:0.2:10; Lam=0.2:0.2:3; H=[];
   for lam=Lam  % solve the differential equations under different parameters
      f=@(t,x)[x(2); -lam*x(1)];
      [t1,x]=ode45(f,t,[1;0]); H=[H x(:,2)];
   end
   surf(Lam,t,H) % draw the surface of the solutions
```

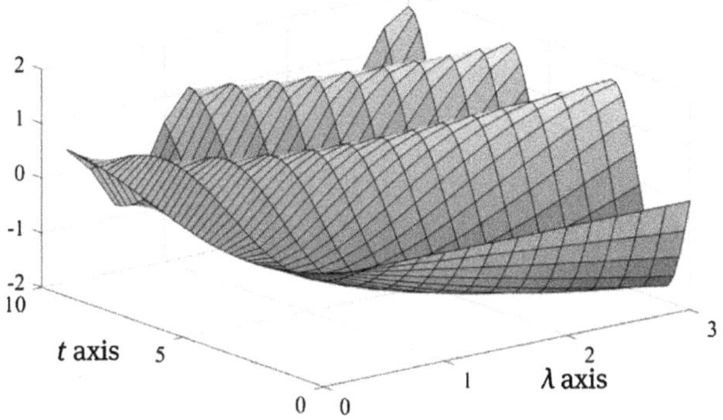

Figure 3.15: Time responses of x_2 at different λ values.

3.4.5 Demonstrations of high-order nonlinear differential equations

Many test examples for initial value problems are provided at the website mentioned in [52]. Some of the examples are of good application illustrations. The simplest and most straightforward example is used here to demonstrate the solution process of differential equations. The interested readers are recommended to try solving more complicated problems in the test set.

Example 3.19. Consider the following first-order explicit differential equation:[52]

$$\mathbf{y}'(t)=\begin{bmatrix} -1.71y_1(t)+0.43y_2(t)+8.32y_3(t)+0.0007 \\ 1.71y_1(t)-8.75y_2(t) \\ -10.03y_3(t)+0.43y_4(t)+0.035y_5(t) \\ 8.32y_2(t)+1.71y_3(t)-1.12y_4(t) \\ -1.745y_5(t)+0.43y_6(t)+0.43y_7(t) \\ -280y_6(t)y_8(t)+0.69y_4(t)+1.71y_5(t)-0.43y_6(t)+0.69y_7(t) \\ 280y_6(t)y_8(t)-1.81y_7(t) \\ -280y_6(t)y_8(t)+1.81y_7(t) \end{bmatrix}$$

with initial values of $\mathbf{y}_0 = [1,0,0,0,0,0,0,0.0057]^{\mathrm{T}}$. Find the numerical solution in the interval $0 \leqslant t \leqslant 321.8122$.

Solutions. If MATLAB is used to solve this problem, the following statements can be written to describe the differential equation with an anonymous function. Then the solver `ode45()` can be called to solve the differential equation. The solution process is fast, and it only needs 0.15 seconds to find the numerical solution. The error tolerances are set to tough values, and 41 521 points are computed. The state space responses are obtained as shown in Figure 3.16. It can be seen that the results are the same, as those

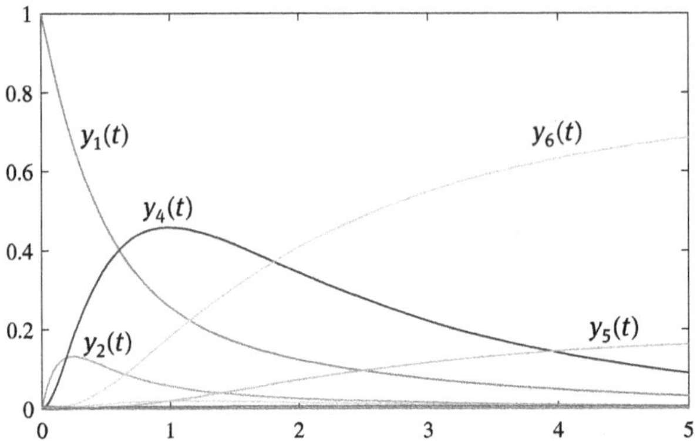

Figure 3.16: The time domain responses (locally zoomed over $0 \leqslant t \leqslant 5$).

provided in [52].

```
>> f=@(t,y)[-1.71*y(1)+0.43*y(2)+8.32*y(3)+0.0007;
              1.71*y(1)-8.75*y(2);
              -10.03*y(3)+0.43*y(4)+0.035*y(5);
              8.32*y(2)+1.71*y(3)-1.12*y(4);
              -1.745*y(5)+0.43*y(6)+0.43*y(7);
              -280*y(6)*y(8)+0.69*y(4)+1.71*y(5)-0.43*y(6)+0.69*y(7);
              280*y(6)*y(8)-1.81*y(7)
              -280*y(6)*y(8)+1.81*y(7)];
   y0=[1;0;0;0;0;0;0;0.0057];  %initial values
   ff=odeset; ff.RelTol=100*eps; ff.AbsTol=100*eps;
   tic, [t,x]=ode45(f,[0,321.8122],y0); toc
   plot(t,x), xlim([0,5])         % time domain responses
```

The step-size curve can also be drawn, as shown in Figure 3.17. It can be seen that at the initial stage, the smallest step-size is 2.9379×10^{-5}. When t is large, relatively large step-

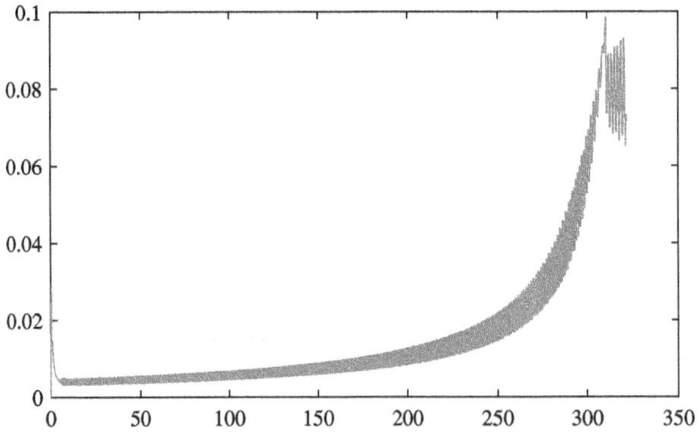

Figure 3.17: The step-sizes during the solution process.

sizes are selected automatically, so as to speed up the solution process, while ensuring the accuracy. It can be seen that the tools provided here are of high efficiency.

In the original reference, this differential equation was regarded as a stiff one. More on stiff differential equation will be presented later in Chapter 5. Therefore it may be hard to solve with usual methods. Here MATLAB is used to solve the equations directly, and the solution process is straightforward. In an extremely short period, the exact solutions of the differential equation can be found, under tough error tolerances. It can be seen that the capabilities in handling differential equations with MATLAB are powerful and reliable.

3.5 Exercises

3.1 Use the algebraic equation solution method to validate the coefficients of the fourth-order Runge–Kutta and Gill's algorithms.

3.2 Write a MATLAB function to implement the 5th and 6th order Adams–Bashforth algorithms, and assess their efficiency:

(1) The 5th order Adams–Bashforth algorithm:

$$\begin{aligned}\boldsymbol{x}_{k+1} = \boldsymbol{x}_k + \frac{h}{720}\Big[&1\,901\boldsymbol{f}(t_k,\boldsymbol{x}_k) - 2774\boldsymbol{f}(t_{k-1},\boldsymbol{x}_{k-1}) + 2\,616\boldsymbol{f}(t_{k-2},\boldsymbol{x}_{k-2})\\ &- 1\,274\boldsymbol{f}(t_{k-3},\boldsymbol{x}_{k-3}) + 251\boldsymbol{f}(t_{k-4},\boldsymbol{x}_{k-4})\Big];\end{aligned}$$

(2) The 6th order Adams–Bashforth algorithm:

$$\begin{aligned}\boldsymbol{x}_{k+1} = \boldsymbol{x}_k + \frac{h}{1440}\Big[&4\,277\boldsymbol{f}(t_k,\boldsymbol{x}_k) - 7\,923\boldsymbol{f}(t_{k-1},\boldsymbol{x}_{k-1}) + 9\,982\boldsymbol{f}(t_{k-2},\boldsymbol{x}_{k-2})\\ &- 7\,298\boldsymbol{f}(t_{k-3},\boldsymbol{x}_{k-3}) + 2\,877\boldsymbol{f}(t_{k-4},\boldsymbol{x}_{k-4}) - 475\boldsymbol{f}(t_{k-5},\boldsymbol{x}_{k-5})\Big].\end{aligned}$$

3.3 Limit cycles are commonly encountered phenomena in nonlinear differential equations. For certain nonlinear differential equations, no matter what the initial values, the phase trajectories may settle down to certain closed paths. The paths are known as limit cycles. Draw the limit cycles for the differential equations. Draw the phase plane trajectory of the differential equations, and make sure that they do form limit cycles if

$$\begin{cases} x'(t) = y(t) + x(t)(1 - x^2(t) - y^2(t)), \\ y'(t) = -x(t) + y(t)(1 - x^2(t) - y^2(t)). \end{cases}$$

3.4 Consider the following nonlinear differential equation. It is pointed out in [22] that the equation may have multiple limit cycles. Use numerical solutions to observe the multiple limit cycles in the equation if

$$\begin{cases} x'(t) = -y(t) + x(t)f\left(\sqrt{x^2(t) + y^2(t)}\right), \\ y'(t) = x(t) + y(t)f\left(\sqrt{x^2(t) + y^2(t)}\right) \end{cases}$$

where $f(r) = r^2 \sin(1/r)$.

3.5 Consider the well-known Rössler equation in chemical reaction processes

$$\begin{cases} x'(t) = -y(t) - z(t), \\ y'(t) = x(t) + ay(t), \\ z'(t) = b + (x(t) - c)z(t). \end{cases}$$

If $a = b = 0.2$ and $c = 5.7$, and the initial values are $x(0) = y(0) = z(0)$, draw the phase space trajectory and its projection on the xy plane. It is suggested to set a, b, and c as additional parameters. If the parameters are changed to $a = 0.2$, $b = 0.5$, and $c = 10$, draw the two- and three-dimensional trajectories of the states.

3.6 Consider the following differential equation:[63]

$$\begin{cases} y'(t) = \tan\phi(t), \\ v'(t) = -\dfrac{g\sin\phi(t)\gamma v^2(t)}{v(t)\cos\phi(t)}, \\ \phi'(t) = -g/v^2(t) \end{cases}$$

where $g = 0.032$, $\gamma = 0.02$. Initial values are $y(0) = 0$, $v(0) = 0.5$. If $\phi(0)$ is selected respectively as 0.3782 and 9.7456, find the numerical solution.

3.7 Chua circuit is an often mentioned differential equation in chaos theory:[79]

$$\begin{cases} x'(t) = \alpha[y(t) - x(t) - f(x(t))], \\ y'(t) = x(t) - y(t) + z(t), \\ z'(t) = -\beta y(t) - \gamma z(t) \end{cases}$$

where $f(x)$ is called Chua diode, having piecewise linear properties

$$f(x) = bx + \frac{1}{2}(a - b)(|x + 1| - |x - 1|),$$

and $a < b < 0$. Write a MATLAB function to describe the differential equation, and draw the phase space trajectory for $\alpha = 15$, $\beta = 20$, $\gamma = 0.5$, $a = -120/7$, $b = -75/7$. The initial values are $x(0) = -2.121304$, $y(0) = -0.066170$, and $z(0) = 2.881090$.

3.8 Solve numerically the differential equation:[25]

$$\begin{cases} y'(x) = -2xy(x)\ln z(x), \\ z'(x) = 2xz(x)\ln y(x) \end{cases}$$

with the initial values $x(0) = 0$, $y(0) = \mathrm{e}$, and $z(0) = 1$. It is known that the analytical solution is $y(x) = \mathrm{e}^{\cos x^2}$ and $z(x) = \mathrm{e}^{\sin x^2}$. Solve the problem with different algorithms and assess the speed and accuracy.

3.9 Consider the following differential equation:[16]

$$\begin{cases} w'(t) = \alpha K_w + \beta y(t) - \gamma x(t)w(t) - \alpha w(t)z(t), \\ x'(t) = -x(t) + \beta y(t) - \gamma x(t)w(t) - y(t)z(t)/K_a, \\ y'(t) = x(t) - \beta y(t) + \gamma x(t)w(t) - y(t)z(t)/K_a, \\ z'(t) = \alpha K_w + x(t) + \alpha w(t)z(t) - y(t)z(t)/K_a \end{cases}$$

where $K_a = 31 \times 10^{-8}$, $K_w = 3.25 \times 10^{-18}$, $\alpha = 55.5 \times 10^6$, $\beta = 10 \times 10^{-6}$, and $\gamma = 1.11 \times 10^6$. If all the initial states are zero, solve the differential equation.

3.10 Lotka–Volterra predator–prey model is given below. Solve the differential equations and draw the relevant curves:

$$\begin{cases} x'(t) = 4x(t) - 2x(t)y(t), \\ y'(t) = x(t)y(t) - 3y(t) \end{cases}$$

where $x(0) = 2$ and $y(0) = 3$.

3.11 Van der Pol differential equation can be mathematical written as follows:

$$\begin{cases} x_1'(t) = x_2(t), \\ x_2'(t) = -\mu(x_1^2(t) - 1)x_2(t) - x_1(t) \end{cases}$$

where $x_1(0) = -0.2$ and $x_2(0) = -0.7$. Draw the surface of state variable $x_1(t)$ with respect to the changes in parameter μ.

3.12 A differential equation set is described as[40]

$$\begin{cases} x'(t) = s\big[qy(t) - x(t)y(t) + x(t)(1 - x(t))\big], \\ y'(t) = h\big[-qy(t) - x(t)y(t) - pz(t)\big], \\ z'(t) = x(t) - z(t) \end{cases}$$

where $h = 8.333$, $p = 0.3$, $q = 0.01$, and $s = 20$. The initial values are known as $x(0) = 1$, $y(0) = 2$ and $z(0) = 0.6$. If $t_{\rm n} = 60$, solve the differential equation system.

3.13 FitzHugh–Nagumo model is given by[40]

$$\begin{cases} v'(t) = a\big(v(t) + w(t) - v^3(t) - 0.5\big), \\ w'(t) = -\big(v(t) - b + cw(t)\big)/a \end{cases}$$

where $a = 3$, $b = 0.7$, and $c = 0.8$. If the initial values are $v(0) = 0.5$ and $w(0) = 0.2$, solve the differential equations with three different methods.

3.14 Consider the following differential equations. Select suitable terminal time to solve the differential equations, and observe the phase plane trajectory during the solution process

$$\begin{cases} x_1'(t) = -x_2(t) - 10x_1^2(t) + 5x_1(t)x_2(t) + x_2^2(t), \\ x_2'(t) = x_1(t) + x_1^2(t) - 25x_1(t)x_2(t) \end{cases}$$

with the initial values $x_1(0) = -0.0914$ and $x_2(0) = -0.1075$.

3.15 Solve the following differential equations[32]

$$\begin{cases} I_1 y_1'(t) = (I_2 - I_3)y_2(t)y_3(t), \\ I_2 y_2'(t) = (I_3 - I_1)y_3(t)y_1(t), \\ I_3 y_3'(t) = (I_1 - I_2)y_1(t)y_2(t) + f(t) \end{cases}$$

where $f(x)$ is a piecewise function

$$f(t) = \begin{cases} 0.25\sin^2 t, & \text{if } 3\pi \leqslant t \leqslant 4\pi, \\ 0, & \text{otherwise.} \end{cases}$$

The constants and initial values are given by $I_1 = 0.5$, $I_2 = 2$, $I_3 = 3$, $y_1(0) = 1$, $y_2(0) = 0$, and $y_3(0) = 0.9$.

3.16 Hodgin–Huxley neural model is given by[40]

$$i = CV'(t) + i_{Na}(t) + i_K(t) + i_L(t)$$

where

$$\begin{cases} i_{Na}(t) = g_{Na}m^3(t)h(t)(v - V_{Na}), \\ i_K(t) = g_K n^4(t)(v - V_K), \\ i_L(t) = g_L(v - V_L), \end{cases}$$

and the constants are $i = 10$, $C = 1$, $V_{Na} = 115$, $V_K = 12$, $V_L = 10.599$, $g_{Na} = 120$, $g_K = 36$, and $g_L = 0.3$. Besides we known the differential equations

$$\begin{cases} m'(t) = a_m(1 - v) - b_m m(t), \\ n'(t) = a_n(1 - v) - b_n n(t), \\ h'(t) = a_h(1 - v) - b_h h(t) \end{cases}$$

where

$$a_m = \frac{0.1(25 - v)}{e^{0.1(25-v)} - 1}, \quad b_m = 4e^{-v/18},$$
$$a_n = \frac{0.01(10 - v)}{e^{0.01(10-v)} - 1}, \quad b_n = 0.125e^{-v/80},$$
$$a_h = 0.07e^{-v/20}, \quad b_h = \frac{1}{e^{0.1(30-v)} + 1},$$

with initial values $V(0) = 0$ and $m(0) = n(0) = h(0) = 0.5$. Assuming that v takes values in the interval $(-80, 80)$, solve the differential equations for different values of v, with $t \in (0, 50)$.

3.17 At the website [52], many test examples are provided as benchmark problems. The differential equation shown in Example 3.19 is only one of them. The readers may visit the website to download certain problems and try to solve the differential equations to test whether accurate Solutions Can Be Found, and to Observe the Efficiency of the Solver.

4 Standard form conversions of ordinary differential equations

The initial value problem solver studied in Chapter 3 can only be used for solving differential equations given in the form of first-order explicit differential equations whose standard form is

$$\boldsymbol{x}'(t) = \boldsymbol{f}(t, \boldsymbol{x}(t)), \quad \text{with given } \boldsymbol{x}(t_0). \tag{4.0.1}$$

This is not adequate for the study of ordinary differential equations. Numerical solvers should be applicable to solve differential equations given in any form. In fact, an alternative solution pattern is to convert the differential equations to be studied into the standard form of first-order explicit differential equations. Then a universal solver such as `ode45()` can be called to find numerical solutions.

It can be seen from the standard forms and solvers that if a system of ordinary differential equations is composed of one or more high-order differential equations, a set of state variables should be selected to convert it first into the first-order explicit differential equations in standard form. In this chapter, the conversion methods are introduced for various differential equations. In Section 4.1, the conversion of a single high-order differential equation is introduced. If differential equations can be successfully converted into the standard form, solvers such as `ode45()` can be called to solve them directly. In Section 4.2, complicated single high-order differential equations are considered for the conversion into standard form. In Section 4.3, conversion from high-order differential equation sets is explored. The major objective is the same as that discussed earlier, that is, to select a set of state variables, and rewrite the differential equation sets into the first-order explicit differential equations. In Section 4.4, high-order matrix differential equations are explored, and numerical algorithms are considered to solve Sylvester and Riccati matrix differential equations. In Section 4.5, for a class of Volterra integro-differential equations with separable variables, conversion methods are presented such that numerical solutions can be found. In fact, solutions of many differential equations will be revisited in the next chapter.

4.1 Conversion method for a single high-order differential equation

If in the high-order differential equation, the explicit form of the highest-order term of the unknown function can be found, a convenient way is suggested to select the state variables, and the high-order differential equation can be converted easily into the standard form. Then solvers such as `ode45()` can be called to find the numerical

https://doi.org/10.1515/9783110675252-004

solutions of the differential equations. If the explicit form does not exist for the original differential equation, algebraic solution techniques can be employed, such that the original differential equation can finally be converted into the standard form and solved numerically.

In this section, some different cases are considered and presented. The final target is to eventually convert the equations into the standard form so that the numerical solution can be found.

4.1.1 Conversion of explicit equations

Assume that a high-order differential equation is given by

$$y^{(n)}(t)=f(t,y(t),y'(t),\dots,y^{(n-1)}(t)) \tag{4.1.1}$$

and the initial values of $y(t)$ and its directives are given as $y(t_0),y'(t_0),\dots,y^{(n-1)}(t_0)$, a set of state variables can be selected. For instance, let $x_1(t)=y(t),x_2(t)=y'(t),\dots,$ $x_n(t)=y^{(n-1)}(t)$. Therefore, the original high-order differential equation can be converted into the following equivalent standard form

$$\begin{cases} x_1'(t)=x_2(t), \\ x_2'(t)=x_3(t), \\ \vdots \\ x_n'(t)=f(t,x_1(t),x_2(t),\dots,x_n(t)) \end{cases} \tag{4.1.2}$$

with initial states $x_1(t_0)=y(t_0),x_2(t_0)=y'(t_0),\dots,x_n(t_0)=y^{(n-1)}(t_0)$. Therefore, function `ode45()` can be called to solve the original differential equation directly.

In fact, there are infinitely many ways to select the state variables. The above mentioned is only one of them. An alternative way to select the state variables is $x_1(t)=y^{(n-1)}(t),x_2(t)=y^{(n-2)}(t),\dots,x_{n-1}(t)=y'(t)$ and $x_n(t)=y(t)$, such that the original differential equation can be converted into the form

$$\begin{cases} x_1'(t)=f(t,x_n(t),x_{n-1}(t),\dots,x_1(t)), \\ x_2'(t)=x_1(t), \\ \vdots \\ x_n'(t)=x_{n-1}(t) \end{cases} \tag{4.1.3}$$

with initial state values $x_1(t_0)=y^{(n-1)}(t_0),x_2(t_0)=y^{(n-2)}(t_0),\dots,x_n(t_0)=y(t_0)$.

Of course, there are other ways of state variable selection. In real applications, if there is no other specific request, the first method is recommended.

Example 4.1. For given initial values $y(0) = -0.2$, $y'(0) = -0.7$, numerically solve the following van der Pol equation, and draw the phase plane trajectories for different values of μ:

$$y''(t) + \mu(y^2(t) - 1)y'(t) + y(t) = 0.$$

Solutions. It has been demonstrated in Example 2.42 that the differential equation has no analytical solution. Numerical method is the only choice to study this differential equation. Since van der Pol equation is not given in the standard form of the first-order explicit differential equations, it should be converted to that form so that the equation can be solved numerically with MATLAB. It is seen from the original equation that it can be rewritten to find the explicit form of $y''(t)$:

$$y''(t) = -\mu(y^2(t) - 1)y'(t) - y(t).$$

Select a set of state variables $x_1(t) = y(t)$, $x_2(t) = y'(t)$, the original equation can easily be converted into the following form:

$$\begin{cases} x_1'(t) = x_2(t), \\ x_2'(t) = -\mu(x_1^2(t) - 1)x_2(t) - x_1(t) \end{cases}$$

with initial states $x_1(0) = -0.2$ and $x_2(0) = -0.7$.

It can be seen that the original differential equation is successfully converted into first-order explicit differential equations.

Here μ is a variable parameter. If for each value of μ, a MATLAB function is written to describe the equation, it may be rather complicated to do so. Therefore μ can be used as an additional parameter, so that the values of μ in MATLAB workspace can be used to define the equations. Alternatively, anonymous functions can be used to define the differential equations.

Let us now consider first the additional parameter method. The following anonymous function can be used to describe the differential equations, and it can be set to accept an additional parameter.

```
>> f=@(t,x,mu)[x(2);
           -mu*(x(1)^2-1)*x(2)-x(1)]; %with additional parameter
```

It can be seen that there is an extra input argument `mu` in the model function. When calling the solver `ode45()`, this variable must be assigned also in the function call. Assuming that the initial state vector is $\boldsymbol{x}(0) = [-0.2, -0.7]^{\mathrm{T}}$, the following statements can be used to solve the differential equation, for $\mu = 1$ and $\mu = 2$, respectively. The time response plots of the states can be obtained, as shown in Figure 4.1.

```
>> x0=[-0.2; -0.7]; tn=20;
   ff=odeset; ff.AbsTol=100*eps; ff.RelTol=100*eps;
```

```
mu=1; f=@(t,x)[x(2); -mu*(x(1)^2-1)*x(2)-x(1)];
[t1,y1]=ode45(f,[0,tn],x0,ff); % solve the equation
mu=2; f=@(t,x)[x(2); -mu*(x(1)^2-1)*x(2)-x(1)];
[t2,y2]=ode45(f,[0,tn],x0,ff); % solve again
plot(t1,y1,t2,y2,'--')            % draw the time responses
```

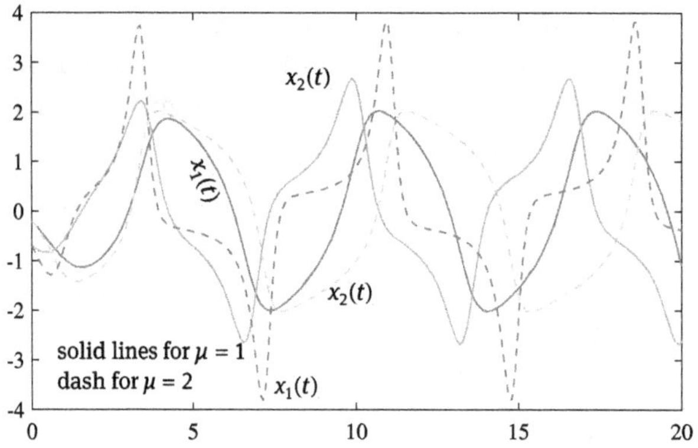

Figure 4.1: Van der Pol equation solutions for different values of μ.

If x_1 and x_2 are selected as the two axes, the phase plane trajectories can be drawn as shown in Figure 4.2. It can be seen from the phase plane trajectories that no matter what the value of μ, the phase portraits may settle down eventually on closed paths. The closed paths are referred to as limit cycles in differential equation theory.

```
>> plot(y1(:,1),y1(:,2),y2(:,1),y2(:,2),'--') % draw phase plane plot
```

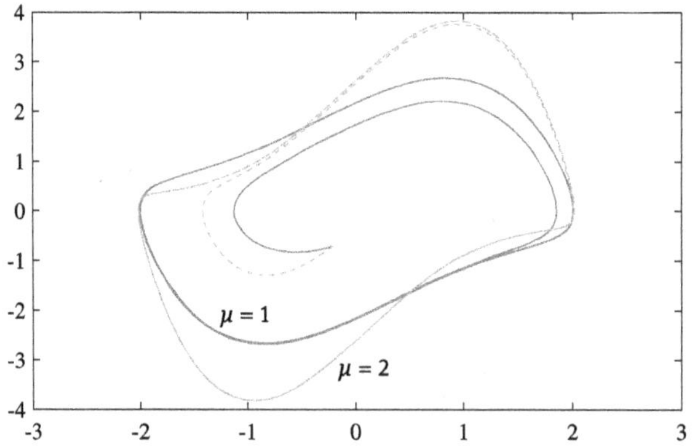

Figure 4.2: Phase portraits of van der Pol equation.

In this example, the initial values provided are located inside the limit cycles. The values of the unknown function may continue growing and finally settle down in the limit cycles in a periodic mode. If the initial positions are located outside of the limit cycles, the unknown function may continue shrinking, and eventually settle down on the limit cycles. Limit cycles will be further studied in Chapter 7.

Example 4.2. Solve the van der Pol equation in Example 4.1 without using additional parameters.

Solutions. If an additional parameter is not used, the value of μ must exist in MATLAB workspace. Then an anonymous function can be defined for the differential equation. Since the variables in MATLAB workspace can be accepted directly by the anonymous function, this method can be used. If the value of μ is changed, the anonymous function should be defined again. The following commands can be used to solve the differential equation with different values of μ. The results obtained are identical to those obtained earlier.

```
>> x0=[-0.2; -0.7]; tn=20;
   ff=odeset; ff.AbsTol=100*eps; ff.RelTol=100*eps;
   mu=1; f=@(t,x)[x(2); -mu*(x(1)^2-1)*x(2)-x(1)];
   [t1,y1]=ode45(f,[0,tn],x0,ff); %solve differential equation
   mu=2; f=@(t,x)[x(2); -mu*(x(1)^2-1)*x(2)-x(1)];
   [t2,y2]=ode45(f,[0,tn],x0,ff); %solve differential equation
   plot(t1,y1,t2,y2,'--')          %draw the solutions
```

Example 4.3. Select the state variables differently for the van der Pol equation in Example 4.1 and solve it again.

Solutions. In Example 4.1, the states were selected as $x_1(x) = y(t)$, $x_2(t) = y'(t)$, such that the original differential equation was converted into standard form. If a new set of states are selected as $x_1(x) = y'(t)$ and $x_2(t) = y(t)$, the original equation can be converted into the following standard form:

$$\begin{cases} x_1'(t) = -\mu(x_2^2(t) - 1)x_1(t) - x_2(t), \\ x_2'(t) = x_1(t). \end{cases}$$

Therefore, the same solver `ode45()` can be called to solve the differential equation. Note that, since the physical meaning of the states is changed, the initial values are also changed. Care must be taken in solving the problems.

Example 4.4. For $\mu = 1\,000$ and $t_n = 3\,000$, solve the equation in Example 4.1 again, and see what happens.

Solutions. There is no need to use an additional parameter in the example, since an anonymous function is adequate for describing the differential equation. The solver ode45() can be used to find the numerical solution of the equations. If the new value of μ is assigned, the terminal time is set to 3 000, and moderate error tolerance is chosen, the following commands can be used to solve the van der Pol equation:

```
>> x0=[2;0]; tn=3000; mu=1000;
   ff=odeset; ff.AbsTol=1e-8; ff.RelTol=1e-8;
   f=@(t,x)[x(2); -mu*(x(1)^2-1)*x(2)-x(1)];
   tic, [t,y]=ode45(f,[0,tn],x0,ff); toc %try to solve the equation
```

After long waiting, the error message “Out of memory. Type HELP MEMORY for your options” appears and the solution process is aborted. Since in the variable-step solution process, at some points the step-size is selected too small to ensure the expected accuracy, this may lead to a huge increase in the total number of points in the solution such that the memory is exhausted. If another solver ode87() is used, similar thing may happen.

```
>> x0=[2;0]; tn=3000; mu=1000;
   ff=odeset; ff.AbsTol=1e-8; ff.RelTol=1e-8;
   f=@(t,x)[x(2); -mu*(x(1)^2-1)*x(2)-x(1)];
   tic, [t,y]=ode87(f,[0,tn],x0,ff); toc  %call another solver
```

This kind of phenomenon is often caused by the so-called stiff differential equations, and the concept and manipulation of stiff differential equations will be further explored in the next chapter.

4.1.2 Solutions of time-varying differential equations

So far the studied examples were all with constant coefficients. In fact, the standard form in (4.0.1) fully supports the descriptions of time-varying differential equations. Examples are used here to show the numerical solutions of time-varying differential equations.

Example 4.5. Solve numerically the time-varying differential equation in Example 2.40 for $t \in (0.2, \pi)$, and assess the accuracy and efficiency of the numerical method. For convenience, the differential equation is given below:

$$x^5 y'''(x) = 2(xy'(x) - 2y(x)), \quad y(1) = 1,\ y'(1) = 0.5,\ y''(1) = -1.$$

Solutions. The following commands can be used to find the analytical solution of the time-varying differential equation:

```
>> syms x y(x)
   d1y=diff(y); d2y=diff(d1y); d3y=diff(d2y);
```

```
y=dsolve(x^5*d3y==2*(x*d1y-2*y),...
         y(1)==1, d1y(1)==0.5, d2y(1)==-1) % analytical solution
```

The analytical solution obtained is

$$y(x) = x^2 - \frac{3\sqrt{2}\mathrm{e}^{\sqrt{2}}}{8}x^2\mathrm{e}^{-\sqrt{2}/x} + \frac{3\sqrt{2}\mathrm{e}^{-\sqrt{2}}}{8}x^2\mathrm{e}^{\sqrt{2}/x}.$$

Dividing both sides of the equation by x^5, the original equation can be converted to the explicit form of $y'''(t)$:

$$y'''(x) = \frac{2(xy'(x) - 2y(x))}{x^5}.$$

To avoid confusion, the independent variable is changed from x to t, and the state variables are selected as $x_1(t) = y(t)$, $x_2(t) = y'(t)$, and $x_3(t) = y''(t)$. The first-order explicit differential equations in standard form can be written as

$$\boldsymbol{x}'(t) = \begin{bmatrix} x_2(t) \\ x_3(t) \\ \dfrac{2}{t^5}\left[tx_2(t) - 2x_1(t)\right] \end{bmatrix}, \quad \boldsymbol{x}(1) = \begin{bmatrix} 2 \\ 0.5 \\ -1 \end{bmatrix}.$$

An anonymous function can be written to express the standardized differential equations. Then the numerical solutions for $t \in (1, \pi)$ can be found. Also the numerical solutions in the interval $t \in (0.2, 1)$ can be found. The two sets of solutions can be joined together, to form the numerical solutions in the interval $t \in (0.2, \pi)$. The numerical solution can be compared with the analytical one, and a norm of the error of 3.575×10^{-11} can be witnessed. It can be seen that the numerical solutions obtained are very accurate. The minimum allowed error tolerance of 100`eps` will be used throughout the book, unless otherwise stated.

```
>> f=@(t,x)[x(2); x(3); 2*(t*x(2)-2*x(1))/t^5];
   x0=[1; 0.5; -1];                      % set initial values
   ff=odeset; ff.AbsTol=100*eps; ff.RelTol=100*eps;
   [tn,xn]=ode45(f,[1,pi],x0,ff);
   [t1,x1]=ode45(f,[1,0.2],x0,ff); % solve differential equation
   tm=[t1(end:-1:2); tn]; xm=[x1(end:-1:2,:); xn];
   ym=xm(:,1); norm(ym-double(subs(y,x,tm)),1) % find the error
```

It can be seen that the interval $t \in (0.2, \pi)$ avoids the singularity at $t = 0$, so the solution obtained is valid.

4.1.3 Singularities in differential equations

In the previous example, the behavior of the differential equation at point $t = 0$ may be special. In this section, the concept of singularities in differential equations is presented. Examples are used to demonstrate the behaviors of the differential equations around their singularities.

Definition 4.1. For linear time-varying differential equations, if the coefficient functions contain singularities, they are referred to as the regular singular points of the differential equations.

For the differential equation studied in Example 4.5, it can be seen that $x = 0$, or $t = 0$, is the regular singular point of the differential equation. Since the point is located outside of the solution interval, there was no impact witnessed. Now an example is given to show the numerical solutions in larger intervals and to illustrate the impact of the singularities.

Example 4.6. Solve the differential equation in Example 4.5 in the interval $t \in (0, \pi)$, and see what happens.

Solutions. The statements in the previous example can still be used, by replacing 0.2 with 0. The differential equation solutions can be found, and the state curves can be drawn, as shown in Figure 4.3. It should be noted that due to the existence of the singularity at $t = 0$, the values of the states around this point are NaN. When t is small, the values of the states tend to infinity. In the plot, only the curves in the range of $y(t) \in (-20, 20)$ are shown.

```
>> f=@(t,x)[x(2); x(3); 2*(t*x(2)-2*x(1))/t^5];
   x0=[1; 0.5; -1];
   ff=odeset; ff.AbsTol=100*eps; ff.RelTol=100*eps;
```

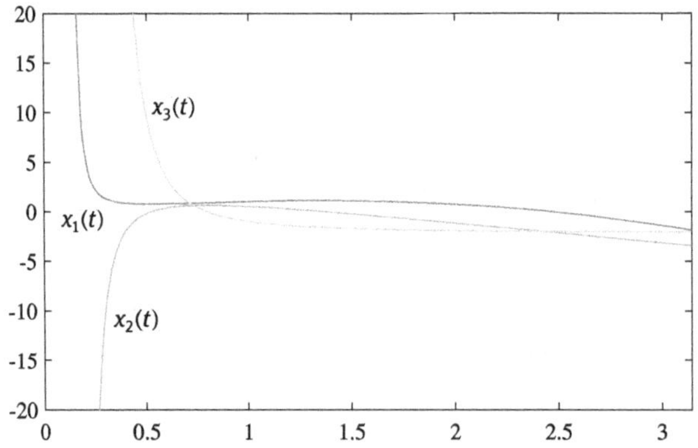

Figure 4.3: Solution of a differential equation having singularities.

```
[tn,xn]=ode45(f,[1,pi],x0,ff);
[t1,x1]=ode45(f,[1,0],x0,ff);
tm=[t1(end:-1:2); tn]; xm=[x1(end:-1:2,:); xn];
plot(tm,ym), ylim([-10,10])
```

Example 4.7. Consider now the following nonlinear differential equation:[58]

$$y(t)y^{(6)}(t)+6y'(t)y^{(5)}(t)+15y''(t)y^{(4)}(t)+10(y'''(t))^2=a\sin^m\lambda t.$$

If $a=2$, $m=3$, $\lambda=2$ and $y(0)=1$, $y^{(i)}(0)=0$, $i=1,2,\dots,5$, find the numerical solution, and draw the state variable curves.

Solutions. It can be seen from the example that if $y(t)=0$, singular behavior may appear in this differential equation. Unfortunately, since the analytical expression of function $y(t)$ is not known, the singularity phenomenon cannot be studied theoretically. Numerical method is the only way to study the behavior of this differential equation.

Since the highest order term in the equation is $y^{(6)}(t)$, its explicit expression can be written as

$$y^{(6)}(t)=\frac{1}{y(t)}[-6y'(t)y^{(5)}(t)-15y''(t)y^{(4)}(t)-10(y'''(t))^2+a\sin^m\lambda t]$$

which implies $y(t)\neq 0$. Introducing the state variables $x_1(t)=y(t)$, $x_2(t)=y'(t)$, $x_3(t)=y''(t)$, $x_4(t)=y'''(t)$, $x_5(t)=y^{(4)}(t)$, and $x_6(t)=y^{(5)}(t)$, the standard form of first-order explicit differential equations can be written as

$$\mathbf{x}'(t)=\begin{bmatrix} x_2(t)\\ x_3(t)\\ x_4(t)\\ x_5(t)\\ x_6(t)\\ \dfrac{1}{x_1(t)}[-6x_2(t)x_6(t)-15x_3(t)x_5(t)-10x_4^2(t)+a\sin^m\lambda t]\end{bmatrix}$$

and $\mathbf{x}(0)=[1,0,0,0,0,0]^{\mathrm{T}}$.

An anonymous function can be used to depict the standard form of the differential equation, and under strict error tolerance, the numerical solution can be found. It can be seen by observing the curves that the first three states are divergent. Only the latter three states are drawn, as shown in Figure 4.4.

```
>> a=2; m=3; lam=2;            %input differential equation parameters
   x0=[1; 0; 0; 0; 0; 0];      %initial state vector
   f=@(t,x)[x(2:6);
      (-6*x(2)*x(6)-15*x(3)*x(5)-10*x(4)^2+a*sin(lam*t)^m)/x(1)];
```

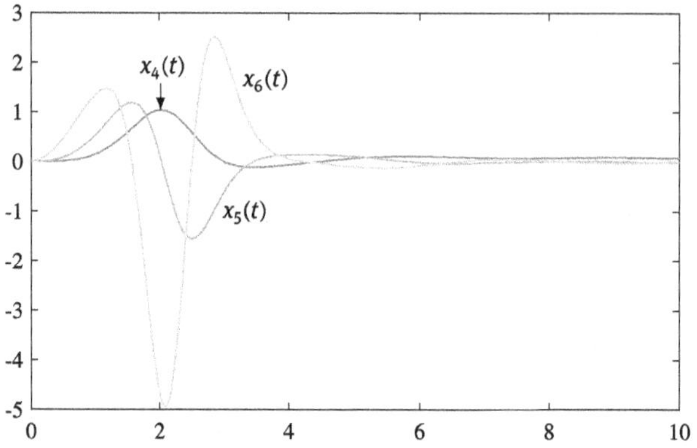

Figure 4.4: The latter three state signals.

```
ff=odeset; ff.AbsTol=100*eps; ff.RelTol=100*eps;
[t x]=ode45(f,[0,10],x0,ff);
plot(t,x(:,4:6))  % the first three states are divergent and omitted
```

In fact, in the solution process, there is a hidden problem that, when finding the explicit form of $y^{(6)}(t)$, it was assumed that $y(t) \neq 0$. We need this assumption, otherwise the numerical solution cannot be found, since it is impossible to show in mathematics that $y(t) \neq 0$. Luckily, in the whole solution process there is no such warning like "Division by zero". It implies that the phenomenon of $y(t) = 0$ did not happen at all in the solution process. The result obtained is correct.

In other words, although the analytical form of $y(t)$ signal is unknown, and it is not possible to decide theoretically whether there are singularities, from simulation it can be concluded that in the solution process $y(t)$ does not equal to zero, and there are no singularities in the differential equation.

4.1.4 State augmentation for constant parameters

If there are constant parameters in the differential equation, the methods illustrated earlier can be used to solve the problem, without any difficulties. In fact, there are many methods to process differential equations with constant parameters. Here a state augmentation is presented to solve the differential equations with constant parameters.

Assume that the mathematical expression of a differential equation is

$$\boldsymbol{x}'(t) = \boldsymbol{f}(t, \boldsymbol{x}(t), a), \tag{4.1.4}$$

and the initial vector is known to be $\boldsymbol{x}(t_0) = \boldsymbol{x}_0$. In the mathematical expression, we deliberately wrote a constant parameter a. If there are n state variables in $\boldsymbol{x}(t)$, an extra one can be introduced, by denoting $x_{n+1}(t) = a$. The augmented state space model can be written as

$$\tilde{\boldsymbol{x}}'(t) = \begin{bmatrix} \boldsymbol{x}(t) \\ x_{n+1}(t) \end{bmatrix} = \begin{bmatrix} \boldsymbol{f}(t, \boldsymbol{x}(t), a) \\ 0 \end{bmatrix} \tag{4.1.5}$$

where the initial values of the augmented state are $\tilde{\boldsymbol{x}}(t_0) = [\boldsymbol{x}^{\mathrm{T}}(t_0), a]^{\mathrm{T}}$. The differential equation can then be solved directly, so that the differential equations with constant parameters can be handled in this manner.

Example 4.8. Use the state augmentation method to solve the van der Pol equation in Example 4.1, where $\mu = 1$.

Solutions. In earlier examples, the state variables are selected as $x_1(t) = y(t)$ and $x_2(t) = y'(t)$. If one wants to set the constant parameter μ as the augmented state $x_3(t) = \mu$, the augmented state space model of van der Pol equation can be rewritten as

$$\begin{cases} x_1'(t) = x_2(t), \\ x_2'(t) = -x_3(t/0(x_1^2(t) - 1)x_2(t) - x_1(t), \\ x_3'(t) = 0, \end{cases}$$

and the initial values of the new states are $x_1(0) = -0.2$, $x_2(0) = -0.7$, and $x_3(0) = 1$. With the first-order explicit differential equations standard form, an anonymous function can be written to express the model, and the solution can then be obtained. The results obtained are identical to those in Figure 4.2, for $\mu = 1$. If can be seen from the solution that $x_3(t)$ is constant. If one wants to study the case when $\mu = 2$, one can simply set the last entry in the initial value vector to 2.

```
>> x0=[-0.2;-0.7;2]; tn=20;        % augmented initial state
   ff=odeset; ff.AbsTol=100*eps; ff.RelTol=100*eps;
   f=@(t,x)[x(2); -x(3)*(x(1)^2-1)*x(2)-x(1); 0];
   [t,y]=ode45(f,[0,tn],x0,ff);  % solve the differential equation
   plot(y(:,1),y(:,2))              % draw phase plane trajectory
```

4.2 Conversions of complicated high-order differential equations

In the previous examples, the differential equations could easily be rewritten in the explicit form of the signal with highest orders. When the differential equations are converted, solvers such as `ode45()` can be used to solve them. In real applications, sometimes the explicit form of the highest order derivative function may not be easily found. Some special treatment should be introduced to convert the differential equations into the expected standard form. Then the differential equations can be solved. Several special differential equations like this will be used as examples to demonstrate the solution process.

4.2.1 Equations containing the square of the highest-order derivative

Suppose the highest order term of the unknown function appears in the square form

$$[y^{(n)}(t)]^2 = f(t, y(t), y'(t), \dots, y^{(n-1)}(t)) \tag{4.2.1}$$

and the initial values $y(t_0), y'(t_0), \dots, y^{(n-1)}(t_0)$ are given. As before, the state variables $x_1(t) = y(t), x_2(t) = y'(t), \dots, x_n(t) = y^{(n-1)}(t)$ can be selected first, and taking the square root of the last term, two different sets of first-order explicit differential equations can be created

$$\begin{cases} x_1'(t) = x_2(t), \\ \quad\vdots \\ x_{n-1}'(t) = x_n(t), \\ x_n'(t) = \sqrt{f(t, y(t), y'(t), \dots, y^{(n-1)}(t))} \end{cases} \tag{4.2.2}$$

and

$$\begin{cases} x_1'(t) = x_2(t), \\ \quad\vdots \\ x_{n-1}'(t) = x_n(t), \\ x_n'(t) = -\sqrt{f(t, y(t), y'(t), \dots, y^{(n-1)}(t))}. \end{cases} \tag{4.2.3}$$

The two state space models comprise the original differential equation. The initial values of the states are

$$\boldsymbol{x}(t_0) = [y(t_0), y'(t_0), \dots, y^{(n-1)}(t_0)]^{\mathrm{T}}. \tag{4.2.4}$$

It can be seen that the two differential equation systems can both be solved directly with MATLAB solvers. Therefore, with anonymous or MATLAB functions, the two equation sets can be described so that they can be solved numerically. Both solutions satisfy the original differential equation.

Example 4.9. Use the numerical method to solve the following differential equation:

$$(y''(t))^2 = 4(ty'(t) - y(t)) + 2y'(t) + 1, \quad y(0) = 0,\ y'(0) = 0.1,$$

and compare these solutions with analytical solutions.

Solutions. Taking square root of the right-hand side of the differential equation, the explicit forms of the original equation can be written as

$$\begin{cases} y''(t) = \sqrt{4(ty'(t) - y(t)) + 2y'(t) + 1}, \\ y''(t) = -\sqrt{4(ty'(t) - y(t)) + 2y'(t) + 1}. \end{cases}$$

Selecting state variables $x_1(t) = y(t)$ and $x_2(t) = y'(t)$, two differential equations in standard form can be established. The first is written as

$$\mathbf{x}'(t) = \begin{bmatrix} x_2(t) \\ \sqrt{4(tx_2(t) - x_1(t)) + 2x_2(t) + 1} \end{bmatrix}.$$

The second is written as

$$\mathbf{x}'(t) = \begin{bmatrix} x_2(t) \\ -\sqrt{4(tx_2(t) - x_1(t)) + 2x_2(t) + 1} \end{bmatrix}.$$

The following MATLAB commands can be used respectively to solve the two differential equations. Two solutions can then be found. It should be noted that in the solution process, tiny imaginary quantities may appear. The bias like this can be neglected by using function `real()`. The genuine numerical solutions can then be found.

```
>> ff=odeset; ff.AbsTol=1e-8; ff.RelTol=1e-8;
   f=@(t,x)[x(2); sqrt(4*(t*x(2)-x(1))+2*x(2)+1)];
   [t1,x1]=ode45(f,[0,1],[0; 0.1],ff); x1=real(x1(:,1));
   f=@(t,x)[x(2); -sqrt(4*(t*x(2)-x(1))+2*x(2)+1)];
   [t2,x2]=ode45(f,[0,1],[0; 0.1],ff);
   x2=real(x2(:,1)); plot(t1,x1,t2,x2)
```

If function `dsolve()` is called, the analytical solutions of the equation can be found, and they can be superimposed on the numerical solutions, as shown in Figure 4.5. It can be found that the first model fits the analytical solution perfectly, while the second has large discrepancies when t is large. If one wants to reduce further the error tolerance, the numerical solutions may not be found.

```
>> syms t y(t); y1d=diff(y);
   y=dsolve((diff(y,2))^2==4*(t*y1d-y)+2*y1d+1,...
```

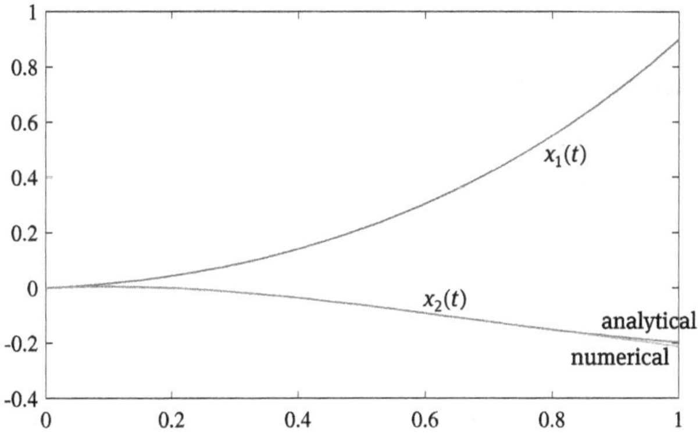

Figure 4.5: Comparisons of analytical and numerical solutions.

```
      y(0)==0, y1d(0)==0.1);                % analytical solution
y1=y(1); y2=y(2); hold on, fplot(y,[0,1]) % draw the curves
```

The analytical solutions of the differential equation are

$$y(t) = \begin{cases} t^4/12 + t^3/6 + \sqrt{30}\,t^2/10 + t/10, \\ t^4/12 + t^3/6 - \sqrt{30}\,t^2/10 + t/10. \end{cases}$$

The results are the most accurate solutions obtained under the double precision using this method. This problem will be revisited in Chapter 5, and accurate results will be found, see Example 5.13.

4.2.2 Equations containing odd powers

In real applications there are even differential equations of complicated form. For instance, the odd power of the highest-order derivative term may exist:

$$[y^n(t)]^{2k+1} = f(t, x_1(t), x_2(t), \dots, x_n(t)). \tag{4.2.5}$$

In the real domain, manipulation of such differential equations looks simpler, compared with the square counterpart, since the root may not have the phenomenon of multiple solutions. The state variables can be selected as usual, for instance, $x_1(t) = y(t)$, $x_2(t) = y'(t)$, ..., $x_n(t) = y^{(n-1)}(t)$. Finally, the first-order explicit differential equation can be written as

$$\mathbf{x}'(t) = \begin{bmatrix} x_2(t) \\ \vdots \\ x_{n-1}(t) \\ \sqrt[2k+1]{f(t, x_1(t), x_2(t), \dots, x_n(t))} \end{bmatrix}. \tag{4.2.6}$$

The next two examples are created by the author based on a given function $y(t) = \mathrm{e}^{-t}$. If one wants to find the analytical solution in a usual way, the solution cannot be easily found. Numerical methods can be tried, and compared with the analytical solution.

Example 4.10. Solve the following initial value problem:

$$(y''(t))^3 + 3y'(t)\sin y(t) - 3y(t)\sin y'(t) = \mathrm{e}^{-3t}, \quad y(0) = 1,\ y'(0) = -1$$

Solutions. If the state variables $x_1(t) = y(t)$ and $x_2(t) = y'(t)$ are selected, it is not hard to write down the first-order explicit differential equation

$$\mathbf{x}'(t) = \begin{bmatrix} x_2(t) \\ \sqrt[3]{-3x_2(t)\sin x_1(t) + 3x_1(t)\sin x_2(t) + \mathrm{e}^{-3t}} \end{bmatrix}$$

with the initial state vector $\boldsymbol{x}_0 = [1, -1]^\mathrm{T}$. Therefore, the following MATLAB commands can be used to solve this differential equation directly. The solutions for the two state variables can be found as shown in Figure 4.6. Since the exact solutions are known as $y(t) = \mathrm{e}^{-t}$, $y'(t) = -\mathrm{e}^{-t}$, the maximum error can be estimated as 6.7722×10^{-12}. It can be seen that the accuracy of the numerical solutions is very high.

```
>> f=@(t,x)[x(2);
        (-3*x(2)*sin(x(1))+3*x(1)*sin(x(2))+exp(-3*t))^(1/3)];
   ff=odeset; ff.AbsTol=100*eps; ff.RelTol=100*eps;
   [t,x]=ode45(f,[0,4],[1; -1],ff); plot(t,x)
   norm(x-[exp(-t) -exp(-t)],1)  % norm of the error matrix
```

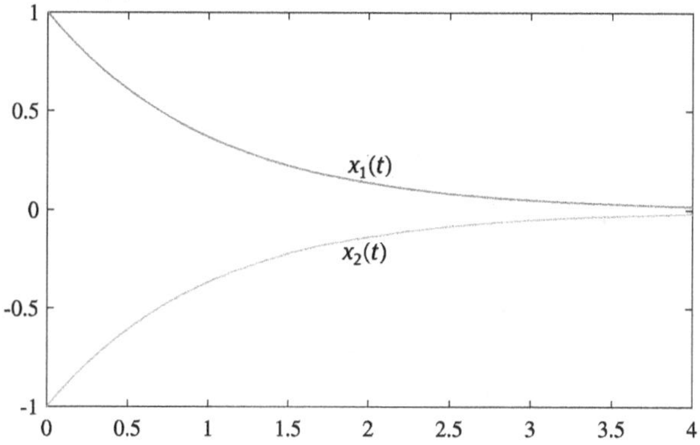

Figure 4.6: Numerical solutions.

4.2.3 Equations containing nonlinear operations

If there exist nonlinear functions of the highest-order derivative of the unknown function, the direct method discussed earlier cannot be used to find the first-order explicit differential equations in standard form. The algebraic equation solution process should be embedded in the differential equation description. Finally, the numerical solutions can be found. An example next will be used to demonstrate how to convert and solve such differential equations.

Example 4.11. Consider a more complicated initial value problem:

$$(y''(t))^3 + 3y''(t)\sin y(t) + 3y'(t)\sin y''(t) = \mathrm{e}^{-3t}, \quad y(0) = 1,\ y'(0) = -1$$

Solutions. It is not possible to find the explicit form of function $y''(t)$ in this example. The state variables can be introduced as usual, $x_1(t) = y(t)$ and $x_2(t) = y'(t)$. Denoting

$p(t) = y''(t)$, the algebraic equation for $p(t)$ can be established as

$$p^3(t) + 3p(t)\sin x_1(t) + 3x_2(t)\sin p(t) - \mathrm{e}^{-3t} = 0.$$

With the numerical method, the solution $p(t)$ can be found, such that the term in the differential equation, $x_2'(t) = p(t)$, can be formulated. The first-order explicit form of the differential equation can be established as

$$\boldsymbol{x}'(t) = \begin{bmatrix} x_2(t) \\ p(t) \end{bmatrix}.$$

With the solver `ode45()`, the numerical solution of the original differential equation can be found.

The standard form of the differential equations cannot be described by anonymous functions, since an algebraic equation solver is embedded. MATLAB function is the only choice to describe such differential equations.

```
function dx=c4exode1(t,x)
f=@(p)p^3+3*p*sin(x(1))+3*x(2)*sin(p)-exp(-3*t);
ff=optimset; ff.Tolx=eps; ff.TolFun=eps;
p=fsolve(f,x(1),ff); dx=[x(2); p];
```

Since the example was created for the analytical solution $y(t) = \mathrm{e}^{-t}$, it cannot be found with the symbolic method in MATLAB with the help of `dsolve()`. Numerical solution to the equation is the only choice. The numerical solution can easily be found with the following statements, and the maximum error is 1.1915×10^{-10}. The elapsed time is 25.92 seconds. The solution process is quite time consuming, since in each step of the solution process, the algebraic equation is solved once. It can be seen that the solution method is quite ineffective.

```
>> ff=odeset; ff.AbsTol=100*eps; ff.RelTol=100*eps;
   tic, [t,x]=ode45(@c4exode1,[0,4],[1; -1],ff); toc
   plot(t,x), norm(x-[exp(-t) -exp(-t)],1)
```

Of course, implicit differential equation solvers will be presented to all the examples in this section, and efficient solvers will be demonstrated in the next chapter.

4.3 Conversions of differential equation sets

Compared with the single high-order differential equations, the conversion and solution process of high-order differential equation sets is even more complicated. In this section, conversion methods will be presented for various differential equations. Some differential equation manipulation may better be performed using the methods to be presented in the next chapter.

4.3.1 Simple explicit differential equation sets

In this section, two differential equations are demonstrated and we show how to convert them into first-order explicit differential equations. Suppose the two equations can be written as the following differential equations:

$$\begin{cases} x^{(m)}(t) = f(t, x(t), x'(t), \dots, x^{(m-1)}(t), y(t), y'(t), \dots, y^{(n-1)}(t)), \\ y^{(n)}(t) = g(t, x(t), x'(t), \dots, x^{(m-1)}(t), y(t), y'(t), \dots, y^{(n-1)}(t)) \end{cases} \tag{4.3.1}$$

where each equation may contain the explicit form of the highest-order derivative of one unknown function. The state variables can still be selected as $x_1(t) = x(t)$, $x_2(t) = x'(t)$, ..., $x_m(t) = x^{(m-1)}(t)$. Then one may continue selecting the state variables as $x_{m+1} = y(t)$, $x_{m+2} = y'(t)$, ..., $x_{m+n}(t) = y^{(n-1)}(t)$. In this case, the original differential equations can be converted into

$$\begin{cases} x_1'(t) = x_2(t), \\ \vdots \\ x_m'(t) = f(t, x_1(t), x_2(t), \dots, x_{m+n}(t)), \\ x_{m+1}'(t) = x_{m+2}(t), \\ \vdots \\ x_{m+n}'(t) = g(t, x_1(t), x_2(t), \dots, x_{m+n}(t)). \end{cases} \tag{4.3.2}$$

The initial values of the state variables can also be set accordingly. The expected first-order explicit differential equations can then be established. An example will be given next to demonstrate the conversion and solution process.

Example 4.12. Assume that the coordinates (x, y) of the Apollo satellite satisfy

$$\begin{cases} x''(t) = 2y'(t) + x(t) - \mu^*(x(t) + \mu)/r_1^3(t) - \mu(x(t) - \mu^*)/r_2^3(t), \\ y''(t) = -2x'(t) + y(t) - \mu^* y(t)/r_1^3(t) - \mu y/r_2^3(t) \end{cases}$$

where $\mu = 1/82.45$, $\mu^* = 1 - \mu$, and

$$r_1(t) = \sqrt{(x(t) + \mu)^2 + y^2(t)}, \quad r_2(t) = \sqrt{(x(t) - \mu^*)^2 + y^2(t)}.$$

The initial values are $x(0) = 1.2$, $x'(0) = 0$, $y(0) = 0$, and $y'(0) = -1.04935751$. Solve the differential equations and draw the (x, y) trajectory of the Apollo satellite.

Solutions. Select a set of state variables as $x_1(t) = x(t)$, $x_2(t) = x'(t)$, $x_3(t) = y(t)$, and $x_4(t) = y'(t)$. The first-order explicit differential equations can be written as

$$\boldsymbol{x}'(t) = \begin{bmatrix} x_2(t) \\ 2x_4(t) + x_1(t) - \mu^*(x_1(t) + \mu)/r_1^3(t) - \mu(x_1(t) - \mu^*)/r_2^3(t) \\ x_4(t) \\ -2x_2(t) + x_3(t) - \mu^* x_3(t)/r_1^3(t) - \mu x_3(t)/r_2^3(t) \end{bmatrix}$$

where

$$r_1(t) = \sqrt{(x_1(t) + \mu)^2 + x_3^2(t)}, \quad r_2(t) = \sqrt{(x_1(t) - \mu^*)^2 + x_3^2(t)}$$

with $\mu = 1/82.45$ and $\mu^* = 1 - \mu$. The initial values of the state variables are $\boldsymbol{x}_0 = [1.2, 0, 0, -1.04935751]^\mathrm{T}$.

Since there are two intermediate variables $r_1(t)$ and $r_2(t)$, a MATLAB function can be used to compute $\boldsymbol{x}'(t)$. The MATLAB function can be written as follows:

```
function dx=apolloeq(t,x)
mu=1/82.45; mu1=1-mu;
r1=sqrt((x(1)+mu)^2+x(3)^2); r2=sqrt((x(1)-mu1)^2+x(3)^2);
dx=[x(2);
    2*x(4)+x(1)-mu1*(x(1)+mu)/r1^3-mu*(x(1)-mu1)/r2^3;
    x(4);
    -2*x(2)+x(3)-mu1*x(3)/r1^3-mu*x(3)/r2^3]; %describe ODE
```

The numerical solution can be obtained with the solver `ode45()`.

```
>> x0=[1.2;0;0;-1.04935751];
   tic, [t,y]=ode45(@apolloeq,[0,20],x0); toc %solution
   length(t), plot(y(:,1),y(:,3)) %draw phase plane trajectory
```

The trajectory of the satellite can be obtained as shown in Figure 4.7. The number of points computed is 689, and the elapsed time is 0.014 seconds.

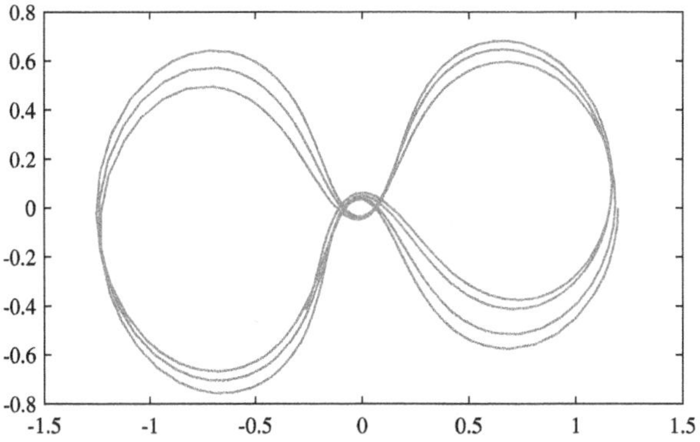

Figure 4.7: An erroneous trajectory found.

In fact, the Apollo satellite trajectory thus found is incorrect. This is because the default error tolerance, `RelTol`, is too large, such that the solver `ode45()` yields the wrong results. To better solve the problem, the error tolerance should be reduced. For instance, it can be educed to 10^{-6}. The following commands can be used to solve the differential equations again:

```
>> ff=odeset; ff.AbsTol=1e-6; ff.RelTol=1e-6;
   tic, [t1,y1]=ode45(@apolloeq,[0,20],x0,ff); toc %solve again
   length(t1), plot(y1(:,1),y1(:,3)) %draw the trajectory
```

The new trajectory obtained is shown in Figure 4.8. The total number of points is increased to 1 873, and the elapsed time is 0.067 seconds. It can be seen that the trajectory is completely different from that under the default setting. The error tolerance can further be decreased, however, the trajectory may look almost the same. In a real solution process, the error tolerance `RelTol` should be set to different values, and one may use such a method to validate the result.

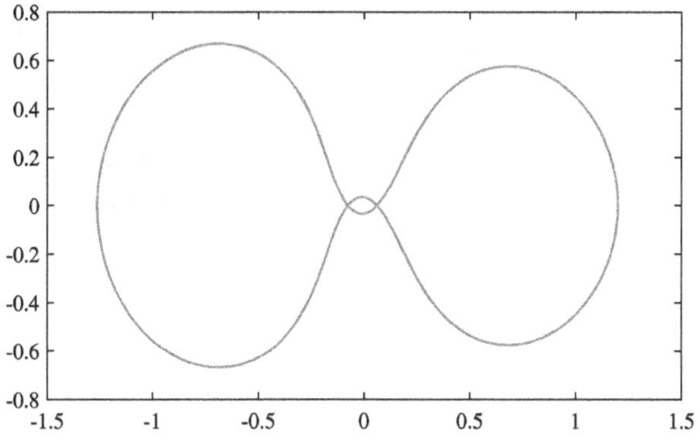

Figure 4.8: The correct Apollo satellite trajectory.

With the following MATLAB commands, the minimum step-size of 1.8927×10^{-4} is witnessed. The step-size curve can be obtained, as shown in Figure 4.9.

```
>> min(diff(t1));                  %find the minimum step-size
   plot(t1(1:end-1),diff(t1)) %draw the step-size curve
```

The significance of using variable-step algorithm can be seen from the step-size curve. It is obvious that when accurate solutions are needed, the step-size can be set to small values; while when the error is already kept small, the step-size can be increased adaptively, so as to speed up the simulation process. In this case, the efficiency of the algorithm is increased.

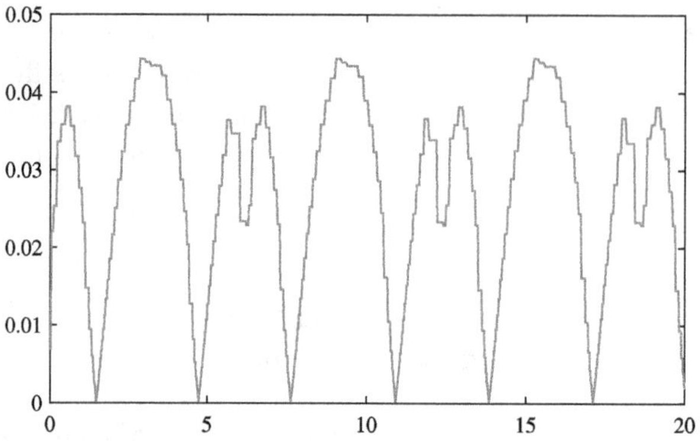

Figure 4.9: The step-size in the simulation process.

It is also seen that most of the time, a large step-size of 0.04 is used. In fixed-step algorithms, no one dares to use such a large step-size in the solution process. To ensure the accuracy of some points, the small step-size of 2×10^{-4} is taken automatically. In other words, at these points, extremely small step-sizes are used to ensure that the error is kept under 10^{-6}. From a fixed-step viewpoint, if one wants to ensure 10^{-6} accuracy, the step-size should not be selected larger than the minimum step-size obtained above. It can be seen the number of computation points is increased to 10^5, 56 times more than for the variable-step algorithm.

The error tolerance can further be set to 100`eps`, and the new step-sizes can be drawn in Figure 4.10. It can be seen that the minimum step-size is automatically set to 5.4904×10^{-6}, and the total number of points is 63 053.

```
>> ff=odeset; ff.AbsTol=100*eps; ff.RelTol=100*eps;
   tic, [t1,y1]=ode45(@apolloeq,[0,20],x0,ff); toc %solve again
   min(diff(t1)), length(t1)   %find the minimum step-size
   plot(t1(1:end-1),diff(t1)) %draw the step-size plot
```

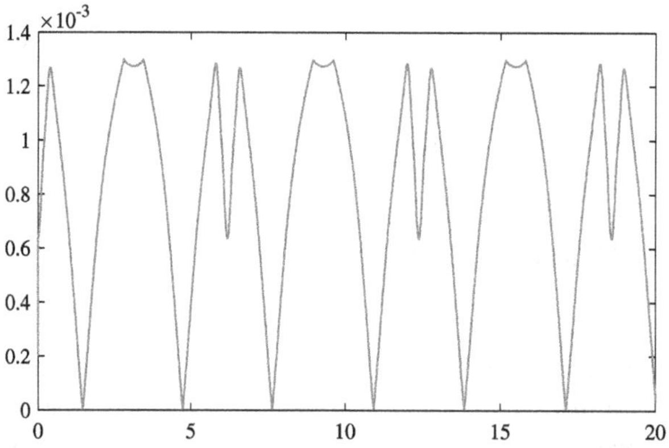

Figure 4.10: The new step-size curve when error tolerance is changed.

It can be imagined that to ensure such precision, the step-size of 5.4904×10^{-6} should be used throughout the fixed-step solution process. It may bring too much computational load for the computers. It can be seen that the variable-step algorithms are far superior to the fixed-step ones.

Example 4.13. In Example 4.12, a MATLAB function was used to express the differential equations. Use an anonymous function to describe the differential equations, and solve them again.

Solutions. In fact, for this type of problem, anonymous functions can be used to describe the differential equations, since intermediate variables can also be described by anonymous functions. Therefore there is no longer any need to use MATLAB files to express differential equation models. An anonymous function can be used to handle such problems. The following commands can be used to solve the differential equations again, and the results are exactly the same.

```
>> mu=1/82.45; mu1=1-mu;
   r1=@(x)sqrt((x(1)+mu)^2+x(3)^2);
   r2=@(x)sqrt((x(1)-mu1)^2+x(3)^2); %intermediate parameters
   f=@(t,x)[x(2);
       2*x(4)+x(1)-mu1*(x(1)+mu)/r1(x)^3-mu*(x(1)-mu1)/r2(x)^3;
       x(4); -2*x(2)+x(3)-mu1*x(3)/r1(x)^3-mu*x(3)/r2(x)^3];
   x0=[1.2; 0; 0; -1.04935751];          %initial values
   ff=odeset; ff.AbsTol=100*eps; ff.RelTol=100*eps;
   tic, [t1,y1]=ode45(f,[0,20],x0,ff); toc %solve differential equation
```

Example 4.14. Consider the seven-body problem. The coordinates $(x_i(t), y_i(t))$ of the seven bodies satisfy the following differential equations:[32]

$$x_i''(t) = \sum_{j\neq i} \frac{m_j}{r_{ij}(t)}(x_j(t) - x_i(t)), \quad y_i''(t) = \sum_{j\neq i} \frac{m_j}{r_{ij}(t)}(y_j(t) - y_i(t)),$$

where $m_i = i$, $i = 1, 2, \dots, 7$, and

$$r_{ij}(t) = \left((x_i(t) - x_j(t))^2 + (y_i(t) - y_j(t))^2\right)^{3/2}.$$

The initial positions and speeds of the bodies are known as $x_1(0) = 3$, $x_2(0) = 3$, $x_3(0) = -1$, $x_4(0) = -3$, $x_5(0) = 2$, $x_6(0) = -2$, $x_7(0) = 2$, $y_1(0) = 3$, $y_2(0) = -3$, $y_3(0) = 2$, $y_4(0) = 0$, $y_5(0) = 0$, $y_6(0) = -4$, $y_7(0) = 4$, $x_6'(0) = 1.75$, $x_7'(0) = -1.5$, $y_4'(0) = -1.25$, $y_5'(0) = 1$, and all the other $x_i'(0) = y_i'(0) = 0$. The simulation interval is $t \in (0, 3)$.

Solutions. Analyzing the original problem, it is found that $r_{ij}(t)$ forms a 7×7 symmetric matrix $\boldsymbol{R}(t)$. Considering the conditions $i \neq j$ in the sum signs, the diagonal terms in matrix $\boldsymbol{R}(t)$ should be set to be very large such as 30^{30}. As before, the state variables can be selected as $\boldsymbol{z}_1(t) = \boldsymbol{x}(t)$, $\boldsymbol{z}_2(t) = \boldsymbol{y}(t)$, $\boldsymbol{z}_3(t) = \boldsymbol{x}'(t)$, $\boldsymbol{z}_4(t) = \boldsymbol{y}'(t)$, and

$z(t) = [z_1(t), z_2(t), z_3(t), z_4(t)]^T$. It is not hard to write the following MATLAB function to express the original differential equations:

```
function dz=seven_body(t,z)
n=7; M=[1:n]'; x=z(1:n); y=z(n+1:2*n);
[xi,xj]=meshgrid(x,x); [yi,yj]=meshgrid(y,y);
R=(xi-xj).^2+(yi-yj).^2; R=R.^(3/2); R=R+1e30*eye(n);
DX=sum(((xj-xi)./R).*M); DY=sum(((yj-yi)./R).*M);
z0=z(2*n+1:end); dz=[z0(:); DX(:); DY(:)];
```

The initial state vector can be entered, while the following commands can be used to initiate the simulation process and draw the trajectories. The trajectories of the seven bodies are as shown in Figure 4.11. The results are exactly the same as those in [32].

```
>> z0=[3;3;-1;-3;2;-2;2; 3;-3;2;0;0;-4;4;
       0;0;0;0;0;1.75;-1.5; 0;0;0;-1.25;1;0;0];
   ff=odeset; ff.RelTol=100*eps; ff.AbsTol=100*eps;
   [t,x]=ode45(@seven_body,[0,3],z0,ff); % solve differential equations
   n=7; plot(x(:,1:n),x(:,n+1:2*n)), length(t)
```

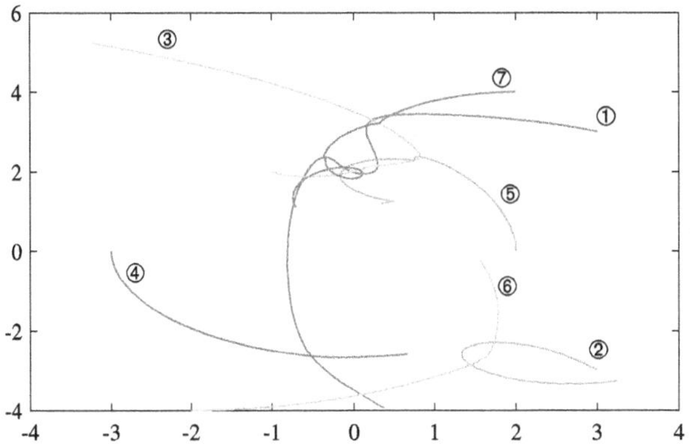

Figure 4.11: Trajectories of the seven bodies.

In fact, one may use many loops to describe the differential equations. However, the method like this is not native to MATLAB programming. The vectorized programming and matrix computation is a better solution, and more suitable for this kind of problem.

Normally, functions like `comet()` cannot be used to draw multiple trajectories in animation form. The animation technique in Volume I can be used and the following commands tried to create the animation display of the seven bodies. A video file `c4stars.avi` can then be made, which can be played on any multimedia player.

```
>> h=plot(x(1,1:n),x(1,n+1:2*n),'o'); axis([-4 4 -4 6])
   vid=VideoWriter('c4stars.avi'); open(vid); %open blank movie file
   for k=2:length(t) %make animation of each move
      set(h,'XData',x(k,1:n),'YData',x(k,n+1:2*n))
      drawnow; hVid=getframe; writeVideo(vid,hVid); %create a frame
   end
   close(vid) %after making a movie, close video file
```

Example 4.15. Solve high-order linear differential equations with constant coefficients in Example 2.15 with the numerical method:

$$\begin{cases} x''(t) - x(t) + y(t) + z(t) = 0, \\ x(t) + y''(t) - y(t) + z(t) = 0, \\ x(t) + y(t) + z''(t) - z(t) = 0 \end{cases}$$

where $x(0) = 1$ and $y(0) = z(0) = x'(0) = y'(0) = z'(0) = 0$.

Solutions. With the recommended method, the state variables can be selected as $x_1(t) = x(t)$, $x_2(t) = x'(t)$, $x_3(t) = y(t)$, $x_4(t) = y'(t)$, $x_5(t) = z(t)$, and $x_6(t) = z'(t)$. It is immediately found that the first-order explicit differential equations can be written as

$$\boldsymbol{x}(t) = \begin{bmatrix} x_2(t) \\ x_1(t) - x_3(t) - x_5(t) \\ x_4(t) \\ -x_1(t) + x_3(t) - x_5(t) \\ x_6(t) \\ -x_1(t) - x_3(t) + x_5(t) \end{bmatrix}$$

and the initial state vector is $\boldsymbol{x}(0) = [1, 0, 0, 0, 0, 0]^{\mathrm{T}}$. With an anonymous function, the standardized differential equations can be defined, and the following commands can be used to solve them. The results are the same as in Figure 2.9.

```
>> f=@(t,x)[x(2); x(1)-x(3)-x(5);
            x(4); -x(1)+x(3)-x(5);
            x(6); -x(1)-x(3)+x(5)];
   x0=[1,0,0,0,0,0]';
   ff=odeset; ff.AbsTol=100*eps; ff.RelTol=100*eps;
   [t0,x]=ode45(f,[0,1],x0,ff); plot(t0,x(:,[1 3 5]))
```

The methods in Example 2.23 can be used to find the analytical solution of the original differential equations. The exact values at each point can also be evaluated, from which the maximum error can be found as 2.9388×10^{-12}. It can be seen that the result obtained is rather accurate.

```
>> syms t x(t) y(t) z(t)
   d1x=diff(x); d1y=diff(y); d1z=diff(z);
   [x1,y1,z1]=dsolve(diff(x,2)-x+y+z==0,x+diff(y,2)-y+z==0,...
                    x+y+diff(z,2)-z==0, x(0)==1, d1x(0)==0,...
                    y(0)==0, d1y(0)==0,z(0)==0, d1z(0)==0);
   x1=simplify(x1); y1=simplify(y1); z1=simplify(z1);
   norm([x(:,[1,3,5])-double(subs([x1,y1,z1],t,t0))],1)
```

4.3.2 Limitations with fixed-step methods

In conventional numerical analysis courses, various fixed-step algorithms are presented. The main reason is that fixed-step algorithms have neat mathematical formulas, and are easy to teach and numerically implement. While in real applications, fixed-step methods are not practical. The Apollo satellite example is used as an example to illustrate the limitations of the fixed-step algorithms.

Example 4.16. Solve the Apollo satellite equation again with the fixed-step fourth-order Runge–Kutta algorithm.

Solutions. Before using a fixed-step algorithm, two questions must be answered: (1) How to select the step-size? and (2) How to ensure accuracy? The trial-and-error method can be used to answer the first question. It can be seen that a smaller step-size may yield more accurate solutions, while the computational load may also be significantly increased.

For the Apollo satellite problem, the step-size of 0.01 can be tried first, and the following commands can be used to solve the differential equations, and the trajectory of the Apollo satellite can be drawn, as shown in Figure 4.12. The elapsed time is 0.053 seconds.

```
>> x0=[1.2; 0; 0; -1.04935751];
   tic, [t1,y1]=ode_rk4(@apolloeq,[0,0.01,20],x0); toc
   plot(y1(:,1),y1(:,3))  % draw the satellite trajectory
```

It is immediately seen that the result is incorrect. A smaller step-size can be tried. For instance, selecting the step-size of 0.001, the differential equations can be solved again. A more accurate trajectory curve can be found, which looks similar to that shown in Figure 4.8. The elapsed time in this case is 0.84 seconds, 13 times of the time needed by a variable-step algorithm.

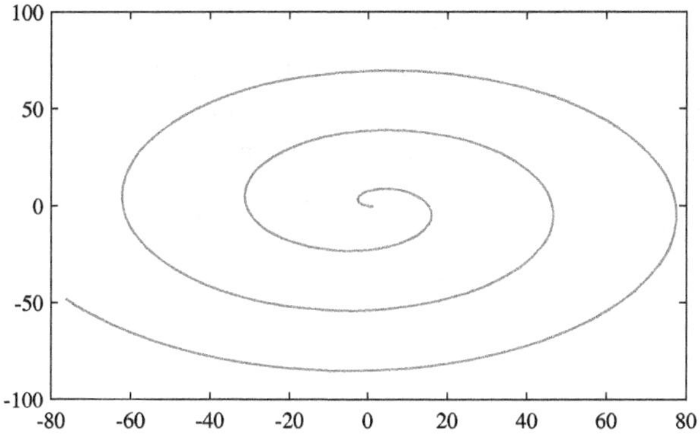

Figure 4.12: Satellite trajectory with $h = 0.01$ (erroneous).

```
>> tic, [t2,y2]=ode_rk4(@apolloeq,[0,0.001,20],x0); toc
   plot(y2(:,1),y2(:,3))  % draw the trajectory
```

If the solutions using the variable-step algorithm `ode45()` and under tough error tolerance are considered as accurate ones, the errors under fixed-step algorithm can be obtained, as shown in Figure 4.13. It can be seen that although the trajectory curves look similar, the actual errors are rather large and not negligible. The errors in the derivative signal are even as large as 4 in magnitude. The errors like this cannot be accepted in real applications.

```
>> ff=odeset; ff.AbsTol=100*eps; ff.RelTol=100*eps;
   tic, [t3,y3]=ode45(@apolloeq,t2,x0,ff); toc % solve again
   plotyy(t2,y3(:,[1 3])-y2(:,[1,3]),t2,y3(:,[2 4])-y2(:,[2,4]))
```

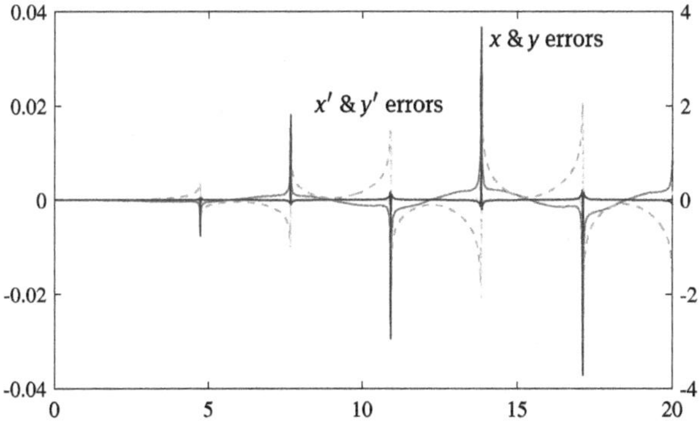

Figure 4.13: Error curves of the fourth-order Runge–Kutta algorithm.

In fact, there may exist very large errors at certain points. Although the shape of the trajectory looks correct, the results may be considered incorrect. In solving differential equations, variable-step algorithms are recommended. There is no need to use the commonly taught fixed-step algorithms in numerical analysis courses. If the tough error-level of 100eps is expected, it is not possible to find the result with a fixed-step method. Besides, since there is no monitoring mechanism in the fixed-step algorithm, it is not possible to select a step-size to ensure certain error bounds. Therefore fixed-step algorithms are unsuitable for dealing with real problems.

4.3.3 Simple implicit differential equations

In the previously studied examples, the explicit form of the highest-order derivative of the unknown function could be found. It was easy to build up the first-order explicit differential equation model. In this section, complicated differential equation sets are discussed, where in each equation the highest-order derivatives of two unknown functions appear simultaneously. Algebraic equation solution must be carried out in describing the first-order explicit differential equations. Examples are given to demonstrate the conversion and solving methods.

Example 4.17. Consider the following differential equations, if $x_1(0) = x_2(0) = 0$,

$$\begin{cases} x_1'(t)\sin x_1(t) + x_2'(t)\cos x_2(t) + x_1(t) = 1, \\ -x_1'(t)\cos x_2(t) + x_2'(t)\sin x_1(t) + x_2(t) = 0. \end{cases}$$

Find their numerical solution.

Solutions. Letting $\boldsymbol{x}(t) = [x_1(t), x_2(t)]^{\mathrm{T}}$, the original differential equations can be expressed in matrix form

$$\boldsymbol{A}(\boldsymbol{x}(t))\boldsymbol{x}'(t) = \boldsymbol{B}(\boldsymbol{x}(t))$$

where

$$\boldsymbol{A}(\boldsymbol{x}(t)) = \begin{bmatrix} \sin x_1(t) & \cos x_2(t) \\ -\cos x_2(t) & \sin x_1(t) \end{bmatrix}, \quad \boldsymbol{B}(\boldsymbol{x}(t)) = \begin{bmatrix} 1 - x_1(t) \\ -x_2(t) \end{bmatrix}.$$

If one can show that $\boldsymbol{A}(\boldsymbol{x}(t))$ is a nonsingular matrix, then the equation can be converted into the following standard form;

$$\boldsymbol{x}'(t) = \boldsymbol{A}^{-1}(\boldsymbol{x}(t))\boldsymbol{B}(\boldsymbol{x}(t)).$$

With the MATLAB solvers, the numerical solutions can be found. In fact, there is no method to strictly show that matrix $\boldsymbol{A}(\boldsymbol{x}(t))$ is nonsingular, one may try using the standard form to solve the problem numerically. If in the solution process, there is no error or warning message indicating $\boldsymbol{A}(\boldsymbol{x}(t))$ is singular, it means that in the solution process $\boldsymbol{A}(\boldsymbol{x}(t))$ is nonsingular. The solutions thus obtained are acceptable. If in the solution process, there were warnings like these, the result obtained may not be of any use. Better solvers should be adopted.

In order to study implicit differential equations, an anonymous function can be taken to describe the differential equations. Therefore, the following commands can be used to solve the considered differential equations:

```
>> f=@(t,x)inv([sin(x(1)) cos(x(2));
            -cos(x(2)) sin(x(1))])*[1-x(1); -x(2)];
   opt=odeset; opt.RelTol=100*eps; opt.AbsTol=100*eps;
   [t,x]=ode45(f,[0,10],[0; 0],opt); plot(t,x)
```

The time responses of the state variables can be drawn as shown in Figure 4.14. It can be seen that no warning or error messages appeared in the solution process. Therefore the solution obtained is correct.

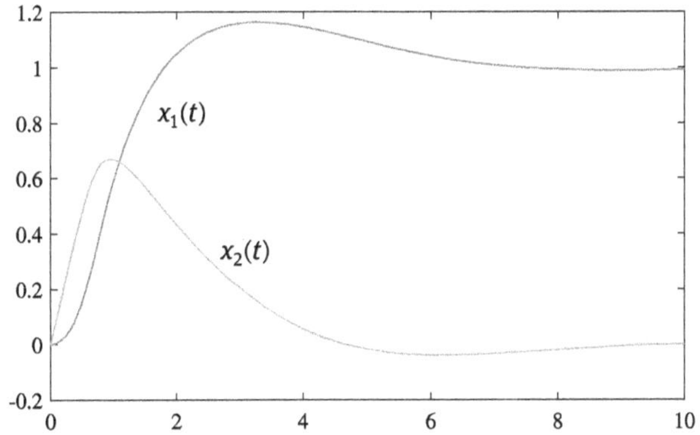

Figure 4.14: The time response of the differential equations.

Example 4.18. Assuming that the differential equations are given as follows:

$$\begin{cases} x''(t) + 2y'(t)x(t) = 2y''(t), \\ x''(t)y'(t) + 3x'(t)y''(t) + x(t)y'(t) - y(t) = 5, \end{cases}$$

convert them to first-order explicit differential equations, and find the numerical solutions.

Solutions. It can be seen that the two equations contain the highest order derivatives $x''(t)$ and $y''(t)$ simultaneously, but the state variables can still be selected as $x_1(t) = x(t)$, $x_2(t) = x'(t)$, $x_3(t) = y(t)$, and $x_4(t) = y'(t)$. The target is to eliminate one of the highest-order derivative term. For this example, solving the first equation, it is found that $y''(t) = y'(t)x(t) + x''(t)/2$. Substituting it into the second equation, $x''(t)$ can be expressed as

$$x''(t) = \frac{2y(t) + 10 - 2x(t)y'(t) - 6x(t)x'(t)y'(t)}{2y'(t) + 3x'(t)},$$

which can be rewritten in the state space form as

$$x_2'(t) = \frac{2x_3(t) + 10 - 2x_1(t)x_4(t) - 6x_1(t)x_2(t)x_4(t)}{2x_4(t) + 3x_2(t)}.$$

Substituting the above result back to the $y''(t)$ equation, it is found that

$$x_4'(t) = \frac{x_3(t) + 5 - x_1(t)x_4(t) + 2x_1(t)x_4^2(t)}{2x_4(t) + 3x_2(t)}.$$

Summarizing the above, the first-order explicit differential equations can be obtained as

$$\boldsymbol{x}'(t) = \begin{bmatrix} x_2(t) \\ \dfrac{2x_3(t) + 10 - 2x_1(t)x_4(t) - 6x_1(t)x_2(t)x_4(t)}{2x_4(t) + 3x_2(t)} \\ x_4(t) \\ \dfrac{x_3(t) + 5 - x_1(t)x_4(t) + 2x_1(t)x_4^2(t)}{2x_4(t) + 3x_2(t)} \end{bmatrix}.$$

In fact, equations like this cannot be easily solved applying the manual method. Symbolic Math Toolbox should be adopted to solve the algebraic equations. For convenience, denote $p_1(t) = x''(t)$ and $p_2(t) = y''(t)$. Therefore, $p_1(t)$ and $p_2(t)$ are, in fact, $x_2'(t)$ and $x_4'(t)$. With the following commands, the solutions of the equation can be found. If can be seen that the results are the same as obtained above.

```
>> syms x1 x2 x3 x4 p1 p2 %declare symbolic variable
   [p1,p2]=solve(p1+2*x4*x1==2*p2,p1*x4+3*x2*p2+x1*x4-x3==5,...
                 [p1,p2]) %solve algebraic equations
```

From the converted standard form, the following commands can be used to describe the differential equations and find their numerical solutions. It can be seen that for these particular differential equations, there are no solutions, since the matrix $\boldsymbol{x}$ obtained this way is composed of `NaN`'s.

```
>> f=@(t,x)[x(2);
      (2*x(3)+10-2*x(1)*x(4)+6*x(1)*x(2)*x(4))/(2*x(4)+3*x(2));
      x(4);
```

```
      (x(3)+5-x(1)*x(4)+2*x(1)*x(4)^2)/(2*x(4)+3*x(2))];
ff=odeset; ff.AbsTol=100*eps; ff.RelTol=100*eps;
[t,x]=ode45(f,[0,10],[1; 0; 1; 0],ff);
plot(t,x)  % solve differential equation and draw solutions
```

4.3.4 Even more complicated nonlinear differential equations

The examples shown earlier are too simple, since manual conversion may help transform them directly into the standard form. If $x^{(m)}(t)$ and $y^{(n)}(t)$ terms appear simultaneously in both equations, corresponding manipulations should be made such that two algebraic equations can be established. Therefore a MATLAB code for solving the two algebraic equations can be written. Solving the equations, a MATLAB function describing the standard differential equations can be constructed, so that the solver `ode45()` can be used to solve them directly. In other words, the algebraic equation solver can be embedded in the differential equation model. Examples will be used to demonstrate the solution procedures.

Example 4.19. Consider a system of more complicated implicit differential equations given below. Assuming that the initial values are $x(0) = y'(0) = 1$, $x'(0) = y(0) = 0$, find the numerical solutions of the differential equations

$$\begin{cases} x''(t)\sin y'(t) + (y''(t))^2 = -2x(t)y(t)\mathrm{e}^{-x'(t)} + x(t)x''(t)y'(t), \\ x(t)x''(t)y''(t) + \cos y''(t) = 3y(t)x'(t)\mathrm{e}^{-x(t)}. \end{cases}$$

Solutions. The state variables can still be selected as $x_1(t) = x(t)$, $x_2(t) = x'(t)$, $x_3(t) = y(t)$, and $x_4(t) = y'(t)$. Then

$$x_1'(t) = x_2(t), \quad x_3'(t) = x_4(t).$$

It is obvious that the method in Example 4.18 cannot be used to solve the algebraic equations, since the explicit expressions of $x_2'(t)$ and $x_4'(t)$ cannot be found. The numerical method can be used solve them from the given $\boldsymbol{x}(t)$.

From the given equations, letting $p_1(t) = x''(t)$ and $p_2(t) = y''(t)$, the algebraic equations can be established as

$$\begin{cases} p_1(t)\sin x_4(t) + p_2^2(t) + 2x_1(t)x_3(t)\mathrm{e}^{-x_2(t)} - x_1(t)p_1(t)x_4(t) = 0, \\ x_1(t)p_1(t)p_2(t) + \cos p_2(t) - 3x_3(t)x_2(t)\mathrm{e}^{-x_1(t)} = 0. \end{cases}$$

The algebraic solution solver can be used to find $p_1(t)$ and $p_2(t)$, after this the results can be assigned to $x_2'(t)$ and $x_4'(t)$ so that

$$x_2'(t) = p_1(t), \quad x_4'(t) = p_2(t).$$

In this way the first-order explicit differential equation model can be written, and then a MATLAB solver can be used to solve the differential equations. Such a procedure is not suitable to implement using anonymous functions, since intermediate variables are needed, which are not supported for anonymous functions. The following MATLAB function can be written to describe the differential equation model:

```
function dy=c4impode(t,x)
dx=@(p)[p(1)*sin(x(4))+p(2)^2+...
            2*x(1)*x(3)*exp(-x(2))-x(1)*p(1)*x(4);
        x(1)*p(1)*p(2)+cos(p(2))-3*x(3)*x(2)*exp(-x(1))];
ff=optimset; ff.Display='off'; ff.TolX=eps;
dx1=fsolve(dx,x([1,3]),ff);          %embedded algebraic equation solver
dy=[x(2); dx1(1); x(4); dx1(2)]; %describe differential equations
```

Note that the continuation notation "..." was used. One has to make sure that operators are used in front of the continuation signs. To see why, consider the following erroneous representation:

```
dx=@(p)[p(1)*sin(x(4))+p(2)^2...
            +2*x(1)*x(3)*exp(-x(2))-x(1)*p(1)*x(4);
```

If the command it written like this, MATLAB may accept the expression in the first line as one expression, and that on the second line as another, led by +. Therefore, the function may be wrongly interpreted, and errors may occur in the function call.

Inside the function, anonymous functions describing the algebraic equations are defined, with $p_1(t)$ and $p_2(t)$ being the unknowns. The two unknowns can be solved for using the solver `fsolve()`. And so $p_1(t)$ and $p_2(t)$ can be found. They can be assigned to $x_2'(t)$ and $x_4'(t)$, respectively, such that the first-order explicit differential equations can be set up. A trick is used in the function, namely, the initial search points $p_1(0) = x_1$ and $p_2(0) = x_3$ are used so as to speed up the algebraic equation solution process.

With the standardized differential equation model, the following commands can be used to solve the differential equations directly. The time responses of the states can be obtained as shown in Figure 4.15. The elapsed time is 12.65 seconds, with 2 217 points computed. The whole process is quite time consuming, since in each step of the differential equation solver, the nonlinear algebraic equation is solved once.

```
>> ff=odeset; ff.AbsTol=100*eps; ff.RelTol=100*eps;
   tic, [t,x]=ode45(@c4impode,[0,2],[1,0,0,1],ff); toc
   plot(t,x), length(t) %draw the curves of the solution
```

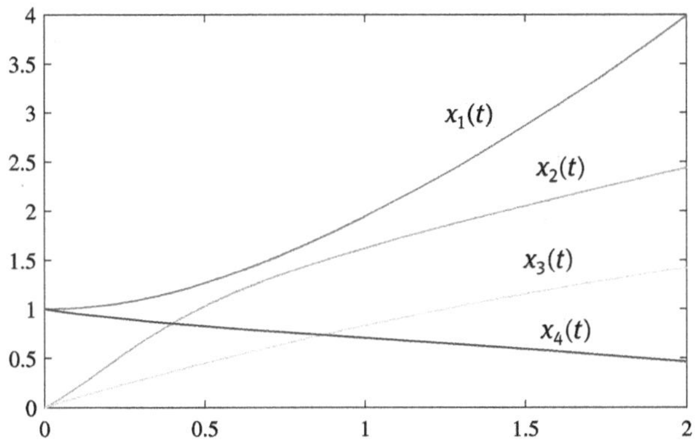

Figure 4.15: Time domain response of implicit differential equation.

4.4 Conversions for matrix differential equations

In certain differential equations, if the unknown functions appear in the matrix form, the equations are referred to as matrix differential equations. For complicated problems, it may be difficult to use the manual method to convert the given differential equations into standard forms. Therefore it is recommended to use computers to find the analytical solutions. If analytical solutions are not available, numerical methods should be used instead. Before the equation can be solved, they should be converted into the standard form. In this chapter, general matrix differential equations are studied first, followed by the solutions of Sylvester and Riccati differential equations.

4.4.1 Conversion and solutions of differential equations in matrix form

In real applications, differential equations in matrix form are usually encountered. For instance, the Lagrangian equations in robotics. The corresponding differential equations in matrix form can be expressed as

$$\boldsymbol{M}\boldsymbol{X}''(t) + \boldsymbol{C}\boldsymbol{X}'(t) + \boldsymbol{K}\boldsymbol{X}(t) = \boldsymbol{F}u(t) \tag{4.4.1}$$

where $\boldsymbol{M}$, $\boldsymbol{C}$, and $\boldsymbol{K}$ are $n\times n$ matrices, while $\boldsymbol{X}$ and $\boldsymbol{F}$ are $n\times 1$ column vectors. Introducing the state vectors $\boldsymbol{x}_1(t) = \boldsymbol{X}(t)$, $\boldsymbol{x}_2(t) = \boldsymbol{X}'(t)$, we get $\boldsymbol{x}_1{}'(t) = \boldsymbol{x}_2(t)$ and $\boldsymbol{x}_2{}'(t) = \boldsymbol{X}''(t)$. It can be seen from (4.4.1) that

$$\boldsymbol{X}''(t) = \boldsymbol{M}^{-1}[\boldsymbol{F}u(t) - \boldsymbol{C}\boldsymbol{X}'(t) - \boldsymbol{K}\boldsymbol{X}(t)]. \tag{4.4.2}$$

Now selecting the state vectors $\boldsymbol{x}(t) = [\boldsymbol{x}_1^{\mathrm{T}}(t), \boldsymbol{x}_2^{\mathrm{T}}(t)]^{\mathrm{T}}$, the state space model can be established as

$$\boldsymbol{x}'(t) = \begin{bmatrix} \boldsymbol{x}_2(t) \\ \boldsymbol{M}^{-1}[\boldsymbol{F}u(t) - \boldsymbol{C}\boldsymbol{x}_2(t) - \boldsymbol{K}\boldsymbol{x}_1(t)] \end{bmatrix}. \tag{4.4.3}$$

It can be seen that the differential equations thus established are already in standard form for the state vector $\boldsymbol{x}(t)$, therefore, the following MATLAB commands can be used to solve them directly. An example is given next to demonstrate the solution process.

Example 4.20. The mathematical model of a double inverted pendulum is[9]

$$\boldsymbol{M}(\boldsymbol{\theta})\boldsymbol{\theta}'' + \boldsymbol{C}(\boldsymbol{\theta},\boldsymbol{\theta}')\boldsymbol{\theta}' = \boldsymbol{F}(\boldsymbol{\theta})$$

where $\boldsymbol{\theta} = [a, \theta_1, \theta_2]^{\mathrm{T}}$, a is the position of the cart, θ_1 and θ_2 are respectively the angles of the upper and lower bars, while the matrices in the inverted pendulum are

$$\boldsymbol{M}(\boldsymbol{\theta}) = \begin{bmatrix} m_c + m_1 + m_2 & (0.5m_1 + m_2)L_1\cos\theta_1 & 0.5m_2L_2\cos\theta_2 \\ (0.5m_1 + m_2)L_1\cos\theta_1 & (m_1/3 + m_2)L_1^2 & 0.5m_2L_1L_2\cos\theta_1 \\ 0.5m_2L_2\cos\theta_2 & 0.5m_2L_1L_2\cos\theta_1 & m_2L_2^2/3 \end{bmatrix},$$

$$\boldsymbol{C}(\boldsymbol{\theta},\boldsymbol{\theta}') = \begin{bmatrix} 0 & -(0.5m_1 + m_2)L_1\theta_1'\sin\theta_1 & -0.5m_2L_2\theta_2'\sin\theta_2 \\ 0 & 0 & 0.5m_2L_1L_2\theta_2'\sin(\theta_1 - \theta_2) \\ 0 & -0.5m_2L_1L_2\theta_1'\sin(\theta_1 - \theta_2) & 0 \end{bmatrix},$$

$$\boldsymbol{F}(\boldsymbol{\theta}) = \begin{bmatrix} u(t) \\ (0.5m_1 + m_2)L_1g\sin\theta_1 \\ 0.5m_2L_2g\sin\theta_2 \end{bmatrix}.$$

For the given parameters in the double pendulum system, $m_c = 0.85\,\mathrm{kg}$, $m_1 = 0.04\,\mathrm{kg}$, $m_2 = 0.14\,\mathrm{kg}$, $L_1 = 0.1524\,\mathrm{m}$, and $L_2 = 0.4318\,\mathrm{m}$, solve numerically the equations and draw the step responses of the signals.

Solutions. It can be seen that the coefficient matrices $\boldsymbol{M}(\theta_1,\theta_2)$, $\boldsymbol{C}(\theta_1,\theta_2)$, and $\boldsymbol{F}(\theta_1,\theta_2)$ are nonlinear functions of the $\boldsymbol{x}$ vector. For instance, the sine and cosine functions of θ_1 make the original differential equation nonlinear. Introducing the additional parameters $\boldsymbol{x}_1 = \boldsymbol{\theta}$, $\boldsymbol{x}_2 = \boldsymbol{\theta}'$, the new state vector $\boldsymbol{x} = [\boldsymbol{x}_1^{\mathrm{T}}, \boldsymbol{x}_2^{\mathrm{T}}]^{\mathrm{T}}$ can be composed. The following MATLAB function can be written to describe the first-order explicit differential equations:

```
function dx=inv_pendulum(t,x,u,mc,m1,m2,L1,L2,g)
M=[mc+m1+m2, (0.5*m1+m2)*L1*cos(x(2)), 0.5*m2*L2*cos(x(3))
   (0.5*m1+m2)*L1*cos(x(2)),(m1/3+m2)*L1^2,0.5*m2*L1*L2*cos(x(2))
   0.5*m2*L2*cos(x(3)),0.5*m2*L1*L2*cos(x(2)),m2*L2^2/3]; %M matrix
C=[0,-(0.5*m1+m2)*L1*cos(x(5))*sin(x(2)),-0.5*m2*L2*x(6)*sin(x(3))
   0, 0, 0.5*m2*L1*L2*x(6)*sin(x(2)-x(3))
   0, -0.5*m2*L1*L2*x(5)*sin(x(2)-x(3)), 0];                %C matrix
F=[u; (0.5*m1+m2)*L1*g*sin(x(2)); 0.5*m2*L2*g*sin(x(3))]; %F matrix
dx=[x(4:6); inv(M)*(F-C*x(4:6))]; %compute x'(t)
```

If a step signal is used to excite the system, the following commands can be used to find the numerical solutions as shown in Figures 4.16 and 4.17:

```
>> opt=odeset; opt.RelTol=100*eps; opt.AbsTol=100*eps;
   u=1; mc=0.85; m1=0.04; m2=0.14; %input relevant parameters
   L1=0.1524; L2=0.4318; g=9.81;    %input system parameters
   f=@(t,x)inv_pendulum(t,x,,u,mc,m1,m2,L1,L2,g);
   [t,x]=ode45(f,[0,0.5],zeros(6,1),ff); %solve equations
   plot(t,x(:,1:3))
   figure; plot(t,x(:,4:6)) %draw x(t) and in new window draw x'(t)
```

It should be noted that since the double inverted pendulum system is naturally unstable, it is meaningless to apply step input to the system in reality. Appropriate control signals should be applied to stabilize the pendulum system.

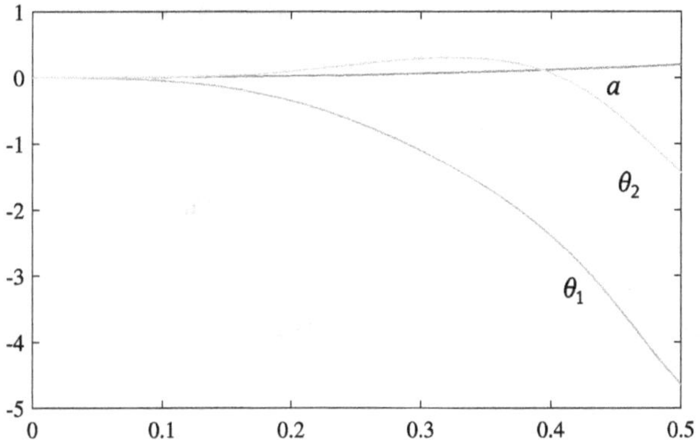

Figure 4.16: Step responses of the double inverted pendulum.

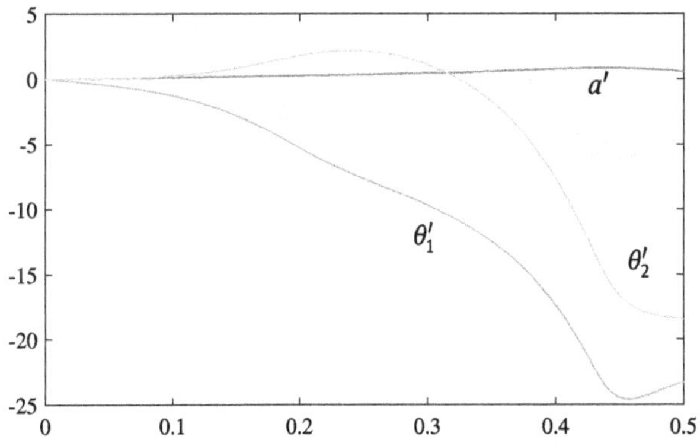

Figure 4.17: Step responses of the derivative signals in the double inverted pendulum.

Besides, if the matrices $\boldsymbol{M}$, $\boldsymbol{C}$, $\boldsymbol{K}$ and $\boldsymbol{F}$ are all independent of $\boldsymbol{X}(t)$, the equation becomes a linear differential equation. Through simple conversions, the following linear state space equation can be found:

$$\begin{bmatrix} x_1'(t) \\ \hdashline x_2'(t) \end{bmatrix} = \begin{bmatrix} 0 & I \\ \hdashline -\boldsymbol{M}^{-1}\boldsymbol{K} & -\boldsymbol{M}^{-1}\boldsymbol{C} \end{bmatrix} \begin{bmatrix} x_1(t) \\ \hdashline x_2(t) \end{bmatrix} + \begin{bmatrix} 0 \\ \hdashline \boldsymbol{M}^{-1}\boldsymbol{F} \end{bmatrix} u(t). \tag{4.4.4}$$

4.4.2 Sylvester differential equations

Sylvester differential equations are commonly encountered matrix equations. The mathematical form and analytical solutions of Sylvester equation are studied in Section 2.5.3. In this section, numerical solutions are presented for Sylvester equations, and demonstrated through examples.

Compared with the analytical solution methods, there are special tricks in solving Sylvester equations numerically. A suitable set of state variables can be chosen to convert the given equations into first-order explicit differential equations. Consider again the mathematical form of a Sylvester differential equation

$$\boldsymbol{X}'(t) = \boldsymbol{A}\boldsymbol{X}(t) + \boldsymbol{X}(t)\boldsymbol{B}, \quad \boldsymbol{X}(0) = \boldsymbol{C} \tag{4.4.5}$$

where $\boldsymbol{X}(t)$ is a matrix that can be expanded in the column-wise format as a column vector, which can then be implemented in MATLAB as $\boldsymbol{x}(t) = \boldsymbol{X}(:)$. With the vector representation, function `X=reshape(x,n,m)` can be used to convert it back to an $n\times m$ matrix. The following MATLAB function can be written to express Sylvester differential equation:

```
function dx=c4msylv(t,x,A,B)
[n1,m1]=size(A); [n2,m2]=size(B);
X=reshape(x,n1,n2); dx=A*X+X*B; dx=dx(:);
```

where $\boldsymbol{A}$ and $\boldsymbol{B}$ are additional parameters. With this function, the Sylvester differential equation can be solved numerically employing solvers. The analytical solutions obtained in Section 2.5.3 can also be used to assess the accuracy and efficiency of the numerical solutions. An example is given next to demonstrate the numerical solutions of Sylvester differential equations.

Example 4.21. Solve numerically the Sylvester differential equation in Example 2.34. Study the accuracy and efficiency of the method by comparing with the analytical solution in the example.

Solutions. Inputting the relevant matrices into MATLAB, an anonymous function can be written as an interface for MATLAB function `c4msylv()` in order to avoid additional parameters. The variables $\boldsymbol{A}$ and $\boldsymbol{B}$ in MATLAB workspace can be extracted directly.

Therefore, with the following statements, the numerical solutions of Sylvester matrix equation can be found. Some of the variables are shown in Figure 4.18. The readers may find the analytical solution in Example 2.34, and compare the accuracy of the solution. This is left as an exercise.

```
>> A=[-1,-2,0,-1; -1,-3,-1,-2; -1,1,-2,0; 1,2,1,1];
   B=[-2,1; 0,-2]; X0=[0,-1; 1,1; 1,0; 0,1];
   f=@(t,x)c4msylv(t,x,A,B);  %an interface from anonymous function
   ff=odeset; ff.AbsTol=100*eps; ff.RelTol=100*eps;
   [t1,x1]=ode45(f,[0,6],X0(:),ff); plot(t1,x1)
```

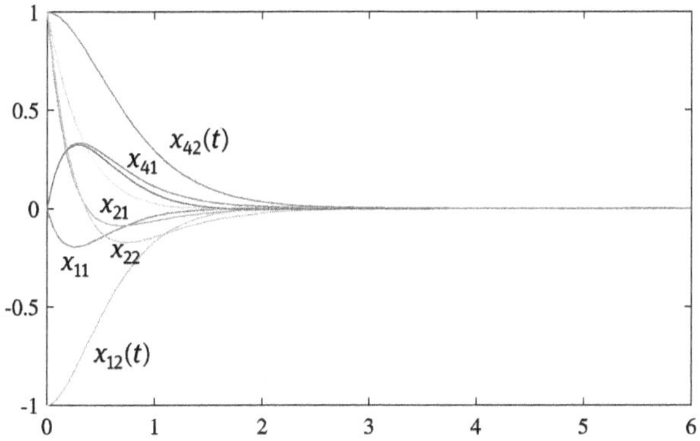

Figure 4.18: Numerical solution of Sylvester matrix equation.

4.4.3 Riccati differential equations

Riccati differential equations are another commonly encountered matrix differential equations. The general form of such equations is

$$\boldsymbol{P}'(t) = \boldsymbol{A}^{\mathrm{T}}\boldsymbol{P}(t) + \boldsymbol{P}(t)\boldsymbol{A} + \boldsymbol{P}(t)\boldsymbol{B}\boldsymbol{P}(t) + \boldsymbol{C} \tag{4.4.6}$$

where $\boldsymbol{B}$ and $\boldsymbol{C}$ are symmetric matrices. It is known at time t_{n} that the terminal value $\boldsymbol{P}(t_{\mathrm{n}})$ is given. A numerical solution in the time interval (t_0, t_{n}) is expected.

To solve such equations numerically, they should be converted into the standard form of first-order explicit differential equations. Then numerical solvers can be applied. In mathematics, the column-wise vector expansion can be denoted as $\mathrm{vec}(\boldsymbol{P}(t))$. In MATLAB, `P(:)` can be used in direct implementation. If the vector is to be transformed back to a matrix, function `reshape()` can be called.

The following MATLAB function may be written to describe Riccati differential equation:

```
function dy=ric_de(t,x,A,B,C)
P=reshape(x,size(A));
Y=A'*P+P*A+P*B*P+C; dy=Y(:); %describe Riccati equation
```

where the given matrices ***A***, ***B***, and ***C*** are fed into the function through additional parameters. In this way, a solver such as `ode45()` can be used to find the numerical solution of Riccati differential equations. Note that, in solvers such as `ode45()`, the terminal time is allowed to be smaller than the starting time.

`[t,p]=ode45(@ric_de,[t_1,0],P_1(:),options,A,B,C)`

Example 4.22. The matrices of a Riccati differential equation and terminal condition are given below. Solve numerically the differential equation if

$$\boldsymbol{A}=\begin{bmatrix}6&6&17\\1&0&-1\\-1&0&0\end{bmatrix},\quad \boldsymbol{B}=\begin{bmatrix}0&0&0\\0&4&2\\0&2&1\end{bmatrix},\quad \boldsymbol{C}=\begin{bmatrix}1&2&0\\2&8&0\\0&0&4\end{bmatrix},\quad \boldsymbol{P}_1(0.5)=\begin{bmatrix}1&0&0\\0&3&0\\0&0&5\end{bmatrix}.$$

Solutions. The matrices should be input first, then the solver can be called to solve the differential equation, and the results are shown in Figure 4.19.

```
>> A=[6,6,17; 1,0,-1; -1,0,0]; B=[0,0,0; 0,4,2; 0,2,1];
   C=[1,2,0; 2,8,0; 0,0,4]; P1=[1,0,0; 0,3,0; 0,0,5];
   ff=odeset; ff.AbsTol=100*eps; ff.RelTol=100*eps;
   [t,p]=ode45(@ric_de,[0.5,0],P1(:),ff,A,B,C);
   plot(t,p)                    % draw the time domain response
```

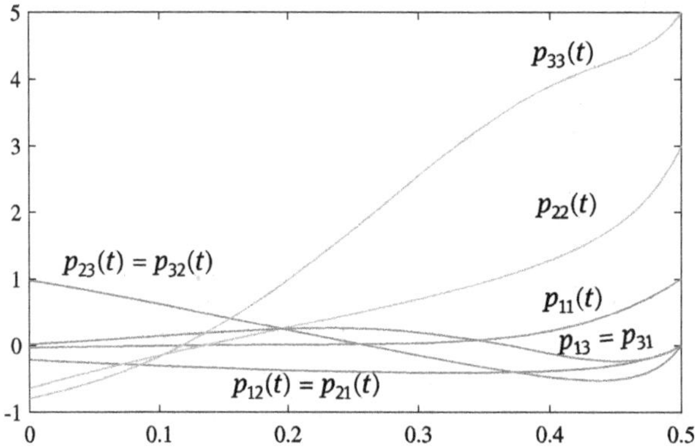

Figure 4.19: Numerical solutions of Riccati differential equation.

For this particular example, if additional parameters are not expected in the solution process, an interface can be designed through an anonymous function, and then the following commands can be used to solve the differential equation directly. Identical results can be obtained in this way.

```
>> f=@(t,x)ric_de(t,x,A,B,C);
   [t,p]=ode45(f,[0.5,0],P1(:),ff);
```

With the above method, the initial state matrix $\boldsymbol{X}(0)$ at time $t = 0$ can be found. From the initial matrix, the differential equation can be solved again, the same results can be found. The maximum error of the terminal matrix recovered in this way is 4.6863×10^{-12}. It can be seen that the original problem can be restored precisely into an initial value problem.

```
>> P0=p(end,:);   %substituting the consistent initial matrix
   [t1,p1]=ode45(@ric_de,[0,0.5],P0(:),ff,A,B,C); %solve again
   norm(p1(end,:)-P1(:)',1)                        %error norm
```

The consistent initial matrix obtained for Riccati equation is

$$\begin{bmatrix} -0.034485182 & -0.21447473 & 0.019514038 \\ -0.21447473 & -0.64287195 & 0.98324799 \\ 0.019514038 & 0.98324799 & -0.79534974 \end{bmatrix}.$$

4.5 Conversions of a class of Volterra integro-differential equations

If in the equations studied, there are both derivative and integral terms of the unknown functions, the equations are referred to as integro-differential equations. In this section, Volterra integro-differential equations with separable variables are discussed.

Definition 4.2. The mathematical form of a first-order Volterra integro-differential equation is given by

$$x'(t) = f(t, x(t)) + \int_0^t K(s, t, x(s))\mathrm{d}s. \tag{4.5.1}$$

Volterra integro-differential equation is named after an Italian mathematician Vito Volterra (1860–1940). In [18], some numerical algorithms are presented for solving certain Volterra integro-differential equations, but the accuracy is not very high.

Similarly, the mathematical definition of high-order Volterra integro-differential equations may also be presented. In fact, through effective transforms, a certain class of Volterra integro-differential equations can be converted into ordinary differential equations or delay differential equations. For instance, if the kernel $K(s,t,x(s))$ in the integral can be factored into a product of functions of t and s, that is,

$$K(s,t,x(s)) = K_1(t)K_2(s,x(s)), \tag{4.5.2}$$

the original equation can be rewritten as

$$x'(t) = f(t,x(t)) + K_1(t)\int_0^t K_2(s,x(s))\mathrm{d}s. \tag{4.5.3}$$

Taking derivatives with respect to t of both sides of (4.5.1) yields

$$x''(t) = f'(t,x(t)) + K_1(t)K_2'(t,x(t)) + K_1'(t)\int_0^t K_2(s,x(s))\mathrm{d}s. \tag{4.5.4}$$

It can be seen that in (4.5.3) and (4.5.4), the same integral term appears. Thus it can be eliminated simply by substituting one equation into another, and the original integro-differential equation can be converted into an ordinary differential equation, so that numerical solvers can be used to study the original integro-differential equations. In this section, the combined method is demonstrated through examples to solve some Volterra integro-differential equations.

Example 4.23. Let us consider a simple Volterra integro-differential equation[18]

$$x'(t) = 1 - \int_0^t x(s)\mathrm{d}s$$

with given initial value $x(0) = 0$. It is known that the analytical solution is $x(t) = \sin t$. Convert it into an ordinary differential equation.

Solutions. For this simple problem, the manual method can be used directly to find the solution. Substituting t into the equation, it is found that $x'(0) = 1$. Taking derivatives of both sides with respect to t, it is seen that

$$x''(t) = -x(t)$$

with initial values $x(0) = 0$ and $x'(0) = 1$. The differential equation has, in fact, multiple analytical solution expressions, namely $x(t) = \sin(2k\pi + t)$, where k is an integer. The solution provided, $x(t) = \sin t$, is merely one of them.

Example 4.24. Solve the following Volterra integro-differential equation:[18]

$$x''(t) = 4 - x(t) - 2x'(t) + 4\int_0^t \mathrm{e}^{t-2s}x(s)\mathrm{d}s \tag{4.5.5}$$

with initial values $x(0) = x'(0) = 1$. It is known that the analytical solution is $x(t) = e^t$. Assess the accuracy of the numerical solution.

Solutions. Although the integro-differential equation contains the second-order derivative, it can be converted in exactly the same way. Consider the integral term in (4.5.5). The part containing t can be moved out of the integral, and it is not hard to derive the integral term explicitly as

$$\int_0^t e^{-2s}x(s)ds = \frac{1}{4e^t}(x''(t) - 4 + x(t) + 2x'(t)). \tag{4.5.6}$$

Now taking the first-order derivative of (4.5.5) with respect to t, there must be an identical term on the right-hand side of the result. Substituting (4.5.6) into the equation, the right-hand side of the equation no longer contains the integral

```
>> syms t s x(t)
   F=4-x-2*diff(x,t)+4*exp(t)*int(exp(-2*s)*x(s),s,0,t)
   F1=subs(diff(F,t),int(exp(-2*s)*x(s),s,0,t),...
      1/4/exp(t)*(diff(x,t,t)-4+x+2*diff(x,t))) % substitute the integral
   simplify(F1), subs(F,t,0)                    % find initial value
```

The left-hand side is $x'''(t)$. The third-order differential equation is found as

$$x'''(t) = x(t) + 4e^{-t}x(t) + x'(t) - x''(t) - 4.$$

The third initial value can also be found from the above expression, as $x''(0) = 4 - 2x'(0) - x(0) = 1$.

This is a normal third-order differential equation with three initial values. Selecting the state variables $x_1(t) = x(t)$, $x_2(t) = x'(t)$, and $x_3(t) = x''(t)$, it can be converted into the following first-order explicit differential equation in standard form

$$\boldsymbol{x}'(t) = \begin{bmatrix} x_2(t) \\ x_3(t) \\ x_1(t) + 4e^{-t}x_1(t) + x_2(t) - x_3(t) - 4 \end{bmatrix}$$

with initial values $\boldsymbol{x}_0 = [1, 1, 1]^{\mathrm{T}}$. With the solver `ode45()`, the numerical solution can immediately be found, and the error of the solution is 6.1937×10^{-14}. The elapsed time is 0.012 seconds. It can be seen that in this way, the Volterra integro-differential equation can be solved effectively.

```
>> f=@(t,x)[x(2); x(3);
            x(1)+4*exp(-t)*x(1)+x(2)-x(3)-4];
   ff=odeset; ff.RelTol=100*eps; ff.AbsTol=100*eps;
   x0=[1; 1; 1]; tic, [t0,y0]=ode45(f,[0,1],x0,ff); toc
   norm(y0(:,1)-exp(t0))
```

In the previous example, the integral term in one equation could be substituted into the other to effectively eliminate the integral term so that the equation can be converted into ordinary differential equations. In some particular cases, this action may lead to singular differential equations. Other methods should be introduced for such problems. This will be demonstrated by the following example.

Example 4.25. Consider the following Volterra integro-differential equation:[18]

$$x'(t) = x(t) + \int_0^t tx^2(s)\mathrm{d}s - 2\mathrm{e}^{-t} + \frac{t}{2}(\mathrm{e}^{-2t} - 1) \tag{4.5.7}$$

with given initial value $x'(0) = 1$. It is known that the analytical solution is $x(t) = \mathrm{e}^{-t}$. Solve the integro-differential equation with the numerical method and assess its accuracy.

Solutions. We use the conversion method discussed earlier:

```
>> syms t s x(t)
   F1=x+t*int(x(s)^2,s,0,t)-2*exp(-t)+t/2*(exp(-2*t)-1);
   F2=simplify(diff(F1,t)) % take first-order derivative
```

It can be seen that the original equation can be converted into

$$x''(t) = 2\mathrm{e}^{-t} + \frac{1}{2}\mathrm{e}^{-2t} - t\mathrm{e}^{-2t} + t^2x(t) + \int_0^t x^2(s)\mathrm{d}s + x'(t) - \frac{1}{2}. \tag{4.5.8}$$

If a substitution method is adopted, that is, by substituting the integral in (4.5.7) into the above equation, the following second-order differential equation can be found:

$$x''(t) = \frac{1}{t}\left[2\mathrm{e}^{-t} - x(t) + 2t\mathrm{e}^{-t} - t^2\mathrm{e}^{-2t} + t^2x^2(t) + tx'(t) + x'(t)\right]. \tag{4.5.9}$$

Unfortunately, if the numerical method is used to handle this differential equation, erroneous results may be obtained. No further comments are given here, and this is left as an exercise for the reader to try solve this equation numerically. It is obvious that the differential equation contains t in the denominator, which implies that it is singular at $t = 0$.

How can we solve the original problem? It can be seen by observing (4.5.8) that the equation has an integral term, but the term is an independent one. It is no longer the product with any other time domain functions. Taking the first-order derivative once more will eliminate the integral term. Evaluating the first-order derivative of the right-hand side with

```
>> F3=simplify(diff(F2,t))
```

the following ordinary differential equation can be derived, which no longer has an integral term in it:

$$x'''(t) = x''(t) + 2tx(t)x'(t) + 2x^2(t) - 2e^{-t} - 2e^{-2t} + 2te^{-2t}.$$

For the third-order differential equation, normally three initial values are needed. It is known that $x(0) = 1$. The other two can be derived with the following statements:

```
>> subs(F1,t,0), subs(F2,0)   % find the initial values
```

It can be seen that $x'(0) = x(0) - 2 = -1$ and $x''(0) = x'(0) + 2 = 1$. With the differential equation and initial values, the following commands can be used to find the numerical solution. It is found that the error of the numerical solution is 1.4462×10^{-13}, and the elapsed time is 0.028 seconds. It can be seen that the solution process is effective.

```
>> f=@(t,x)[x(2:3);
         x(3)+2*t*x(1)*x(2)+2*x(1)^2-2*exp(-t)-...
            2*exp(-2*t)+2*t*exp(-2*t)];  % describe ODE
   ff=odeset; ff.RelTol=100*eps; ff.AbsTol=100*eps;
   tic, [t,x]=ode45(f,[0,1],[1;-1;1],ff); toc % solution
   plot(t,x), norm(x(:,1)-exp(-t))              % find error
```

It is worth pointing out that the conversion method introduced here has certain limitations. It cannot be used to handle typical Volterra integro-differential equations. Only the case if the kernel can be separated may be handled by this method. The interested readers may try other numerical integral methods to compute the kernel integral. The readers may also study the code in [21].

4.6 Exercises

4.1 Solve the following differential equation and draw the $y(t)$ curve:

$$y'''(t) + ty(t)y''(t) + t^2y'(t)y^2(t) = e^{-ty(t)}$$

where $y(0) = 2$ and $y'(0) = y''(0) = 0$.
Is there an analytical solution? When the fixed-step fourth-order Runge–Kutta algorithm is used to solve the differential equation, what should be the suitable step-size to ensure expected precision? Use existing MATLAB functions to solve the problem, and assess the speed and accuracy.

4.2 Find the analytical and numerical solutions of the following differential equations:

$$\begin{cases} x''(t) = -2x(t) - 3x'(t) + e^{-5t}, \\ y''(t) = 2x(t) - 3y(t) - 4x'(t) - 4y'(t) - \sin t \end{cases}$$

where $x(0) = 1$, $x'(0) = 2$, $y(0) = 3$, and $y'(0) = 4$.

4.3 Consider the Duffing differential equation

$$x''(t) + \mu_1 x'(t) - x(t) + 2x^3(t) = \mu_2 \cos t, \quad \text{where } x_1(0) = \gamma,\ x_2(0) = 0.$$

(1) If $\mu_1 = \mu_2 = 0$, find the numerical solutions of the differential equation. If $\gamma = [0.1 : 0.1 : 2]$, draw the phase plane trajectories for different initial values;

(2) If $\mu_1 = 0.01$ and $\mu_2 = 0.001$, selecting $\gamma = 0.99, 1.01$, draw the phase plane trajectories for different initial values;

(3) If $x_2(0) = 0.2$, draw the phase plane trajectories for different values of γ.

4.4 Select the state variables to convert the following differential equations into first-order explicit ones. Solve the nonlinear differential equation with MATLAB and draw the phase plane or phase space trajectories for

(1) $$\begin{cases} x''(t) = -x(t) - y(t) - (3x'(t))^2 + (y'(t))^3 + 6y''(t) + 2t, \\ y'''(t) = -y''(t) - x'(t) - \mathrm{e}^{-x(t)} - t \end{cases}$$

where $x(1) = 2$, $x'(1) = -4$, $y(1) = -2$, $y'(1) = 7$ and $y''(1) = 6$;

(2) $$\begin{cases} x''(t) - 2x(t)z(t)x'(t) = 3t^2x^2(t)y(t), \\ y''(t) - \mathrm{e}^{y(t)}y'(t) = 4t^2x(t)z(t), \\ z''(t) - 2tz'(t) = 2t\mathrm{e}^{x(t)y(t)} \end{cases}$$

where $z'(1) = x'(1) = y'(1) = 2$, $z'(1) = x(1) = y(1) = 3$;

(3) $$\begin{cases} x^{(4)}(t) - 8\sin ty(t) = 3t - \mathrm{e}^{-2t}, \\ y^{(4)}(t) + 3t\mathrm{e}^{-5t}x(t) = 12\cos t \end{cases}$$

where $x(0) = y(0) = 0$, $x'(0) = y'(0) = 0.3$, $x''(0) = y''(0) = 1$, and $x'''(0) = y'''(0) = 0.1$.

4.5 Find the analytical and numerical solutions of the following differential equations:

$$\begin{cases} x''(t) = -2x(t) - 3x'(t) + \mathrm{e}^{-5t}, & x(0) = 1, x'(0) = 2, \\ y''(t) = 2x(t) - 3y(t) - 4x'(t) - 4y'(t) - \sin t, & y(0) = 3, y'(0) = 4. \end{cases}$$

4.6 For the given differential equation model, if $u(0) = 1$, $u'(0) = 2$, $v'(0) = 2$ and $v(0) = 1$, select a set of states to convert the equation into first-order explicit differential equation. Solve the equation and draw the $u(t)$ and $v(t)$ trajectories if

$$\begin{cases} u''(t) = -u(t)/r^3(t), \\ v''(t) = -v(t)/r^3(t) \end{cases}$$

where $r(t) = \sqrt{u^2(t) + v^2(t)}$.

4.7 A system of differential equations is given below,[53] where $u_1(0) = 45$, $u_2(0) = 30$, $u_3(0) = u_4(0) = 0$, and g= 9.81. Solve it numerically and draw the time responses of the states:

$$\begin{cases} u_1'(t) = u_3(t), \\ u_2'(t) = u_4(t), \\ 2u_3'(t) + \cos(u_1(t) - u_2(t))u_4'(t) = -\mathrm{g}\sin u_1(t) - \sin(u_1(t) - u_2(t))u_4^2(t), \\ \cos(u_1(t) - u_2(t))u_3'(t) + u_4'(t) = -\mathrm{g}\sin u_2(t) + \sin(u_1(t) - u_2(t))u_3^2(t). \end{cases}$$

4.8 Solve the following initial value problem:[62]

$$\begin{cases} u''(t) + 5v'(t) + 7u(t) = \sin t, \\ v''(t) + 6v'(t) + 4u'(t) = 3u(t) + v(t) = \cos t \end{cases}$$

where $u(0) = 1$, $u'(0) = 2$, $v(0) = 3$, and $v'(0) = 4$.

4.9 Solve the following initial value problem:[62]

$$\begin{cases} s''(t) + 0.042s'(t) + 0.961s(t) = \theta'(t) + 0.063\theta(t), \\ u''(t) + 0.087u'(t) = s'(t) + 0.025s(t), \\ v'(t) = 0.973(u(t) - v(t)), \\ w'(t) = 0.433(v(t) - w(t)), \\ x'(t) = 0.508(w(t) - x(t)), \\ \theta'(t) = -0.396(x(t) - 47.6) \end{cases}$$

where $s(0) = s'(0) = u'(0) = \theta(0) = 0$, $u(0) = 50$, and $v(0) = w(0) = x(0) = 75$. If t increases indefinitely, what may be the final limit of $v(t)$?

4.10 Consider the following double inverted pendulum model:[40]

$$\begin{cases} \theta_1'(t) = \dfrac{1}{6mL^2}\dfrac{2p_1(t) - 3p_2(t)\cos\Delta\theta(t)}{16 - 9\cos^2\Delta\theta(t)}, \\ \theta_2'(t) = \dfrac{1}{6mL^2}\dfrac{8p_1(t) - 3p_1(t)\cos\Delta\theta(t)}{16 - 9\cos^2\Delta\theta(t)}, \\ p_1'(t) = -\dfrac{1}{mL^2}\left[\theta_1'(t)\theta_2'(t)\sin\Delta\theta(t) + \dfrac{3\mathrm{g}}{L}\sin\theta_1(t)\right], \\ p_2'(t) = -\dfrac{1}{mL^2}\left[-\theta_1'(t)\theta_2'(t)\sin\Delta\theta(t) + \dfrac{\mathrm{g}}{L}\sin\theta_2(t)\right] \end{cases}$$

where $\Delta\theta(t) = \theta_1(t) - \theta_2(t)$, $\mathrm{g} = 9.81$, $m = L = 1$, $\theta_1(0) = \pi/4$, $\theta_2(0) = 2\pi/4$, and $p_1(0) = p_2(0) = 0$. Solve this system of differential equations. Is this differential equation system a genuine implicit one? Is there a method to convert it into a first-order explicit differential equation? If there is, solve the equation again and compare the results.

4.11 Assume that a simplified drone model is[40]

$$\begin{cases} x''(t) = (\cos\phi(t)\sin\theta(t)\cos\psi(t) + \sin\phi(t)\sin\psi(t))U_1/m, \\ y''(t) = (\cos\phi(t)\sin\theta(t)\cos\psi(t) - \sin\phi(t)\sin\psi(t))U_1/m, \\ z''(t) = \cos\phi(t)\cos\theta(t)U_1/m - g, \\ \phi''(t) = \theta'(t)\psi'(t)(I_{yy} - I_{zz})/I_{xx} + U_2/I_{xx}, \\ \theta''(t) = \phi'(t)\psi'(t)(I_{zz} - I_{yy})/I_{yy} + U_3/I_{yy}, \\ \psi''(t) = \phi'(t)\theta'(t)(I_{xx} - I_{yy})/I_{zz} + U_4/I_{zz} \end{cases}$$

where the given constants are $I_{xx} = I_{yy} = 0.0081$, $I_{zz} = 0.0142$, $m = 1$, $g = 9.81$, $b = 54.2\times10^{-6}$, $d = 1.1\times10^{-6}$, and $L = 0.24$. Assuming that $f_i = 2031.4$, $i = 1, 2, 3, 4$, it is found that $\omega_i = 2\pi f_i$. The constants $U_1 = b(\omega_1^2+\omega_2^2+\omega_3^2+\omega_4^2)$, $U_2 = bL(-\omega_2^2+\omega_4^2)$, $U_3 = bL(\omega_2^2 - \omega_4^2)$, and $U_4 = b(-\omega_1^2 + \omega_2^2 - \omega_3^2 + \omega_4^2)$ can also be found. If the initial values of the states and their first-order derivatives are all zero, solve the differential equations numerically.

4.12 Convert the following differential equations into the standard forms:

$$\begin{cases} x^2y''(x) + 2xy'(x) = 2y(x)z^2(x) + \lambda^2x^2y(x)[y^2(x) - 1], \\ x^2z''(x) = z(x)[z^2(x) - 1] + x^2y^2(x)z(x). \end{cases}$$

4.13 Convert the following differential equations into first-order explicit differential equations:

$$\begin{cases} p_1'(t) = p_2^2(t)\cos q_1(t)/\sin^3 q_1(t), \\ p_2'(t) = 0, \\ q_1'(t) = 1, \\ q_2'(t) = p_2(t)/\sin^2 q_1(t). \end{cases}$$

4.14 Find the analytical and numerical solutions of the following differential equations. Draw the trajectory of (x, y), and assess the accuracy of the numerical solutions if

$$\begin{cases} (2x''(t) - x'(t) + 9x(t)) - (y''(t) + y'(t) + 3y(t)) = 0, \\ (2x''(t) + x'(t) + 7x(t)) - (y''(t) - y'(t) + 5y(t)) = 0. \end{cases}$$

The initial values are given as $x(0) = x'(0) = 1$ and $y(0) = y'(0) = 0$.

4.15 Find the numerical solutions of the following implicit differential equations, with $x_1(0) = 1$, $x_1'(0) = 1$, $x_2(0) = 2$ and $x_2'(0) = 2$. Draw the trajectory of the solutions if

$$\begin{cases} x_1'(t)x_2''(t)\sin(x_1(t)x_2(t)) + 5x_1''(t)x_2'(t)\cos(x_1^2(t)) + t^2x_1(t)x_2^2(t) = e^{-x_2^2(t)}, \\ x_1''(t)x_2(t) + x_2''(t)x_1'(t)\sin(x_1^2(t)) + \cos(x_2''(t)x_2(t)) = \sin t. \end{cases}$$

4.16 Assess the precision of the numerical solution for the Sylvester matrix differential equation studied in Example 4.21.

4.17 If the initial values are $x(0) = 1$ and $x'(0) = -1$, solve the differential equation in (4.5.9) numerically. Observe whether the exact numerical solution can be found, why? It is known that the exact solution is $x(t) = \mathrm{e}^{-t}$.

4.18 Convert the following Volterra integro-differential equations into standard ordinary differential equations,[18] and find their numerical solutions. With the given analytical solutions, assess the accuracy and efficiency:

(1) $x'(t) = 1 + x(t) - t\mathrm{e}^{-t^2} - 2\int_0^t ts\mathrm{e}^{-x^2(s)}\mathrm{d}s$, $x(0) = 0$, with analytical solution $x(t) = t$;

(2) $x'(t) = x(t) + \dfrac{1}{1500}x^2(t) + \dfrac{1}{3000}\int_{-1}^{1} \mathrm{e}^{2(t-s)}x^2(s)\mathrm{d}s$, $x(0) = 1$, with analytical solution $x(t) = \mathrm{e}^{-t}$.

5 Special differential equations

It can be seen from the presentations and examples in the previous two chapters that ordinary differential equations of various forms can normally be converted to first-order explicit differential equations and then solved numerically by MATLAB solvers, such as `ode45()`. It is also shown that the solvers such as `ode45()` may not work, as in the case of stiff equations. Therefore dedicated solvers should be introduced to solve stiff differential equations. Besides, differential-algebraic equations, implicit differential equations, and many others should be considered. Their solutions should also be considered to fill the gap left by the `ode45()` solver.

In Section 5.1, stiffness phenomenon is demonstrated, followed by the introduction of dedicated solvers for stiff differential equations. Some thought on stiffness detection and problems in fixed-step methods will be shown. In Section 5.2, the solvers of implicit differential equations are introduced. The so-called implicit differential equations are those which cannot be converted into first-order explicit differential equations. The mathematical formula and consistent initial value computation problems are introduced. Also implicit equations with multiple solutions are discussed. In Section 5.3, solutions of differential-algebraic equations are addressed. The semi-explicit differential-algebraic equations are studied first, then the implicit equation based methods are discussed for the solution of differential-algebraic equations. In Section 5.4, switched differential equations are presented. The zero-crossing detection and even response handling problems are discussed, and nonlinear switched differential equations are explored. In Section 5.5, linear stochastic differential equations are considered. A discretization method is proposed for linear stochastic differential equations.

5.1 Stiff differential equations

The so-called stiff differential equations are those which cannot be handled well with conventional explicit differential equation solvers. The usual phenomenon is that the analytical solutions, if any, of differential equations are smooth, but in the numerical solution, artificial blurs are introduced, due to the selection of step-size for numerical algorithms. Therefore, if a variable-step mechanism is adopted, the step-size must be set to an extremely small value to find suitable solutions. However, this may lead to a significant increase in computational load which may be too heavy to solve the problem on any computer. Care must be taken to find dedicated solvers for stiff differential equations. In this section, the stiffness phenomenon is demonstrated through an example. Then numerical solvers are introduced to handle stiff differential equations. At last, stiff solvers are compared with some conventional solvers in certain examples.

https://doi.org/10.1515/9783110675252-005

5.1.1 Time constants in linear differential equations

It can be seen in Chapter 2 that the analytical solutions of linear differential equations are weighted sums of exponential functions. For stable differential equations, the exponential functions are decaying. These exponential functions are referred to as transient solutions. Normally, a time constant is usually used as a specification to describe the speed of the transient solution decaying process. The definition of the time constant is proposed next.

Definition 5.1. For a first-order linear differential equation with constant coefficient, $u'(t) + \lambda u(t) = 0$, the analytical solution is $u(t) = Ce^{-\lambda t}$, where $1/\lambda$ is referred to as the time constant of the differential equation.

Definition 5.2. It has been claimed that the analytical solution of a linear differential equation with constant coefficients is a weighted sum of exponential functions. The smallest constant of the exponential functions can be regarded as the time constant of the system.

5.1.2 Demonstrations of stiff phenomena

In the 1950s, the phenomena and solution methods of stiff differential equations attracted the attention of scholars and researchers in the numerical analysis community. There were some important international symposia dedicated to the stiff differential equations problems.[72] Here an example is used to present the stiffness behavior and its impact on numerical solutions.

Example 5.1. Consider the initial value problem of the differential equation

$$y'(x) = -\alpha[y(x) - \cos x], \quad y(0) = 0.$$

Find the analytical solution and draw the curve for $\alpha = 50$.

Solutions. With the foundations studied in Chapter 2, the following statements can be written to solve this differential equation directly, and draw the curve within the interval $(0, \pi/2)$, as shown in Figure 5.1:

```
>> syms x y(x) alpha
   y0=dsolve(diff(y)==-alpha*(y(x)-cos(x)),y(0)==0);
   y0=simplify(y0), y1=subs(y0,alpha,50); % analytical solution
   fplot(y1,[0,pi/2])                     % draw the solution
```

The mathematical form of the analytical solution is

$$y_0(t) = \frac{\alpha}{\alpha^2+1}(\sin x + \alpha\cos x) - \frac{\alpha^2}{\alpha^2+1}e^{-\alpha x}.$$

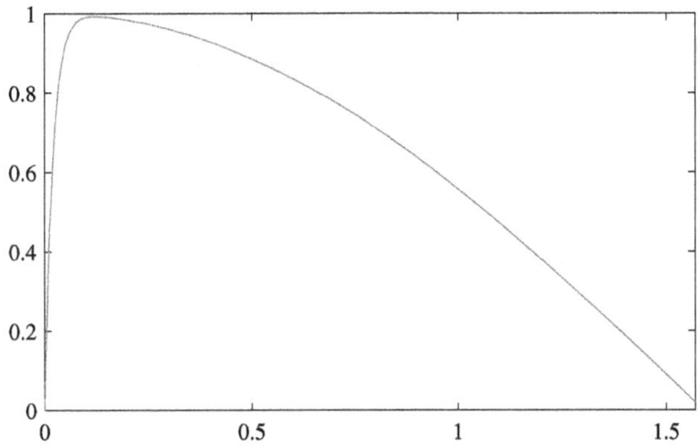

Figure 5.1: The analytical solution curve.

If $\alpha = 50$, a particular solution is

$$y_1(x) = \frac{50\sqrt{2501}}{2501}\cos\left(x - \operatorname{atan}\left(\frac{1}{50}\right)\right) - \frac{2500}{2501}\mathrm{e}^{-50x}.$$

It can be seen from the analytical solution and curve that the e^{-50x} term vanishes rapidly. Its impact lasts very shortly, then the remaining response is the persistent cosine function. When x is increasing, the solution of the differential equation is almost a periodic function. Here only a very small interval of x is selected to draw the curve, the periodic behavior cannot be witnessed in the plot. It can be seen from the analytic solution that the solution is smooth. When x is relatively large, the solution is almost a cosine function.

In Example 5.1, the time constant of the differential equation is $1/50 = 0.02$. If the time constant of the system is small, while the step-size is similar or larger than its scale, there might be oscillations in the numerical results. One can be imagine that if the time constant is not 0.02, but a much smaller value, such as 10^{-7}, the impact of the transient response may not be captured by conventional numerical algorithms. Therefore the correct numerical solution may not be found.

Example 5.2. Use a fixed-step algorithm to solve numerically the differential equation in Example 5.1, and observe the behavior.

Solutions. Since the original system is already in standard form, an anonymous function can be written to describe the differential equation. The step-sizes of $h = 0.04$ and $h = 0.03$ can be selected. With Euler's method, the solution can be found, as shown in Figure 5.2. It is obvious that the solutions are incorrect.

```
>> f=@(x,y)-50*(y-cos(x)); x0=0;
   h=0.04; [x1,y1]=ode_euler(f,[0,h,pi/2],x0);
```

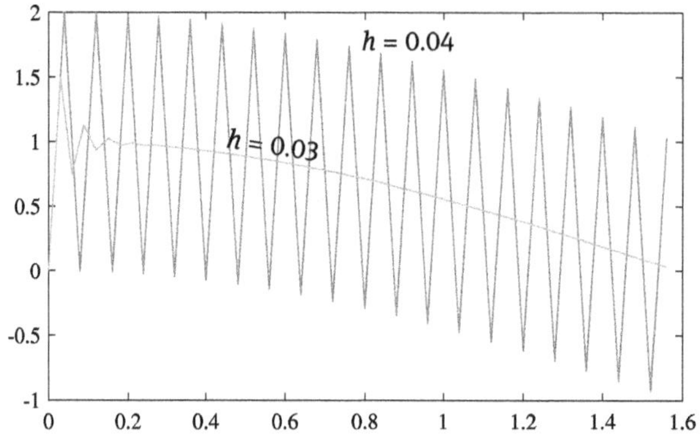

Figure 5.2: The erroneous solutions.

```
   h=0.03; [x2,y2]=ode_euler(f,[0,h,pi/2],x0);
   plot(x1,y1,x2,y2)  % draw the solutions for different step-sizes
```

If the fourth-order Runge–Kutta algorithm is used, with slightly larger step-sizes, there may also be large errors, as in Figure 5.3; when $h = 0.03$, the error is large. A further increase of the step-size may make the numerical process unstable.

```
>> h=0.04; [x1,y1]=ode_rk4(f,[0,h,pi/2],x0);
   h=0.03; [x2,y2]=ode_rk4(f,[0,h,pi/2],x0);
   plot(x1,y1,x2,y2)  % fourth-order Runge–Kutta algorithm
```

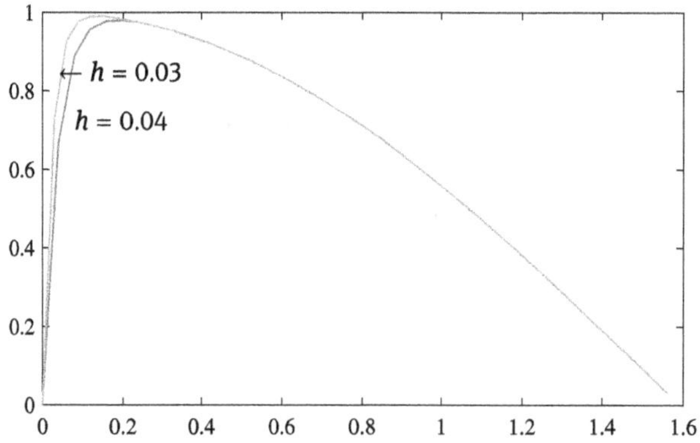

Figure 5.3: Erroneous result with Runge–Kutta algorithm.

Of course, for this specific problem, if a variable-step algorithm such as `ode45()` is used directly, accurate solutions can be found, indicating that at the initial stage, the step-size is assigned to a tiny quantity such that the correct solution can be found. On the other hand, since the time constant is 1/50, the stiffness of the differential equation is not very serious. If it is 1/1 000 or even smaller, the frequently used single-step algorithms may also fail. Dedicated solvers for stiff differential equations are expected.

It can be seen from the phenomena that the sudden change in the solution may mislead the numerical algorithm, such that blurs or large errors may occur. The errors continue misleading the subsequent solution process, such that erroneous solutions are found. The differential equation itself does not have these problems. This kind of equation is also known as a stiff differential equation.

It is also said that although the differential equations are referred to as stiff, the equations themselves are not stiff. The initial values in certain regions are stiff.[26]

5.1.3 Direct solution of stiff differential equations

Many stiff differential equations are not suitable to solve with function `ode45()`, since in certain points variable-step algorithms must select extremely small step-size so as to satisfy the specified error tolerance. The step-sizes are so small and the number of points selected is so large that computer memory is exhausted, and the solution processes are aborted.

For stiff differential equations, some dedicated solvers are provided, such as the variable-order solver `ode15s()`, trapezoidal rule based solver `ode23t()`, trapezoidal rule with backward difference formula based solver `ode23tb()`, and so on. The syntaxes of these functions are the same, as that of `ode45()`. Therefore, if one wants to solve a stiff differential equation, one just needs to substitute the solver `ode45()` with a stiff solver.

Example 5.3. When the van der Pol equation was discussed earlier, μ was selected as $\mu = 1\,000$, where $t \in (0, 3\,000)$. Solve this differential equation again.

Solutions. It was pointed out in Example 4.4 that `ode45()` function cannot be used to find the numerical solution to this differential equation. Similar to the earlier example, the following MATLAB commands can be used, and within 2.55 seconds the numerical solution is found.

```
>> ff=odeset; ff.RelTol=100*eps; ff.AbsTol=100*eps; % control options
   x0= [2;0]; tn=3000; mu=1000;
   f=@(t,x)[x(2); -mu*(x(1)^2-1)*x(2)-x(1)];  % differential equation
   tic, [t,y]=ode15s(f,[0,tn],x0,ff); toc      % solve differential equation
   length(t), plot(t,y(:,1));
   figure; plot(t,y(:,2))                      % draw the two states
```

It can be seen that the solution can easily be found with the stiff differential equation solver. The time responses of the two states can be drawn, as shown in Figures 5.4 and 5.5. It can be seen that there are sudden changes in the solutions $x_1(t)$ and $x_2(t)$ at certain points. Therefore, when $\mu = 1000$, van der Pol equation is a typical stiff differential equation. Dedicated solvers should be used for this equation. It is the most exact numerical solution under the double precision framework.

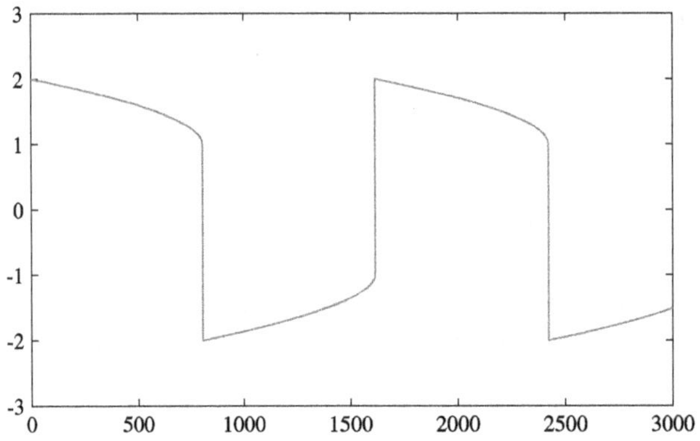

Figure 5.4: Solution $y(t)$ of van der Pol equation for $\mu = 1000$.

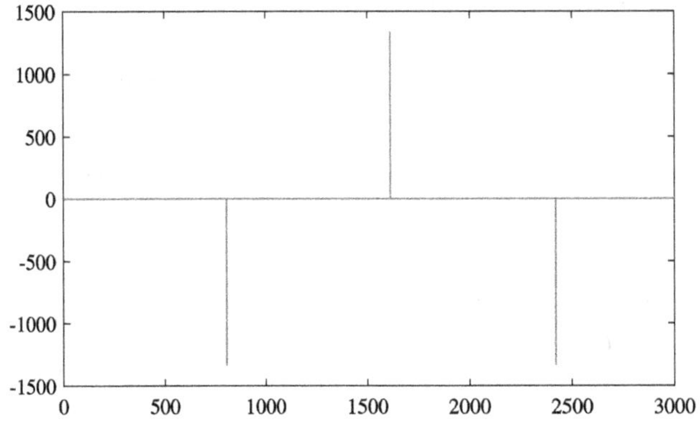

Figure 5.5: Solution $y'(t)$ of van der Pol equation for $\mu = 1000$.

It can be seen that at the same points, the curves of $y(t)$ and $y'(t)$ are extremely steep. In fact, if the regions are zoomed-in, one can observe gradual change in the tiny intervals. In other smooth intervals, the step-size can be selected relatively large.

In the solution process, the step-size is as shown in Figure 5.6. In order to keep the error small, at certain points the step-size must be set to small values such as 2.1073×10^{-9}, while at other points the step-size may be selected as large as 4. A total of 24 441

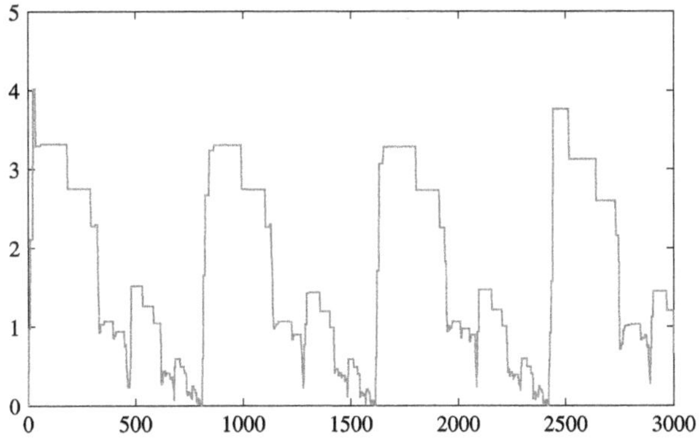

Figure 5.6: The step-size curve.

points are computed.

```
>> plot(t(1:end-1),diff(t))   % draw the step-size curve
   min(diff(t)), length(t)    % find the minimum step-size
```

Example 5.4. Use stiff differential equation solvers to solve again the problem in Example 3.19.

Solutions. Exactly the same code can be used in solving the problem in Example 3.19. Just replace the solver `ode45()` by the dedicated stiff differential equation solver `ode15s()`. If can be seen that the solution is exactly the same as that in Example 3.19. No discrepancy can be witnessed from the curves.

```
>> f=@(t,y)[-1.71*y(1)+0.43*y(2)+8.32*y(3)+0.0007;
            1.71*y(1)-8.75*y(2);
            -10.03*y(3)+0.43*y(4)+0.035*y(5);
            8.32*y(2)+1.71*y(3)-1.12*y(4);
            -1.745*y(5)+0.43*y(6)+0.43*y(7);
            -280*y(6)*y(8)+0.69*y(4)+1.71*y(5)-0.43*y(6)+0.69*y(7);
            280*y(6)*y(8)-1.81*y(7)
            -280*y(6)*y(8)+1.81*y(7)];
   y0=[1;0;0;0;0;0;0;0.0057]; % set the initial values
   ff=odeset; ff.RelTol=100*eps; ff.AbsTol=100*eps;
   tic, [t,x]=ode15s(f,[0,321.8122],y0); toc
   plot(t,x), xlim([0,5])     % solve differential equation and draw curves
```

With a stiff differential equation solver, the step-size curve can be drawn as in Figure 5.7. With the dedicated solver, the elapsed time is only 0.056 seconds, much smaller than that for `ode45()`. The number of points computed is only 119. At the

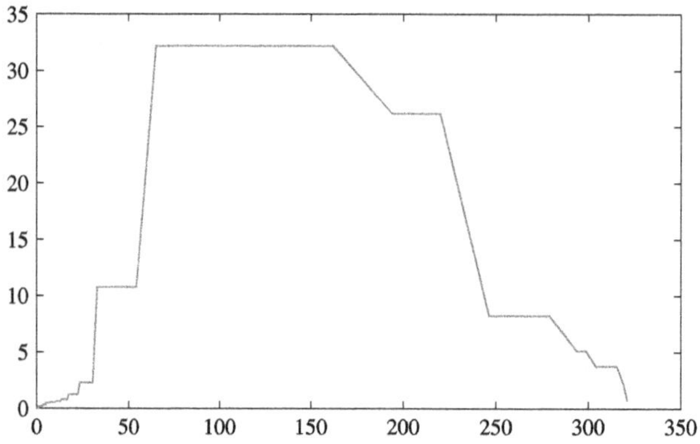

Figure 5.7: Step-size in the entire solution process.

initial stage, the minimum step-size is only 2.6756×10^{-4}, and later, a huge step-size of 30 is allowed. It can be seen that the dedicated solver is of high efficiency.

```
>> plot(t(1:end-1),diff(t)), min(diff(t)) %draw the step-size curve
```

5.1.4 Stiffness detection

The stiffness of a differential equation cannot be found by merely observing the equation itself. For linear differential equations, the differences in the characteristic roots can be used to judge whether a system is stiff or not. For nonlinear differential equations, there are no such simple judgement methods. In [26], a stiffness detection method is illustrated and shown to have some theoretical value. In real applications, it is not really necessary to judge the differential equations in this way, since it is too complicated to decide.

In reality, if there is a differential equation to solve, the stiffness may not be considered initially. The high precision solvers such as `ode45()` can be applied directly. There are two cases where the differential equation may be stiff. In those cases, the stiffness should be considered, and dedicated solvers can be selected instead:

(1) If the solution process is extraordinarily too time consuming, and no solution can be found after a long waiting time, the differential equation may be stiff. In that case, the solution process should be aborted by pressing simultaneously the Ctrl and C keys. The solvers such as `ode15s()` can be tried instead.

(2) If the numerical solutions obtained are found with blurs and strongly oscillating, the original differential equation may be stiff. The solver `ode15s()` should be used instead, and one can see whether the same thing happens. If the solution obtained is smooth, it is quite probable that the original differential equation is stiff. The dedicated stiff differential equation solver should be used to find the solution.

In fact, there are many problems which were considered as stiff, but under the standard of MATLAB, the stiffness was not reflected. Ordinary solvers are suggested to find the numerical solutions. There is no need to solve problems with stiff solvers, since accuracy may be sacrificed. Of course, for certain problems where ordinary solvers cannot be used, stiff solvers must be employed, as seen in Example 5.6.

Example 5.5. In the classical textbooks on numerical solutions of ordinary differential equations,[73] the following equation is regarded as stiff:

$$\boldsymbol{y}'(t)=\begin{bmatrix}-21 & 19 & -20\\ 19 & -21 & 20\\ 40 & -40 & -40\end{bmatrix}\boldsymbol{y}(t),\quad \boldsymbol{y}(0)=\begin{bmatrix}1\\0\\-1\end{bmatrix}.$$

Solve the differential equation with MATLAB.

Solutions. The analytical solution of the equation can be found directly evaluating the matrix exponential with Symbolic Math Toolbox commands in MATLAB.

```
>> syms t;
   A=[-21,19,-20; 19,-21,20; 40,-40,-40]; %input the matrices
   y0=[1; 0; -1];
   y=expm(A*t)*y0 %find the analytical solution with matrix exponential
```

The analytical solution of the equation can be written as

$$\boldsymbol{y}(t)=\begin{bmatrix}0.5\mathrm{e}^{-2t}+0.5\mathrm{e}^{-40t}(\cos 40t+\sin 40t)\\ 0.5\mathrm{e}^{-2t}-0.5\mathrm{e}^{-40t}(\cos 40t+\sin 40t)\\ \mathrm{e}^{-40t}(\sin 40t-\cos 40t)\end{bmatrix}.$$

Now consider the numerical methods. Seeing the original problem, an anonymous function can be used to describe the differential equation, then the following MATLAB commands can be used to find the numerical solution. Note that the $\boldsymbol{A}$ matrix must be converted into double precision data type, and symbolic matrix should not be involved in the numerical solution process, otherwise the solver may not be used.

```
>> f=@(t,x)A*x; %anonymous function for the differential equation
   opt=odeset; opt.RelTol=100*eps; opt.AbsTol=100*eps;
   tic, [t,y]=ode45(f,[0,1],[1;0;-1],opt); toc
   x1=exp(-2*t); x2=exp(-40*t).*cos(40*t);
   x3=exp(-40*t).*sin(40*t); %analytical solution
   y1=[0.5*x1+0.5*x2+0.5*x3, 0.5*x1-0.5*x2-0.5*x3, -x2+x3];
   plot(t,y,t,y1,'--'), norm(y-y1,1), length(t)
```

The numerical and analytical solutions of the equation are as shown in Figure 5.8. The two curves just coincide. The maximum difference between the numerical and

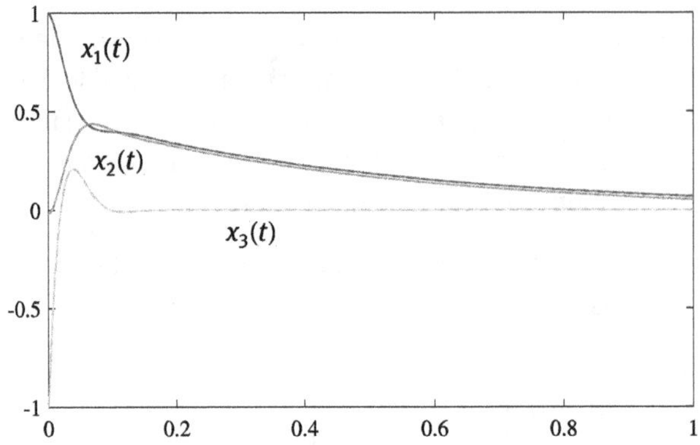

Figure 5.8: Solution comparison.

analytical solution is 1.9295×10^{-13}, and the number of points computed is 4 937, with the elapsed time of 0.0345 seconds.

The following statements can be used to draw the step-size curve, as shown in Figure 5.9, with the minimum step-size of 4.6476×10^{-6}:

```
>> plot(t(1:end-1),diff(t)), min(diff(t))
```

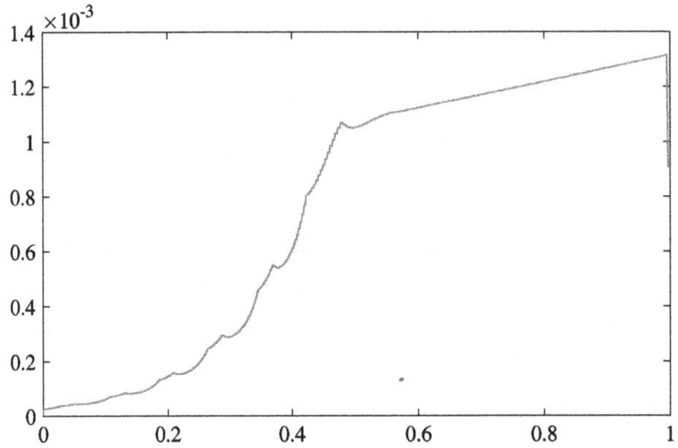

Figure 5.9: Step-size plot in the solution process.

If a stiff solver is used to solve such a problem, the elapsed time is only about 0.023 seconds, and the number of points computed is only 1 540. Compared with the analytical solution, the norm of the error is 3.4377×10^{-11}. It can be seen that with the dedicated solvers, the number of points is significantly reduced, but the cost is heavier, since the error is evidentally increased, which means that the accuracy is sacrificed. It is shown in this example that for such problems, it is not really necessary to introduce dedicated stiff solvers in handling the differential equation like this.

```
>> tic, [t,y]=ode15s(f,[0,1],[1;0;-1],opt); toc
   x1=exp(-2*t); x2=exp(-40*t).*cos(40*t);
   x3=exp(-40*t).*sin(40*t); %analytical solutions
   y1=[0.5*x1+0.5*x2+0.5*x3, 0.5*x1-0.5*x2-0.5*x3, -x2+x3];
   plot(t,y,t,y1,'--'), norm(y-y1), length(t) %find the error
```

For this specific problem, it can be seen that the accuracy of the numerical solution is high, and the solution process is fast. The stiffness issue seems to be not reflected in the solution process. This is because the variable-step solvers in MATLAB adjust the step-size automatically, according the assigned error tolerance. Therefore the stiffness issue cannot be felt. It is also seen that the difference in time constant is only 40/2=20 times; this is a relatively small number under the double precision standard. Therefore ordinary solvers are sufficient in handling this problem.

Example 5.6. Consider the following ordinary differential equation:

$$\begin{cases} y_1'(t) = 0.04(1-y_1(t)) - (1-y_2(t))y_1(t) + 0.0001(1-y_2(t))^2, \\ y_2'(t) = -10^4 y_1(t) + 3\,000(1-y_2(t))^2 \end{cases}$$

where the initial values are $y_1(0) = 0$ and $y_2(0) = 1$. If the interval of interest is $t \in (0, 100)$, select a suitable algorithm to find the numerical solution of the equations.

Solutions. Based on the given ordinary differential equations, the following anonymous function can be written. The following MATLAB commands can be tried:

```
>> f=@(t,y)[0.04*(1-y(1))-(1-y(2))*y(1)+0.0001*(1-y(2))^2; ...
          -10^4*y(1)+3000*(1-y(2))^2]; %describe the differential equation
   ff=odeset; ff.RelTol=100*eps; ff.AbsTol=100*eps;
   tic,[t2,y2]=ode45(f,[0,100],[0;1],ff); toc
   length(t2), plot(t2,y2)          %draw the solution curve
```

After 1.49 seconds of waiting, the numerical solution can be found as shown in Figure 5.10. It can be seen that the ordinary solver `ode45()` needs too much time, and the number of points computed is as high as 481 289. Now the dedicated stiff solver is tried as follows:

```
>> [min(diff(t2)), max(diff(t2))]
   plot(t2(1:end-1), diff(t2)) %changes in the step-size
```

It can be seen that since the precision requirement is rather high, very small step-sizes must be explored. The actual step-size curve for this problem is as shown in Figure 5.11. It is seen that during the solution process the step-size is alternating in an interval around 0.0003. This means that the variable-step method is trying to find

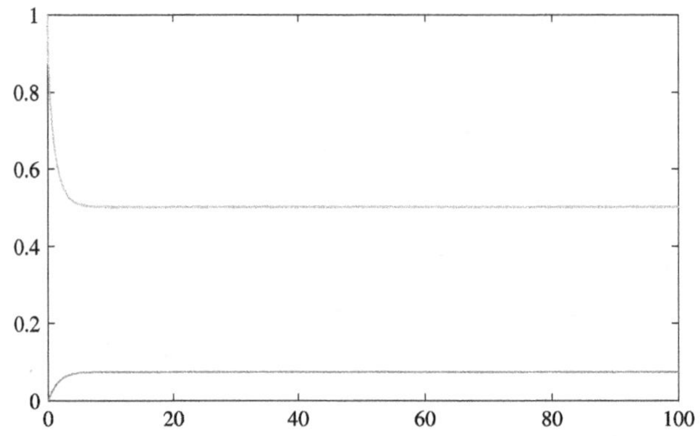

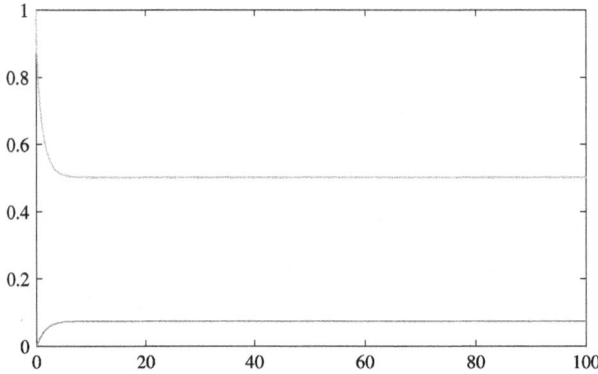

Figure 5.10: Solution of the 4/5th order Runge–Kutta–Felhberg algorithm.

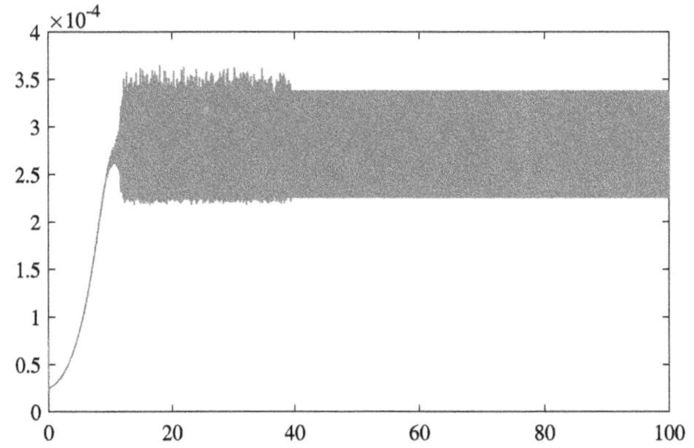

Figure 5.11: The changes in the step-sizes.

an appropriate step-size for the problem. At the initial stage, the step-size is switching in a small neighborhood around 2×10^{-5}.

If the `ode15s()` is used to replace `ode45()`, the elapsed time is 0.25 seconds, and the number of points computed is 2404. It can be seen that the time needed is significantly reduced, and the efficiency is boosted by 5.96 times. It can be seen that the curves are almost identical with those in Figure 5.10. No error can be witnessed from the curves.

```
>> tic,[t1,y1]=ode15s(f,[0,100],[0;1],ff); toc
   length(t1), plot(t1,y1) % change a solver and try again
```

The step-size curve of the new solver can be drawn, using the above code, as shown in Figure 5.12. Compared with the time response curves, it is seen that, when the time response is smooth, the step-size of a stiff solver may reach $h = 9 \sim 10$. Therefore the efficiency of the solver is significantly higher than that of the ordinary solvers. To ensure precision requirements, the initial step-size is selected as small as 8.4294×10^{-9}.

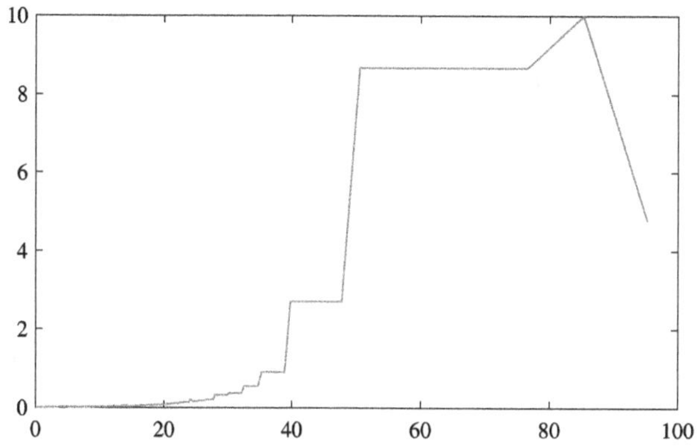

Figure 5.12: The step-size with the solver `ode15s()`.

Luckily, the phenomena like this only lasted a very small period of time, and did not affect the efficiency too much.

```
>> min(diff(t1))                       % compute the minimum error
   plot(t1(1:end-1), diff(t1)) % draw the step-size curve
```

5.1.5 Fixed-step solution of stiff differential equations

In the earlier demonstrations, variable-step solvers provided in MATLAB were recommended. Here for a stiff differential equation, the fixed-step algorithm is taken. Let us see what may happen.

Example 5.7. Solve again the differential equation in Example 5.3 with a fixed-step Runge–Kutta algorithm.

Solutions. If a fixed-step algorithm is to be used, the solver `ode_rk4()` implemented from the fourth-order Runge–Kutta algorithm can be adopted to solve the differential equation with the following commands:

```
>> A=[-21,19,-20; 19,-21,20; 40,-40,-40]; %input the matrices
   f=@(t,x)A*x; % describe differential equation and solve it
   tic, [t,y]=ode_rk4(f,[0,0.001,1],[1;0;-1]); toc
   x1=exp(-2*t); x2=exp(-40*t).*cos(40*t);
   x3=exp(-40*t).*sin(40*t);   % find the analytical solution
   y1=[0.5*x1+0.5*x2+0.5*x3, 0.5*x1-0.5*x2-0.5*x3, -x2+x3];
   plot(t,y1-y), norm(y-y1,1) % draw error curve and find 2-norm
```

The error curves of the state variables are shown in Figure 5.13. It can be seen that the 2-norm of the error signal is 2.6184×10^{-6}.

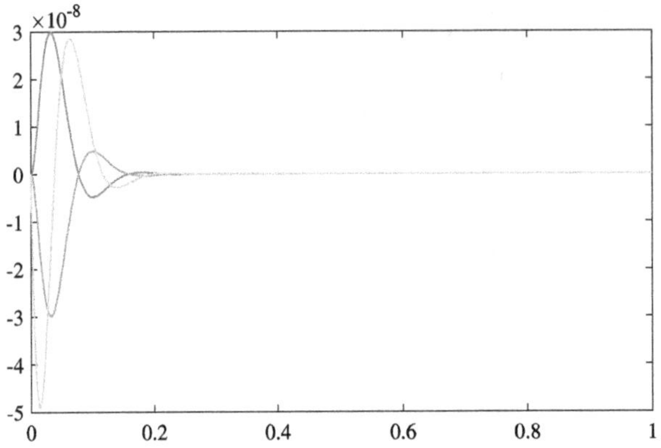

Figure 5.13: The errors when $h = 0.001$ is selected.

It is seen from the results that obviously, since the step-size is selected too large, the solution is not precise. The step-size should be reduced. However, if the step-size is reduced to 10^{-7}, the total elapsed time is increased to 22.21 seconds, more than 500 times that of the `ode45()` solver, and the 2-norm of the error is still as high as 8.57×10^{-8}. Therefore variable-step algorithms should be adopted in actual differential equation solutions.

It can be seen from the curves that the differences in the changes among the three curves are not very significant, the system is not really a stiff one, under the current standard in the software. In the past, due to the limitations in the computer support, it was mistakenly considered as a stiff one.

In fact, this is a good example to exhibit the benefit of variable-step algorithms. In fixed-step algorithms, if a step-size is chosen, it is used throughout the solution process, without considering the errors caused in a step and error prorogations. On the other hand, if a variable-step is chosen, then each step in the whole solution process is monitored. If the error is too small, the step-size is increased so as to speed up the solution process; if the error is large, the step-size is automatically reduced such that the expected accuracy is retained. This cannot be completed with any fixed-step algorithms. Therefore, unless absolutely necessary, the fixed-step algorithms usually taught in numerical analysis courses should not be chosen in solving differential equations. Variable-step algorithms should be adopted instead.

5.2 Implicit differential equations

The so-called implicit differential equations are those which cannot be converted directly into the first-order explicit differential equations in the form shown in (3.1.1). For instance, the differential equations studied in Example 4.19 are such. In the original example, we tried to convert the equation into the standard form of (3.1.1), but in each step of the solution process, an algebraic equation had to be solved. This is obviously not a good choice for implicit differential equations, since the solution process is extremely slow. In this section, direct methods are introduced for dealing with implicit differential equations.

5.2.1 Mathematical description of implicit differential equations

In this section the mathematical form of implicit differential equations is provided, with necessary explanations. Examples are then used to demonstrate the solution of implicit differential equations.

Definition 5.3. The mathematical model of an implicit differential equation is

$$\boldsymbol{F}(t,\boldsymbol{x}(t),\boldsymbol{x}'(t)) = \boldsymbol{0} \tag{5.2.1}$$

with given $\boldsymbol{x}(t_0) = \boldsymbol{x}_0$ and $\boldsymbol{x}'(t_0) = \boldsymbol{x}'_0$.

It can be seen from the standard form that the explicit expression illustrated earlier is no longer needed. The implicit expression of the differential equations is needed to be described in MATLAB. This equation format is much simpler and more flexible. It appears from the mathematical model that not only the regular $\boldsymbol{x}_0$, but also the initial value of the first-order derivative is needed. This is usually not a usual condition. An example is given here to show how to describe an implicit differential equation with MATLAB.

Example 5.8. Express the differential equation in Example 4.19 in the standard form of an implicit differential equation. For ease of presentation, the original model is given below:

$$\begin{cases} x''(t)\sin y'(t) + (y''(t))^2 = -2x(t)y(t)\mathrm{e}^{-x'(t)} + x(t)x''(t)y'(t), \\ x(t)x''(t)y''(t) + \cos y''(t) = 3y(t)x'(t)\mathrm{e}^{-x(t)}. \end{cases}$$

The initial values of the equations are $x(0) = y'(0) = 1$ and $x'(0) = y(0) = 0$.

Solutions. Selecting the state variables $x_1(t) = x(t)$, $x_2(t) = x'(t)$, $x_3(t) = y(t)$, and $x_4(t) = y'(t)$, the original differential equations can be described by the following implicit differential equations in standard form:

$$\begin{bmatrix} x_1'(t) - x_2(t) \\ x_2'(t)\sin x_4(t) + (x_4'(t))^2 + 2x_1(t)x_3(t)\mathrm{e}^{-x_2(t)} - x_1(t)x_2'(t)x_4(t) \\ x_3'(t) - x_4(t) \\ x_1(t)x_2'(t)x_4'(t) + \cos x_4'(t) - 3x_3(t)x_2(t)\mathrm{e}^{-x_1(t)} \end{bmatrix} = \boldsymbol{0}.$$

From the given initial values, the initial value vector can be established as $\boldsymbol{x}(0) = [1, 0, 0, 1]^{\mathrm{T}}$. Therefore the following anonymous function can be established to express the implicit differential equations:

```
>> f=@(t,x,xd)[xd(1)-x(2);
        xd(2)*sin(x(4))+xd(4)^2+2*x(1)*x(3)*exp(-x(2))-x(1)*xd(2)*x(4);
        xd(3)-x(4);
        x(1)*xd(2)*xd(4)+cos(xd(4))-3*x(3)*x(2)*exp(-x(1))];
```

It can be seen that the description is neat and simple. There is no need to convert it into the form in Example 4.19, with algebraic equation embedded. Simple commands can be used to solve the implicit differential equations directly.

Comparing the explicit differential equations and the implicit one given here, there are several differences:

(1) The standard form here can be used to describe the complicated implicit differential equations, with an algebraic equation embedded. Since no algebraic equations are solved each time, the solution efficiency may be boosted.
(2) In standard implicit differential equations, not only $\boldsymbol{x}(t_0)$, but also an extra vector $\boldsymbol{x}'(t_0)$ is expected. The latter should be found before the solution of the differential equations can be carried out. In fact, observing the standard implicit form in (5.2.1), it can be found that if t_0 and $\boldsymbol{x}(t_0)$ are substituted into (5.2.1), the consistent $\boldsymbol{x}'(t_0)$ can be found directly. For implicit differential equations, the algebraic equation needs to be solved only once, and it will not affect the solution efficiency.

It can be seen from the above presentation that the whole solution process can be divided into two parts:

(1) The consistent initial value of the first-order derivative $\boldsymbol{x}'(t_0)$ can be found using the implicit differential equations.
(2) The entire implicit differential equation can be solved using the given consistent initial values.

Since the initial values have a significant impact on the accuracy of the solutions, consistent initial values should be very accurate, otherwise they may yield large errors. The solution procedures of implicit differential equations will be systematically illustrated next.

5.2.2 Consistent initial value transformation

In the earlier versions of MATLAB, implicit differential equation algorithms and solvers were not available, regular differential equation solvers such as `ode45()` were used, with algebraic equation solvers embedded. In the current versions, implicit differential equation solvers can be used directly. Here MATLAB based solvers are presented.

Implicit differential equations are different from explicit ones. Before the solution process, the initial values of $\boldsymbol{x}(t_0)$ and $\boldsymbol{x}'(t_0)$ should both be provided. They cannot be arbitrarily assigned. At most n components in them should be assigned independently, while the others should be solved for from the implicit differential equations. Otherwise, conflicting initial values may be found.

Here the first step in solving implicit differential equations is introduced. We will shoe how to use the given implicit differential equation model to find the consistent first-order derivative initial values $\boldsymbol{x}'(t_0)$.

It can be seen in mathematics that by substituting t_0 and $\boldsymbol{x}(t_0)$ into (5.2.1), the following algebraic equation appears:

$$\boldsymbol{F}(t_0, \boldsymbol{x}(t_0), \boldsymbol{x}'(t_0)) = \boldsymbol{0}. \tag{5.2.2}$$

There are various of ways of solving such algebraic equations. Volume IV of this book also provided some solvers. If the user does not want to solve the equations, the solver `decic()` provided in MATLAB can be used to find the consistent initial values, with the syntax

$[\boldsymbol{x}_0^*, \boldsymbol{x}'^*_0]$=`decic(fun,`$t_0$`,`$\boldsymbol{x}_0$`,`$\boldsymbol{x}_0^{\rm F}$`,`$\boldsymbol{x}_0'$`,`$\boldsymbol{x}'^{\rm F}_0$`),` %find consistent initial values
$[\boldsymbol{x}_0^*, \boldsymbol{x}'^*_0]$=`decic(fun,`$t_0$`,`$\boldsymbol{x}_0$`,`$\boldsymbol{x}_0^{\rm F}$`,`$\boldsymbol{x}_0'$`,`$\boldsymbol{x}'^{\rm F}_0$`,options)`
$[\boldsymbol{x}_0^*, \boldsymbol{x}'^*_0, f_0]$=`decic(fun,`$t_0$`,`$\boldsymbol{x}_0$`,`$\boldsymbol{x}_0^{\rm F}$`,`$\boldsymbol{x}_0'$`,`$\boldsymbol{x}'^{\rm F}_0$`,options)`

where `fun` is the function handle for the implicit differential equation. It can be a regular MATLAB function or an anonymous function, with input arguments of t, $\boldsymbol{x}(t)$, and $\boldsymbol{x}'(t)$. Examples were used earlier to show the description of implicit differential equations.

In the solution process, the consistent $\boldsymbol{x}'^*_0$ should be found first by calling function `decic()`. In the function call, the input arguments $\boldsymbol{x}_0$ and $\boldsymbol{x}_0'$ are the initial values. They can be specific or initial values for the algebraic equation solution process. The $\boldsymbol{x}_0^{\rm F}$ and $\boldsymbol{x}'^{\rm F}_0$ are both n-dimensional column vectors, indicating which values in the two initial value vectors should be retained. If the value is 1, then the corresponding initial value should be retained, otherwise, it should be found with an algebraic equation solver. The total number of 1's in the two vectors should not exceed n. After the solution process, consistent initial values $\boldsymbol{x}_0^*$ and $\boldsymbol{x}'^*_0$ are retained. The `options` is the control option which could be set by function `odeset()`. The members such as `RelTol` can be selected to define the expected recision. It should be noted that in the solution process, `RelTol` member should not be set to too small numbers, otherwise consistent initial values may not be found.

If no error messages appear in the solution process, the returned arguments $\boldsymbol{x}_0^*$ and $\boldsymbol{x}'^*_0$ are consistent initial values, and f_0 is the norm of the error when the consistent values are substituted back to (5.2.2).

Examples will be given next to demonstrate the consistent initial value computation methods.

Example 5.9. Find the consistent initial values x_0' for the implicit differential equation in Example 4.19.

Solutions. For this specific example, the whole vector x_0 is given. It should be input into the computer, and x_0^F should be set to a vector of ones, indicating that each element in vector x_0 should be retained. Since x_0' is unknown, it can be chosen randomly, or set in other forms, but vector ${x'}_0^F$ should be set to a zero vector, or an empty vector, indicating that each element in x_0' should be computed such that consistent initial values can be found. The following commands should be employed to find the consistent initial values:

```
>> f=@(t,x,xd)[xd(1)-x(2);
      xd(2)*sin(x(4))+xd(4)^2+2*exp(-x(2))*x(1)*x(3)-x(1)*xd(2)*x(4);
      xd(3)-x(4);
      x(1)*xd(2)*xd(4)+cos(xd(4))-3*exp(-x(1))*x(3)*x(2)];
   x0=[1;0;0;1]; x0F=ones(4,1);        % retain x0
   xd0=rand(4,1); xd0F=zeros(4,1); % initial first-order derivatives are needed
   [x0,xd0,f0]=decic(f,0,x0,x0F,xd0,xd0F)  % use x0 to determine x0'
```

With the above function call, the consistent $x_0' = [0, 1.6833, 1, -0.5166]^T$ can be found and the norm of the error vector is 1.1102×10^{-16}, indicating that the solution is successful.

In fact, the solver `decic()` provided in MATLAB occasionally fails to find consistent initial values, and returns an error message "Convergence failure in DECIC". An outer loop structure can be designed to find the consistent initial values.

```
function [x0,xd0,f0]=decic_new(f,t0,x0,x0F,varargin)
n=length(x0);
[xd0,xd0F,a,b,tol]=default_vals(...
      {rand(n,1),zeros(n,1),0,1,eps},varargin{:});
while (1)
   xd0=rand(n,1); x0=a+(b-a).*x0;
   try     % find consistent values
      [x0,xd0,f0]=decic(f,t0,x0,x0F,xd0,xd0F);
   catch, continue; end
   if abs(f0)<tol, break; end % if found then terminate the loop
end
```

The syntax of the function is

$[\boldsymbol{x}_0^*, \boldsymbol{x'}_0^*, f_0]$=decic_new(fun, t_0, $\boldsymbol{x}_0$, $\boldsymbol{x}_0^{\rm F}$, $\boldsymbol{x}_0'$, $\boldsymbol{x'}_0^{\rm F}$, ϵ)

where ϵ is the error tolerance, and the other input arguments are the same as those defined earlier. For finding consistent initial values, the latter three arguments can be omitted, and default values are used. To ensure high precision, the initial value of ϵ is set to eps. Sometimes this cannot be achieved. A slightly larger number should be used instead.

Function default_vals() is used to set the default values of the function. This function is a common one, which also appears in other volumes in the series. For convenience, the listing is given below.

```
function varargout=default_vals(vals,varargin)
if nargout~=length(vals), error('number of arguments mismatch');
else, nn=length(varargin)+1; % assign default values
   varargout=varargin; for i=nn:nargout, varargout{i}=vals{i};
end, end, end
```

With consistent initial values, the next step is to solve the implicit differential equations.

5.2.3 Direct solution of implicit differential equations

With the consistent $\boldsymbol{x}_0$ and $\boldsymbol{x}_0'$, and the standard form of the implicit differential equations, the solver ode15i() provided in MATLAB can be called to solve them:

$[\boldsymbol{t}, \boldsymbol{x}]$=ode15i(fun,tspan, $\boldsymbol{x}_0^*$, $\boldsymbol{x'}_0^*$, options)

where the definition of tspan is the same as that studied earlier. It can be the interval $[t_0, t_n]$, or a user specified time vector $\boldsymbol{t}$. The control options options is also the same as those discussed earlier.

Examples are given next to demonstrate the solution process of implicit differential equations.

Example 5.10. Solve the implicit differential equations in Example 4.19.

Solutions. In Examples 5.8 and 5.9, the standard form the implicit differential equations and the consistent initial values were found. With the following MATLAB commands, the differential equations can be described with an anonymous function, and the consistent initial values are found. Finally, the equations can be solved directly. The total elapsed time is 0.94 seconds, and the number of points computed is 6 334. The efficiency is significantly higher than in Example 4.19, since the elapsed time there was 12.65 seconds, with 2 217 points computed. The elapsed time was 13.5 times more than here.

```
>> f=@(t,x,xd)[xd(1)-x(2);
               xd(2)*sin(x(4))+xd(4)^2+...
                  2*x(1)*x(3)*exp(-x(2))-x(1)*xd(2)*x(4);
               xd(3)-x(4);
               x(1)*xd(2)*xd(4)+cos(xd(4))-3*x(3)*x(2)*exp(-x(1))];
   ff=odeset; ff.AbsTol=100*eps; ff.RelTol=100*eps;
   x0=[1,0,0,1]'; x0F=ones(4,1); tic
   [x0,xd0,f0]=decic_new(f,0,x0,x0F) % consistent initial value
   [t,x]=ode15i(f,[0,2],x0,xd0,ff); toc
   plot(t,x), length(t)                % solution and plot
```

In Example 4.19, the main reason for the large time consumption was that in each differential equation solution step, the algebraic equation was solved once. While in the algorithm here, the algebraic equation is solved only once, and the information can be used to solve the implicit differential equation. Therefore the solver here is more efficient.

Example 5.11. Solve the implicit differential equations in Example 5.9 again. Compare the results with the those in Example 4.17. For convenience, the original differential equation model is recalled as follows:

$$\begin{cases} x_1'(t)\sin x_1(t) + x_2'(t)\cos x_2(t) + x_1(t) = 1, \\ -x_1'(t)\cos x_2(t) + x_2'(t)\sin x_1(t) + x_2(t) = 0 \end{cases}$$

with initial values $x_1(0) = x_2(0) = 0$.

Solutions. The standard form the implicit differential equation can be written as

$$\begin{bmatrix} x_1'(t)\sin x_1(t) + x_2'(t)\cos x_2(t) + x_1(t) - 1 \\ -x_1'(t)\cos x_2(t) + x_2'(t)\sin x_1(t) + x_2(t) \end{bmatrix} = \mathbf{0}.$$

An anonymous function can be used to describe the implicit differential equation, and find the consistent initial values. The information can then be used to solve the equations. It can be seen that the results are identical to those in Example 4.17. The elapsed time is only 0.29 seconds, with the number of points being 2 948.

```
>> f=@(t,x,xd)[xd(1)*sin(x(1))+xd(2)*cos(x(2))+x(1)-1;
              -xd(1)*cos(x(2))+xd(2)*sin(x(1))+x(2)];
   x0=[0;0]; x0F=ones(2,1); tic
   [x0,xd0,f0]=decic_new(f,0,x0,x0F); % consistent initial values
   ff=odeset; ff.AbsTol=100*eps; ff.RelTol=100*eps;
   [t,x]=ode15i(f,[0,10],x0,xd0,ff);  % solve the equation
   plot(t,x); length(t), toc           % draw the solution curves
```

Example 5.12. Solve the following differential equations:

$$\begin{cases} y_1'(t) = -1.71y_1(t) + 0.43y_2(t) + 8.32y_3(t) + 0.0007, \\ y_2'(t) = 1.71y_1(t) - 8.75y_2(t), \\ y_3'(t) = -10.03y_3(t) + 0.43y_4(t) + 0.035y_5(t), \\ y_4'(t) = 8.32y_2(t) + 1.71y_3(t) - 1.12y_4(t), \\ y_5'(t) = -1.745y_5(t) + 0.43y_6(t) + 0.43y_7(t), \\ y_6'(t) = -280y_6(t)y_8(t) + 0.69y_4(t) + 1.71y_5(t) - 0.43y_6(t) + 0.69y_7(t), \\ y_7'(t) = 280y_6(t)y_8(t) - 1.81y_7(t), \\ y_8'(t) = -y_7'(t) \end{cases}$$

where $y_1(0) = 1$, $y_2(0) = y_3(0) = \cdots = y_7(0) = 0$, and $y_8(0) = 0.0057$.

Solutions. At first glance, it seems that this is an explicit differential equation. Yet in the last term, the quantities $y_7'(t)$ and $y_8'(t)$ are both contained. Therefore, in fact, this is an implicit differential equation. If the equation is to be solved, the standard model should be established:

$$\begin{bmatrix} y_1'(t) + 1.71y_1(t) - 0.43y_2(t) - 8.32y_3(t) - 0.0007 \\ y_2'(t) - 1.71y_1(t) + 8.75y_2(t) \\ y_3'(t) + 10.03y_3(t) - 0.43y_4(t) - 0.035y_5(t) \\ y_4'(t) - 8.32y_2(t) - 1.71y_3(t) + 1.12y_4(t) \\ y_5'(t) + 1.745y_5(t) - 0.43y_6(t) - 0.43y_7(t) \\ y_6'(t) + 280y_6(t)y_8(t) - 0.69y_4(t) - 1.71y_5(t) + 0.43y_6(t) - 0.69y_7(t) \\ y_7'(t) - 280y_6(t)y_8(t) + 1.81y_7(t) \\ y_8'(t) + y_7'(t) \end{bmatrix} = \mathbf{0}.$$

With an anonymous function, the implicit differential equation can be described first, and then function `decic()` can be called to find the consistent $\boldsymbol{x}'(0)$. The implicit differential equation can be solved directly.

```
>> f=@(t,y,yd)[yd(1)+1.71*y(1)-0.43*y(2)-8.32*y(3)-0.0007;
     yd(2)-1.71*y(1)+8.75*y(2);
     yd(3)+10.03*y(3)-0.43*y(4)-0.035*y(5);
     yd(4)-8.32*y(2)-1.71*y(3)+1.12*y(4);
     yd(5)+1.745*y(5)-0.43*y(6)-0.43*y(7);
     yd(6)+280*y(6)*y(8)-0.69*y(4)-1.71*y(5)+0.43*y(6)-0.69*y(7);
     yd(7)-280*y(6)*y(8)+1.81*y(7);
     yd(8)+yd(7)];                          % describe the equation
   ff=odeset; ff.AbsTol=100*eps; ff.RelTol=100*eps;
   x0=[1;0;0;0;0;0;0;0.0057]; x0F=ones(8,1); tic
```

```
[x0,xd0,f0]=decic_new(f,0,x0,x0F) % consistent initial values
[t,x]=ode15i(f,[0,10],x0,xd0,ff); toc
plot(t,x), length(t)               % solve the equations
```

The consistent values are $y_1'(0) = -1.7093$ and $y_2'(0) = 1.7100$, with the rest of the first-order values being zero. The norm of the error is 1.4496×10^{-16}. It is seen that the initial values are very accurate. The total elapsed time is 0.38 seconds, with the number of points being 2 558. The solutions of the differential equation are as shown in Figure 5.14.

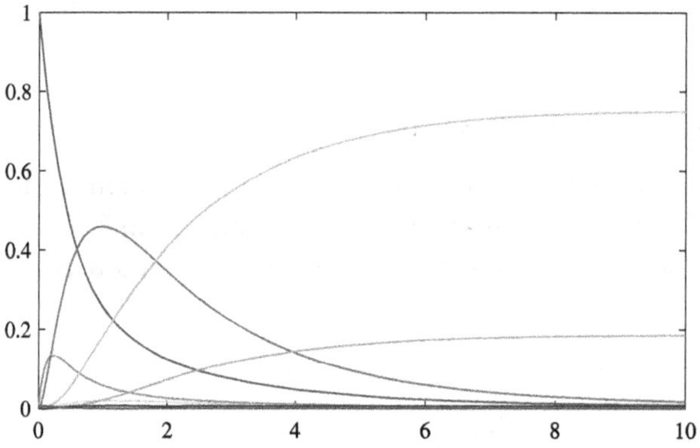

Figure 5.14: Numerical solution of the implicit differential equations.

It can be seen from the results that the efficiency in the implicit differential equation solution process is significantly higher than that in Example 4.3.4 of Section 4.3.3. Therefore for implicit differential equations, if possible, the implicit differential equation solver is the top choice.

5.2.4 Implicit differential equations with multiple solutions

When solving for consistent initial values of implicit differential equations, algebraic equations are involved. For these algebraic equations, the uniqueness of the solutions should also be considered. It means that multiple consistent initial values may be found, whereas each set of consistent initial values may yield a numerical solution for the original implicit differential equations. Examples are presented next to find the multiple solutions of implicit differential equations.

Example 5.13. Consider the differential equation in Example 4.9. For convenience, the differential equation is given below. Solve the following equation:

$$(y''(t))^2 = 4(ty'(t) - y(t)) + 2y'(t) + 1$$

where $y(0) = 0$ and $y'(0) = 0.1$.

Solutions. In Example 4.9, attempts were made to derive the explicit form. There were two branches in the first-order explicit differential equations. Numerical methods were adopted in solving them. Now the implicit differential equation solver can be used to solve again the original problem.

Introducing the variables $x_1(t) = y(t)$ and $x_2(t) = y'(t)$, the standard form of the implicit differential equation can be written as

$$\begin{bmatrix} x_1'(t) - x_2(t) \\ (x_2'(t))^2 - 4(tx_2(t) - x_1(t)) - 2x_2(t) - 1 \end{bmatrix} = \mathbf{0}$$

with $x_1(0) = 0$ and $x_2(0) = 0.1$. With an anonymous function to describe the consistent initial values, the numerical solution of the differential equation can finally be found. The following commands can be used:

```
>> f=@(t,x,xd) [xd(1)-x(2);
          xd(2)^2-4*(t*x(2)-x(1))-2*x(2)-1];
   x0=[0; 0.1]; x0F=[1;1]; tic
   [x0,xd0,f0]=decic_new(f,0,x0,x0F) % get consistent conditions
   [t1,x1]=ode15i(f,[0,1],x0,xd0,ff); toc
   plot(t1,x1(:,1)), length(t)         % draw the solutions
```

It can be seen that the consistent initial values are $x_1'(0) = 0.1$ and $x_2'(0) = -1.0954$, with the error being 2.2204×10^{-16}. From this initial values, after 0.045 seconds of time, 122 points are computed, the solution is the lower branch shown in Figure 4.5.

In Example 4.9, the precision of the lower branch is extremely low. The difference between the numerical and analytical solution can be distinguished easily from the curves. Here the analytical solution is found again, and compared with the numerical solutions. The norm of the error vector is reduced to 1.7955×10^{-13}. It can be seen that the accuracy is boosted in the new solutions.

```
>> syms t y(t); y1d=diff(y);
   y=dsolve((diff(y,2))^2==4*(t*y1d-y)+2*y1d+1,...
            y(0)==0, y1d(0)==0.1); % analytical solution
   y2=y(2); y0=double(subs(y2,t,t1)); norm(x1(:,1)-y0)
```

In fact, repeating the above code, especially repeating it for finding the consistent initial values, when function `decic_new()` is called several times, another set of consistent values $x_1'(0) = 0.1$ and $x_2'(0) = 1.0954$ can be found. It can be seen that in the two initial values, the absolute values of $x_2'(0)$ are the same, satisfying the square characteristic property in the $x_2'(t)$ term. It can be seen that the error found is 1.7955×10^{-13}. With such an initial value, the numerical solution of the other branch can be found. From practical computations, it can be seen that the accuracy is much higher than that obtained in the previous sections.

It can be seen that the initial value problems of differential equations may have multiple solutions. Therefore in the solution process, the consistent value solver `decic_new()` should be executed several times. If different initial values can be found, the differential equations may have multiple solutions. The solutions on different branches can be obtained from different initial values.

5.3 Differential-algebraic equations

It can be seen from the previous examples that the solutions of some ordinary differential equations can be found if equations can be converted into first-order explicit differential equations. In real applications, if some of the differential equations degenerate into algebraic equations, the previously illustrated methods cannot be used. Special dedicated methods should be used to handle differential-algebraic equations. The solution approach for differential-algebraic equations is introduced in this section. The concept of differential-algebraic equation index is also addressed. Index reduction methods for systems with higher indexes are introduced next, followed by numerical solution methods.

5.3.1 General form of differential-algebraic equations

Differential-algebraic equation (DAE) is a special case of differential equations in science and engineering. If in some differential equations certain terms are degenerated into algebraic constraints, the solution of the differential equations may become complicated. Conventional differential equation solution algorithms cannot be applied directly. In this section, the definition and general form of differential-algebraic equations are presented, and then numerical solution methods are illustrated.

Definition 5.4. The so-called differential-algebraic equation set is such that some of the equations are given in the form of differential equations, while others are given by algebraic equations, which can be regarded as the algebraic constraints among the state variables. This type of equations cannot be solved directly with the differential equation solvers. The general form of differential-algebraic equations is

$$\begin{cases} \boldsymbol{F}(t,\boldsymbol{x}(t),\boldsymbol{x}'(t)) = \boldsymbol{0}, \\ \boldsymbol{g}(t,\boldsymbol{x}(t)) = \boldsymbol{0} \end{cases} \tag{5.3.1}$$

where $\boldsymbol{x}(t_0) = \boldsymbol{x}_0$. The number of equations in $\boldsymbol{F}(t,\boldsymbol{x}(t),\boldsymbol{x}'(t))$ is smaller than the number of unknowns, where each equation must contain at least one first-order derivative term. The total number of equations in the two sets of $\boldsymbol{F}(\cdot)$ and $\boldsymbol{G}(\cdot)$ equals to the number of unknowns.

Definition 5.5. A class of commonly used semi-explicit differential-algebraic equations are given by

$$M(t,x(t))x'(t) = f(t,x(t)) \tag{5.3.2}$$

where $M(t,x(t))$ is also known as a mass matrix, which is singular.

If MATLAB is used to solve differential-algebraic equations, the right-hand side vector function $f(t,x(t))$ and the left-hand side matrix $M(t,x(t))$ should be expressed separately. If it can be shown that $M(t,x(t))$ is a square nonsingular matrix, the two sides should be left-multiplied by $M^{-1}(t,x(t))$, so that the semi-explicit differential-algebraic equation can be transformed to

$$x'(t) = M^{-1}(t,x(t))f(t,x(t)) \tag{5.3.3}$$

and the usual methods can be used to solve the resulting differential equation.

For genuine differential-algebraic equations, matrix $M(t,x(t))$ is singular. Special methods for this kind of equations should be studied.

5.3.2 Indices of differential-algebraic equations

The index of a differential-algebraic equation is a very important quantity. Its size may indicate the easiness of its numerical solution process. In (5.3.1), the equations containing derivative terms are dynamic functions, while the rest are static functions. For a differential-algebraic equation, it is not an easy thing to find the index. Experience and mathematical formulation are needed for that. Here only a rough judgement method for the index is illustrated.

Normally, the more static functions are involved in the differential-algebraic equations, the higher the index, so the harder are numerical computations. For semi-explicit differential-algebraic equations, the index sometimes equals to the difference of the size of the matrix $M(t,x(t))$ and its rank. Before really solving the differential-algebraic equations, the index should be reduced in some methods.

5.3.3 Semi-explicit differential-algebraic equations

Index-1 semi-explicit differential-algebraic equations can be solved directly with MATLAB solvers such as `ode45()`. In conventional control options, the member `Mass` can be used. It can be a constant matrix or a function handle. Examples are given next to show the direct solution method of differential-algebraic equations.

Example 5.14. Consider the following differential-algebraic equations:

$$\begin{cases} x_1'(t) = -0.2x_1(t) + x_2(t)x_3(t) + 0.3x_1(t)x_2(t), \\ x_2'(t) = 2x_1(t)x_2(t) - 5x_2(t)x_3(t) - 2x_2^2(t), \\ 0 = x_1(t) + x_2(t) + x_3(t) - 1 \end{cases}$$

for given $x_1(0) = 0.8$ and $x_2(0) = x_3(0) = 0.1$. Find their numerical solution.

Solutions. It can be seen that the last equation is an algebraic one. It can be regarded as the algebraic constraint among the three state variables. The matrix form of the differential-algebraic equation can be written as

$$\begin{bmatrix} 1 & 0 & 0 \\ 0 & 1 & 0 \\ 0 & 0 & 0 \end{bmatrix} \begin{bmatrix} x_1'(t) \\ x_2'(t) \\ x_3'(t) \end{bmatrix} = \begin{bmatrix} -0.2x_1(t) + x_2(t)x_3(t) + 0.3x_1(t)x_2(t) \\ 2x_1(t)x_2(t) - 5x_2(t)x_3(t) - 2x_2^2(t) \\ x_1(t) + x_2(t) + x_3(t) - 1 \end{bmatrix}.$$

The following MATLAB function can be used to express the model:

```
>> f=@(t,x)[-0.2*x(1)+x(2)*x(3)+0.3*x(1)*x(2);
        2*x(1)*x(2)-5*x(2)*x(3)-2*x(2)*x(2);
        x(1)+x(2)+x(3)-1]; % right-hand side of the equation
```

Matrix $\boldsymbol{M}$ can be entered into MATLAB workspace, and the following commands can be written. The differential-algebraic equation can be solved, and the solutions are shown in Figure 5.15. The elapsed time of the solution process is 0.51 seconds, and 2 268 points are computed.

```
>> M=[1,0,0; 0,1,0; 0,0,0];
   ff=odeset; ff.AbsTol=100*eps; ff.RelTol=100*eps;
   ff.Mass=M; x0=[0.8; 0.1; 0.1]; % mass matrix and initial values
   tic, [t,x]=ode15s(f,[0,20],x0,ff); toc
   plot(t,x), length(t)            % draw solutions
```

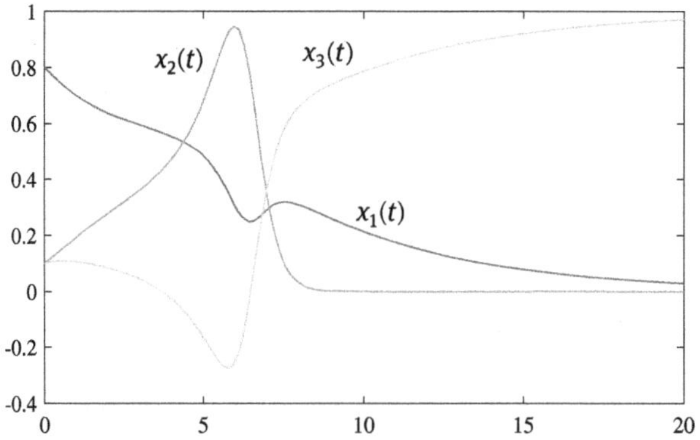

Figure 5.15: Numerical solutions of differential-algebraic equations.

In fact, some simple differential-algebraic equations can be converted into ordinary differential equations. For instance, it is found from the constraints that $x_3(t) = 1 - x_1(t) - x_2(t)$. Substituting this expression into other differential equations, it is found that

$$\begin{cases} x_1'(t) = -0.2x_1(t) + x_2(t)[1 - x_1(t) - x_2(t)] + 0.3x_1(t)x_2(t), \\ x_2'(t) = 2x_1(t)x_2(t) - 5x_2(t)[1 - x_1(t) - x_2(t)] - 2x_2^2(t). \end{cases}$$

An anonymous function can be established in describing the differential equations:

```
>> f=@(t,x)[-0.2*x(1)+x(2)*(1-x(1)-x(2))+0.3*x(1)*x(2);
            2*x(1)*x(2)-5*x(2)*(1-x(1)-x(2))-2*x(2)*x(2)];
```

After a transformation like this, the following commands can be used to solve the differential equations numerically, so as to find the solutions of the original differential-algebraic equations. The solutions obtained in this way are exactly the same as those obtained earlier.

```
>> fDae=@(t,x)[-0.2*x(1)+x(2)*(1-x(1)-x(2))+0.3*x(1)*x(2);...
            2*x(1)*x(2)-5*x(2)*(1-x(1)-x(2))-2*x(2)*x(2)];
   ff=odeset; ff.AbsTol=100*eps; ff.RelTol=100*eps; ff.Mass=M;
   x0=[0.8; 0.1]; [t1,x1]=ode45(fDae,[0,20],x0);
   plot(t1,x1,t1,1-sum(x1'))
```

Note that even though the solver `ode45()` is used here, no error or warning messages appear.

Example 5.15. Use a differential-algebraic equation solver to solve again the implicit differential equations studied in Example 5.12.

Solutions. Slightly modifying the last equation, it is easy to find the matrix form of the original differential equations;

$$\boldsymbol{M}\boldsymbol{y}'(t) = \begin{bmatrix} -1.71y_1(t) + 0.43y_2(t) + 8.32y_3(t) + 0.0007 \\ 1.71y_1(t) - 8.75y_2(t) \\ -10.03y_3(t) + 0.43y_4(t) + 0.035y_5(t) \\ 8.32y_2(t) + 1.71y_3(t) - 1.12y_4(t) \\ -1.745y_5(t) + 0.43y_6(t) + 0.43y_7(t) \\ -280y_6(t)y_8(t) + 0.69y_4(t) + 1.71y_5(t) - 0.43y_6(t) + 0.69y_7(t) \\ 280y_6(t)y_8(t) - 1.81y_7(t) \\ 0 \end{bmatrix}$$

where the mass matrix is

$$M=\begin{bmatrix}1&0&0&0&0&0&0&0\\0&1&0&0&0&0&0&0\\0&0&1&0&0&0&0&0\\0&0&0&1&0&0&0&0\\0&0&0&0&1&0&0&0\\0&0&0&0&0&1&0&0\\0&0&0&0&0&0&1&0\\0&0&0&0&0&0&1&1\end{bmatrix}.$$

If the mass matrix is fed into MATLAB, the following MATLAB commands can be used to solve the differential equations directly:

```
>> f=@(t,y)[-1.71*y(1)+0.43*y(2)+8.32*y(3)+0.0007;
            1.71*y(1)-8.75*y(2);
            -10.03*y(3)+0.43*y(4)+0.035*y(5);
            8.32*y(2)+1.71*y(3)-1.12*y(4);
            -1.745*y(5)+0.43*y(6)+0.43*y(7);
            -280*y(6)*y(8)+0.69*y(4)+1.71*y(5)-0.43*y(6)+0.69*y(7);
            280*y(6)*y(8)-1.81*y(7); 0];
   ff=odeset; ff.AbsTol=100*eps; ff.RelTol=100*eps;
   M=eye(8); M(8,7)=1; ff.Mass=M; x0=[1;0;0;0;0;0;0;0.0057];
   tic, [t,x]=ode45(f,[0,10],x0,ff); toc
   plot(t,x), length(t)  %solve the differential equations
```

The elapsed time is 0.57 seconds, and the number of points computed is 9 913. It can be seen from the solution efficiency that the speed of the solution process is slightly lower than that of the implicit differential equation solver. Therefore, implicit differential equation solvers are better for this type of problems.

For this particular problem, the mass matrix is clearly singular. There is no way to find M^{-1}, and convert the equation into a first-order explicit differential equation. Therefore, `ode45()` solver cannot be called in this way to solve the problems:

```
>> f0=@(t,y)inv(M)*f(t,y); ff.Mass=[];
   tic, [t1,x1]=ode45(f0,[0,10],x0,ff); toc
   plot(t1,x1), length(t1)  %solve the differential equation
```

The numerical solution can be found directly for the differential equation. The total elapsed time is 0.24 seconds, which is obviously better than that obtained previously. The number of points computed is 9 913, the same as with the differential-algebraic equation solver, and more than with the implicit differential equation solver.

5.3.4 Limitations of the direct solver

The limitations of the differential-algebraic equation solver provided in MATLAB have been pointed out earlier. One of them is that the equation must be index-1, the other is that it must be a semi-explicit differential-algebraic equation. The two conditions must be satisfied, otherwise the equations cannot be solved in this way. Examples are presented here. In the next section, the solution methods for such differential-algebraic equations will be explored.

Example 5.16. Consider a complicated differential-algebraic equation:[31]

$$\begin{cases} k'(t) = \dfrac{1}{20}(c(t) - k(t)), \\ c'(t) + \dfrac{1}{15}p'(t) = -\dfrac{1}{75}(p(t) - 99.1), \\ M'(t) = \mu(t) - m(t), \\ \dfrac{p(t)}{P(t)} = 3.35 - 0.075m(t) + 0.001m^2(t), \\ P^2(t) = 49.58^2 - \left(\dfrac{\mu(t)}{1.2k(t)}\right)^2, \\ M(t) = 20P(t), \\ \mu(t) = 15 + 5\tanh(t-10) \end{cases}$$

where the initial values of the states are

$$\begin{aligned} p(0) &= 99.1, \quad k(0) = 0.25, \quad \mu(0) = 15 + 5\tanh(-10), \\ c(0) &= 0.2500145529515559, \quad P(0) = 734.0477598381585, \\ m(0) &= 9.998240239254898, \quad M(0) = 36.70238799190793. \end{aligned}$$

The solution interval is $t \in (0, 40)$. Solve the equation.

Solutions. It can be seen from the given model that the dynamical signals $k(t)$, $c(t)$, $p(t)$, $M(t)$, $P(t)$, $m(t)$, and $\mu(t)$ can be regarded as the state variables. If the given signal $\mu(t)$ is substituted into the differential equation, it can be seen that there are 7 state variables, and there are also 7 equations. Let $x_1(t) = k(t)$, $x_2(t) = c(t)$, $x_3(t) = p(t)$, $x_4(t) = M(t)$, $x_5(t) = m(t)$, $x_6(t) = P(t)$ and $x_7(t) = \mu(t)$. The differential-algebraic equation can easily be set up as

$$\begin{bmatrix} 1 & 0 & 0 & 0 & 0 & 0 & 0 \\ 0 & 1 & 1/15 & 0 & 0 & 0 & 0 \\ 0 & 0 & 0 & 1 & 0 & 0 & 0 \\ 0 & 0 & 0 & 0 & 0 & 0 & 0 \\ 0 & 0 & 0 & 0 & 0 & 0 & 0 \\ 0 & 0 & 0 & 0 & 0 & 0 & 0 \\ 0 & 0 & 0 & 0 & 0 & 0 & 0 \end{bmatrix} \mathbf{x}'(t) = \begin{bmatrix} (x_2(t) - x_1(t))/20 \\ -(x_3(t) - 99.1)/75 \\ x_7(t) - x_5(t) \\ x_3(t) - [3.35 - 0.075x_5(t) + 0.001x_5^2(t)]x_6(t) \\ -x_6^2(t) + 49.58^2 - [x_7(t)/(1.2x_1(t))]^2 \\ x_4(t) - 20x_6(t) \\ -x_7(t) + 15 + 5\tanh(t-10) \end{bmatrix}.$$

It is natural to write the following commands to solve the differential-algebraic equation directly:

```
>> M=[1 0 0 0 0 0 0; 0 1 1/15 0 0 0 0;
      0 0 0 1 0 0 0;  zeros(4,7)];
   f=@(t,x)[(x(2)-x(1))/20; -(x(3)-99.1)/75; x(7)-x(5);
      x(3)-(3.35-0.075*x(5)+0.001*x(5)^2)*x(6);
      -x(6)^2+49.58^2-(x(7)/1.2/x(1))^2;
      x(4)-20*x(6); -x(7)+15+5*tanh(t-10)];
   ff=odeset; ff.AbsTol=100*eps; ff.RelTol=100*eps; ff.Mass=M;
   x0=[0.25; 0.2500145529515559; 99.1;
       734.0477598381585; 9.998240239254898;
       36.70238799190793; 15+5*tanh(-10)];
   tic, [t,x]=ode15s(f,[0,40],x0,ff); toc  % solve the differential equation
```

Unfortunately, the error message “This DAE appears to be of index greater than 1” is displayed and the solution process is aborted. Since the rank of $\boldsymbol{M}$ is 3, and there are 7 state variables, the differential-algebraic equation is of index 4, which cannot be solved by many solvers. Manual methods may be used to reduced the index. For instance, $\mu(t)$ is a given function, the signal can be computed directly without assigning a state for it. Besides, from a simple algebraic relationship $M(t) = 20P(t)$, one redundant state $P(t)$ can be eliminated. Of course, the manual index reduction like this can only reduce the index to 2, rather than 1. The semi-explicit differential-algebraic equation solver discussed earlier still cannot be used.

It can be seen from the example that there is a major limitation in the solver that the equation must have index 1. Later the solutions for differential-algebraic equations of index larger than 1 will be explored.

Example 5.17. In [31, 41], a seemingly simpler differential-algebraic equation is given by

$$\begin{cases} p'(t) = u(t), \\ q'(t) = v(t), \\ mu'(t) = -p(t)\lambda(t), \\ mv'(t) = -q(t)\lambda(t) - mg, \\ 0 = p^2(t) + q^2(t) - l^2 \end{cases} \tag{5.3.4}$$

where $g = 9.81$, m and l are constants. The model describes the motion of a pendulum of mass m and length l.

This equation is a differential-algebraic equation of index 3. It cannot be solved directly with solvers such as `ode45()`. The numerical methods will be explored later.

5.3.5 Implicit solvers for differential-algebraic equations

It can be seen from the previous examples that MATLAB provides a differential-algebraic equation solver applicable only for index-1 equations, with semi-explicit form. The solution process has limitations.

Now consider the implicit differential equation shown in (5.2.1). It can be seen from the mathematical model that it does not require each model have derivative terms. Therefore even if some of the equations are degenerated to algebraic constraints, the modeling structure can still be used. In this section, examples are given to explore the solution of differential-algebraic equations with implicit differential equation solvers. Equations having higher indices or those which cannot be expressed by semi-explicit differential-algebraic equations are also explored.

Example 5.18. Solve again the differential-algebraic equation in Example 5.14 with an implicit differential equation solver.

Solutions. The original differential-algebraic equation model can be written in the standard form of an implicit differential equation as follows:

$$\begin{bmatrix} x_1'(t) + 0.2x_1(t) - x_2(t)x_3(t) - 0.3x_1(t)x_2(t) \\ x_2'(t) - 2x_1(t)x_2(t) + 5x_2(t)x_3(t) + 2x_2^2(t) \\ x_1(t) + x_2(t) + x_3(t) - 1 \end{bmatrix} = \mathbf{0}.$$

The following commands can be used to describe the equation:

```
>> f=@(t,x,xd)[xd(1)+0.2*x(1)-x(2)*x(3)-0.3*x(1)*x(2);
         xd(2)-2*x(1)*x(2)+5*x(2)*x(3)+2*x(2)^2;
         x(1)+x(2)+x(3)-1];
```

If the implicit differential equation solver `ode15i()` provided in MATLAB is used, the consistent initial values should be found first. If one selects $\boldsymbol{x}_0 = [0.8, 0.1, 2]^{\mathrm{T}}$, and initial derivative values are $\boldsymbol{x}_0' = [1,1,1]^{\mathrm{T}}$, and the following MATLAB commands can be issued:

```
>> ff=odeset; ff.AbsTol=100*eps; ff.RelTol=100*eps;
   x0=[0.8;0.1;2]; x0F=[1;1;1]; xd0=[1;1;1]; xdF=[0;0;0];
   [x0,xd0]=decic(f,0,x0,x0F,xd0,xdF)  % consistent initial values
```

While the error message "Try freeing 1 fixed component" is returned, indicating that three values in $\boldsymbol{x}_0$ cannot be assigned. One of them should be set to a free value. For instance, the initial value of x_3 needs to be set to a free value. The vector $\boldsymbol{x}_0$ can be kept unchanged, we only need to modify its indicator. The following commands can be used to compute consistent initial values, and directly solve the differential-algebraic equation. For this particular example, the elapsed time is 0.4 seconds and the number of points is 2 916. They are similar to the previous case using other methods.

```
>> x0F=[1;1;0];                    % the last number is set to 0
   [x0,xd0]=decic_new(f,0,x0,x0F)  % consistent initial values
   tic, [t,x]=ode15i(f,[0,20],x0,xd0,ff); toc
   plot(t,x), length(t)            % implicit differential equation solution
```

The obtained consistent initial values are

$$\boldsymbol{x}_0 = [0.8, 0.1, 0.1]^{\mathrm{T}} \text{ and } \boldsymbol{x}_0' = [-0.126, 0.09, 1]^{\mathrm{T}}.$$

Example 5.19. Use the differential-algebraic equation solver to solve the implicit differential equation in Example 4.17.

Solutions. In Example 4.19, the inverse of matrix $\boldsymbol{A}(\boldsymbol{x})$ was used directly to convert the model into a first-order explicit differential equation, so that solvers such as `ode45()` could be used it. The numerical solution of the differential equation could then be found. Since the assumption that $\boldsymbol{A}(\boldsymbol{x})$ is a nonsingular matrix was made, the solution process could not be very rigorous. Therefore, the equation should be solved again without making such an assumption.

It can be seen that two anonymous functions can be used to describe the equation and the mass matrix:

```
>> f=@(t,x)[1-x(1);-x(2)];
   M=@(t,x)[sin(x(1)),cos(x(2));-cos(x(2)),sin(x(1))];
```

The following commands can be used to solve the differential-algebraic equation:

```
>> ff=odeset; options.Mass=M;
   ff.RelTol=100*eps; ff.AbsTol=100*eps;       % precision setting
   [t,x]=ode45(f,[0,10],[0;0],ff); plot(t,x)  % solve equation
```

The result obtained is exactly the same as that shown in Figure 5.15. Since there is no analytical solution for the original differential equation, there is no method to decide which solution is more accurate.

Example 5.20. Solve the differential-algebraic equation in Example 5.16 with the implicit differential equation solver.

Solutions. The code for a semi-explicit differential-algebraic equation description in Example 5.16 can still be used. Then it can be converted into an implicit differential equation model with an anonymous function. The following commands can be used:

```
>> M=[1 0 0 0 0 0 0; 0 1 1/15 0 0 0 0;
      0 0 0 1 0 0 0;  zeros(4,7)];
   f0=@(t,x)[(x(2)-x(1))/20; -(x(3)-99.1)/75; x(7)-x(5);
      x(3)- (3.35-0.075*x(5)+0.001*x(5)^2)*x(6);
      -x(6)^2+49.58^2-(x(7)/1.2/x(1))^2;
      x(4)-20*x(6); -x(7)+15+5*tanh(t-10)];
   f=@(t,x,xd)M*xd-f0(t,x);
   ff=odeset; ff.AbsTol=100*eps; ff.RelTol=100*eps;
   x0=[0.25; 0.2500145529515559; 99.1;
       734.0477598381585; 9.998240239254898;
       36.70238799190793; 15+5*tanh(-10)];
   x0F=ones(7,1); xd0=rand(7,1); xd0F=zeros(7,1);
   [x0,xd0]=decic(f,0,x0,x0F,xd0,xd0F) %consistent initial values
```

When the above statements are employed to find the consistent initial values, an error message "Try freeing 4 fixed components" is displayed. Again it shows that the index of the differential-algebraic equation is 4. If the latter 4 components in $\boldsymbol{x}(0)$ are assumed as free values, the following statements can be used to solve the differential-algebraic equation. Note that the differential equation model is a stiff one, and the error tolerance cannot be assigned to very small values, otherwise the solution process may not converge. According to [31], the relative tolerance is set to 10^{-4}.

```
>> x0F(4:7)=0;    %free some initial values
   [x0,xd0]=decic_new(f,0,x0,x0F,xd0,xd0F,1e-14)
   ff=odeset; ff.AbsTol=1e-10; ff.RelTol=1e-4;
   tic, [t,x]=ode15i(f,[0,40],x0,xd0,ff); toc
```

The consistent initial values are obtained as follows. The curves of $k(t)$ and $m(t)$ are obtained as shown in Figure 5.16. It can be seen that the results are the same as those

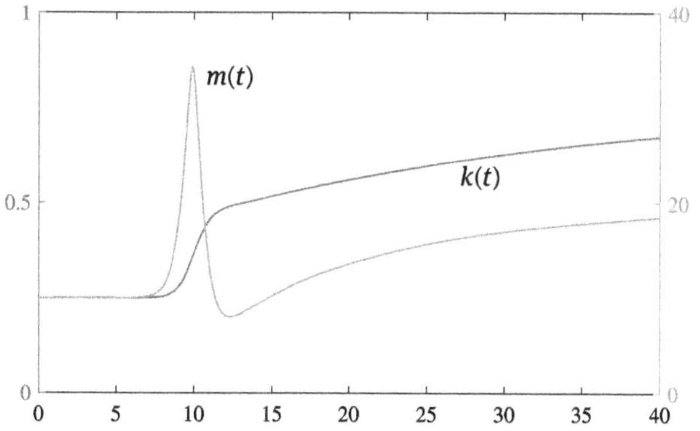

Figure 5.16: Solution of the differential–algebraic equation.

in Figure [31]. The elapsed time is 2.74 seconds.

$$\begin{aligned} \boldsymbol{x}(0) &= [0.25, 0.25001, 99.1, 734.0478, 9.9982, 36.7024, 10]^{\mathrm{T}}, \\ \boldsymbol{x}'(0) &= [0, -0.0358, 0.5366, 0.00176, 0.1860, 0.4817, 0.2071]^{\mathrm{T}}. \end{aligned}$$

Example 5.21. Consider again the differential-algebraic equation in Example 5.16. Since the index is 4, numerical solution process is made more complicated. The index can be appropriately reduced to 2. Solve the differential-algebraic equation in this manner.

Solutions. The idea of index reduction for this problem was discussed in Example 5.16. That is, we should select state variables

$$x_1(t) = k(t), \quad x_2(t) = c(t), \quad x_3(t) = p(t), \quad x_4(t) = M(t), \quad x_5(t) = m(t).$$

The other two functions need not be selected as states, since they are either a given function or can be expressed simply by other state variables. That is, $P(t) = 0.05x_4(t)$ and $\mu(t) = 10 + 5\tanh(t-10)$. Therefore the new implicit differential equation can be written as

$$\begin{bmatrix} x_1'(t) - (x_2(t) - x_1(t))/20 \\ x_2'(t) + x_3'(t)/15 + (x_3(t) - 99.1)/75 \\ x_4'(t) - 15 - 5\tanh(t-10) + x_5(t) \\ x_3(t) - 0.05[3.35 - 0.075x_5(t) + 0.001x_5^2(t)]x_4(t) \\ -(0.05x_4(t))^2 + 49.58^2 - [(15 + 5\tanh(t-10))/(1.2x_1(t))]^2 \end{bmatrix} = \mathbf{0}.$$

The index is reduced to 2. The latter two initial values can be set to free values, so that the following commands can be issued to solve the differential equation again. The results are the same as those in Example 5.20. The elapsed time is 2.52 seconds, similar to that in the previous example.

```
>> f=@(t,x,xd)[xd(1)-(x(2)-x(1))/20;
          xd(2)+xd(3)/15+(x(3)-99.1)/75;
          xd(4)-15-5*tanh(t-10)+x(5);
          x(3)-0.05*[3.35-0.075*x(5)+0.001*x(5)^2]*x(4);
          -(0.05*x(4))^2+49.58^2-[(15+5*tanh(t-10))/(1.2*x(1))]^2];
   x0=[0.25; 0.2500145529515559; 99.1;
       734.0477598381585; 9.998240239254898];
   x0F=[1;1;1;0;0]; xd0=rand(5,1); x0dF=0*x0F;
   [x0,xd0]=decic_new(f,0,x0,x0F,xd0,x0dF,1e-14)
   ff=odeset; ff.AbsTol=1e-10; ff.RelTol=1e-4;
   tic, [t,x]=ode15i(f,[0,40],x0,xd0,ff); toc % solve equation
```

Good news is that when the index is reduced, a smaller relative error tolerance can be used, also it may become very time-consuming. For instance, if the relative error tolerance is set to 10^{-5}, the elapsed time is 19.25 seconds, while if it is set to 10^{-6}, the time needed is increased to 209.41 seconds.

The consistent initial values are as follows. Note that the initial values for the state variables are the same as those in the previous example, but the initial derivative vector is different:

$$\begin{aligned}\boldsymbol{x}(0) &= [0.25, 0.25, 99.1, 734.0478, 9.9982]^{\mathrm{T}},\\ \boldsymbol{x}'(0) &= [0, -0.0121, 0.1813, 0.0018, 0.0079]^{\mathrm{T}}.\end{aligned}$$

Example 5.22. Consider again the differential-algebraic equation in Example 5.17. If the initial values are $p(0) = 1$, $q(0) = u(0) = v(0) = \lambda(0) = 0$, and the parameters are $m = l = 1$, solve the differential-algebraic equation.

Solutions. It was indicated in Example 5.17 that, due to the existence of the $p^2(t) + q^2(t) - l^2 = 0$ term, the differential equation cannot be converted into a semi-explicit differential-algebraic equation. Therefore an implicit differential equation solver can be tried by selecting the state variables $x_1(t) = p(t)$, $x_2(t) = q(t)$, $x_3(t) = u(t)$, $x_4(t) = v(t)$, and $x_5(t) = \lambda(t)$. The standard form of the implicit differential equation can be written as

$$\begin{bmatrix} x_1'(t) - x_3(0) \\ x_2'(t) - x_4(0) \\ x_3'(t) + x_1(t)x_5(t) \\ x_4'(t) + x_2(t)x_5(t) + 9.81 \\ x_1^2(t) + x_2^2(t) - 1 \end{bmatrix} = \mathbf{0}.$$

With the implicit differential equation model, an anonymous function can be written to express it, and the consistent initial values can be found with the following commands:

```
>> f=@(t,x,xd)[xd(1)-x(3); xd(2)-x(4);
     xd(3)+x(1)*x(5); xd(4)+x(2)*x(5)+9.81; x(1)^2+x(2)^2-1];
   x0=[1;0;0;0;0]; x0F=ones(5,1);
   xd0=rand(5,1); xd0F=0*x0F;
   [x0,xd]=decic(f,0,x0,x0F,xd0,xd0F)
```

Since the original differential equation is, in fact, a differential-algebraic equation, an error message "Try freeing 1 fixed component" is displayed. Which of the values is to be freed? The values of the second up to the fifth state are tried, and the same error message is displayed. If the first one is set free

```
>> x0F=[0; 1; 1; 1; 1];                        % set the first one free
   [x0,xd]=decic(f,0,x0,x0F,xd0,xd0F) % find consistent initial values
```

the consistent initial values are found as

$$x(0) = [1, 0, 0, 0, 0]^{\mathrm{T}}, \quad x_0'(0) = [0, 0, 0, -9.81, 0.27086]^{\mathrm{T}}.$$

With the consistent initial values, the following commands can be tried to obtain more accurate values. But it is found that in the solution process, the value of $\lambda'(0)$ is not unique.

```
>> [x0,xd,f0]=decic_new(f,0,x0,x0F) % find consistent initial values
```

With the obtained consistent initial values, the differential-algebraic equation can be solved. For this particular problem, the solution process may fail, since it is singular at the initial time instance.

```
>> ff=odeset; ff.AbsTol=100*eps; ff.RelTol=100*eps; ff.Mass=M;
   tic, [t,x]=ode15i(f,[0,10],x0,xd0,ff);
   toc % solve implicit differential equation
```

5.3.6 Index reduction for differential-algebraic equations

It has been mentioned that if the index is larger than 1, some of the solution methods may fail. An index reduction technique should be introduced to convert the differential equations. For instance, derivatives can be taken of an algebraic equation, such that a new differential-algebraic equation can be found. Examples are shown next to demonstrate the index reduction and solution method.

Example 5.23. Consider the differential-algebraic equation in Example 5.17. Reduce its index and then find numerical solutions.

Solutions. Taking the first-order derivative of the algebraic equation $p^2(t)+q^2(t)-l^2 = 0$ with respect to t, it is found that

$$2p(t)p'(t) + 2q(t)q'(t) = 0.$$

Substituting (5.3.4) into the above, then

$$p(t)u(t) + q(t)v(t) = 0.$$

The new equation can be used to replace the original algebraic constraint in (5.3.4), and the index of the equation is reduced to 2. The $p^2(t)+q^2(t)-l^2 = 0$ constraint vanishes. If the differential-algebraic equations are tried to be solved, this still cannot be accomplished. A further derivative should be computed, to further reduce the index. Taking the first-order derivative again, we get

$$\begin{aligned} &p'(t)u(t) + p(t)u'(t) + q'(t)v(t) + q(t)v'(t) \\ &\quad = u^2(t) - p^2(t)\lambda(t)/m + v^2(t) - q^2(t)\lambda(t)/m - gq(t)/m \\ &\quad = u^2(t) + v^2(t) - (p^2(t) + q^2(t))\lambda(t)/m - gq(t)/m \\ &\quad = u^2(t) + v^2(t) - l^2\lambda(t)/m - gq(t)/m. \end{aligned}$$

A new differential-algebraic equation can be set up from (5.3.4). If the states in Example 5.22 are selected, the standard form of the implicit differential equation can be set up as

$$\begin{bmatrix} x_1'(t) - x_3(0) \\ x_2'(t) - x_4(0) \\ x_3'(t) + x_1(t)x_5(t) \\ x_4'(t) + x_2(t)x_5(t) + 9.81 \\ x_3^2(t) + x_4^2(t) - x_5(t) - 9.81x_2(t) \end{bmatrix} = \mathbf{0}.$$

With the standard form of the implicit differential equation, the following statements can be used to solve the equation:

```
>> f=@(t,x,xd)[xd(1)-x(3); xd(2)-x(4);
     xd(3)+x(1)*x(5); xd(4)+x(2)*x(5)+9.81;
     x(3)^2+x(4)^2-x(5)-9.81*x(2)];
   x0=[1;0;0;0;0]; x0F=[ones(4,1); 0]; %free one initial value
   [x0,xd0,f0]=decic_new(f,0,x0,x0F)    %find consistent initial values
   ff=odeset; ff.RelTol=100*eps; ff.AbsTol=100*eps;
   tic, [t,x]=ode15i(f,[0,3],x0,xd0,ff); toc %solve equations
   max(abs(x(:,1).^2+x(:,2).^2-1)), plot(t,x(:,[1,2]))
```

It can be seen that the maximum error in the algebraic constraint is 6.1655×10^{-11}. The time responses of the states obtained are as shown in Figure 5.17. The elapsed time is 7.23 seconds, and the total number of points is 60 632. It can be seen that by index reduction, the original problem can be solved, and the results are satisfactory.

Example 5.24. Solve the equivalent differential-algebraic equations in Example 5.23 with the semi-explicit differential-algebraic equation solver.

Solutions. It is not difficult to write the standard form of the semi-explicit differential-algebraic equation as follows:

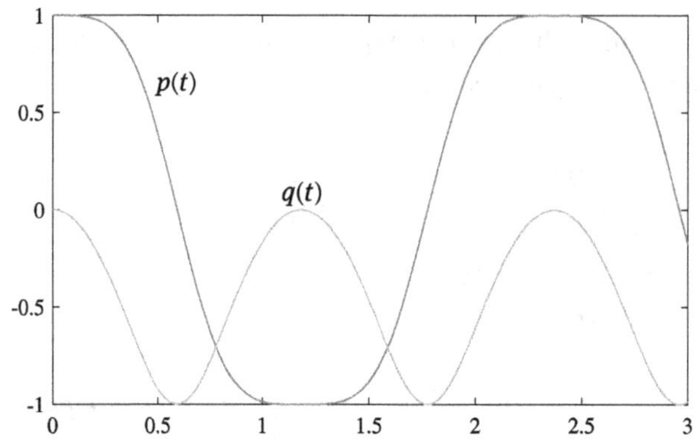

Figure 5.17: Numerical solution when the index is reduced.

$$\begin{bmatrix} 1 & 0 & 0 & 0 & 0 \\ 0 & 1 & 0 & 0 & 0 \\ 0 & 0 & 1 & 0 & 0 \\ 0 & 0 & 0 & 1 & 0 \\ 0 & 0 & 0 & 0 & 0 \end{bmatrix} \boldsymbol{x}'(t) = \begin{bmatrix} x_3(0) \\ x_4(0) \\ x_1(t)x_5(t) \\ x_2(t)x_5(t) + 9.81 \\ x_3^2(t) + x_4^2(t) - x_5(t) - 9.81x_2(t) \end{bmatrix}.$$

The following commands can be used to solve the semi-explicit differential-algebraic equations. Unfortunately, after a period of waiting, an error message is displayed, indicating that the original equation diverges at a certain point. The solution process is unsuccessful.

```
>> M=eye(5); M(5,5)=0;
   f=@(t,x)[x(3); x(4); x(1)*x(5); x(2)*x(5)+9.81;
            x(3)^2+x(4)^2-x(5)-9.81*x(2)];
   x0=[1;0;0;0;0]; eps0=100*eps;
   ff=odeset('RelTol',eps0,'AbsTol',eps0,'Mass',M);
   tic, [t,x]=ode45(f,[0,3],x0,ff); toc
   max(abs(x(:,1).^2+x(:,2).^2-1)), plot(t,x(:,[1,2]))
```

It can be concluded from the example that for differential-algebraic equations it is not necessary to change them manually to the semi-explicit form. An implicit form can be used to find numerical solutions. If in the solution process it is prompted that the equations are singular, index reduction techniques may be introduced by taking derivatives of the algebraic constraints. If the model is still singular, derivatives should be computed again to further reduce the index. Implicit differential equation solvers can be used to directly solve the differential equations.

5.4 Switched differential equations

Switched systems is an important research field in control systems theory.[48] The so-called switched system is a system composed of several subsystems, and these subsystems are selected by certain switching laws.

Definition 5.6. The multiple models of the differential equation of switched systems can be expressed as

$$\boldsymbol{x}'(t) = \boldsymbol{f}_i(t, \boldsymbol{x}(t), \boldsymbol{u}(t)), \quad i = 1, \dots, m, \tag{5.4.1}$$

where $\boldsymbol{x}(t_0)$ is given and $\boldsymbol{u}(t)$ is the external input signal.

Certain switching laws are defined such that the entire model is switching under the control of switching laws, among the subsystems. With the switched system theory, a controller can be designed to stabilize the whole system, under certain switching laws. The subsystems $\boldsymbol{f}_i(\cdot)$ can be unstable.

In this section, numerical solutions of linear switched systems are demonstrated. Then the concept of zero-crossing detection is proposed. Finally, numerical solutions of nonlinear switched differential equations are studied.

5.4.1 Linear switched differential equations

In the theoretical references such as,[66, 78, 81] we commonly see demonstrative examples to allow switching among different subsystems. In this section, linear switched differential equation is formulated with physical explanations. Then examples are used to demonstrate the numerical solution methods.

Definition 5.7. The state space form of a linear switched system model is

$$\begin{cases} \boldsymbol{x}'(t) = \boldsymbol{A}_{\sigma(t,\boldsymbol{x}(t))}\boldsymbol{x}(t) + \boldsymbol{B}_{\sigma(t,\boldsymbol{x}(t))}\boldsymbol{u}(t), \\ \boldsymbol{y}(t) = \boldsymbol{C}_{\sigma(t,\boldsymbol{x}(t))}\boldsymbol{x}(t) + \boldsymbol{D}_{\sigma(t,\boldsymbol{x}(t))}\boldsymbol{u}(t) \end{cases} \tag{5.4.2}$$

where $\sigma(t, \boldsymbol{x}(t)) \in \{1, 2, \dots, m\}$ is the switching law.

Here the physical meaning of the so-called switching law is that if the time and states satisfy the ith preset condition, the system is switched to the ith subsystem model. It is not hard to see that if the switching law can be expressed clearly in mathematics, anonymous or MATLAB functions can then be used to describe the entire state space model such that numerical methods can be used to solve switched systems directly. A simple example is given next to demonstrate the numerical solutions of the switched linear differential equation.

Example 5.25. Assume that the system model is given by $\boldsymbol{x}'(t) = \boldsymbol{A}_i\boldsymbol{x}(t)$ where

$$\boldsymbol{A}_1 = \begin{bmatrix} 0.1 & -1 \\ 2 & 0.1 \end{bmatrix}, \quad \boldsymbol{A}_2 = \begin{bmatrix} 0.1 & -2 \\ 1 & 0.1 \end{bmatrix}.$$

It can be seen that the two subsystems are unstable. If $x_1(t)x_2(t) < 0$, that is, the states are located in quadrants II and IV in the phase plane, the coefficient is switched to subsystem $\boldsymbol{A}_1$, while if $x_1(t)x_2(t) \geqslant 0$, that is, if the states are located in quadrants I and III, the coefficient matrix is switched to subsystem $\boldsymbol{A}_2$. If the initial states are $x_1(0) = x_2(0) = 5$, solve the differential equation.

Solutions. According to the system model and switching laws, it is easy to use an anonymous function to describe the switched system as

```
>> A1=[0.1 -1; 2 0.1]; A2=[0.1 -2; 1 0.1];
   f=@(t,x)(x(1)*x(2)<0)*A1*x+(x(1)*x(2)>=0)*A2*x;
```

Therefore, the following statements can be used to solve the switched differential equation directly. The time responses of the state variables are obtained as shown in Figure 5.18.

```
>> ff=odeset; ff.RelTol=100*eps; ff.AbsTol=100*eps;
   [t,x]=ode45(f,[0,30],[5;5],ff); plot(t,x) %solve switched system
```

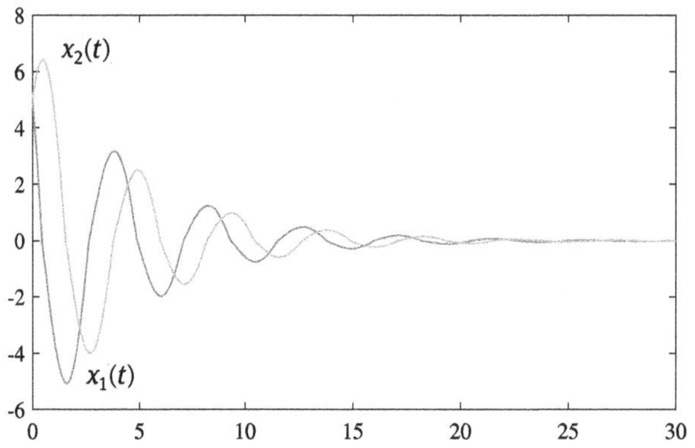

Figure 5.18: Time responses of the switched system.

The phase plane trajectory of the system can also be drawn as shown in Figure 5.19. It can be seen that although the two subsystems are unstable, under appropriate switching laws, the entire switched system is stable.

```
>> plot(x(:,1),x(:,2))    %phase plane trajectory
```

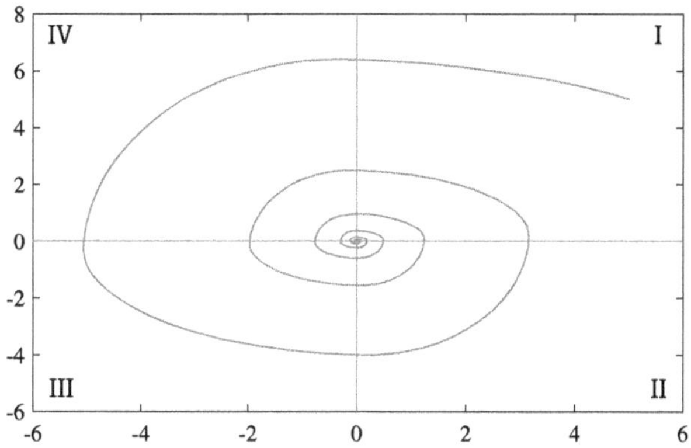

Figure 5.19: Phase plane trajectory of the switched system.

5.4.2 Zero-crossing detection and event handling

It can be seen from the previous example that the switching laws between the two subsystems are respectively $x_1(t)x_2(t) > 0$ and $x_1(t)x_2(t) \leqslant 0$. The two conditions are mutual exclusive. If one of them is satisfied, the other is not. Since the signals $x_1(t)$ and $x_2(t)$ are continuous, there must be a certain point when $x_1(t)x_2(t) = 0$ is satisfied. This is the so-called zero-crossing point. If the zero-crossing point can be detected accurately, the simulation process may run correctly. Now let us have a look on the importance of zero-crossing detection.

Example 5.26. Draw the curve of the function $y = |\sin t^2|$, $t \in (0, \pi)$.

Solutions. It is natural to select a step-size, say, $h = 0.02$ seconds to compute the given function. The following MATLAB statements can be used to draw the curve of the function as shown in Figure 5.20:

```
>> t=0:0.02:pi; y=abs(sin(t.^2)); plot(t,y)
```

It is obvious that the values of the function at some particular points are wrong. If the function values are close to 0, say at points A, B, and C, the function value should be decreased to 0, and then increase gradually. It should take a turning at the suspending point A. If there is a mechanism to find the exact point when the function value is 0, then that point can be inserted into vector $\boldsymbol{t}$ such that the correct curve can be drawn. The method of finding the exact point is the so-called zero-crossing detection. In variable-step differential equation solvers, if the error tolerance is set to a small enough value, the zero-crossing detection facilities will be automatically implemented, while in fixed-step algorithms, zero-crossing detection is not ensured. Therefore the phenomena in Figure 5.20 may be inevitable. Therefore, a variable-step

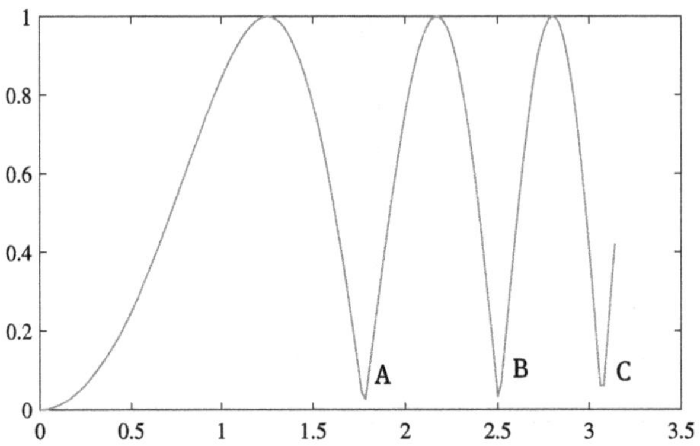

Figure 5.20: The $y = |\sin t^2|$ curve and zero-crossing demonstration.

algorithm with a small error tolerance must be employed in solving switched differential equations.

The user may select the zero-crossing points of the signal which is to be monitored. If the user is interested in a particular signal, then the point at which the signal equals zero is regarded as a zero-crossing point. The case when the signal equals zero can be regarded as an event. In the differential equation solution process, the user may define responses to certain events. For instance, the intermediate data at the zero-crossing points or solution processes may be controlled by setting these events.

If one wants to detect zero-crossing points, and make certain responses for such points, responding functions should be written. Responding functions are MATLAB functions. Meanwhile, the `Events` member of the control should be set to the function handle. The event setting and responding will be demonstrated next through examples.

If an event detection mechanism is used, the syntax of the solver `ode45()` should be adjusted to

$[\boldsymbol{t},\boldsymbol{x},\boldsymbol{t}_{\rm e},\boldsymbol{x}_{\rm e},\boldsymbol{i}_{\rm e}]$=`ode45(Fun,`$[t_0,t_n]$,$\boldsymbol{x}_0$`,options)`

where $\boldsymbol{t}_{\rm e}$, $\boldsymbol{x}_{\rm e}$, and $\boldsymbol{i}_{\rm e}$ return respectively the time instance the event happens, and the state values and crossing direction when the event happens.

Example 5.27. Consider the switched system model in Example 5.25. If in the plot in Figure 5.18, the $x_1(t)x_2(t) = 0$ event points are needed to be superimposed, the expression $|x_1(t)x_2(t)| = 0$ can be set as the condition for the zero-crossing points. Superimpose all the zero-crossing points on the plot.

Solutions. Under the double precision framework, the condition $|x_1(t)x_2(t)| = 0$ is too strict from the numerical viewpoint. Therefore, condition $|x_1(t)x_2(t)| < \epsilon$ is usually used instead, for instance, with $\epsilon = 10^{-11}$. For this particular problem, if ϵ is selected too small, the zero-crossing point may not be detected. For better describing zero-crossing points, logic expressions are used to describe the value of the zero-crossing conditions.

If the information at zero-crossing points needs to be intercepted, a response function in MATLAB is written. The input arguments are t and $\boldsymbol{x}$, and there can be additional parameters. The returned argument is `values`, returning the values of the zero-crossing points. Besides, other arguments can be returned, such as `istermimnal`, which can be used to control the solution process, with 1 for termination of the solution process; argument `direction` is used to express the direction of the zero-crossing. Normally it is set to 1. If set to zero, the two directions are both included in the events, which means that a zero-crossing event may be detected twice.

```
function [values,isterminal,direction]=event_fun(t,x)
values=abs(x(1)*x(2))>=1e-11; isterminal=0; direction=1;
```

With such a response function, the following commands can be used to solve the original problem again. The time responses of the states are drawn, superimposed with zero-crossing points, as shown in Figure 5.21. It can be seen that all 27 zero-crossing points are detected. The returned vector `ie` is a vector of 1s, indicating that zero-crossings in the positive direction are detected. It is as one has expected.

```
>> A1=[0.1 -1; 2 0.1]; A2=[0.1 -2; 1 0.1];
   f=@(t,x)(x(1)*x(2)<0)*A1*x+(x(1)*x(2)>=0)*A2*x;
   ff=odeset; ff.RelTol=100*eps; ff.AbsTol=100*eps;
   ff.Events=@event_fun;               % event response function handle
   [t,x,te,xe,ie]=ode45(f,[0,30],[5;5],ff);
   plot(t,x,te,xe,'o'), length(te) % switched equation solutions
```

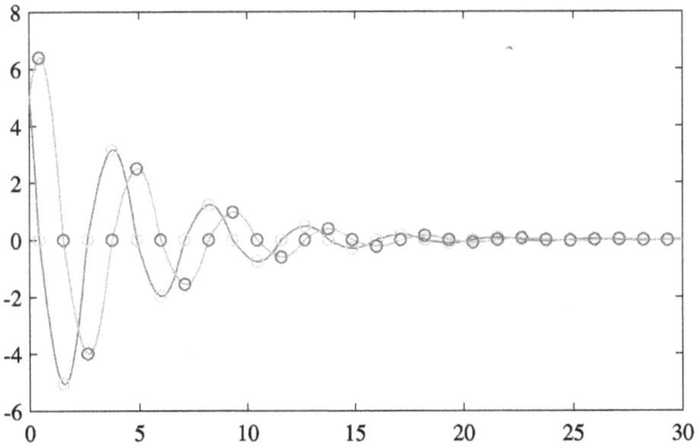

Figure 5.21: State response curves with zero-crossing marks.

5.4.3 Nonlinear switched differential equations

In this section, functions such as `ode45()` are used mainly to solve differential equations. The simulation method for switched differential equations is not restricted to linear systems. It can be applied directly for nonlinear switched differential equations.

Definition 5.8. The mathematical model of a nonlinear switched differential equation is

$$\boldsymbol{x}'(t) = \boldsymbol{f}_{\sigma(t,\boldsymbol{x}(t))}(t, \boldsymbol{x}(t)) \tag{5.4.3}$$

where $\sigma(t, \boldsymbol{x}(t)) \in \{1, 2, \ldots, m\}$ are the control laws.

Similar to the ideas discussed earlier, a MATLAB function can be used to express the switched differential equation. Then solvers such as `ode45()` can be used to solve these differential equations directly. The examples given next demonstrate the numerical solution process of nonlinear switched differential equations.

Example 5.28. Consider the kinetic model of a wheeled mobile robot:[81]

$$\begin{cases} x_1'(t) = u_1(t)\cos\theta(t), \\ x_2'(t) = u_1(t)\sin\theta(t), \\ \theta'(t) = u_2(t). \end{cases}$$

The column vector $\boldsymbol{x}(t) = [x_1(t), x_2(t), \theta(t)]^{\mathrm{T}}$ is composed such that the robot model may reach the final states at $\boldsymbol{x}(t_{\mathrm{n}}) = \mathbf{0}$. The following switching laws are considered:

Condition $\mathscr{D}_1$ is set to $|x_3(t)| > \|\boldsymbol{x}(t)\|/2$. If this condition is satisfied, then the control law is set to

$$\boldsymbol{u}_{\mathrm{I}}(t) = \begin{bmatrix} u_1(t) \\ u_2(t) \end{bmatrix} = \begin{bmatrix} -4y_2(t) - 6y_3(t)/y_1(t) - y_3(t)y_1(t) \\ -y_1(t) \end{bmatrix}.$$

If Condition $\mathscr{D}_1$ is not satisfied, the control law is set to

$$\boldsymbol{u}_{\mathrm{II}}(t) = \begin{bmatrix} -y_2(t) - y_3(t)\operatorname{sgn}(y_2(t)y_3(t)) \\ -\operatorname{sgn}(y_2(t)y_3(t)) \end{bmatrix}$$

where $\operatorname{sgn}(\cdot)$ is the sign function

$$\operatorname{sgn}(\alpha) = \begin{cases} 1, & \text{if } \alpha \geqslant 0, \\ -1, & \text{otherwise} \end{cases}$$

and

$$\begin{cases} y_1(t) = \theta(t), \\ y_2(t) = x_1(t)\cos\theta(t) + x_2(t)\sin\theta(t), \\ y_3(t) = x_1(t)\sin\theta(t) - x_2(t)\cos\theta(t). \end{cases}$$

If the initial values are $x_1(0) = x_2(0) = 5$, $\theta(0) = \pi$, solve the switched differential equation.

Solutions. This example is rather complicated. It is not suitable to describe the system with anonymous functions. A MATLAB function should be written instead. For MATLAB functions, the input arguments are t and $\boldsymbol{x}$. The intermediate variables $y_1(t)$, $y_2(t)$, and $y_3(t)$ can be computed first. Then from Condition $\mathscr{D}_1$, the control laws can be implemented with a conditional structure to compute $\boldsymbol{u}_{\mathrm{I}}(t)$ and $\boldsymbol{u}_{\mathrm{II}}(t)$, and finally to compose the derivative of the states $\boldsymbol{x}'(t)$. Based on the above consideration, the following MATLAB function can be written:

```
function dx=c5mwheels(t,x)
c=cos(x(3)); s=sin(x(3));
y=[x(3); x(1)*c+x(2)*s; x(1)*s-x(2)*c];
if abs(x(3))>norm(x)/2
   u=[-4*y(2)-6*y(3)/y(1)-y(3)*y(1); -y(1)];
else
   sgn=-1; if y(2)*y(3)>=0, sgn=1; end
   u=[-y(2)-y(3)*sgn; -sgn];
end
dx=[u(1)*c; u(1)*s; u(2)]; %compute the derivatives of the states
```

It can be seen that the following commands can be used to solve the nonlinear differential equations directly. Under the control laws, the phase plane plot obtained is shown in Figure 5.22. Note that the solution process is rather time consuming (it takes 16.4 seconds).

```
>> ff=odeset; ff.RelTol=1e-6; ff.AbsTol=1e-7;
   tic, [t,x]=ode45(@c5mwheels,[0,2],[5; 5; pi],ff); toc
   plot(x(:,1),x(:,2))          %draw phase portrait
```

It is a pity that, although the author faithfully described the theory and MATLAB implementations, and different control options were tried, the expected solution could never be reached.

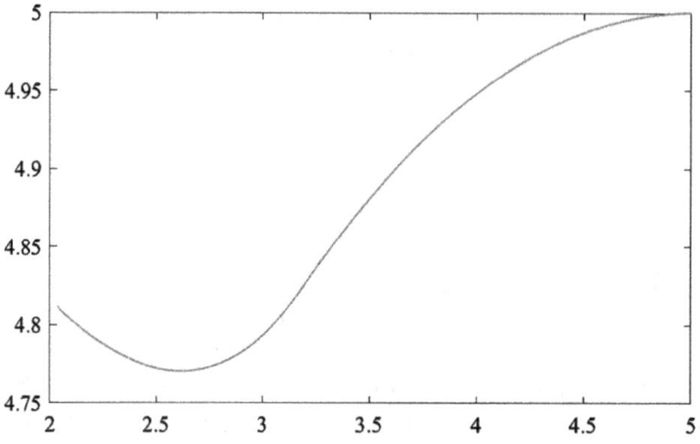

Figure 5.22: Phase portrait of the switched differential equation.

5.4.4 Discontinuous differential equations

The so-called discontinuous differential equations are those where certain parameters may have discontinuities or sudden jumps. In normal cases, in order to solve discontinuous differential equations well, the relative error tolerances should not be assigned to large quantities, such as 10^{-5}. Besides, stiff equation solvers are not recommended. An example of a discontinuous system is demonstrated next.

Example 5.29. Consider the following discontinuous differential equation model:[32]

$$y''(t) + 2Dy'(t) + \mu \operatorname{sgn}(y'(t)) + y(t) = A \cos \omega t$$

where $D = 0.1$, $\mu = 4$, $A = 2$, and $\omega = \pi$. The initial values are $y(0) = 3$ and $y'(0) = 4$.

Solutions. Due to the existence of the $\mu \operatorname{sgn}(y'(t))$ term, used in modeling the Coulomb friction, the coefficient of $y'(t)$, that is, the velocity, has jumps in the Coulomb friction in the range from −4 to 4. The jump scale is large such that the differential equation is discontinuous. Many numerical algorithms cannot be used in handling such jumps.

Now a stiff solver is used to solve the problem. Letting $x_1(t) = y(t)$ and $x_2(t) = y'(t)$, the original differential equation can be rewritten into the standard form of first-order explicit differential equations:

$$\boldsymbol{x}'(t) = \begin{bmatrix} x_2(t) \\ A\cos\omega t - 2Dx_2(t) - \mu \operatorname{sgn}(x_2(t)) - x_1(t) \end{bmatrix}.$$

The following commands can be tried to solve the differential equations:

```
>> A=2; w=pi; D=0.1; mu=4;
   f=@(t,x)[x(2); A*cos(w*t)-2*D*x(2)-mu*sign(x(2))-x(1)];
   x0=[3,4]; [t,x]=ode15s(f,[0,10],x0); % solve the differential equation
```

Unfortunately in the solution process, a warning "Warning: Failure at t=2.035200e+00" is displayed, indicating that at $t = 2.0352$, the solution process was aborted. The stiff solver failed to work. If other stiff solvers are used, similar situation happens.

Now we try to use the solver `ode45()` to tackle the problem again. After 11.02 seconds of waiting, the solution in Figure 5.23 is found, which is the same as that in [32], meaning that the solver in MATLAB is reliable in solving such differential equations.

```
>> ff=odeset; ff.RelTol=1e-6; ff.AbsTol=1e-6;
   tic, [t,x]=ode45(f,[0,10],x0,ff); toc
   plot(t,x,t,A*cos(w*t),'--'); length(t) %solutions and inputs
```

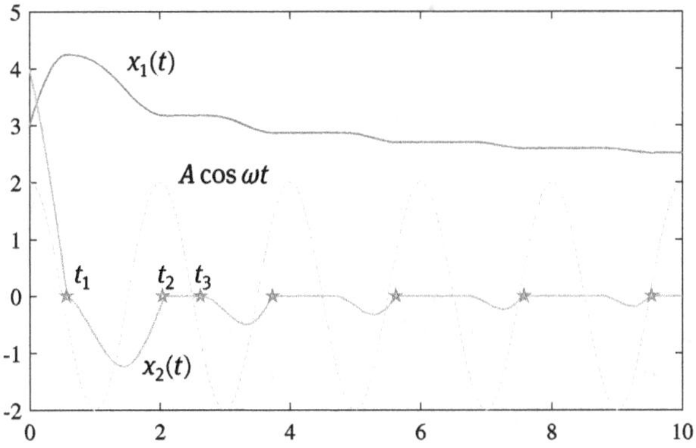

Figure 5.23: Solution results for a discontinuous system.

Under the setting here, the number of points computed is as high as 5 211 569. If a smaller relative error tolerance is adopted, the time needed may further increase, and the solutions may not even be found.

It can be seen from the result that at $t_1 = 0.5628$, the first jump happens, yet the solvers `ode15s()` and `ode45()` both successfully found the solution at this point. At the second jump $t_2 = 2.0352$, the solver `ode15s()` failed and aborted. The `ode45()` solver automatically changed into a small step-size mode, until $t_3 = 2.6281$, when the jump period terminates. Large step-size computation is resumed until the next jump. The changes of step-size in the entire solution process are as shown in Figure 5.24. The minimum step-size is 1.9440×10^{-7}. One can clearly see the behavior of the solver in the jumping regions.

5.5 Linear stochastic differential equations

The stochastic differential equations studied here are mainly deterministic differential equations subject to stochastic process inputs. Compared with deterministic differential equations, stochastic differential equations are more complicated in research. In

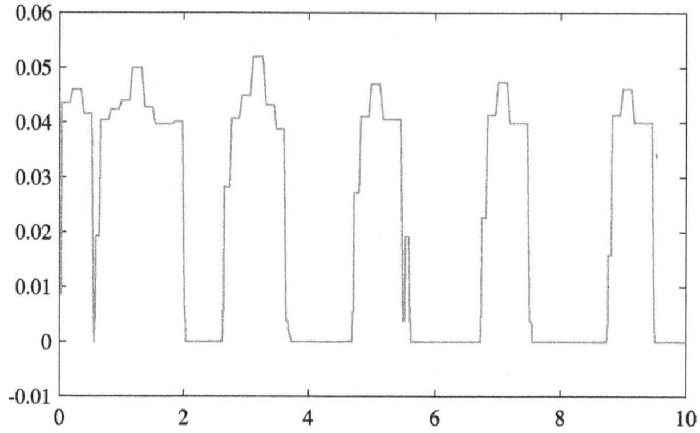

Figure 5.24: Changes of the step-size curves.

mathematical descriptions and numerical solutions, the following changes in conventional viewpoints are made:

(1) If the input signals are described by a stochastic process, the output signals of the differential equation are also stochastic processes. The derivative representation is no longer suitable. Differentials should be used instead.
(2) Since the input signal is stochastic, random signals can be used to drive the system for numerical solution and analysis. One-time simulation results are not of any use. Necessary statistical analysis for the simulation results should be introduced.

In this section, the mathematical model of a linear stochastic differential equation is presented. A transfer function model is also given. Examples are used to indicate that a conventional simulation method, even if the system is merely driven by signals from a random number generator, is not correct. Discretization of linear stochastic differential equations is presented.

In this section, only linear stochastic differential equations are studied. In Chapter 9, numerical solutions of nonlinear stochastic differential equations will be further explored.

5.5.1 Transfer function model for linear stochastic differential equations

In this section, we discuss the mathematical form of continuous stochastic differential equations and their transfer function representation. The output signal of the system subject to Gaussian white noise is also presented.

Definition 5.9. The mathematical form of a linear stochastic differential equation is

$$\begin{aligned} &\mathrm{d}^n y(t) + a_1 \mathrm{d}^{n-1} y(t) + \cdots + a_{n-1} \mathrm{d}y(t) + a_n y(t) \\ &\quad = b_0 \mathrm{d}^m \xi(t) + b_1 \mathrm{d}^{m-1} \xi(t) + \cdots + b_{m-1} \mathrm{d}\xi(t) + b_m \xi(t) \end{aligned} \tag{5.5.1}$$

where $\xi(t)$ is a stochastic process.

Theorem 5.1. *From the given linear stochastic differential equation model, the corresponding transfer function can be written down as*

$$G(s) = \frac{b_0 s^m + b_1 s^{m-1} + \cdots + b_{m-1}s + b_m}{s^n + a_1 s^{n-1} + \cdots a_{n-1}s + a_n}. \tag{5.5.2}$$

In control system theory, the transfer function of a system is defined as the ratio of the Laplace transforms of the output and input signals. The transfer function can be regarded as the gain of the system in the s-domain.

Theorem 5.2. *If the input signal $\xi(t)$ is Gaussian white noise with zero mean and variance σ^2, the output signal is also Gaussian, with zero mean and a variance of*

$$\sigma_y^2 = \frac{\sigma^2}{2\pi \mathrm{j}} \int_{-\mathrm{j}\infty}^{\mathrm{j}\infty} G(s)G(-s)\mathrm{d}s. \tag{5.5.3}$$

The output signal $y(t)$ is known as colored noise.

5.5.2 Erroneous methods in continuous stochastic system simulation

In many academic research fields, continuous stochastic simulations are carried out where the differential equations are driven by random number generators, so as to find the simulation results. In fact, the method is incorrect. An example is given next to show the phenomenon.

Example 5.30. Consider the simple first-order stochastic differential equation

$$a\,\mathrm{d}y(t) = -y(t)\mathrm{d}t + \xi(t)\mathrm{d}t$$

where the initial value for the equation is $y(0) = 0$. In the system with $a = 1$, the input signal $\xi(t)$ is $\mathrm{N}(0, \sigma^2)$ white noise. Find the probability density function of the output signal.

Solutions. It is known from stochastic differential equation theory that[4] the output signal $y(t)$ is also Gaussian, with zero mean and variance $\sigma_y^2 = \sigma^2/2$.

Now let us observe the simulation analysis of such a differential equation. Although derivatives cannot be used mathematically for such equations, in the solution process, they can be regarded as "derivatives":

$$y'(t) = -\frac{1}{a}y(t) + \frac{1}{a}\xi(t).$$

Let us assume that $\sigma = 1$. It is natural to generate signal $\xi(t)$ as a set of random numbers from $\mathrm{N}(0, 1)$, then solve the differential equation. Fixed-step Euler's method can be used, and some different step-sizes are selected. Let us observe the simulation results.

```
>> H=[0.0001, 0.001 0.01 0.1]; N=30000; v=[];
   for h=H    % simulation with different step-sizes
      Xi=randn(1,N+1); f=@(t,y)-y+Xi(fix(t/h)+1);
      [t,y]=ode_euler(f,[0,h,(N-1)*h],0); v=[v,cov(y)]
   end
```

It can be seen that for different step-size h, different variances of the output signal are found as follows:

$$\boldsymbol{v} = \left[1.2264 \times 10^{-5}, 4.146 \times 10^{-4}, 0.0042, 0.0683\right]$$

This is obviously wrong, since the theoretical one is a fixed variance of 1/2. It is obvious that this simulation method has problems.

It can also be shown in theory that with the simulation method, the variance of the output signal is not a constant. Assume that within each step-size, the input signal is a constant e_k, having normal distribution $\mathrm{N}(0,1)$, then discretizing the system model, the following system model can be obtained:

$$y_{k+1} = \mathrm{e}^{-h/a} y_k + (1 - \mathrm{e}^{-h/a})\sigma e_k$$

where h is the step-size. It is found that

$$\mathrm{E}[y_{k+1}^2] = \mathrm{e}^{-2h/a} + 2\sigma \mathrm{e}^{-h/a}\mathrm{E}[e_k y_k] + \sigma^2(1 - \mathrm{e}^{-h/a})^2\mathrm{E}[e_k^2].$$

If the input and output signals are both stationary processes, then $\mathrm{E}[y_{k+1}^2] = \mathrm{E}[y_k^2] = \sigma_y^2$. Since the two signals y_k and e_k are mutually independent, $E[y_k e_k] = 0$. It can be shown that

$$\frac{\sigma^2(1 - \mathrm{e}^{-h/a})^2}{(1 - \mathrm{e}^{-2h/a})} = \frac{\sigma^2(1 - \mathrm{e}^{-h/a})}{1 + \mathrm{e}^{-h/a}}.$$

If $h \to 0$, using power series to approximate respectively the numerator and denominator, it is found that

$$\sigma_y^2 = \frac{h/a + o[(h/a)^2]}{2 + o(h/a)}\sigma^2 \xrightarrow{h \to 0} \frac{h}{2a}\sigma^2. \tag{5.5.4}$$

It can be seen that the variance of the output function is independent of h. It is obvious that the result is incorrect, indicating that common signal generator, which was adopted in many documents, cannot be used at all as the input signal for the continuous system, since the intensity of the signal is not adequate. It can be concluded that if $h \to 0$, the random numbers should be multiplied by $\sqrt{1/h}$, such that the variance of the output signal may be close to the theoretical value.

5.5.3 Discretizing stochastic linear systems

Assume that the mathematical model of a linear stochastic differential equation is given by

$$\mathrm{d}\boldsymbol{x}(t) = \boldsymbol{A}\boldsymbol{x}(t)\mathrm{d}t + \boldsymbol{B}[\boldsymbol{d}(t) + \boldsymbol{\gamma}(t)]\mathrm{d}t, \quad \boldsymbol{y}(t) = \boldsymbol{C}\boldsymbol{x}(t) \tag{5.5.5}$$

where $\boldsymbol{A}$, $\boldsymbol{B}$, and $\boldsymbol{C}$ are compatible matrices. Signal $\boldsymbol{d}(t)$ is deterministic, while $\boldsymbol{\gamma}(t)$ is a Gaussian white signal, satisfying

$$\mathrm{E}[\boldsymbol{\gamma}(t)] = 0, \quad \mathrm{E}[\boldsymbol{\gamma}(t)\boldsymbol{\gamma}^{\mathrm{T}}(t)] = \boldsymbol{V}_\sigma\delta(t-\tau). \tag{5.5.6}$$

Defining a variable $\boldsymbol{\gamma}_{\mathrm{c}}(t) = \boldsymbol{B}\boldsymbol{\gamma}(t)$, it can be shown that $\boldsymbol{\gamma}_{\mathrm{c}}(t)$ is also a Gaussian white noise, satisfying

$$\mathrm{E}[\boldsymbol{\gamma}_{\mathrm{c}}(t)] = 0, \quad \mathrm{E}[\boldsymbol{\gamma}_{\mathrm{c}}(t)\boldsymbol{\gamma}_{\mathrm{c}}^{\mathrm{T}}(t)] = \boldsymbol{V}_{\mathrm{c}}\delta(t-\tau) \tag{5.5.7}$$

where $\boldsymbol{V}_{\mathrm{c}} = \boldsymbol{B}\boldsymbol{V}_\sigma\boldsymbol{B}^{\mathrm{T}}$ is a covariance matrix. Equation (5.5.5) can be regarded as

$$\boldsymbol{x}'(t) = \boldsymbol{A}\boldsymbol{x}(t) + \boldsymbol{B}\boldsymbol{d}(t) + \boldsymbol{\gamma}_{\mathrm{c}}(t), \quad \boldsymbol{y}(t) = \boldsymbol{C}\boldsymbol{x}(t). \tag{5.5.8}$$

The analytical solution of the state variable can be written as

$$\boldsymbol{x}(t) = \mathrm{e}^{-\boldsymbol{A}t}\boldsymbol{x}(t_0) + \int_{t_0}^{t} \mathrm{e}^{\boldsymbol{A}(t-\tau)}d(\tau)\boldsymbol{B}\mathrm{d}\tau + \int_{t_0}^{t} \boldsymbol{\gamma}_{\mathrm{c}}(t)\mathrm{d}\tau. \tag{5.5.9}$$

Assuming that $t_0 = kh$, $t = (k+1)h$, where h is the step-size, and that within each step, the deterministic input signal $d(t)$ is a constant for $\Delta t \leqslant t \leqslant (k+1)h$, one has $\boldsymbol{d}(t) = \boldsymbol{d}(kh)$. The discretized form of (5.5.9) can be written as

$$\boldsymbol{x}[(k+1)h] = \boldsymbol{F}\boldsymbol{x}(kh) + \boldsymbol{G}\boldsymbol{d}(kh) + \boldsymbol{\gamma}_{\mathrm{d}}(kh), \quad \boldsymbol{y}(kh) = \boldsymbol{C}\boldsymbol{x}(kh) \tag{5.5.10}$$

where

$$\boldsymbol{F} = \mathrm{e}^{\boldsymbol{A}h}, \quad \boldsymbol{G} = \int_0^h \mathrm{e}^{\boldsymbol{A}(h-\tau)}\boldsymbol{B}\mathrm{d}\tau \tag{5.5.11}$$

and

$$\boldsymbol{\gamma}_{\mathrm{d}}(kh) = \int_{kh}^{(k+1)h} \mathrm{e}^{\boldsymbol{A}[(k+1)h-\tau]}\boldsymbol{\gamma}_{\mathrm{c}}(t)\mathrm{d}\tau = \int_0^h \mathrm{e}^{\boldsymbol{A}t}\boldsymbol{\gamma}_{\mathrm{c}}[(k+1)h-\tau]\mathrm{d}\tau. \tag{5.5.12}$$

It can be seen that the matrices $\boldsymbol{F}$ and $\boldsymbol{G}$ are the same as the discretized ones in the deterministic system. They can be found easily. If the system contains stochastic signals, the discretized form is different from that of the deterministic systems. It can be shown that $\boldsymbol{y}_{\rm d}(t)$ is also a Gaussian white noise, satisfying

$$\mathrm{E}[\boldsymbol{y}_{\rm d}(kh)] = 0, \quad \mathrm{E}[\boldsymbol{y}_{\rm d}(kh)\boldsymbol{y}_{\rm d}^{\rm T}(jh)] = \boldsymbol{V}\delta_{kj} \tag{5.5.13}$$

where $\boldsymbol{V} = \int_0^h \mathrm{e}^{\boldsymbol{A}t}\boldsymbol{V}_{\rm c}\mathrm{e}^{\boldsymbol{A}^{\rm T}t}\mathrm{d}t$. With the Taylor series technique, it is found that

$$\boldsymbol{V} = \int_0^h \sum_{k=0}^{\infty} \frac{\boldsymbol{R}_k(0)}{k!} t^k \mathrm{d}t = \sum_{k=0}^{\infty} \boldsymbol{V}_k \tag{5.5.14}$$

where $\boldsymbol{R}_k(0)$ and $\boldsymbol{V}_k$ can be recursively evaluated from[67]

$$\begin{cases} \boldsymbol{R}_k(0) = \boldsymbol{A}\boldsymbol{R}_{k-1}(0) + \boldsymbol{R}_{k-1}(0)\boldsymbol{A}^{\rm T}, \\ \boldsymbol{V}_k = \dfrac{h}{k+1}(\boldsymbol{A}\boldsymbol{V}_{k-1} + \boldsymbol{V}_{k-1}\boldsymbol{A}^{\rm T}), \end{cases} \tag{5.5.15}$$

with initial values $\boldsymbol{R}_0(0) = \boldsymbol{R}(0) = \boldsymbol{V}_{\rm c}$ and $\boldsymbol{V}_0 = \boldsymbol{V}_{\rm c}h$. It is known from singular value decomposition that matrix $\boldsymbol{V}$ can be written as $\boldsymbol{V} = \boldsymbol{U}\boldsymbol{\Gamma}\boldsymbol{U}^{\rm T}$, where $\boldsymbol{U}$ is an orthogonal matrix, while $\boldsymbol{\Gamma}$ is a diagonal matrix whose diagonal elements are nonzero. With Cholesky decomposition, $\boldsymbol{V} = \boldsymbol{D}\boldsymbol{D}^{\rm T}$. It is found that $\boldsymbol{y}_{\rm d}(kh) = \boldsymbol{D}\boldsymbol{e}(kh)$, where $\boldsymbol{e}(kh)$ is an $n \times 1$ vector, and $\boldsymbol{e}(kh) = [e_k, e_{k+1}, \dots, e_{k+n-1}]^{\rm T}$, such that each component e_k has the standard normal distribution, i. e., $e_k \sim \mathrm{N}(0, 1)$. The recursive discretized form can be written as

$$\begin{cases} \boldsymbol{x}[(k+1)h] = \boldsymbol{F}\boldsymbol{x}(kh) + \boldsymbol{G}\boldsymbol{d}(kh) + \boldsymbol{D}\boldsymbol{e}(kh), \\ \boldsymbol{y}(kh) = \boldsymbol{C}\boldsymbol{x}(kh). \end{cases} \tag{5.5.16}$$

Based on the above algorithm, the following MATLAB function can be written to discretize linear stochastic differential equations:

```
function [F,G,D,C]=sc2d(G,sig,T)
G=ss(G); G=balreal(G); Gd=c2d(G,T);
A=G.a; B=G.b; C=G.c; i=1;
F=Gd.a; G=Gd.b; V0=B*sig*B'*T; Vd=V0; V1=Vd;
while (norm(V1)<eps)
   V1=T/(i+1)*(A*V0+V0*A'); Vd=Vd+V1; V0=V1; i=i+1;
end
[U,S,V0]=svd(Vd); V0=sqrt(diag(S)); Vd=diag(V0); D=U*Vd;
```

The syntax of the function is $[\boldsymbol{F},\boldsymbol{G},\boldsymbol{D},\boldsymbol{C}]$=sc2d$(G,\sigma,h)$, where G is the system model, σ is the covariance matrix of the input signal, h is the sample time, and $(\boldsymbol{F}, \boldsymbol{G}, \boldsymbol{D}, \boldsymbol{C})$ are the corresponding matrices in the discretized state space model.

In the simulation process, a set of pseudorandom numbers can be generated to form the vector $\boldsymbol{e}(kh)$. Then the state variables $\boldsymbol{x}[(k+1)h]$ can be computed to find the output $y[(k+1)h]$.

Example 5.31. Consider the linear stochastic differential equation

$$\mathrm{d}^4y(t)+10\mathrm{d}^3y(t)+35\mathrm{d}^2y(t)+50\mathrm{d}y(t)+24y(t)=\mathrm{d}^3\xi(t)+7\mathrm{d}^2\xi(t)+24\mathrm{d}\xi(t)+24\xi(t),$$

and assume that the stochastic differential equation has zero mean. If a white noise signal is used to excite the system, use the simulation method to find the statistical properties of the output signal.

Solutions. The transfer function model can be written directly from

$$G(s)=\frac{s^3+7s^2+24s+24}{s^4+10s^3+35s^2+50s+24}.$$

Assuming that the sample time is selected as $h=0.02$ seconds, the following state can be used to discretize the model:

```
>> G=tf([1,7,24,24],[1,10,35,50,24]); % transfer function
   [F,G0,D,C]=sc2d(G,1,0.02)           % system discretization
```

The discretized state space model can be written as

$$\boldsymbol{F}=\begin{bmatrix}0.9838 & -0.00673 & 0.0132 & 0.00129\\ 0.00673 & 0.9883 & 0.07022 & 0.00364\\ 0.0132 & -0.07022 & 0.8653 & -0.0257\\ 0.00129 & -0.0036401 & -0.0257 & 0.9684\end{bmatrix},\quad \boldsymbol{G}_0=\begin{bmatrix}0.01823\\ -0.00355\\ -0.00757\\ -0.000718\end{bmatrix},$$

$$\boldsymbol{D}=\begin{bmatrix}-0.12893 & 0.00088028 & -4.6919\times10^{-6} & 4.6917\times10^{-10}\\ 0.0251 & -0.0012 & -1.3573\times10^{-5} & 2.3791\times10^{-9}\\ 0.05356 & 0.002635 & -5.2322\times10^{-6} & -8.8812\times10^{-10}\\ 0.00508 & 0.0005 & 3.1358\times10^{-6} & 9.5176\times10^{-9}\end{bmatrix}.$$

For the obtained discrete state space model, the following commands can be used to perform simulation in MATLAB, where the total number of simulation points is selected as $n=30\,000$. If this number is small, then statistical analysis results may be meaningless.

```
>> n=30000; e=randn(n+4,1); e=e-mean(e);
   y=zeros(n,1); x=zeros(4,1); d0=0;
   for i=1:n, x=F*x+G0*d0+D*e(i:i+3); y(i)=C*x; end
   T=0.02; t=0:T:(n-1)*T;
   plot(t,y), v=norm(G) % draw output signal and compute 2-norm
```

It can be seen that the $\mathcal{H}_2$ of the output is $v = 0.6655$. The time response curve is shown in Figure 5.25. It can be seen that the output signals are in a mess, since the output is stochastic. Therefore one curve of the output signal is usually meaningless. Statistical analysis should be performed instead.

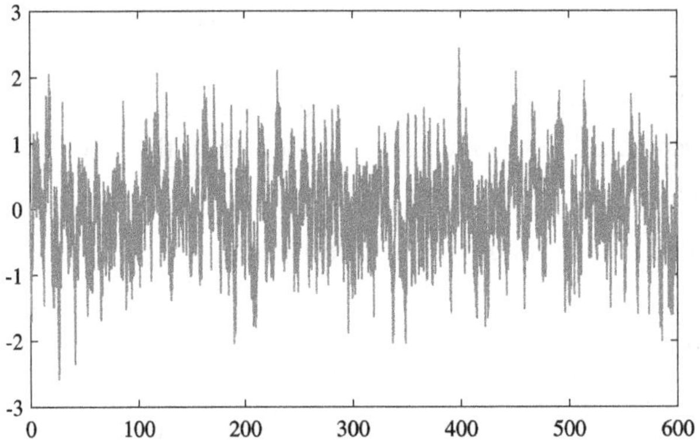

Figure 5.25: Time responses of the output signal.

Since the variance of the output signal is known to be v^2, it can be shown in theory that[4] the probability density function of the output signal is

$$p(y) = \frac{1}{\sqrt{2\pi}v} e^{-y^2/(2v^2)}.$$

Statistical methods can also be used to obtain the histogram of the response data. Specifically, the range of the output $(-2.5, 2.5)$ is divided equally into subintervals of width $w = 0.2$. The total number of points falling into each subinterval can be found, and then, dividing by nw, the numerical probability density function from the data can be found and superimposed on the theoretical probability density function curve, as shown in Figure 5.26. The results from simulation match the theoretical results very well.

```
>> w=0.2; x=-2.5:w:2.5;
   y1=hist(y,x); bar(x,y1/n/w); % draw histogram
   x1=-2.5:0.05:2.5;
   y2=1/sqrt(2*pi)/v*exp(-x1.^2/2/v^2); % compute analytical PDF
   line(x1,y2)  % superimposed probability density function
```

Simulation of nonlinear stochastic differential equations are far more complicated than for other differential equations. They are not discussed here. In Chapter 9, block diagram based simulation methods for nonlinear stochastic differential equations will be presented.

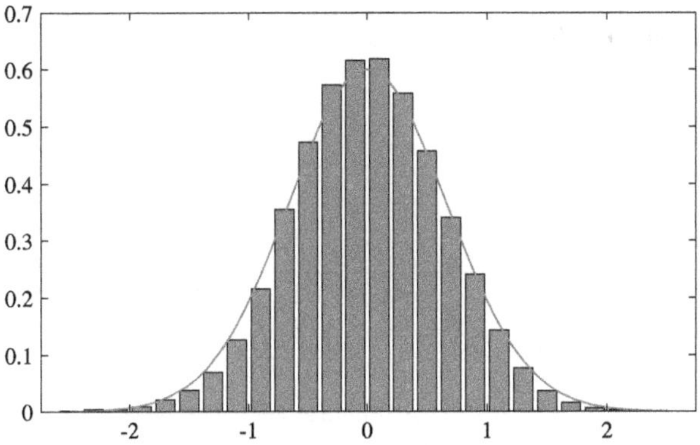

Figure 5.26: System responses under stochastic inputs.

5.6 Exercises

5.1 The simplified dynamic model of a catalytic fluidized bed is[42]

$$\begin{cases} x'(t) = 1.30(y_2(t) - x(t)) + 2.13 \times 10^6 k y_1(t), \\ y_1'(t) = 1.88 \times 10^3 (y_3(t) - y_1(t)(1+k)), \\ y_2'(t) = 1752 - 269 y_2(t) + 267 x(t), \\ y_3'(t) = 0.1 + 320 y_1(t) - 321 y_2(t) \end{cases}$$

where $x(0) = 761$, $y_1(0) = 0$, $y_2(0) = 600$, $y_3(0) = 0.1$, and $k = 0.006\mathrm{e}^{20.7-15\,000/x(t)}$. If $t \in (0, 100)$, find the numerical solution of the stiff differential equation.

5.2 Solve the following linear stiff differential equation:[23]

$$\boldsymbol{y}'(t) = \begin{bmatrix} -2a & a & & & & \\ 1 & -2 & 1 & & & \\ 0 & 1 & -2 & 1 & & \\ & & \ddots & \ddots & \ddots & \\ & & & 1 & -2 & 1 \\ & & & & b & -2b \end{bmatrix} \boldsymbol{y}(t) + \begin{bmatrix} 0 \\ 0 \\ 0 \\ \vdots \\ 0 \\ b \end{bmatrix}$$

where $a = 900$ and $b = 1\,000$. If the initial vector $\boldsymbol{y}(0)$ is zero and the order of the coefficient matrix is $n = 9$, solve the stiff differential equation in the interval $t \in (0, 120)$.

5.3 Solve the following nonlinear stiff differential equation:[23]

$$\begin{cases} y_1'(t) = y_2(t), \\ y_2'(t) = y_3(t), \\ y_3'(t) = y_4(t), \\ y_4'(t) = (y_1^2(t) - \sin y_1(t) - \Gamma^4) + \left(\dfrac{y_2(t)y_3(t)}{y_1^2(t)+1} - 4\Gamma^3\right) \\ \qquad + (1 - 6\Gamma^2)y_3(t) + (10\mathrm{e}^{-y_4^2(t)} - 4\Gamma)y_4(t) + 1 \end{cases}$$

where $\Gamma = 100$ and $t \in (0, 1)$. Assume that the initial state vector is zero.

5.4 Solve the following stiff differential equation:[33]

$$\begin{cases} y_1'(x) = -2\,000(\cos xy_1(x) + \sin xy_2(x) + 1), \\ y_2'(x) = -2\,000(-\sin xy_1(x) + \cos xy_2(x) + 1) \end{cases}$$

where $y_1(0) = 1$, $y_2(0) = 0$ and $t \in [0, \pi/2]$.

5.5 Numerically solve the following linear stiff differential equation:[42]

$$\begin{cases} x_1'(t) = -k_1x_1(t) + k_2y(t), \\ x_2'(t) = -k_4x_2(t) + k_3y(t), \\ y'(t) = k_1x_1(t) + k_4x_2(t) - (k_2 + k_3)y(t) \end{cases}$$

where $x_1(0) = y(0) = 0$, $x_2(0) = 1$, and $k_1 = 8.430327\times10^{-10}$, $k_2 = 2.9002673\times10^{11}$, $k_3 = 2.4603642 \times 10^{10}$, $k_4 = 8.760058 \times 10^{-6}$, $t \in (0, 100)$. In fact, since this is a linear differential equation, function `dsolve()` can be used to find the quasi-analytical solution. Assess the accuracy of the numerical solutions.

5.6 Solve the following stiff differential equation:[33]

$$\begin{cases} y_1'(x) = -0.04y_1(x) + 10^4y_2(x)y_3(x), \\ y_2'(x) = 0.04y_1(x) - 10^4y_2(x)y_3(x) - 3\times10^7y_2^2), \\ y_3'(x) = 3\times10^7y_2^2) \end{cases}$$

where $y_1(0) = 1$, $y_2(0) = 0$, $y_3(0) = 0$, and $t \in [0, 0.3]$.

5.7 The mathematical model of a chemical reaction is[33]

$$\begin{cases} y_1'(t) = -Ay_1(t) - By_1(t)y_3(t), \\ y_2'(t) = Ay_1(t) - MCy_2(t)y_3(t), \\ y_3'(t) = Ay_1(t) - By_1(t)y_3(t) - MCy_2(t)y_3(t) + Cy_4(t), \\ y_4'(t) = By_1(t)y_3(t) - Cy_4(t) \end{cases}$$

where $A = 7.89\times10^{-10}$, $B = 1.1\times10^7$, $C = 1.13\times10^3$, $M = 10^6$, $y_1(0) = 1.76\times10^{-3}$, $y_2(0) = y_3(0) = y_4(0) = 0$, and $t \in (0, 1\,000)$. Solve the stiff differential equation. If the interval is increased to $t \in (0, 10^{13})$, solve the differential equation again.

5.8 Use regular and stiff differential equation solvers to solve the following differential equation:[32]

$$\begin{cases} y_1'(x) = 77.27\left[y_2(x) + y_1(x)(1 - 8.375\times 10^{-6}y_1(x) - y_2(x))\right],\\ y_2'(x) = \dfrac{1}{77.27}\left[y_3(x) - (1 + y_1(x))y_2(x)\right],\\ y_3'(x) = 0.161(y_1(x) - y_3(x)) \end{cases}$$

where $y_1(0) = 1$, $y_2(0) = 2$, $y_3(0) = 3$, and $t \in [0, 0.3]$.

5.9 For the following differential equation,[53] if $u_1(0) = 45$, $u_2(0) = 30$, $u_3(0) = u_4(0) = 0$, and $\mathrm{g} = 9.81$, solve it and draw the time responses of the states:

$$\begin{cases} u_1'(t) = u_3(t),\\ u_2'(t) = u_4(t),\\ 2u_3'(t) + \cos(u_1(t) - u_2(t))u_4'(t) = -\mathrm{g}\sin u_1(t) - \sin(u_1(t) - u_2(t))u_4^2(t),\\ \cos(u_1(t) - u_2(t))u_3'(t) + u_4'(t) = -\mathrm{g}\sin u_2(t) + \sin(u_1(t) - u_2(t))u_3^2(t). \end{cases}$$

5.10 Solve the following differential-algebraic equation:[11]

$$\begin{cases} x_1'(t) = x_3(t) - 2x_1(t)y_2(t),\\ x_2'(t) = x_4(t) - 2x_2(t)y_2(t),\\ x_3'(t) = -2x_1(t)y_1(t),\\ x_4'(t) = -\mathrm{g} - 2x_2(t)y_1(t),\\ x_1^2(t) + x_2^2(t) = 1,\\ x_1(x)x_3(t) + x_2(t)x_4(t) = 0, \end{cases}$$

with known initial values $\boldsymbol{x}(0) = [1, 0, 0, 0]^{\mathrm{T}}$, $\boldsymbol{y}(t) = \boldsymbol{0}$. The constant $\mathrm{g} = 9.81$, and the interval of interest is $t \in (0, 6)$. Find the numerical solution of the differential-algebraic equation.

5.11 Solve the following time-varying differential-algebraic equation:[3]

$$\begin{cases} x_1'(t) = (\alpha - 1/(2-t))x_1(t) + (2-t)\alpha z(t) + (3-t)/(2-t),\\ x_2'(t) = (1-\alpha)x_1(t)/(t-2) - x_2(t) + (\alpha - 1)z(t) + 2\mathrm{e}^{-t},\\ (t+2)x_1(t) + (t^2 - 4)x_2(t) - (t^2 + t - 2)\mathrm{e}^t = 0 \end{cases}$$

where $x_1(0) = x_2(0) = 1$ and α is a constant. The analytical solutions are $x_1(t) = x_2(t) = \mathrm{e}^t$, $z(t) = -\mathrm{e}^t/(2-t)$.

5.12 Solve the following implicit differential equation:[40]

$$\begin{cases} 0 = y_1'(t) - y_2(t), \\ 0 = (m_1 + m_2)y_2'(t) - m_2Ly_6'(t)\sin y_5(t) - m_2Ly_6^2(t)\cos y_5(t), \\ 0 = y_3'(t) - y_4(t), \\ 0 = (m_1 + m_2)y_4'(t) + m_2Ly_6'(t)\cos y_5(t) - m_2Ly_6'(t)\sin y_5(t) + (m_1 + m_2)g, \\ 0 = y_5'(t) - y_6(t), \\ 0 = -Ly_2'(t)\sin y_5(t) - Ly_4'(t)\cos y_5(t) - L^2y_6^2(t) + gL\cos y_5(t) \end{cases}$$

where $m_1 = m_2 = 0.1$, $L = 1$, $g = 9.81$, $\boldsymbol{y}_0 = [0, 4, 2, 20, -\pi/2, 2]^{\mathrm{T}}$, and the time interval is $t \in (0, 4)$.

5.13 In Example 5.21, an index 4 differential-algebraic equation has been analyzed, and the index has been reduced to 2. Try to further reduce the index and see whether it can be reduced to an index-1 differential-algebraic equation. Solve the equation and see whether it can be solved with a high efficiency method.

5.14 At the website [52], many benchmark problems on stiff differential equations and differential-algebraic equations are provided. The readers may visit that website to download relevant problems, and test whether the problems can be solved with MATLAB. What are the accuracy and efficiency when solving these problems?

5.15 Solve the following linear switched differential equation:

$$\begin{cases} x_1'(t) = f(x_1(t)) + x_2(t), \\ x_2'(t) = -x_1(t) \end{cases}$$

where $x_1(0) = x_2(0) = 5$. Function $f(x_1(t))$ is given piecewise:

$$f(x_1(t)) = \begin{cases} -4x_1(t), & \text{if } x_1(0) > 0, \\ 2x_1(t), & \text{if } -1 \leqslant x_1(0) \leqslant 0. \\ -x_1(t) - 3, & \text{if } x_1(0) < -1. \end{cases}$$

5.16 Consider a linear switched system with state feedback[46]

$$\boldsymbol{x}'(t) = \boldsymbol{A}_\sigma \boldsymbol{x}(t) + \boldsymbol{B}_\sigma u(t), \quad u(t) = \boldsymbol{k}_\sigma \boldsymbol{x}(t)$$

where $\sigma \in \{1, 2\}$. The two subsystems are respectively

$$\boldsymbol{A}_1 = \begin{bmatrix} 1 & 0 \\ 1 & 1 \end{bmatrix}, \quad \boldsymbol{A}_2 = \begin{bmatrix} 1 & 1 \\ 0 & 1 \end{bmatrix}, \quad \boldsymbol{B}_1 = \begin{bmatrix} 1 \\ 0 \end{bmatrix}, \quad \boldsymbol{B}_2 = \begin{bmatrix} 0 \\ 1 \end{bmatrix},$$

and the two state feedback vectors are $\boldsymbol{k}_1 = [6, 9]$ and $\boldsymbol{k}_2 = [9, 6]$. It is known that the condition to switch from subsystem 1 to 2 is $|x_1(t)| = 0.5|x_2(t)|$, and the condition to switch from subsystem 2 to 1 is $|x_1(t)| = 2|x_2(t)|$. If the initial states are $\boldsymbol{x}_0 = [100, 100]^{\mathrm{T}}$, solve the switched system and draw the phase plane trajectory.

5.17 Solve the following discontinuous differential equation,[32] with initial value of $y(0) = 0.3$:

$$y'(t) = \begin{cases} t^2 + 2y^2(t), & \text{if } (t + 0.05)^2 + \big(y(t) + 0.15\big)^2 \leqslant 1, \\ 2t^2 + 3y^2(t) - 2, & \text{if } (t + 0.05)^2 + \big(y(t) + 0.15\big)^2 > 1. \end{cases}$$

5.18 Consider a linear feedback control system with unit negative feedback, as shown in Figure 5.27, where the plant and controller models are given as

$$G(s) = \frac{s^3 + 7s^2 + 24s + 24}{s^4 + 10s^3 + 35s^2 + 50s + 24}, \quad G_c(s) = \frac{s + 0.1}{0.1s + 1}.$$

If a Gaussian white noise with variance 1 is applied to the system as signal $u(t)$, simulate the system and find the probability density function of signal $e(t)$, as well as the variance. (Hint: From feedback control theory it is known that the equivalent transfer function from $r(t)$ to $e(t)$ can be derived from $\widetilde{G}(s) = 1/(1 + G(s)G_c(s))$. Discretize the model and start a simulation study.)

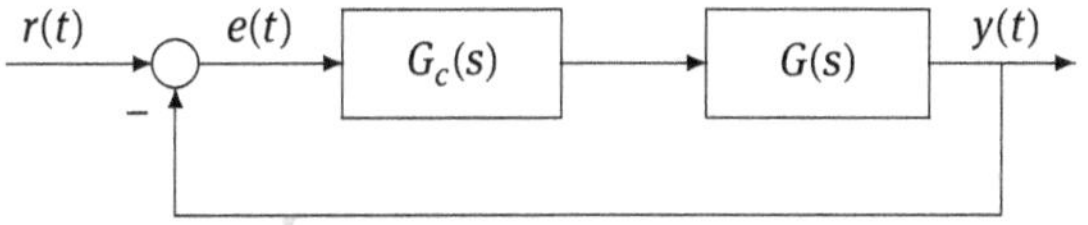

Figure 5.27: Block diagram of a typical feedback control system.

6 Delay differential equations

In the ordinary differential equations studied so far, the standard form of $\boldsymbol{x}'(t) = \boldsymbol{f}(t, \boldsymbol{x}(t))$ was used. All the signals in the equation happen exactly at time t. If the signals contain values from the past, the equations are referred to as delay differential equations. In this chapter, various delay differential equations are discussed, including standard delay differential equations, neutral-type delay differential equations, and delay differential equations with variable delays.

In Section 6.1, the delay differential equations with delay constants are introduced first. The solver provided in MATLAB is used to solve the delay differential equations directly. The cases of zero initial value problems and constant history functions are respectively considered. In Section 6.2, the solutions of differential equations with complicated delay forms are presented. Three types of delays are mainly considered, that is, time-dependent delays, state-dependent delays, and generalized delays. In Section 6.3, neutral-type delay differential equations are discussed. Examples are introduced to demonstrate various neutral-type delay differential equations. In Section 6.4, the Volterra integro-differential equations of certain types are discussed and explored.

6.1 Numerical solutions of delay differential equations with constant delays

As the ordinary differential equations studied earlier, delay differential equations are also commonly encountered dynamic modeling problems in science and engineering. In this section, the general form of delay differential equations is given first. Then a MATLAB based solution method is used to find solutions of the equations, and observe the impact of the history information on delay differential equations.

Definition 6.1. If in a differential equation, the information of the unknown signal $\boldsymbol{x}(t)$ at the current time instance t and the information in the past are involved, the differential equation is referred to as a delay differential equation (DDE).

6.1.1 Ordinary and delay differential equations

In Example 1.6, the latent period of a disease was introduced, so that the original differential equations were changed to delay differential equations. Another example is that, in practical engineering applications, sensors are used to measure a certain signal. If these sensors have their own time delay, the current measured data may be the actual value of the signal at a previous time. Strictly speaking, this is also a delay differential equation model. In some applications, the delays may be too short, and they can be neglected. Therefore the delay differential equations can be

https://doi.org/10.1515/9783110675252-006

approximated by differential equations. In some specific fields, the delays cannot be neglected. Therefore the modeling of delay differential equations is considered, and the solution methods for delay differential equations are expected.

Delay differential equations have various formats. The most commonly used ones are those with delay constants. In many literature sources, delay differential equations are also referred to as functional differential equations (FDEs). Apart from the initial values of the states, sometimes the information of the unknown variables when $t \leqslant t_0$ are also expected. This information is also known as history functions. In this section, numerical methods for this type of delay differential equation are mainly presented.

If the delay differential equation only contains the delay terms at constant time instances, one may consider converting those which can be manually converted into differential equations with no delays. In this section, an example is used to demonstrate this conversion process.

Example 6.1. Consider the delay differential equation with history functions[2]

$$u''(x) = -\frac{1}{16}\sin u(x) - (x+1)u(x-1) + x$$

where $0 \leqslant x \leqslant 2$. It is also known that

$$u(x) = x - \frac{1}{2}, \quad -1 \leqslant x \leqslant 0, \quad u(2) = -\frac{1}{2}.$$

Solutions. Since the delay term $u(x-1)$ exists, the equation here cannot be solved with the solvers discussed earlier. Special manipulation is needed. Here two cases are considered for this differential equation.

(1) Let $t = x$ and $u_1(t) = u(x)$, then, when $-1 \leqslant x \leqslant 0$, $u(x-1) = u(t-1) = (t-1) - 1/2 = t - 3/2$. It can be seen that

$$u_1''(t) = -\frac{1}{16}\sin u_1(t) - (t+1)\left(t - \frac{3}{2}\right) + t.$$

(2) Let $t = x - 1$, that is, $x = t + 1$, and $u_2(t) = u(x) = u(t+1)$, then

$$u''(x) = u''(t+1) = -\frac{1}{16}\sin u(t+1) - (t+2)u_1(t) + t + 1.$$

Substituting $u_2(t)$ back into the equation,

$$u_2''(t) = -\frac{1}{16}\sin u_2(t) - (t+2)u_1(t) + t + 1.$$

The above two equations are now both defined in the interval $0 \leqslant t \leqslant 1$. Therefore the following differential equations can be found directly:

$$\begin{cases} u_1''(t) = -\dfrac{1}{16}\sin u_1(t) - (t+1)\left(t - \dfrac{3}{2}\right) + t, \\ u_2''(t) = -\dfrac{1}{16}\sin u_2(t) - (t+2)u_1(t) + t + 1 \end{cases}$$

with boundary conditions $u_1(0) = -1/2$, $u_2(1) = -1/2$, $u_1(1) = u_2(0)$, and $u_1'(1) = u_2'(0)$. It can be seen that in the manually transformed formula, an auxiliary function is introduced and the delay term is eliminated. A high-order differential equation is found.

In fact, the manual method can be used to eliminate delay terms, if there are only a finite number of delay terms, such that ordinary differential equations can be found. Unfortunately, this kind of manual conversion may be too tedious and error-prone. A dedicated solver is needed for delay differential equations.

6.1.2 Delay differential equation with zero history functions

The mathematical forms of delay differential equations with constant delays are introduced in this section. Then a solver provided in MATLAB for delay differential equations is described, and the solutions of delay differential equations with zero history information are illustrated.

Definition 6.2. The general form of a delay differential equation with constant delays is given by

$$\boldsymbol{x}'(t) = \boldsymbol{f}(t, \boldsymbol{x}(t), \boldsymbol{x}(t-\tau_1), \boldsymbol{x}(t-\tau_2), \ldots, \boldsymbol{x}(t-\tau_m)) \tag{6.1.1}$$

where $\tau_i \geqslant 0$ are the delay constants of the state variables $\boldsymbol{x}(t)$.

In the mathematical description, $\boldsymbol{x}(t)$ contains the state variables at the current time instance t, while $\boldsymbol{x}(t-\tau_i)$ can be understood as the state variable vector τ_i seconds ago. Assuming that there are m delay constants, $\boldsymbol{\tau} = [\tau_1, \tau_2, \ldots, \tau_m]$, and the delay constants are given values, based on the information, the values of the first-order derivative $\boldsymbol{x}'(t)$ at time instance t can be computed. Similar to the cases in other differential equation solutions, the standard form of the delay differential equations should be written by the user, in the format understandable in MATLAB.

An implicit Runge–Kutta algorithm solver is provided in MATLAB, named `dde23()`; the solver can be used to solve delay differential equations with the following syntax:

`sol=dde23(`f_1`,`$\boldsymbol{\tau}$`,`f_2`,[`t_0`,`t_n`],options)`

where, as before, `options` are the controls in the solver, whose initial members can be extracted with function `ddeset()`. The function is similar to that in the `odeset()` solver. The member names are similar, for instance, `AbsTol`, `RelTol`, `OutputFcn`, and `Events`. The definitions are the same as for the differential equation solvers.

In the statements, f_1 is a MATLAB function to describe the delay differential equation, whose syntax will be demonstrated later through examples. Function f_2 is used to express the state variables when $t \leqslant t_0$. If they are functions, then MATLAB function handles are used. If they are constants, then constant vectors can be used directly.

The returned argument `sol` is a structured variable whose `sol.x` member is the time vector $\boldsymbol{t}$ and the member `sol.y` is the matrix $\boldsymbol{x}$ composed of the solutions at different time instances. The format is different from that used in the solver `ode45()`. The data in `sol.y` is arranged in rows. It can be seen that the syntaxes of the functions are not quite unified. It is expected that in later versions they can be unified.

In describing the delay differential equations, apart from the regular scalar t and state vector $\boldsymbol{x}$, a matrix $\boldsymbol{Z}$ is also used as the input argument. The kth column $\boldsymbol{Z}(:,k)$ stores the values of the state at the delay constant τ_k, i. e., $\boldsymbol{x}(t-\tau_k)$.

Examples are used next to demonstrate the solutions of simple delay differential equations and control parameters, so as to provide useful suggestions for delay differential equation solutions.

Example 6.2. For the given delay differential equations with constant delays

$$\begin{cases} x'(t) = 1 - 3x(t) - y(t-1) - 0.2x^3(t-0.5) - x(t-0.5), \\ y''(t) + 3y'(t) + 2y(t) = 4x(t) \end{cases}$$

where, when $t \leqslant 0$, $x(t) = y(t) = y'(t) = 0$, find their numerical solutions.

Solutions. It can be seen that the values of the variables $x(t)$ and $y(t)$ at time instances $t, t-1$, and $t-0.5$ are involved. Therefore, dedicated delay differential equation solvers are expected. To find the numerical solutions, the differential equations must be first converted to the standard forms of the first-order explicit differential equations.

A straightforward method for the conversion is to introduce a set of state variables $x_1(t) = x(t)$, $x_2(t) = y(t)$, and $x_3(t) = y'(t)$. Then the first-order explicit differential equations can be written as follows:

$$\begin{cases} x_1'(t) = 1 - 3x_1(t) - x_2(t-1) - 0.2x_1^3(t-0.5) - x_1(t-0.5), \\ x_2'(t) = x_3(t), \\ x_3'(t) = 4x_1(t) - 2x_2(t) - 3x_3(t). \end{cases}$$

In this equation, two delay constants $\tau_1 = 1$ and $\tau_2 = 0.5$ are involved. It can be seen from the first equation that the first delay τ_1 corresponds to the first column of the state matrix $\boldsymbol{Z}$. Therefore $x_2(t-\tau_1)$ is the first column, second row element of matrix $\boldsymbol{Z}$, denoted as $\boldsymbol{Z}(2,1)$. If the state variable $x_1(t-\tau_2)$ is used, the element $\boldsymbol{Z}(1,2)$ should be extracted. Therefore the standard form of the delay differential equation can be written as

$$\begin{cases} x_1'(t) = 1 - 3x_1(t) - \boldsymbol{Z}(2,1) - 0.2\boldsymbol{Z}^3(1,2) - \boldsymbol{Z}(1,2), \\ x_2'(t) = x_3(t), \\ x_3'(t) = 4x_1(t) - 2x_2(t) - 3x_3(t). \end{cases}$$

The following anonymous function can be written to describe the delay differential equations:

```
>> f=@(t,x,Z)[1-3*x(1)-Z(2,1)-0.2*Z(1,2)^3-Z(1,2);
              x(3);
              4*x(1)-2*x(2)-3*x(3)];
```

Since it is known that the history information of the unknown signal $\boldsymbol{x}(t)$ at time instances $t \leqslant 0$ is all zero, this kind of problem is also known as a problem with zero history functions. The zero history information can be described by a zero vector f_2. Therefore the following statements can be tried to solve the delay differential equations. Unfortunately, the statements cannot be used in finding the numerical solutions of the delay differential equations.

```
>> ff=ddeset; ff.RelTol=100*eps; ff.AbsTol=100*eps;
   tau=[1 0.5]; x0=zeros(3,1);     %set the two delay constants
   tic, tx=dde23(f,tau,x0,[0,10],ff); toc %solve equation
   plot(tx.x,tx.y), length(tx.x) %draw the state variables
```

It seems that there is no problem in the above statements. After an extremely long period, nothing is found. What happens in this kind of phenomena? In the previous presentations, the readers became familiar with the error tolerance `100*eps`. Unfortunately, such tough error tolerance cannot be used in solving delay differential equations. The error tolerances must be increased for delay differential equation solvers. The setting of the error tolerance and is impact on solution efficiency will be explored through examples.

Example 6.3. Consider the above example again. Solve the delay differential equation with different error tolerance and assess the efficiency.

Solutions. With the following MATLAB commands, the numerical solution of the delay differential equation can be found as shown in Figure 6.1:

```
>> f=@(t,x,Z)[1-3*x(1)-Z(2,1)-0.2*Z(1,2)^3-Z(1,2);
              x(3);
              4*x(1)-2*x(2)-3*x(3)]; %delay differential equation
   ff=ddeset; ee=100*eps; ff.RelTol=ee; ff.AbsTol=ee;
   tau=[1 0.5]; x0=zeros(3,1);     %setting the delay vector
   tic, tx=dde23(f,tau,x0,[0,10],ff); toc %solve the equation
   plot(tx.x,tx.y), length(tx.x) %draw the state variables
```

Note that in the returned variables tx.$\boldsymbol{y}$, the results are stored in rows. If the signal $y(t)$ in the original equation is expected, the variable tx.$\boldsymbol{y}$`(2,:)` should be used, rather than tx.$\boldsymbol{y}$`(:,2)`.

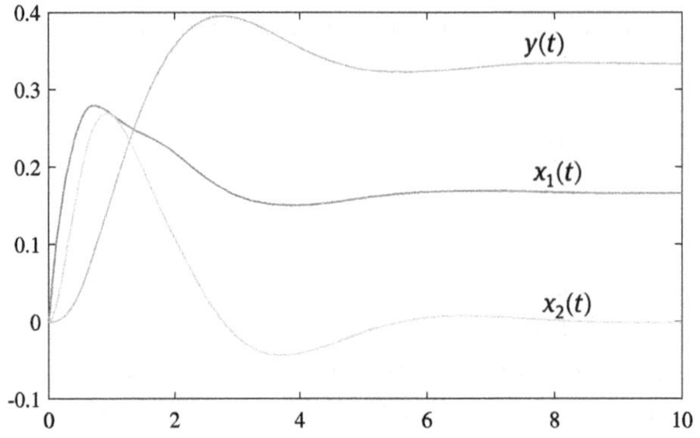

Figure 6.1: Numerical solutions of delay differential equations.

Since the error tolerance is set too tough, the solution process is rather time consuming. The total elapsed time is 89.9 seconds, with 60 006 points computed. Different tolerance `ee` values are tried and compared in Table 6.1.

Table 6.1: Error tolerance and elapsed time.

tolerance `ee`	2.2204×10^{-14}	10^{-13}	10^{-12}	10^{-11}	10^{-10}	10^{-9}	10^{-8}	default
elapsed time	89.9	60.16	11.41	2.84	0.78	0.18	0.10	0.041
number of points	60 006	36 339	16 871	7 833	3 639	1 692	789	70

In the `dde23()` function call, in order to speed up the computation process, the expected precision must be reduced. For instance, under the default setting, the elapsed time is 0.041 seconds, with only 70 points computed. If the obtained curve is superimposed on the exact one, no differences can be witnessed.

```
>> tic, tx=dde23(f,tau,x0,[0,10]); toc %solve the equations
   line(tx.x,tx.y), length(tx.x)          %draw the state variables
```

Although the relationship of error tolerance and elapsed time is summarized, this information is obtained from a particular example. It is not really of any significance, but the trends are meaningful. In normal cases, the value of `ee` can be set to 10^{-10}, and used to solve the delay differential equations. If the solutions can be found immediately, the error tolerance should be reduced slightly, so as to find solutions of higher accuracy. If the solutions cannot be obtained in a long time, the error tolerance should be increased so as to find the approximate solutions. In latter discussions, if not specially indicated, the error tolerance is uniformly set to 10^{-10}.

Example 6.4. In a typical feedback control system shown in Figure 6.2, the plant model $G(s)$ and the controller $G_c(s)$ are described by the following transfer function

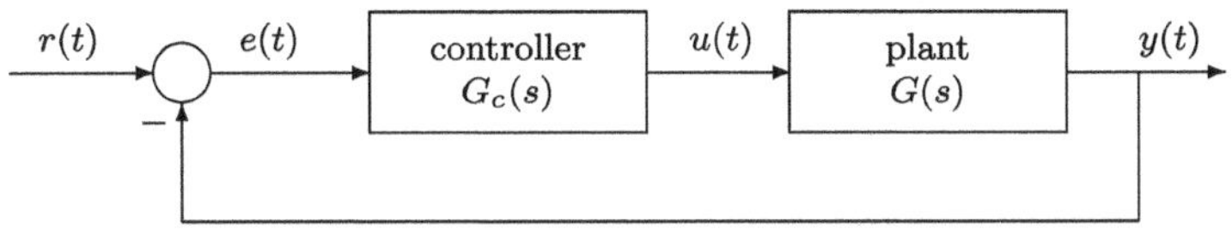

Figure 6.2: Typical feedback control system.

models:

$$G(s) = \frac{e^{-s}}{(s+1)(s+2)} = \frac{e^{-s}}{s^2+3s+2}, \quad G_c(s) = 1.66 + \frac{0.91}{s}.$$

Draw the step responses of the system under zero initial values.

Solutions. With the properties of Laplace transform, the transfer function models can be converted back to differential equations. Since there exists a delay in the plant model, the corresponding model is a delay differential equation. Let us observe first the differential equation model from signal $u(t)$ to signal $y(t)$. It is easy to find that the corresponding delay differential equation model can be written directly as

$$y''(t) + 3y'(t) + 2y(t) = u(t-1).$$

Now let us see the mathematical model of $u(t)$, directly derived as

$$\begin{aligned} u'(t) &= 1.66(r(t) - y(t))' + 0.91(r(t) - y(t)) \\ &= 1.66\delta(t) - 1.66y'(t) + 0.91 - 0.91y(t) \end{aligned}$$

where $\delta(t)$ is a Dirac function. Dirac function is named after a British theoretical physicist Paul Adrien Maurice Dirac (1902–1984), whose definition is the derivative of the step signal. It can be understood as having infinite magnitude at $t = 0$, while the function value is zero elsewhere.

Selecting the state variables $x_1(t) = y(t)$, $x_2(t) = y'(t)$, and $x_3(t) = u(t)$, the standard form of the explicit differential equations can be written as

$$\mathbf{x}'(t) = \begin{bmatrix} x_2(t) \\ x_3(t-1) - 3x_2(t) - 2x_1(t) \\ 1.66\delta(t) - 1.66x_2(t) + 0.91 - 0.91x_1(t) \end{bmatrix}.$$

Although function `dirac()` is provided in MATLAB, it can only be used in symbolic computation, and the function value is `Inf` at time $t = 0$. Therefore the function cannot be effectively used in simulation. An alternative is to assume the width of the signal being 10^{-4} and the magnitude of 10^4. It is an approximate Dirac function $\delta(t)$, and the MATLAB expression can be expressed as `1e4-1e4*heaviside(t-1e-4)`. An anonymous function can be written to describe the delay differential equations. Then the equations can be solved, and the output signal is obtained as shown in Figure 6.3.

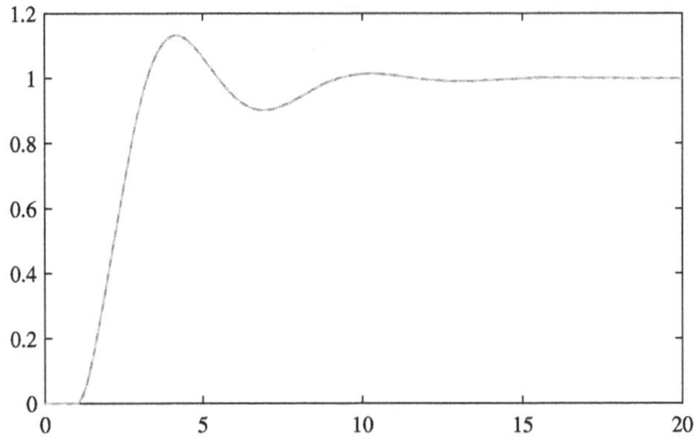

Figure 6.3: Output curve of the delay control system.

For the same problem, Control System Toolbox function can be used to solve the problem directly. The two sets of curves coincide.

```
>> f=@(t,x,Z)[x(2); Z(3)-3*x(2)-2*x(1);
      1.66*(1e4-1e4*heaviside(t-1e-4))-1.66*x(2)+0.91-0.91*x(1)];
   ff=ddeset; ee=1e-10; ff.RelTol=ee; ff.AbsTol=ee;
   tau=1; x0=zeros(3,1);                    % delay constants and initial values
   tic, tx=dde23(f,tau,x0,[0,20],ff); toc  % solve the equation
   s=tf('s'); G=exp(-s)/(s^2+3*s+2); Gc=1.66+0.91/s;
   [y,t]=step(feedback(G*Gc,1),20);        % linear system step response
   plot(t,y,tx.x,tx.y(1,:),'--')           % comparisons of the two methods
```

Note that in the conversion from transfer functions to differential equations, we cannot forget the $\delta(t)$ function term, otherwise wrong results are obtained.

6.1.3 Nonzero history functions

In the previous examples, it is always assumed that when $t \leqslant 0$, the history function is $\boldsymbol{x}(t) = \boldsymbol{0}$. In real applications, when $t \leqslant 0$, the state variables are not always so simple. They can be given constants or functions.

In the solving statement, if $\boldsymbol{x}_0$ is a constant vector, it means that, when $t \leqslant t_0$, the history values of $\boldsymbol{x}$ are kept as constant $\boldsymbol{x}_0$. Care must be taken in understanding such differential equations. An example is given to demonstrate the constant history function and its impact on the delay differential equations.

Example 6.5. Consider again the delay differential equations in Example 6.2. If the history values of the three states are respectively $x_1(t) = -1$, $x_2(t) = 2$, and $x_3(t) = 0$, when $t \leqslant 0$, solve the delay differential equations.

Solutions. Compared with the statements in Example 6.2, the only difference is that the zero vector is changed to vector $\boldsymbol{x}_0$. The solutions of the delay differential equations can be found directly, as shown in Figure 6.4. The elapsed time is 1.24 seconds.

```
>> f=@(t,x,Z)[1-3*x(1)-Z(2,1)-0.2*Z(1,2)^3-Z(1,2);
              x(3);
              4*x(1)-2*x(2)-3*x(3)]; %delay differential equations
   ff=odeset; ee=1e-10; ff.RelTol=ee; ff.AbsTol=ee;
   tau=[1 0.5]; x0=[-1; 2; 0];   %set the delay vector
   tic, tx=dde23(f,tau,x0,[0,10],ff); toc %solve the equations
   plot(tx.x,tx.y), length(tx.x) %draw the state variables
```

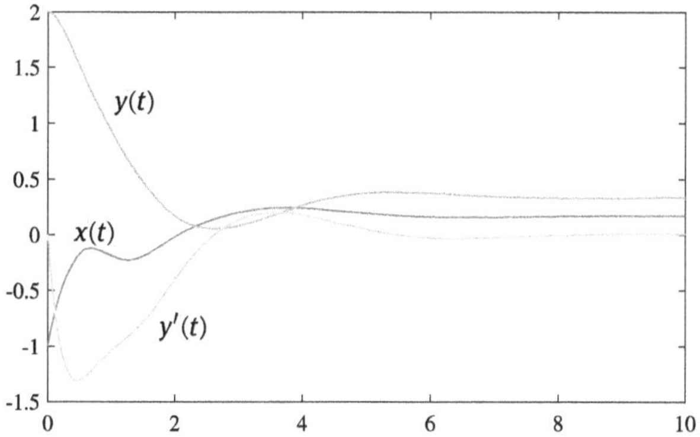

Figure 6.4: Numerical solution of the delay differential equations.

If the history values are not constants, but rather functions of time, or other state related functions, anonymous or MATLAB functions can be used to describe the "pre-history" functions, and express them in function handle f_2. Then the delay differential equations can be solved numerically. For the users, solving these differential equations is no more complicated than solving the zero initial value problems, since the history functions can be described directly. Apart from the history functions, the other statements are exactly the same. Examples are used next to demonstrate the solutions of delay differential equations with time-varying history functions.

Example 6.6. Consider again the delay differential equation studied in Example 6.2. If the history functions of the three state variables are known, as $x_1(t) = \mathrm{e}^{2.1t}$, $x_2(t) = \sin t$, and $x_3(t) = \cos t$, when $t \leqslant 0$, solve again the delay differential equations.

Solutions. An anonymous function can be used to depict the history functions when $t \leqslant 0$. The following statements can then be used to directly solve the delay differential equations with history functions, as shown in Figure 6.5. The elapsed time of code

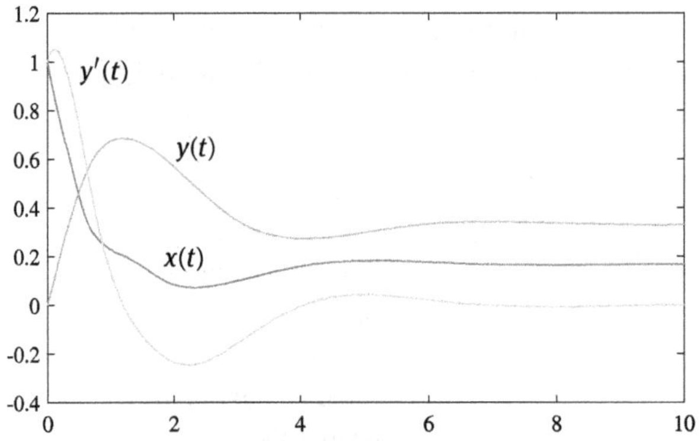

Figure 6.5: Numerical solutions of the delay differential equations.

execution is 1.01 seconds, which is similar to that of the previous example, meaning that the computer burden is not increased.

```
>> f=@(t,x,Z)[1-3*x(1)-Z(2,1)-0.2*Z(1,2)^3-Z(1,2);
              x(3); 4*x(1)-2*x(2)-3*x(3)];
   f2=@(t,x)[exp(2.1*t); sin(t); cos(t)]; %describing history functions
   lags=[1 0.5];  %specify the delay vector
   ff=ddeset; ff.RelTol=1e-10; ff.AbsTol=1e-10;
   tic, tx=dde23(f,lags,f2,[0,10],ff); toc, plot(tx.x,tx.y)
```

It can be seen from the examples that if $t \leqslant t_0$, the definition of the history function is different, the same delay differential equations may yield completely different results. Therefore in solving these delay differential equations, the definition of the initial value and history functions must be noted. Besides, although the definitions may be different, the initial values are not the same, the final values of the solutions of these delay differential equation are virtually the same.

6.2 Differential equations with variable delays

The delay differential equations studied earlier are only a special form of the delay differential equations, since only constant delays were involved. In fact, in many applications, the delay instants are not constant. They can be variable delays, stochastic delays, and state-dependent delays. They can also have other complicated forms. In this section, the mathematical forms of variable delay differential equations are proposed, then the solver provided in MATLAB is addressed. Examples are used next to demonstrate the numerical solutions of delay differential equations with various delays.

6.2.1 Variable delay models

The mathematical form of variable delay differential equations is given first in this section, then MATLAB solutions of the problems are presented.

Definition 6.3. The mathematical form of a variable delay differential equation is given by

$$x'(t) = f(t, x(t), x(\tau_1(t, x(t))), x(\tau_2(t, x(t))), \dots, x(\tau_m(t, x(t)))) \tag{6.2.1}$$

where $0 \leqslant \tau_i(t, x(t)) \leqslant t$ are functions of time t, or even functions of the state variables $x(t)$. It should be noted that the delay should be described as $\tau_1(t, x(t))$, not $t - \tau_1(t, x(t))$.

Three types of delays are explored in particular, and the solution methods are also discussed:

(1) Time-dependent delays. If the delays can be expressed as $t - \tau_i(t)$, with $\tau_i(t) \geqslant 0$, they are referred to as time-dependent delays.
(2) State dependent delays. If the delays can be written as $t-\tau_i(x(t))$, with $\tau_i(x(t)) \geqslant 0$, they are referred to as state-dependent delays.
(3) Generalized delays. The delays in the general form in Definition 6.3 are referred to as generalized delays. The delay form is no longer described by $t - \tau_i$, and it is a function of $x(t)$ at the time instance $\tau_i(t, x(t))$. For instance, $x(\alpha t)$ with $\alpha \leqslant 1$ can be regarded as a generalized delay.

All these forms of delays can be handled with the MATLAB solver `ddesd()`, where function handles are allowed to describe the delays. Of course, function `ddesd()` can be used to directly take care of the three types of delays. They can be handled individually with the solvers in this section.

The syntax of function `ddesd()` is

`sol=ddesd(`f`,`f_τ`,`f_2`,[`t_0`,`t_n`],options)`

where f is the function handle for describing the first-order explicit differential equations; f_τ is the function handle for describing the delay functions; f_2 is the function handle for describing history functions. All these functions can be MATLAB or anonymous functions.

6.2.2 Time-dependent delays

If the delay quantity $\tau_i(t)$ is a function of t, anonymous or MATLAB functions can be used to directly describe the delays, and then the general solver `ddesd()` can be called directly for these delay differential equations. Examples are given next to demonstrate the solutions of time-dependent delay differential equations with zero history functions. Then the nonzero history function problems are considered.

Example 6.7. If the history functions for each state variables are zero, solve the following variable delay differential equations:

$$\begin{cases} x_1'(t) = -2x_2(t) - 3x_1(t - 0.2|\sin t|), \\ x_2'(t) = -0.05x_1(t)x_3(t) - 2x_2(t - 0.8) + 2, \\ x_3'(t) = 0.3x_1(t)x_2(t)x_3(t) + \cos(x_1(t)x_2(t)) + 2\sin 0.1t^2. \end{cases}$$

Solutions. It is obvious that the delay differential equations contain time-dependent delays, that it, with delay at $t - 0.2|\sin t|$ time instance, the solver `dde23()` is no longer usable. Assuming that the first delay is $0.2|\sin t|$, and the second delay is constant 0.8, while the history functions for $t \leqslant 0$ are all zero, the following commands can be used to directly solve the variable delay differential equations, and the solution is shown in Figure 6.6.

```
>> tau=@(t,x)[t-0.2*abs(sin(t)); t-0.8];  % variable delay description
   f=@(t,x,Z)[-2*x(2)-3*Z(1,1);
         -0.05*x(1)*x(3)-2*Z(2,2)+2;        % delay differential equations
         0.3*x(1)*x(2)*x(3)+cos(x(1)*x(2))+2*sin(0.1*t^2)];
   ff=ddeset; ff.RelTol=1e-10; ff.AbsTol=1e-10;
   sol=ddesd(f,tau,zeros(3,1),[0,10],ff); % solve the equations
   plot(sol.x,sol.y)                        % draw the solutions
```

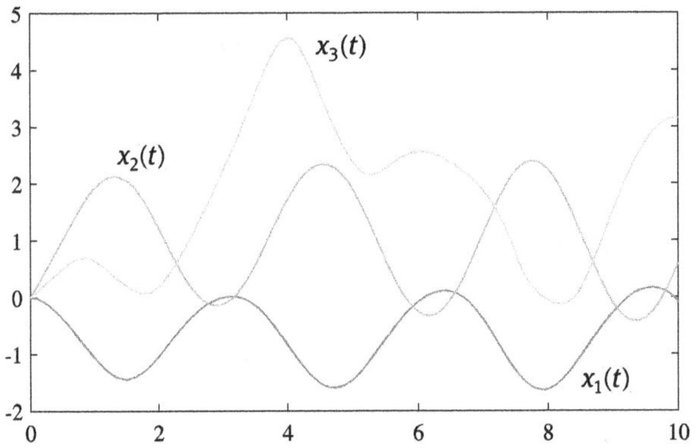

Figure 6.6: Numerical solutions of variable delay differential equations.

It should be noted that since an anonymous function is used to describe the delays, the second delay should be expressed as 0.8. It must be written as $t - 0.8$, otherwise erroneous results may be obtained, since 0.8 may be misunderstood as $x_2(0.8)$, which is the state value at time 0.8, not the correct $x_2(t-0.8)$. Besides, although the delays are functions of time t, they are independent of states, so when expressing the delay function, the argument $\boldsymbol{x}$ should still be used to hold the place, otherwise an error message may be displayed.

Now let us see the solutions with nonzero history functions. As illustrated before, if the history function is expressed in a constant vector $\boldsymbol{x}_0$, the machine understands that when $t \leqslant t_0$, the values of the state are kept the same as the constants $\boldsymbol{x}_0$. If the history functions are given time domain functions, anonymous or MATLAB functions should be used to describe history functions directly. Examples are demonstrated next for such problems.

Example 6.8. For the previous delay differential equations, if the history functions are given below, solve the equation:

$$x_1(t) = \sin(t+1), \quad x_2(t) = \cos t, \quad x_3(t) = \mathrm{e}^{3t}, \quad t \leqslant 0.$$

Solutions. An anonymous function can be written to describe the history functions. Then the following commands can be used to solve the delay differential equations, and the solutions are shown in Figure 6.7.

```
>> tau=@(t,x)[t-0.2*abs(sin(t)); t-0.8];  %variable delay description
   f=@(t,x,Z)[-2*x(2)-3*Z(1,1);
         -0.05*x(1)*x(3)-2*Z(2,2)+2;       %delay differential equations
         0.3*x(1)*x(2)*x(3)+cos(x(1)*x(2))+2*sin(0.1*t^2)];
   f2=@(t,x)[sin(t+1); cos(t); exp(3*t)]; %history functions
   ff=ddeset; ff.RelTol=1e-10; ff.AbsTol=1e-10;
   sol=ddesd(f,tau,f2,[0,10],ff); plot(sol.x,sol.y)
```

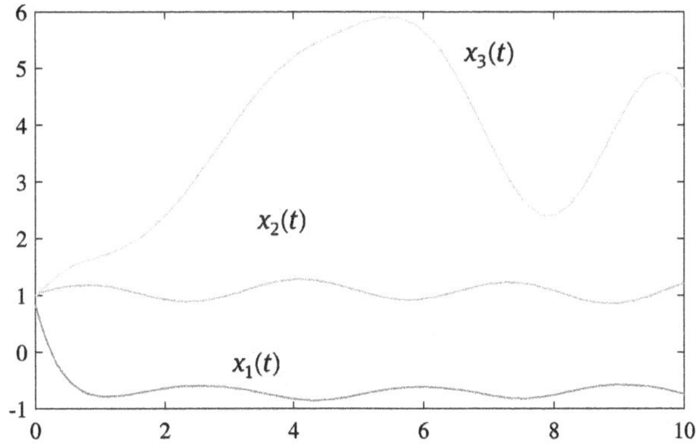

Figure 6.7: Numerical solutions of differential equations with nonzero history functions.

Example 6.9. If the history values for $t < 0$ are all zero, and only at time $t = 0$ a nonzero initial value is defined, solve the differential equations from the previous example.

Solutions. If the history functions are nonzero only at the time instance $t = 0$, the following commands can be used to describe the history functions, where the dot product of *t*==0 ensures that at time $t = 0$, the initial values can be found. At earlier times, the history values are forced to be converted to zeros.

```
>> f2=@(t,x)[sin(t+1); cos(t);
            exp(3*t)].*(t==0); %zero history nonzero initial values
```

Therefore the following MATLAB commands can be used to describe again the differential equations, and the solutions found are shown in Figure 6.8.

```
>> tau=@(t,x)[t-0.2*abs(sin(t)); t-0.8];  %variable delay descriptions
   f=@(t,x,Z)[-2*x(2)-3*Z(1,1);
          -0.05*x(1)*x(3)-2*Z(2,2)+2;           %delay differential equations
          0.3*x(1)*x(2)*x(3)+cos(x(1)*x(2))+2*sin(0.1*t^2)];
   ff=ddeset; ff.RelTol=1e-10; ff.AbsTol=1e-10;
   sol=ddesd(f,tau,f2,[0,10],ff); plot(sol.x,sol.y)
```

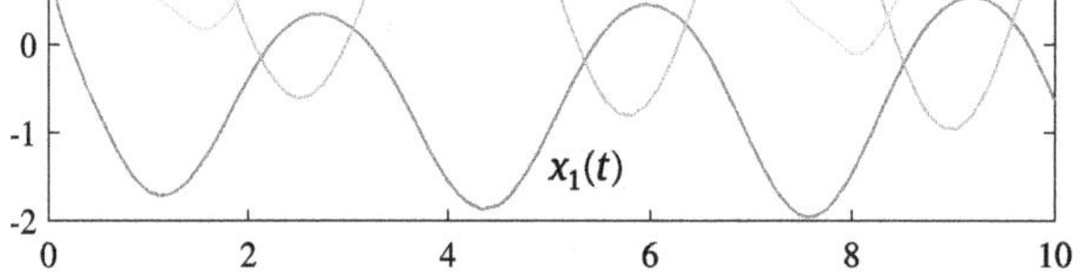

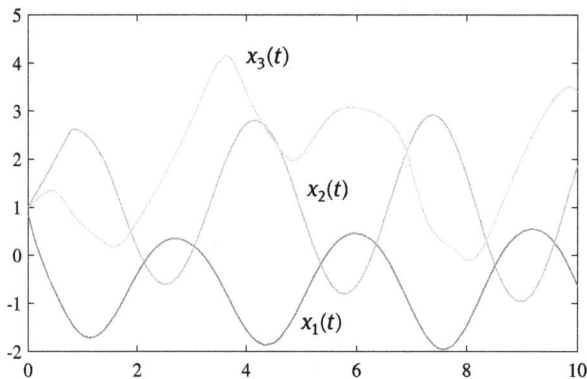

Figure 6.8: Numerical solutions of differential equations with nonzero initial values.

It can be seen that, since the definitions of history values are different, there exist significant differences in the solutions. The history functions are important and affect the transient solutions. Therefore, before solving the delay differential equations, the definition of the history functions must be understood, to avoid ambiguous solutions.

6.2.3 State-dependent delays

In certain applications, the delay quantities are functions of the states, and such differential equations are referred to as state-dependent delay differential equations. In [34], some examples with state-dependent delays are listed. As introduced before, if the delay function handles can be constructed, the solver `ddesd()` can be used in solving the differential equations directly. By default, the function handles of the delay quantities are functions of time t and state $\boldsymbol{x}(t)$. Therefore, there is no special treatment needed to describe the delays. Examples are shown next to demonstrate the solutions of differential equations with state-dependent delays.

Example 6.10. Solve the following simple delay differential equation with state-dependent delay:[34]

$$x'(t) = -x(t - |x(t)|)$$

where, when $t \leqslant 0$, the history function is given by the following piecewise function:

$$y(t) = \begin{cases} -1, & t <= -1, \\ 3(t+1)^{1/3}/2 - 1, & -1 < t \leqslant -7/8, \\ 10t/7 + 1, & -7/8 < t <= 0, \end{cases}$$

while the analytical solution is $y(t) = t + 1$.

Solutions. The differential equation has one delay term whose mathematical expression is $\tau = t - |x(t)|$, which is based on the state variable $x(t)$. The delay differential equation can be easily described with an anonymous function. For history functions, since they are piecewise functions, an anonymous function can also be used. Therefore the following statements can be employed to solve the delay differential equation. The norm of the error is 1.0649×10^{-15}. It can be seen that the accuracy is rather high.

```
>> tau=@(t,x)t-abs(x);  % only one time-dependent delay
   f=@(t,x,Z)-Z;         % differential equation is simple
   x0=@(t,x)-1*(t<=-1)+(10*t/7+1)*(t>-7/8)+...
         3*(t+1)^(3/2)/2*(-1<t & t<=-7/8); % piecewise history functions
   ff=ddeset; ff.RelTol=1e-10; ff.AbsTol=1e-10;
   sol=ddesd(f,tau,x0,[0,10],ff); t=sol.x; y=sol.y;
   plot(t,y,t,t+1), norm(y-t-1) % draw plot and assess accuracy
```

Example 6.11. Consider the delay differential equation in Example 6.7. If $t \leqslant 0$, the history functions of the states are respectively $x_1(t) = 1$ and $x_2(t) = x_3(t) = 0$. The differential equation can be rewritten in the form

$$\begin{cases} x_1'(t) = -2x_2(t - |x_3(t)|) - 3x_1(t - 0.2|\sin t|), \\ x_2'(t) = -0.05x_1(t)x_3(0.85t) - 2x_2(t - 0.8) + 2, \\ x_3'(t) = 0.3x_1(t)x_2(t)x_3(0.85t) + \cos(x_1(t - |x_3(t)|)x_2(t)) + 2\sin 0.1t^2. \end{cases}$$

Solutions. It can be seen from the given delay differential equations that there are 4 delays: $\tau_1 = t - |x_3(t)|$, $\tau_2 = t - 0.2|\sin t|$, $\tau_3 = 0.85t$, and $\tau_4 = t - 0.8$. Therefore the original delay differential equations can be rewritten as

$$\begin{cases} x_1'(t) = -2\mathbf{Z}(2,1) - 3\mathbf{Z}(1,2), \\ x_2'(t) = -0.05x_1(t)\mathbf{Z}(3,3) - 2\mathbf{Z}(2,4) + 2, \\ x_3'(t) = 0.3x_1(t)x_2(t)\mathbf{Z}(3,3) + \cos(\mathbf{Z}(1,1)x_2(t)) + 2\sin 0.1t^2. \end{cases}$$

Anonymous functions can be used to model the delay functions and differential equations. The following commands can then be used to solve the differential equations, with the results shown in Figure 6.9. Since there is no analytical solution, the accuracy cannot be assessed. The validation process can only be made by changing the algorithms or error tolerances, and seeing whether the results are consistent or not. For this example, the results are validated.

```
>> tau=@(t,x)[t-abs(x(3)); t-0.2*sin(t); 0.85*t; t-0.8];
   f=@(t,x,Z)[-2*Z(2,1)-3*Z(1,2);
         -0.05*x(1)*Z(3,3)-2*Z(2,4)+2;          % delay differential equation
         0.3*x(1)*x(2)*Z(3,3)+cos(Z(1,1)*x(2))+2*sin(0.1*t^2)];
   ff=ddeset; ff.RelTol=1e-10; ff.AbsTol=1e-10;
   sol=ddesd(f,tau,[1;0;0],[0,10],ff); plot(sol.x,sol.y)
```

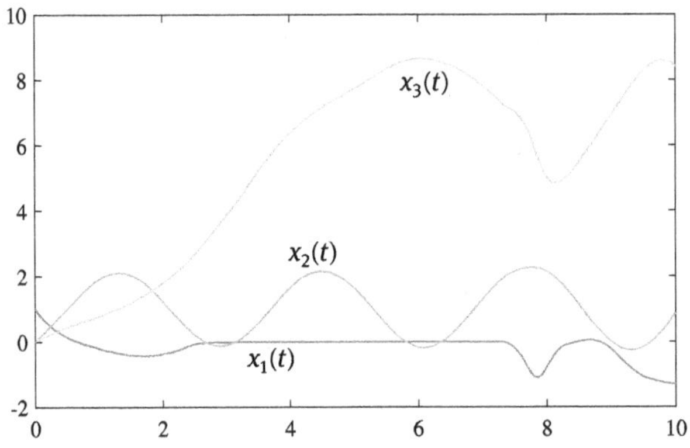

Figure 6.9: Numerical solutions of variable delay differential equations.

6.2.4 Delay differential equations with generalized delays

If some of the delays cannot be expressed as $\boldsymbol{x}(t-\tau_i)$, but they can be expressed as $\boldsymbol{x}(g(t))$, where $g(t) \leqslant t$, or using other complicated forms, the problems like this are known as differential equations with generalized delays. Differential equations with generalized delays can also be solved directly with function `ddesd()`. Here examples are given to demonstrate the solution of differential equations with generalized delays.

Example 6.12. Solve the differential equation with generalized delay[18]

$$x'(t) = [x(t/u(t))]^{u(t)}$$

where $u(t) = (1+2t)^2$ and $x(0) = 1$, with $t \in (0,1)$. The analytical solution is $x(t) = \mathrm{e}^t$.

Solutions. For this specific example, since $(1+2t)^2 \geqslant 1$, we have $t/(1+2t)^2 \leqslant t$. The delay here is a generalized one. Besides, since the differential equation does need the information for $t < 0$, the given $x(0)$ is sufficient for the original delay differential equation.

Anonymous functions can be used directly to describe the differential equation and the delay term. Then the following commands can be used to find the numerical solutions. The error can then be evaluated by comparing the analytical solutions, and its norm 2.7766×10^{-10}. It can be seen that the solution is reliable.

```
>> tau=@(t,x)t/(1+2*t)^2;                   % generalized delay
   f=@(t,x,Z)Z^((1+2*t)^2); x0=1;           % delay differential equation
   ff=ddeset; ff.RelTol=1e-10; ff.AbsTol=1e-10;
   sol=ddesd(f,tau,x0,[0,1],ff);             % solve the DDE
   t=sol.x; y=sol.y; plot(t,y,t,exp(t))      % draw the solutions
```

Example 6.13. Slightly modifying the differential equations in Example 6.7, the differential equations can be described as follows. Solve the following differential equations with generalized delays, where $\alpha = 0.77$:

$$\begin{cases} x_1'(t) = -2x_2(t) - 3x_1(t - 0.2|\sin t|), \\ x_2'(t) = -0.05x_1(t)x_3(t) - 2x_2(\alpha t) + 2, \\ x_3'(t) = 0.3x_1(t)x_2(t)x_3(t) + \cos(x_1(t)x_2(t)) + 2\sin 0.1t^2. \end{cases}$$

Solutions. It can be seen that the function contains the $x_2(0.77t)$ term, not $t - 0.77t$, therefore this equation is with a generalized delay. The value of the state x_2 at $0.77t$ is involved. The following anonymous function can be used to describe the delay terms:

```
>> tau=@(t,x)[t-0.2*abs(sin(t)); 0.77*t]; % describe the functions
```

With the delay description, the remaining commands are the same as those given above. The following commands can be used to directly solve the delay differential equations, with the results shown in Figure 6.10:

```
>> f=@(t,x,Z)[-2*x(2)-3*Z(1,1);
          -0.05*x(1)*x(3)-2*Z(2,2)+2;     % explicit form
          0.3*x(1)*x(2)*x(3)+cos(x(1)*x(2))+2*sin(0.1*t^2)];
   ff=ddeset; ff.RelTol=1e-10; ff.AbsTol=1e-10;
   sol=ddesd(f,tau,zeros(3,1),[0,10],ff);
   plot(sol.x,sol.y)      % the solutions and plots
```

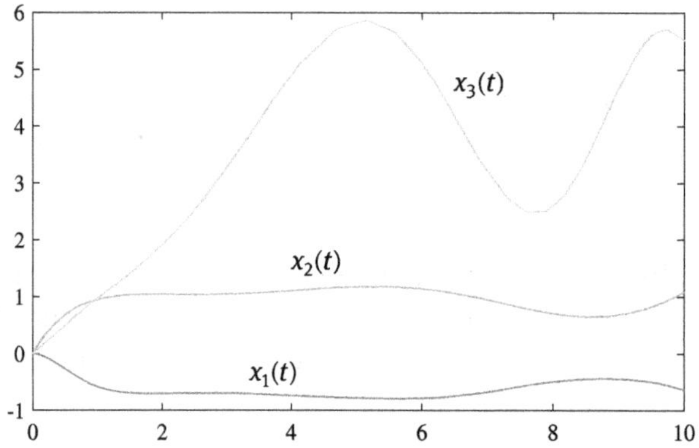

Figure 6.10: Numerical solution of variable delay differential equation.

If the delay vectors contain the values of the unknown functions and their derivatives at future instances t, such as in Example 6.13 where the value $\alpha = 1.1$ indicates that the future values of the unknowns are involved, there is no algorithm which can be used to solve such differential equations. Still the user may use the following expression for the delay function:

```
tau=@(t,x)[t-0.2*abs(sin(t)); 1.1*t]
```

Unfortunately, even function `ddesd()` is unable to solve this kind of problems, since the state $\boldsymbol{x}(t)$ at time $1.1t$ is not previously known. The equations do not have numerical solutions. The values of $\boldsymbol{x}(t)$ are automatically used to substitute for the values in $\boldsymbol{x}(1.1t)$.

6.3 Neutral-type delay differential equations

It can be seen from the delay differential equations studied so far that a differential equation only contains the derivatives of the state variable at the time instance t only.

The previous values of the derivatives are not involved. In delay differential equations research, if the derivatives of the state variables contain the values from the past, such differential equations are referred to as neutral-type delay differential equations. In this section, numerical methods for various neutral-type delay differential equations are presented.

6.3.1 Neutral-type equations

Compared with regular delay differential equations, neutral-type delay differential equations are more complicated to solve. A solver for neutral-type delay differential equations is presented next.

Definition 6.4. The general form of neutral-type delay differential equations can be expressed as

$$\boldsymbol{x}'(t) = \boldsymbol{f}(t, \boldsymbol{x}(t), \boldsymbol{x}(t-\tau_{p_1}), \ldots, \boldsymbol{x}(t-\tau_{p_m}), \boldsymbol{x}'(t-\tau_{q_1}), \ldots, \boldsymbol{x}'(t-\tau_{q_k})) \tag{6.3.1}$$

where the delays of the states and their derivatives are involved. Two vectors $\boldsymbol{\tau}_1 = [\tau_{p_1}, \tau_{p_2}, \ldots, \tau_{p_m}]$ and $\boldsymbol{\tau}_2 = [\tau_{q_1}, \tau_{q_2}, \ldots, \tau_{q_k}]$ are used.

From the given definition, it can be seen that τ_p and τ_q can be constant vectors or functions $\tau_p(t, \boldsymbol{x}(t))$ and $\tau_q(t, \boldsymbol{x}(t))$. These functions can be time-dependent, state-dependent, or generalized. The mathematical form of the delays refers to the presentation given earlier.

Neutral-type delay differential equations can be solved directly with the solver `ddensd()`, with the syntax

`sol=ddensd(`f`,`$\boldsymbol{\tau}_1$`,`$\boldsymbol{\tau}_2$`,`f_2`,[`t_0`,`t_n`],options)`

If the delays in the differential equations are not fixed constants, the format in function `ddesd()` can be used to express $\boldsymbol{\tau}_1$ and $\boldsymbol{\tau}_2$ as function handles. They can be described by anonymous or MATLAB functions.

Example 6.14. Solve a simple neutral-type delay differential equation:[6]

$$y'(t) = y(t) + y'(t-1)$$

where, when $t \leqslant 0$, $y(t) = 1$ and $t \in (0, 3)$. The analytical solution of the delay differential equation is known as

$$y(t) = \begin{cases} \mathrm{e}^t, & \text{if } 0 \leqslant t \leqslant 1, \\ \mathrm{e}^t + (t-1)\mathrm{e}^{t-1}, & \text{if } 1 < t < 2, \\ \mathrm{e}^t + \mathrm{e}^{t-1} + (t-2)(t+2\mathrm{e})\mathrm{e}^{t-2}/2, & \text{if } 2 \leqslant t \leqslant 3. \end{cases}$$

Solutions. The following commands can be used to directly describe the neutral-type delay differential equations. The delay information can also be described, such that the error of the numerical solutions is found to be 3.2820×10^{-4}. The solutions are found as shown in Figure 6.11.

```
>> f=@(t,y,z1,z2)y+z2; tau1=[]; tau2=1; y0=1;
   ff=ddeset; ff.RelTol=1e-10; ff.AbsTol=1e-10;
   tic, sol=ddensd(f,tau1,tau2,y0,[0,3],ff); toc; t=sol.x;
   z=exp(t)+(t-1).*exp(t-1).*(t>=1 & t<2)+...
      (exp(t-1)+(t-2).*(t+2*exp(1)).*exp(t-2)/2).*(t>=2);
   y=sol.y; norm(z-y), plot(t,y,t,z) % solve the DDE
```

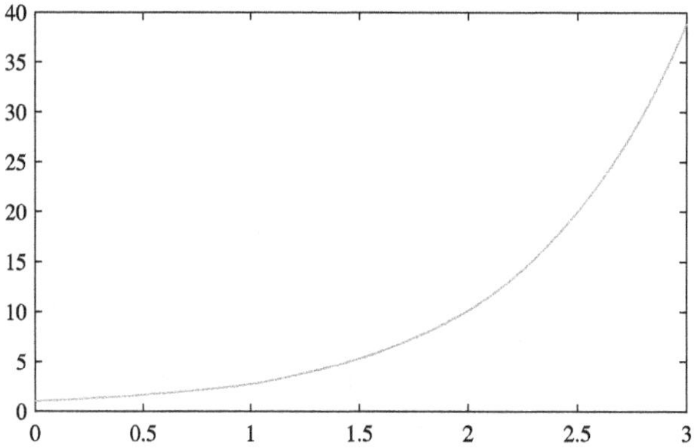

Figure 6.11: Numerical and analytical solutions of neutral-type delay differential equations.

In the solution process, the error message "Warning: `DDENSD` is intended only for modest accuracy. `RelTol` has been increased to 1e–05 (`ddensd()`" indicates that the error tolerance cannot be assigned to extremely small values. For this specific problem, the solution is satisfactory.

Example 6.15. Solve the following neutral-type delay differential equation:

$$\boldsymbol{x}'(t) = \boldsymbol{A}_1\boldsymbol{x}(t-0.15) + \boldsymbol{A}_2\boldsymbol{x}'(t-0.5) + \boldsymbol{B}u(t)$$

where the input signal is $u(t) \equiv 1$, and the known matrices are

$$\boldsymbol{A}_1 = \begin{bmatrix} -13 & 3 & -3 \\ 106 & -116 & 62 \\ 207 & -207 & 113 \end{bmatrix}, \quad \boldsymbol{A}_2 = \begin{bmatrix} 0.02 & 0 & 0 \\ 0 & 0.03 & 0 \\ 0 & 0 & 0.04 \end{bmatrix}, \quad \boldsymbol{B} = \begin{bmatrix} 0 \\ 1 \\ 2 \end{bmatrix}.$$

Solutions. Since the equation contains simultaneously the terms $\boldsymbol{x}'(t)$ and $\boldsymbol{x}'(t-0.5)$, the solver `dde23()` cannot be used. The solver `ddensd()` should be used in the solution process. Here the delay of the state variable is $\tau_1 = 0.15$, while the derivative delay is $\tau_2 = 0.5$. The following anonymous function can be used to describe the neutral-type delay differential equation. Then the following commands can be used to solve the equation, and the curves of the states are obtained as shown in Figure 6.12.

```
>> A1=[-13,3,-3; 106,-116,62; 207,-207,113];
   A2=diag([0.02,0.03,0.04]); %input the given matrices
   B=[0; 1; 2]; u=1; x0=zeros(3,1);
   f=@(t,x,z1,z2)A1*z1+A2*z2+B*u;
   ff=ddeset; ff.RelTol=1e-5; ff.AbsTol=1e-10;
   sol=ddensd(f,0.15,0.5,x0,[0,15],ff); %delay differential equation
   plot(sol.x,sol.y) %equation solutions and plots
```

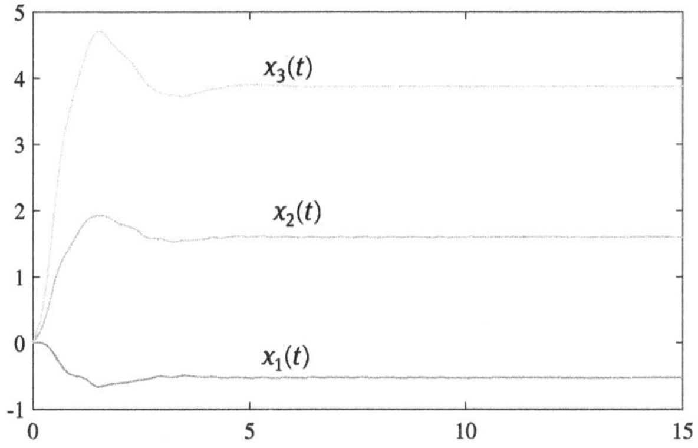

Figure 6.12: Numerical solutions of delay differential equation.

Example 6.16. Consider now the neutral-type delay differential equations with non-zero history functions[6]

$$\begin{cases} y_1'(t) = y_1(t)(1 - y_1(1 - 0.42) - 2.9y_1'(t - 0.42)) - \dfrac{y_1^2(t)y_2(t)}{y_1^2(t) + 1}, \\ y_2'(t) = \left(\dfrac{y_1^2(t)}{y_1^2(t) + 1} - 0.1\right)y_2(t) \end{cases}$$

and, for $t \leqslant 0$, $y_1(t) = 0.33 - 0.1t$, $y_2(t) = 2.22 + 0.1t$. In the interval $t \in (0, 6)$, find the numerical solution of the neutral-type delay differential equations.

Solutions. It can be seen that $\tau_p = \tau_q = 0.42$. Therefore the following statements can be used to solve the delay differential equations, and the results are as shown in Figure 6.13. Since the magnitudes of the solutions differ significantly, the plots with two vertical axes are used.

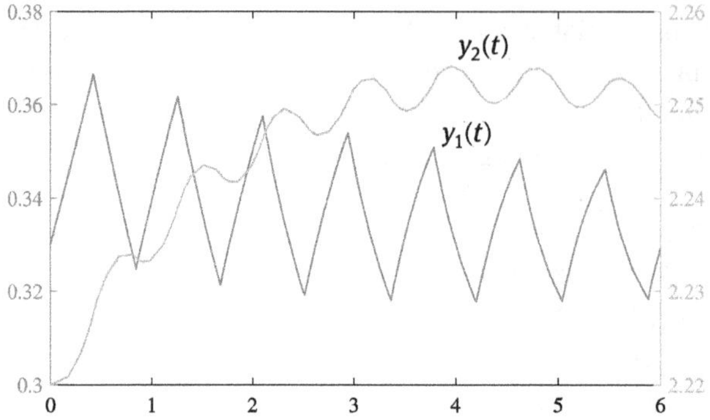

Figure 6.13: Numerical solutions of neutral-type differential equations.

```
>> tau_p=0.42; tau_q=0.42;   % set the two sets of delays
   f0=@(t,y) [0.33-0.1*t; 2.22+0.1*t];  % history functions
   f=@(t,y,Z,z) [y(1)*(1-Z(1)-2.9*z(1))-y(1)^2*y(2)/(y(1)^2+1);
                 (y(1)^2/(y(1)^2+1)-0.1)*y(2)]; % describe equations
   ff=ddeset; ff.RelTol=1e-5; ff.AbsTol=1e-10;
   sol=ddensd(f,tau_p,tau_q,f0,[0,6],ff);      % solve the equations
   plotyy(sol.x,sol.y(1,:),sol.x,sol.y(2,:)) % draw the solutions
```

6.3.2 Neutral-type differential equations with variable delays

It can be seen from the syntaxes of function `ddensd()` that it can be used to directly solve neutral-type variable delay differential equations. The two delays should be first described by MATLAB or anonymous functions. Then the function can be used to directly solve the differential equations. Examples are given next to illustrate the numerical solutions of neutral-type variable delay differential equations.

Example 6.17. The neutral-type delay differential equations in Example 6.16 are changed here to the following form with variable delays:

$$\begin{cases} y_1'(t) = y_1(t)(1 - y_1(t - |y_2(t)|) - 2.9y_1'(t - |\sin y_1(t)|)) - \dfrac{y_1^2(0.6t)y_2(t)}{y_1^2(t) + 1}, \\ y_2'(t) = \left(\dfrac{y_1^2(0.6t)}{y_1^2(t) + 1} - 0.1\right) y_2(t) \end{cases}$$

where, for $t \leqslant 0$, $y_1(t) = 0.33 - 0.1t$ and $y_2(t) = 2.22 + 0.1t$. In the interval $t \in (0, 6)$, solve the neutral-type delay differential equations.

Solutions. It can be seen from the mathematical model that the two delays contained in the states are $t - |y_2(t)|$ and $0.6t$. There is one delay in the derivative, $t - |\sin y_1(t)|$. Therefore anonymous functions can be used to describe the delays. An anonymous function can be used to describe also the original differential equation model. The original problem can be solved with the following MATLAB commands, and the solution is shown in Figure 6.14. Selecting different solvers and error tolerances, it can be seen that consistent curves are obtained.

```
>> tau_p=@(t,y)[t-abs(y(2)), 0.6*t];
   tau_q=@(t,y)[t-abs(sin(y(1)))];       % set the delay vectors
   f0=@(t,y)[0.33-0.1*t; 2.22+0.1*t]; % history function
   f=@(t,y,Z,z)[y(1)*(1-Z(1,1)-2.9*z(1))-Z(1,2)^2*y(2)/(y(1)^2+1);
               (Z(1,2)^2/(y(1)^2+1)-0.1)*y(2)]; % describe equations
   ff=ddeset; ff.RelTol=1e-5; ff.AbsTol=1e-10;
   sol=ddensd(f,tau_p,tau_q,f0,[0,6],ff);      % solve the equations
   plotyy(sol.x,sol.y(1,:),sol.x,sol.y(2,:)) % draw the solutions
```

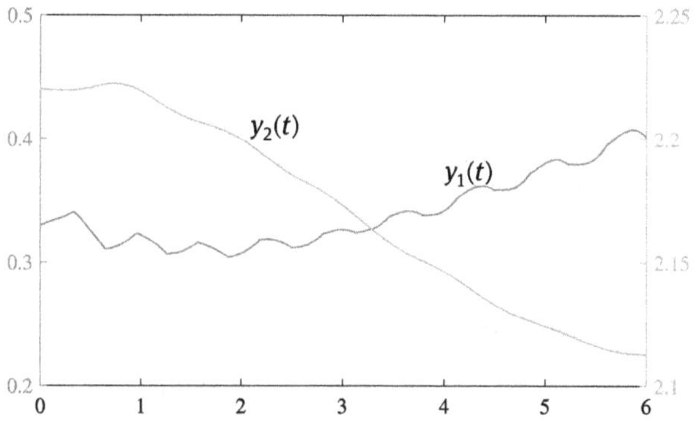

Figure 6.14: Neutral-type delay differential equation solutions.

Of course, the function can be used to solve delay differential equations of non-neutral type. However, there are limitations, since the error tolerance cannot be set to too small numbers. In this consideration it may not be as good as function `ddesd()`. Therefore it is not recommended to use this function for non-neutral-type problems.

Example 6.18. Consider again the variable delay differential equation shown in Example 6.13, and solve it with the neutral-type solver.

Solutions. Function `ddensd()` can be considered to solve the differential equations. The delay vector of the derivative of the unknowns should be assigned to an empty matrix `[]`. The following commands can be used to solve the original differential equation. The results are obtained and they are the same those in Example 6.6, indicating that `ddensd()` function can be used in solving the original problems.

```
>> f=@(t,x,Z,z)[-2*x(2)-3*Z(1,1);
       -0.05*x(1)*x(3)-2*Z(2,2)+2;
```

```
    0.3*x(1)*x(2)*x(3)+cos(x(1)*x(2))+2*sin(0.1*t^2)];
ff=ddeset; ff.RelTol=1e-5; ff.AbsTol=1e-10;
tau=@(t,x)[t-0.2*abs(sin(t)); 0.77*t]; % describing the delays
sol=ddensd(f,tau,[],zeros(3,1),[0,10],ff);
plot(sol.x,sol.y)                          % draw solutions
```

6.4 Volterra differential equations with delays

The general form of a first-order Volterra integro-differential equation was presented in Section 4.5, and the second-order Volterra equation solution was demonstrated with examples. In fact, with the idea of a delay differential equation, $f(t, x(t))$ may be considered having more complicated forms of the unknown functions, such as $x(t/2)$. With similar transforms, Volterra integro-differential equation can be converted into delay differential equations. An example is given next to demonstrate integro-differential equations and their solution methods.

Example 6.19. Consider the integro-differential equation in [18]:[1]

$$x'(t) = \frac{5}{2}t - \frac{t}{2}e^{4x(t/2)} + t\int_0^t se^{x(s)}ds. \tag{6.4.1}$$

When $t \leqslant 0$, $x(t) = 0$. The analytical solution is $x(t) = t^2$. Solve the equation.

Solutions. It is known from Example 4.25 that for this system, the integral term cannot be eliminated by the substitution method. The original equation should be differentiated twice such that the integral term can be removed. The third-order delay differential equation can be derived. The formulation process can be carried out under the symbolic framework, so that no manual work is needed.

```
>> syms t x(t) s
   F=5*t/2-t/2*exp(4*x(t/2))+t*int(s*exp(x(s)),s,0,t);
   F2=diff(F,t);                  % 2nd-order derivative
   F3=simplify(diff(F2,t)) % 3rd-order derivative and simplification
```

The original equation is converted into the following third-order delay differential equation:

$$x'''(t) = 3te^{x(t)} - 2e^{4x(t/2)}x'(t/2) + t^2e^{x(t)}x'(t) - \frac{t}{2}e^{4x(t/2)}x''(t/2) - 2te^{4x(t/2)}(x'(t/2))^2.$$

1 In the reference, the integral was erroneously written as se^s. It is modified to $se^{x(s)}$, otherwise the analytical solution is not $x(t) = t^2$.

To solve this equation, three initial values are needed. The given $x(0) = 1$ is just one of them, and the other two can be found by the following commands:

```
>> subs(F,t,0), subs(F2,t,0)
```

The two other initial values are $x'(0) = 0$ and $x''(0) = 5/2 - e^{4x(0)}/2 = 2$.

It can be seen from the delay differential equation that there is a generalized delay $\tau = t/2$. Selecting the state variables as $x_1(t) = x(t)$, $x_2(t) = x'(t)$, and $x_3(t) = x''(t)$, and denoting $\boldsymbol{z}(t) = \boldsymbol{x}(t/2)$, the standard form of the delay differential equation can be written as

$$\boldsymbol{x}'(t) = \begin{bmatrix} x_2(t) \\ x_3(t) \\ 3te^{x_1(t)} - 2e^{4z_1(t)}z_2(t) + t^2e^{x_1(t)}x_2(t) - te^{4z_1(t)}z_3(t)/2 - 2te^{4z_1(t)}z_2^2(t) \end{bmatrix}.$$

With the delay differential equation and initial states (in this example, history information is not used), the following commands can be written to compute the numerical solutions. Compared with the analytical solution, it can be seen that the error norm is 3.0899×10^{-11}, and the elapsed time is 0.096 seconds. It can be seen that the method here is quite efficient.

```
>> f=@(t,x,Z)[x(2:3);
        3*t*exp(x(1))-2*exp(4*Z(1))*Z(2)+t^2*exp(x(1))*x(2)-...
          t*exp(4*Z(1))*Z(3)/2-2*t*exp(4*Z(1))*Z(2)^2];
   tau=@(t,x)t/2;
   ff=ddeset; ff.RelTol=1e-13; ff.AbsTol=1e-13;
   tic, sol=ddesd(f,tau,[0;0;2],[0,1],ff); toc
   t=sol.x; y=sol.y; plot(t,y), norm(y(1,:)-t.^2)
```

6.5 Exercises

6.1 Consider the following delay differential equation:[32]

$$\begin{cases} y_1'(t) = -y_1(t)y_2(t-1) + y_2(t-10), \\ y_2'(t) = y_1(t)y_2(t-1) - y_2(t), \\ y_3'(t) = y_2(t) - y_2(t-10) \end{cases}$$

where, when $t \leqslant 0$, $y_1(t) = 5$, $y_2(t) = 0.1$, and $y_3(t) = 1$. Solve the delay differential equation for $t \in (0, 40)$.

6.2 Solve the following delay differential equation:

$$y'(t) = \frac{by(t-\tau)}{1 + y^n(t-\tau)} - ay(t)$$

where $a = 0.1$, $b = 0.2$, $n = 10$, and $r = 20$, in the interval $t \leqslant 1000$.

6.3 Solve the following immunological delay differential equations:[32]

$$\begin{cases} V'(t) = (h_1 - h_2F(t))V(t), \\ C'(t) = \xi(m(t))h_3F(t-\tau)V(t-\tau) - h_5(C(t)-1), \\ F'(t) = h_4(C(t)-F(t)) - h_8F(t)V(t) \end{cases}$$

where $\tau = 0.5$, $h_1 = 2$, $h_2 = 0.8$, $h_3 = 104$, $h_4 = 0.17$, $h_5 = 0.5$, $h_7 = 0.12$, and $h_8 = 8$. Parameter h_6 can be set to 10 or 300. If $t \leqslant 0$, $V(t) = \max(0, 10^{-6} + t)$ and $C(t) = F(t) = 1$. Function $\xi(m(t))$ is defined by a piecewise function

$$\xi(m(t)) = \begin{cases} 1, & \text{if } m(t) \leqslant 0.1, \\ 10(1-m(t))/9, & \text{if } 0.1 \leqslant m(t) \leqslant 0.1 \end{cases}$$

where $m(t)$ satisfies the differential equation $m'(t) = h_6V(t) - h_7m(t)$, if $t \leqslant 0$, $m(t) = 0$.

6.4 Solve the following delay differential equations:[63]

(1) $y'(t) = \dfrac{2y(t-2)}{1+y^{9.65}(t-2)} - y(t)$, where, if $t \leqslant 0$, $y(t) = 0.5$, and $t \in [0, 100]$;

(2) $y'(t) = ry(t)\left(1 - \dfrac{1}{m}y(t-0.74)\right)$, where $r = 3.5$, $m = 19$, and $t \in [0, 40]$.

6.5 Solve the following delay differential equations:[63]

$$\begin{cases} y_1'(t) = y_5(t-1) + y_3(t-1), \\ y_2'(t) = y_1(t-1) + y_2(t-0.5), \\ y_3'(t) = y_3(t-1) + y_1(t-0.5), \\ y_4'(t) = y_5(t-1)y_4(t-1), \\ y_5'(t) = y_1(t-1) \end{cases}$$

where, when $t \leqslant 0$, $y_1(t) = y_4(t) = y_5(t) = \mathrm{e}^{t+1}$, $y_2(t) = \mathrm{e}^{t+0.5}$, and $y_3(t) = \sin(t+1)$. The solution interval is $t \in [0, 1]$.

6.6 Consider the following complicated delay differential equations:[6]

$$\begin{cases} y_1'(t) = -5\times 10^4 y_1(t)y_2(t) - 10^5 y_1(t)y_4(t), \\ y_2'(t) = -5\times 10^4 y_1(t)y_2(t) + 9\times 10^4 y_1(t-y_5(t))y_2(t-y_5(t))H(t-35), \\ y_3'(t) = 5\times 10^4 y_1(t)y_2(t), \\ y_4'(t) = -10^5 y_1(t)y_4(t) - 2\times 10^{-3}y_4(t) + 10^6 y_1(t-y_6(t))y_2(t-y_6(t))H(t-197), \\ y_5'(t) = \dfrac{y_1(t)y_2(t) + y_3(t)}{y_1(t-y_5(t))y_2(t-y_5(t)) + y_3(t-y_5(t))}H(t-35), \\ y_6'(t) = \dfrac{10^{-12} + y_2(t) + y_3(t)}{10^{-12} + y_2(t-y_6(t)) + y_3(t-y_6(t))}H(t-197), \end{cases}$$

with $H(\cdot)$ being the Heaviside function. When $t \leqslant 0$, the history functions are $y_1(t) = 5\times 10^{-6}$, $y_2(t) = 10^{-15}$, and $y_3(t) = y_4(t) = y_5(t) = y_6(t) = 0$. If $t \in (0, 300)$, solve the delay differential equations.

6.7 Solve the following delay differential equations:[6]

$$y'(t) = -y(t) + y(t-20) + \frac{1}{20}\cos\left(\frac{t}{20}\right) + \sin\left(\frac{t}{20}\right) - \sin\left(\frac{t-20}{20}\right)$$

where, when $t \leqslant 0$, $y(t) = \sin(t/20)$, and $t \in (0, 1000)$.

6.8 Consider the following delay differential equations:

$$y'(t) = \frac{by(t-\tau)}{1+y^n(t-\tau)} - ay(t)$$

where $b = 0.2$, $a = 0.1$, $n = 10$, $\tau = 20$, and $t \in (0, 1000)$. Solve the delay differential equations and draw the relationship between $y(t)$ and $y(t-\tau)$.

6.9 Solve the following variable delay differential equations:[6]

$$y'(t) = \frac{t-1}{t} y(t - \ln t - 1)y(t), \quad t \geqslant 1$$

where, when $0 \leqslant t \leqslant 1$, $y(t) = 1$.

6.10 Solve the following delay differential equations:[6]

$$y'(t) = -y(t) + y(t-2)\left(2.5 - 1.5\frac{y^{2.5}(t-2)}{1000^{2.5}}\right)$$

where $0 \leqslant t \leqslant 40$. When $-2 \leqslant t \leqslant 0$, $y(t) = 999$.

6.11 Solve the following delay differential equation:[6]

$$y'(t) = -y(t) + y(\alpha(t)) + \frac{t}{20}\cos\frac{t}{20} + \sin\frac{t}{20} - \sin\frac{\alpha(t)}{20}$$

where $0 \leqslant t \leqslant 1000$ and $\alpha(t) = t - 1 + \sin t$. If $-20 \leqslant t \leqslant 0$, $y(t) = \sin(t/20)$.

6.12 Solve the following delay differential equation:[63]

$$y'(t) = \begin{cases} -0.4r(1-t)y(t), & 0 \leqslant t \leqslant 1-c, \\ -ry(t)(0.4(1-t) + 10 - e^{\mu}y(t)), & 1-c < t \leqslant 1, \\ -ry(t)(10 - e^{\mu}y(t)), & 1 < t \leqslant 2, \\ -re^{\mu}y(t)(y(t-1) - y(t)), & 2-c < t \end{cases}$$

where, when $t \leqslant 0$, $y(t) = 10$. The constants are $c = 1/\sqrt{2}$, $r = 0.5$, and $\mu = r/10$. Find the numerical solution of the delay differential equation for $0 \leqslant t \leqslant 10$.

6.13 Solve the following delay differential equation:[18]

$$x'(t) = -x(t - 1 - e^{-t}) + \cos t + \sin(t - 1 - e^{-t})$$

where, when $t \leqslant 0$, $x(t) = \sin t$. The analytical solution of the equation is $x(t) = \sin t$; assess the precision of the numerical solution.

6.14 Solve Volterra delay integro-differential equation:[18]

$$x'(t) = x(t-0.1)\int_{-0.1}^{t-0.1} x(s)\mathrm{d}s$$

where, when $t < 0$, $x(t) = 0$, and $x(0) = 1$.

7 Properties and behaviors of ordinary differential equations

Studies of the solutions of differential equations were fully discussed in earlier chapters. Sometimes, the analysis of the properties and behaviors of differential equations is also needed. For instance, from a practical viewpoint, if a continuous model is described by differential equations, the stability of the model is a very important property. If a system is unstable, it should not be running alone. Some precautions are needed, such that a stable system can be created.

Section 7.1 in this chapter studies the stability issues of differential equations. The definition of stability is presented first, then the direct and indirect methods for assessing stability of linear differential equations with constant coefficients are proposed. The Lyapunov function based stability assessment methods for nonlinear systems are demonstrated. In Section 7.2, some special behaviors of differential equations, such as limit cycles, periodicity, and chaos, are studied. The graphical study of Poincaré mapping and section is also demonstrated. In Section 7.3, the concept of equilibrium points and linearization of nonlinear differential equations are presented. Approximation analysis based linearized models are illustrated. In Section 7.4, the bifurcation issue is introduced.

7.1 Stability of differential equations

Stability is the most important property of differential equations. In this section the stability judgement methods for differential equations with constant coefficients are introduced first. Illustrations are presented for the classical Routh–Hurwitz stability criterion, and we introduce Lyapunov stability and selection of Lyapunov function. Examples are used to demonstrate how to investigate stability of time-varying and nonlinear differential equations.

7.1.1 Stability of linear differential equations with constant coefficients

There are two major ways for describing linear differential equations with constant coefficients. One of them is with high-order linear differential equations (see Definition 2.15)

$$\begin{aligned}\frac{\mathrm{d}^n y(t)}{\mathrm{d}t^n} &+ a_1\frac{\mathrm{d}^{n-1} y(t)}{\mathrm{d}t^{n-1}} + a_2\frac{\mathrm{d}^{n-2} y(t)}{\mathrm{d}t^{n-2}} + \cdots + a_{n-1}\frac{\mathrm{d}y(t)}{\mathrm{d}t} + a_n y(t)\\ &= b_1\frac{\mathrm{d}^m u(t)}{\mathrm{d}t^m} + b_2\frac{\mathrm{d}^{m-1} u(t)}{\mathrm{d}t^{m-1}} + \cdots + b_m\frac{\mathrm{d}u(t)}{\mathrm{d}t} + b_{m+1}u(t).\end{aligned} \tag{7.1.1}$$

https://doi.org/10.1515/9783110675252-007

The other is through first-order explicit differential equations, also known as state space equations (see Definition 2.18)

$$\begin{cases} \boldsymbol{x}'(t) = \boldsymbol{A}\boldsymbol{x}(t) + \boldsymbol{B}\boldsymbol{u}(t), \\ \boldsymbol{y}(t) = \boldsymbol{C}\boldsymbol{x}(t) + \boldsymbol{D}\boldsymbol{u}(t). \end{cases} \tag{7.1.2}$$

With the state variable selection method in Chapter 4, differential equations in (7.1.1) can be converted into state space expressions in (7.1.2). It is also illustrated in Chapter 2 that the analytical solution of linear differential equations with constant coefficients can be written as (see Theorem 2.12)

$$\boldsymbol{x}(t) = \mathrm{e}^{\boldsymbol{A}(t-t_0)}\boldsymbol{x}(t_0) + \int_{t_0}^{t} \mathrm{e}^{\boldsymbol{A}(t-\tau)}\boldsymbol{B}\boldsymbol{u}(\tau)\,\mathrm{d}\tau. \tag{7.1.3}$$

Definition 7.1. For given differential equations, if a small change in the input signal yields a small change in the state variables, the system is stable. In other words, if the state variables are finite, the differential equation is stable; while if the state variables tend to infinity, the differential equation is unstable.

Theorem 7.1. *If the input signal $u(t)$ is bounded, the necessary and sufficient condition of stability of the linear differential equations with constant coefficients is $\mathscr{R}[p_i] < 0$ for all $i = 1, 2, \ldots, n$, where p_i are the eigenvalues of the coefficient matrix $\boldsymbol{A}$.*

It can be seen from the current viewpoints that if the coefficients of the linear differential equations are known, the stability judgement is an easy thing, since an advanced calculator or any computer mathematical language can be used to find all the eigenvalues of the coefficient matrix $\boldsymbol{A}$. What the user needs is merely to see whether there exists an eigenvalue whose real part is larger than 0. If there is none, the differential equation is stable, otherwise it is unstable, since its corresponding signal tends to infinity when $t \to \infty$. The differential equation is then divergent. In particular, if there are eigenvalues whose real part are zero, there are oscillations with constant magnitudes as $t \to \infty$. The differential equation is then critically stable.

Example 7.1. Assess the stability of the following differential equation:

$$\boldsymbol{x}'(t) = \begin{bmatrix} 2 & 1 & -2 & -1 & 0 & -1 \\ -1 & 0 & 2 & 1 & 0 & 1 \\ 4 & 0 & -5 & -2 & -1 & -1 \\ 2 & -3 & -2 & -2 & 1 & -1 \\ -5 & 1 & 3 & 1 & -2 & 1 \\ -4 & 2 & 4 & 3 & 1 & 0 \end{bmatrix} \boldsymbol{x}(t), \quad \boldsymbol{x}(0) = \begin{bmatrix} 1 \\ 0 \\ 0 \\ 0 \\ 0 \\ 0 \end{bmatrix}.$$

Solutions. It is simple to judge whether this differential equation is stable or not with direct methods. The coefficient matrix $\boldsymbol{A}$ can be loaded into MATLAB workspace first, then the eigenvalues can be found. For convenience, symbolic computation is adopted in computing the eigenvalues.

```
>> A=[2,1,-2,-1,0,-1; -1,0,2,1,0,1; 4,0,-5,-2,-1,-1;
      2,-3,-2,-2,1,-1; -5,1,3,1,-2,1; -4,2,4,3,1,0];
   s=eig(sym(A)).'      % the transpose of the eigenvalue vector
```

It can be seen that the eigenvalues are $-2, -2, -2, 1, -1 \pm \mathrm{j}$. Due to the existence of the $s = 1$ eigenvalue, with positive real part, the differential equation is unstable.

Example 7.2. Judge the stability of the differential equation in Example 7.1 from the analytical solution or time domain response.

Solutions. In fact, by the method in Chapter 2, it is not hard to compute the analytical solution as $\boldsymbol{x}(t) = \mathrm{e}^{\boldsymbol{A}t}\boldsymbol{x}(0)$.

```
>> A=[2,1,-2,-1,0,-1; -1,0,2,1,0,1; 4,0,-5,-2,-1,-1;
      2,-3,-2,-2,1,-1; -5,1,3,1,-2,1; -4,2,4,3,1,0];
   x0=[1; 0; 0; 0; 0; 0];    % initial state vector
   syms t real;
   x=expm(A*t)*x0             % analytical solution of the differential equation
```

The analytical solution of the differential equation is obtained as

$$\boldsymbol{x}(t) = \begin{bmatrix} \mathrm{e}^{t} - \mathrm{e}^{-2t} + \mathrm{e}^{-t}\cos t \\ \mathrm{e}^{-2t} - \mathrm{e}^{-t}\cos t \\ \mathrm{e}^{t} - 2\mathrm{e}^{-2t} + \mathrm{e}^{-t}\cos t \\ 2\mathrm{e}^{-t}\cos t - t\mathrm{e}^{-2t} - 2\mathrm{e}^{-2t} + \mathrm{e}^{-t}\sin t \\ 3\mathrm{e}^{-2t} - \mathrm{e}^{t} + t\mathrm{e}^{-2t} - 2\mathrm{e}^{-t}\cos t - \mathrm{e}^{-t}\sin t \\ 3\mathrm{e}^{-2t} - \mathrm{e}^{t} + t\mathrm{e}^{-2t} - 2\mathrm{e}^{-t}\cos t \end{bmatrix}.$$

It can be seen that since the e^{t} term exists, when $t \to \infty$, the analytical solution $\boldsymbol{x}(t)$ tends to infinity, while the terms such as $t\mathrm{e}^{-2t}$ tend to zero. Therefore the stability of the differential equation is determined by the existence of eigenvalues with positive real parts. If there is one such eigenvalue, the differential equation is unstable. The time domain response of the state variables can be drawn, with the following statements:

```
>> fplot(x,[0,pi]) % the time domain response in a certain region
```

The curves of the state variables are obtained as shown in Figure 7.1. It can be seen that, when t increases, some of the states increase as exponential functions. They tend to infinity quickly. Therefore the differential equation is unstable. It is also seen from the analytical solution that the states $x_2(t)$ and $x_4(t)$ do not contain the e^t components, the two signals are stable, while the whole differential equation is not.

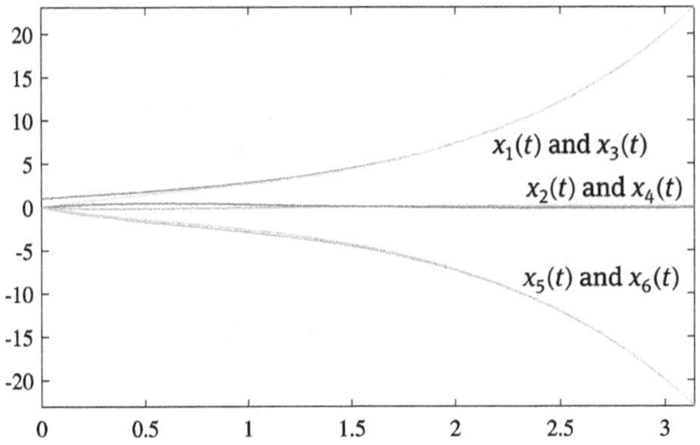

Figure 7.1: The time domain responses of the states.

7.1.2 Routh–Hurwitz stability criterion

A simple direct stability judgement method is illustrated based on the eigenvalues of the coefficient matrices. Hundreds of years ago, there were no tools such as computers, so the researches were not able to find the eigenvalues of a high-order matrix, and the direct method discussed above could not be used. They had to invent different indirect methods to assess the stability. A British mathematician Edward John Routh (1831–1907) proposed such a method. He constructed a table from the characteristic polynomial coefficients, and used this table to assess the stability.[60] The method is an indirect one, known as Routh criterion. A German mathematician Adolf Hurwitz (1859–1919) in 1898 proposed a matrix-based stability judgement method. Since the stability judgement matrix is very similar to Routh table, it is known as Routh–Hurwitz stability criterion.

The classical Routh–Hurwitz stability criterion is presented first in this section, and examples are used to assess the stability of linear differential equations with constant coefficients.

Theorem 7.2. *If the characteristic polynomial is given by*

$$p(s) = a_0 s^n + a_1 s^{n-1} + a_2 s^{n-2} + \cdots + a_{n-1} s + a_n, \tag{7.1.4}$$

we can construct the following Hurwitz judgement matrix:

$$H = \begin{bmatrix} a_1 & a_3 & a_5 & a_7 & \cdots & 0 \\ a_0 & a_2 & a_4 & a_6 & \cdots & 0 \\ 0 & a_1 & a_3 & a_5 & \cdots & 0 \\ 0 & a_0 & a_2 & a_4 & \cdots & 0 \\ \vdots & \vdots & \vdots & \vdots & \ddots & 0 \\ 0 & 0 & 0 & 0 & \cdots & a_n \end{bmatrix}. \tag{7.1.5}$$

*If the determinants of all the upper-left submatrices of matrix **H** are all positive, the differential equation is stable:*

$$\Delta_1 = a_1 > 0, \quad \Delta_2 = \begin{vmatrix} a_1 & a_3 \\ a_0 & a_2 \end{vmatrix} = a_1a_2 - a_0a_3 > 0, \tag{7.1.6}$$

$$\Delta_3 = \begin{vmatrix} a_1 & a_3 & a_5 \\ a_0 & a_2 & a_4 \\ 0 & a_1 & a_3 \end{vmatrix} = a_3\Delta_2 - a_1(a_1a_4 - a_0a_5) > 0, \ldots \tag{7.1.7}$$

The so-defined Hurwitz matrix is a rectangular matrix. Since only the major submatrices are involved, the first n rows are sufficient. The generation of Hurwitz matrix is demonstrated with examples, with their applications in stability analysis.

Example 7.3. Assess the stability of the differential equation in Example 7.1 with Routh–Hurwitz criterion.

Solutions. Matrix $\boldsymbol{A}$ can be entered into MATLAB workspace first, then the characteristic polynomial can be established

```
>> A=[2,1,-2,-1,0,-1; -1,0,2,1,0,1; 4,0,-5,-2,-1,-1;
     2,-3,-2,-2,1,-1; -5,1,3,1,-2,1; -4,2,4,3,1,0];
   syms s real; p=charpoly(sym(A),s) %compute polynomial
```

The characteristic polynomial of matrix $\boldsymbol{A}$ can be computed directly with the above statements as

$$p(s) = s^6 + 7s^5 + 18s^4 + 18s^3 - 4s^2 - 24s - 16.$$

Of course, since the signs are changing in the polynomial matrix, it is sufficient to say that the differential equation is unstable. Now, the Hurwitz matrix can be constructed with the following statements:

```
>> p=charpoly(sym(A)); n=length(p)-1; H=zeros(n); d=[];
   for i=1:(n+1)/2      %compose Hurwitz matrix with a loop
     H(2*i-1,i-1+(2:2:n+1)/2)=p(2:2:n+1);
```

```
    H(2*i,i-1+(2:2:n+2)/2)=p(1:2:n+1);
  end, H=H(1:n,1:n)  % extract Hurwitz matrix
  for i=1:n, d=[d,det(H(1:i,1:i))]; end, d
```

The Hurwitz matrix thus composed is

$$H = \begin{bmatrix} 7 & 18 & -24 & 0 & 0 & 0 \\ 1 & 18 & -4 & -16 & 0 & 0 \\ 0 & 7 & 18 & -24 & 0 & 0 \\ 0 & 1 & 18 & -4 & -16 & 0 \\ 0 & 0 & 7 & 18 & -24 & 0 \\ 0 & 0 & 1 & 18 & -4 & -16 \end{bmatrix}.$$

The determinants of the major submatrices can be found as

$$\boldsymbol{d} = [7, 108, 1\,972, 26\,768, 128\,000, -2\,048\,000].$$

It can be seen that there is one negative determinant in $\boldsymbol{d}$, therefore the differential equation is unstable. This is consistent with the conclusion in Example 7.1. The judgement method here is much more complicated than the direct method. It is not possible to find which of the states are stable, and which are not, with the indirect method. Therefore indirect methods are not recommended.

Example 7.4. Consider the fourth-order characteristic polynomial $s^4+ps^3+qs^2+rs+h = 0$. Show that the stability conditions for the differential equation are:[60]

$$p > 0, \quad pq-r > 0, \quad (pq-r)r-p^2h > 0, \quad h > 0.$$

Solutions. With the following commands, the Hurwitz matrix can be established, from which the determinants of the four major submatrices can be found. Compared with the code presented earlier, the zero matrix $\boldsymbol{H}$ is converted to a symbolic one before the loop structure, otherwise error messages are given. The other statements are exactly the same as those in the previous example.

```
>> syms p q r h; p=[1 p q r h];
   n=length(p)-1; H=sym(zeros(n)); d=[];
   for i=1:(n+1)/2
      H(2*i-1,i-1+(2:2:n+1)/2)=p(2:2:n+1);
      H(2*i,i-1+(2:2:n+2)/2)=p(1:2:n+1);
   end, H=H(1:n,1:n)
   for i=1:n    % compute the major determinants of Hurwitz matrix
      d=[d,prod(factor(det(H(1:i,1:i))))];
   end, d
```

It can be seen that the Hurwitz matrix is

$$\boldsymbol{H} = \begin{bmatrix} p & r & 0 & 0 \\ 1 & q & h & 0 \\ 0 & p & r & 0 \\ 0 & 1 & q & h \end{bmatrix}$$

and the factorized major determinants are

$$\boldsymbol{d} = [p, pq - r, -hp^2 + qpr - r^2, h(-hp^2 + qpr - r^2)].$$

To ensure that the differential equation is stable, the four terms must be positive. The first three terms are the same as expected. Now let us have a look at the fourth term $(-hp^2 + qpr - r^2)h$. Since the expression inside the parentheses is the third major determinant d_3, therefore $d_3 h > 0$ can be reduced to $h > 0$.

7.1.3 Lyapunov function and Lyapunov stability

The definitions of positive and negative definiteness of functions are introduced in this section first, followed by the definition of Lyapunov function. Finally, the Lyapunov stability and its criterion are presented.

Definition 7.2. If a scalar function $V(\boldsymbol{x})$ satisfies $V(\boldsymbol{x}) > 0$ when $\boldsymbol{x} \neq \boldsymbol{0}$, and $V(\boldsymbol{x}) = 0$ when $\boldsymbol{x} = \boldsymbol{0}$, function $V(\boldsymbol{x})$ is referred to as a positive definite function; if $V(\boldsymbol{x}) \geqslant 0$, then $V(\boldsymbol{x})$ is a semipositive definite function.

Definition 7.3. If a function $V(\boldsymbol{x})$ satisfies $V(\boldsymbol{x}) < 0$ when $\boldsymbol{x} \neq \boldsymbol{0}$, and $V(\boldsymbol{x}) = 0$ if $\boldsymbol{x} = \boldsymbol{0}$, $V(\boldsymbol{x})$ is referred to as a negative definite function; if $V(\boldsymbol{x}) \leqslant 0$, then $V(\boldsymbol{x})$ is a seminegative definite function.

Example 7.5. Show that the following function is positive definite:

$$V(x_1, x_2) = a^2 x_1^2 + b^2 x_2^2 + abx_1 x_2 \cos(x_1 + x_2).$$

Solutions. For any a, b, x_1, and x_2, we have $abx_1 x_2 \cos(x_1 + x_2) \geqslant -|ax_1||bx_2|$, therefore, it is easily seen that $V(x_1, x_2)$ is a positive definite function:

$$\begin{aligned} V(x_1, x_2) &\geqslant a^2 x_1^2 + b^2 x_2^2 - |ax_1||bx_2| \\ &= a^2 x_1^2/2 + b^2 x_2^2/2 + a^2 x_1^2/2 + b^2 x_2^2/2 - |ax_1||bx_2| \\ &= a^2 x_1^2/2 + b^2 x_2^2/2 + (|ax_1| - |bx_2|)^2 \geqslant 0. \end{aligned}$$

Definition 7.4. If a given function $V(\boldsymbol{x}(t))$ is positive definite, and its derivative $\mathrm{d}V(\boldsymbol{x}(t))/\mathrm{d}t$ is negative definite, then $V(\boldsymbol{x}(t))$ is referred to as a Lyapunov function.

Lyapunov function is named after a Russian mathematician Aleksandr Mikhailovich Lyapunov (1857–1918). The function is a very important function in the theoretical study of differential equations stability. Lyapunov stability judgement theorem can only be used to assess stability of autonomous differential equations, which are defined as follows.

Definition 7.5. If a differential equation does not explicitly have the time t term, it is referred to as an autonomous one. Autonomous differential equations are briefly denoted as $\boldsymbol{x}'(t) = \boldsymbol{f}(\boldsymbol{x}(t))$.

In other words, autonomous systems require that the parameters in the model are all time-invariant. If there exist time-varying parameters, Lyapunov criterion cannot be used in assessing the stability of the system.

Theorem 7.3. *If function $\boldsymbol{x}(t)$ in Definition 7.4 is the state described by $\boldsymbol{x}'(t) = \boldsymbol{f}(\boldsymbol{x}(t))$, then having a Lyapunov function for it is sufficient for the Lyapunov stability of the autonomous differential equation.*

Note that the condition here is just a sufficient condition. There are no methods known of how to construct Lyapunov functions in Lyapunov stability assessment method. This theorem is also known as the second Lyapunov method, or Lyapunov direct assessment method. Lyapunov stability assessment is based on the art of Lyapunov function construction. If a Lyapunov function can be constructed, the differential equation is shown to be stable. If a Lyapunov function cannot be constructed, or a positive definite $V(\boldsymbol{x}(t))$ is composed, but when it is substituted into the differential equation, $V'(\boldsymbol{x}(t)) < 0$ cannot be shown, this does not necessarily imply that the differential equation is unstable.

7.1.4 Autonomous conversion of time-varying differential equations

Many theorems are based on the autonomous differential equations, that is, t is not explicitly contained in the differential equations. In real applications, if a time-varying (or nonautonomous) differential equation is to be studied, a new state can be introduced such that the differential equation can be augmented to become an autonomous system. An example is given next to show how to convert a time-varying differential equation into an autonomous one.

Example 7.6. Assume that a time-varying differential equation is given by[44]

$$y'''(t) = y'(t)y(t) - 2t(y''(t))^2.$$

Convert it into an autonomous one.

Solutions. In normal cases, the state variables can be selected as $x_1(t) = v(t)$, $x_2(t) = v'(t)$, and $x_3(t) = v''(t)$ such that the original equation can be converted into a first-order explicit differential equation. While such differential equations explicitly contain t, if an autonomous differential equation is expected, which does not explicitly contain t, an augmented state $x_4(t) = t$ can be introduced such that the following autonomous differential equation can be established:

$$\boldsymbol{x}'(t) = \begin{bmatrix} x_2(t) \\ x_3(t) \\ x_2(t)x_1(t) - 2x_4(t)(x_3(t))^2 \\ 1 \end{bmatrix}$$

where $x_4(0) = 0$, and the initial values of other states are exactly the same as those defined with ordinary methods. Now the converted differential equation is an autonomous one.

7.1.5 Stability assessment of nonlinear systems by examples

In this section, simple nonlinear differential equations are used to demonstrate stability judgement, and validate the assessment with examples.

Consider the following autonomous nonlinear differential equations:

$$\begin{cases} x_1'(t) = f(x_1(t), x_2(t)), \\ x_2'(t) = g(x_1(t), x_2(t)). \end{cases} \tag{7.1.8}$$

The following positive definite test function can be constructed:

$$V(t) = x_1^2(t) + x_2^2(t). \tag{7.1.9}$$

Differentiating this function, and substituting the differential equation into the results, it is found that

$$\begin{aligned} V'(t) &= 2x_1(t)x_1'(t) + 2x_2(t)x_2'(t) \\ &= 2x_1(t)f(x_1(t), x_2(t)) + 2x_2(t)g(x_1(t), x_2(t)). \end{aligned} \tag{7.1.10}$$

If one is able to show that $V'(t) < 0$, the differential equations can then be shown to be stable; if $V'(t)$ cannot be proved negative definite, another test function should be chosen. Examples are shown for the stability analysis of nonlinear differential equations.

Example 7.7. Consider the following differential equations:

$$\begin{cases} x_1'(t) = x_2(t) - x_1(t)(x_1^2(t) + x_2^2(t)), \\ x_2'(t) = -x_1(t) + x_2(t)(x_1^2(t) + x_2^2(t)). \end{cases}$$

Show that they are stable.

Solutions. Choosing the positive definite test function according to (7.1.9), the following commands can be tried to compute its derivative:

```
>> syms x1 x2
   f=x2-x1*(x1^2+x2^2); g=-x1-x2*(x1^2+x2^2);
   Vd=2*x1*f+2*x2*g; simplify(Vd)  % find the derivative
```

Through MATLAB simplification command, it is found that $V'(t) = -2(x_1^2(t) + x_2^2(t))^2$. It can be seen that $V'(t)$ is negative definite, indicating that the system is stable.

Example 7.8. Show that the following system of differential equations is stable:[47]

$$\begin{cases} x_1'(t) = -x_1(t)(x_1^2(t) + x_2^2(t))(1 - \cos\ln(x_1^2(t) + x_2^2(t)) - \sin\ln(x_1^2(t) + x_2^2(t))), \\ x_2'(t) = -x_2(t)(x_1^2(t) + x_2^2(t))(1 - \cos\ln(x_1^2(t) + x_2^2(t)) - \sin\ln(x_1^2(t) + x_2^2(t))). \end{cases}$$

Solutions. As in the previous example, the following MATLAB code can be written:

```
>> syms x1 x2; D=(x1^2+x2^2);
   f=-x1*D*(1-cos(log(D))-sin(log(D)));
   g=-x2*D*(1-cos(log(D))-sin(log(D)));
   Vd=2*x1*f+2*x2*g; Vd=simplify(Vd) % find the derivative
```

After simplification, an intermediate result is obtained, which is the same as that in [47]:

$$V'(t) = 2\left(x_1^2(t) + x_2^2(t)\right)^2 \left(\sqrt{2}\sin\left(\frac{\pi}{4} + \ln\left(x_1^2(t) + x_2^2(t)\right)\right) - 1\right).$$

If it can be shown that the expression in the parenthesis is always smaller than zero, then theoretically the system will be stable. In [47] it seems that $V'(t) < 0$ was apparently proved, but the author of this book is suspicious about the result. Luckily, with the powerful MATLAB graphical facilities provided, the surface of $V'(t)$ can be drawn directly, as shown in Figure 7.2. One may pick a point t_0 on the surface and see that $V'(t_0) = 1.311 > 0$, which is sufficient to indicate that $V'(t) < 0$ does not always hold.

```
>> fsurf(Vd,[-2,2])  % the surface of the derivative function
```

But even if this happens, it is not necessarily the case that the system is unstable. From the theoretical viewpoint, one can only blame the improper choice of the function $V(t)$. Other similar functions must be tried. No conclusion can be made so far with Lyapunov criterion.

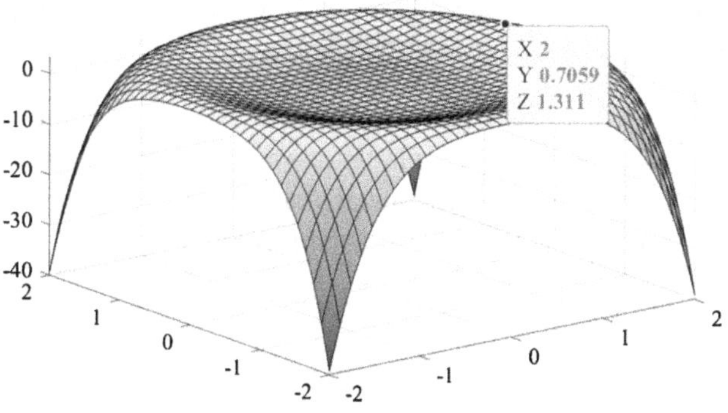

Figure 7.2: The surface of function $V'(t)$.

Example 7.9. Now let us see a frequently used example of second-order differential equation in stability related textbooks:

$$\begin{cases} x'(t) = -y(t) + x(t)(x^2(t) + y^2(t) - 1), \\ y'(t) = x(t) + y(t)(x^2(t) + y^2(t) - 1). \end{cases}$$

How an we check the stability of such a system?

Solutions. If a positive definite function $V(t) = x^2(t) + y^2(t)$ is chosen, the following commands can be used to compute $V'(t)$ directly:

```
>> syms t x(t) y(t)
   f=-y+x*(x^2+y^2-1); g=x+y*(x^2+y^2-1);
   V=x^2+y^2; Vd=prod(factor(2*x*f+2*y*g)) % function derivative
```

It can be seen that the result is $V'(t) = 2(x^2(t) + y^2(t))(x^2(t) + y^2(t) - 1)$, where $2(x^2(t) + y^2(t)) \geqslant 0$ always holds. The final result is determined by the $(x^2(t) + y^2(t) - 1)$ term. If this term is $\leqslant 0$, then $V'(t)$ is negative definite such that the original system is stable. Unfortunately, $(x^2(t) + y^2(t) - 1) < 0$ does not always hold, because it cannot be guaranteed that the coordinate $(x(t), y(t))$ always lies inside a unit circle. Therefore, no useful clue can be found regarding the stability of the system.

7.1.6 Stability assessment of complicated systems with simulation methods

It can be seen in the earlier examples that, in order to construct a Lyapunov function, the structure and parameters in the examples were deliberately chosen. For instance, if the sign of a term in Example 7.8 is altered, the function $V(t)$ cannot be constructed. Therefore the direct use of Lyapunov judgement is quite restricted. For practical stability judgement, simulation methods cannot be used in assessing the system stability.

Example 7.10. Consider again Example 7.9. Use the numerical method to assess the stability of the differential equations.

Solutions. If initial points $(x(0), y(0))$ are select randomly in the interval $[-5, 5]$, the differential equation can be solved, and the phase plane trajectory can be drawn as shown in Figure 7.3. It can be seen that the system is unstable. If a system is unstable, no conclusion can be found with merely using Lyapunov stability judgement.

```
>> f=@(t,x)[-x(2)+x(1)*(x(1)^2+x(2)^2-1);
            x(1)+x(2)*(x(1)^2+x(2)^2-1)];
   ff=odeset; ff.RelTol=1e-10; ff.AbsTol=1e-10;
   for i=1:100, i, x0=-5+10*rand(2,1); %random initial value
      [t,x]=ode15s(f,[0,100],x0,ff); line(x(:,1),x(:,2))
   end
```

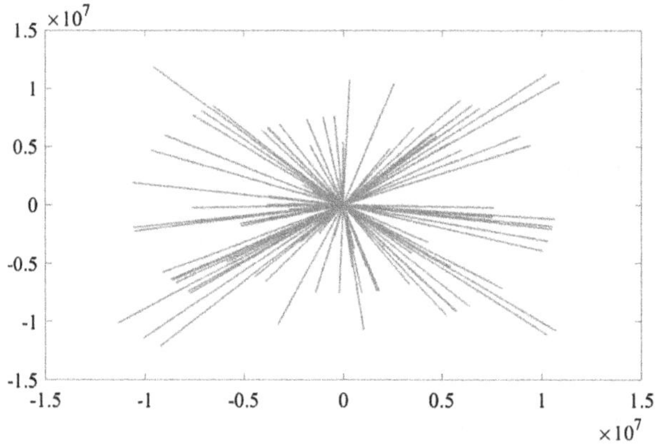

Figure 7.3: Phase plane trajectory of an unstable system.

In fact, if a simulation example shows that a system is stable, there should be an infinite number of cases supporting this fact. This is usually very difficult to implement in practice. If one wants to show that the system is unstable, one such example is sufficient. In the previous example, there are many such examples, they are enough to say that the system is unstable.

Example 7.11. Selecting initial values in a large range, assess the stability of the differential equations in Example 7.8.

Solutions. The theoretical formulation of the stability test depends heavily the selection of the test function. Especially if a system is unstable, there is no way to validate that with Lyapunov criterion. Therefore, the simulation method can be used to assess that. For example, 100 random initial values are generated in $[-500, 500]$ to find the numerical solutions of differential equations. All the solutions converge to fixed values, and there is no case where the differential equation is divergent. For the sake of safety, 10 000 more simulations are carried out, without a single case where the system is divergent. Therefore it is sufficiently safe to say that the system is stable. In Figure 7.4, the zoomed responses in the interval (0, 0.0001) are shown, and the simulation region is (0, 100).

```
>> D=@(x)(x(1)^2+x(2)^2);
   f=@(t,x)[-x(1)*D(x)*(1-cos(log(D(x)))-sin(log(D(x))));
            -x(2)*D(x)*(1-cos(log(D(x)))-sin(log(D(x))))];
   ff=odeset; ff.RelTol=100*eps; ff.AbsTol=100*eps;
   for i=1:100, i, x0=-500+1000*rand(2,1); %large range trial
      [t,x]=ode15s(f,[0,100],x0,ff); line(t,x), drawnow
   end, xlim([0,0.0001])  %display transient response
```

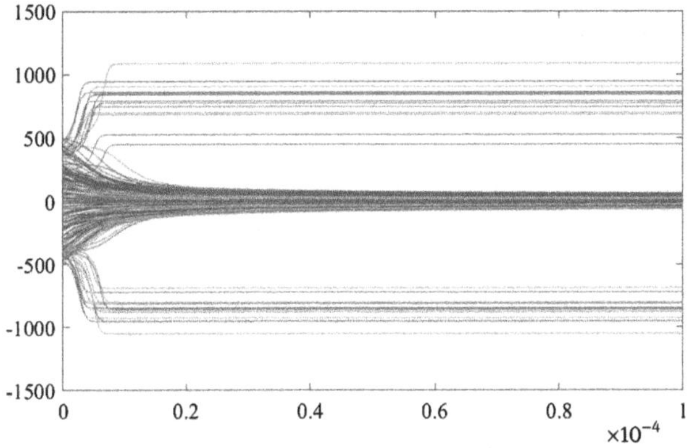

Figure 7.4: Zoomed time domain responses.

Example 7.12. Consider the following differential equations, and assess its stability:

$$\begin{cases} x_1'(t) = -4x_1(t) - 2x_1(t)x_2(t)\sin|x_1(t)|, \\ x_2'(t) = x_1(t)x_2(t) + 3x_2(t)\mathrm{e}^{-x_2(t)}. \end{cases}$$

Solutions. For this non-deliberately composed differential equation, if a theoretical method such as Lyapunov method is used, it may be almost impossible to give a clear answer. Therefore the simulation method can be tried. With the following statements, the responses in Figure 7.5 can be drawn. It can be seen that there are several cases where the curves tend to infinity. Therefore the differential equation is unstable.

```
>> f=@(t,x)[-4*x(1)-2*x(1)*x(2)*sin(abs(x(1)));
            x(1)*x(2)+3*x(2)*exp(-x(2))];
```

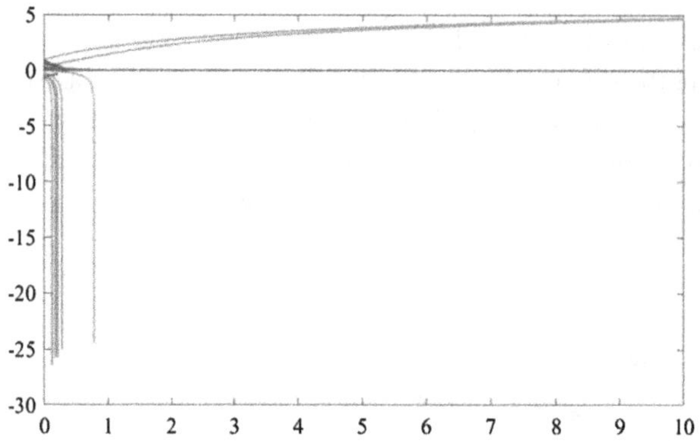

Figure 7.5: Time responses with several diverging cases.

```
ff=odeset; ff.RelTol=100*eps; ff.AbsTol=100*eps;
for i=1:10, i, x0=-1+2*rand(2,1); %random initial values
   [t,x]=ode15s(f,[0,10],x0,ff); line(t,x), drawnow
end
```

It can be seen that the simulation method is a practical one. Especially when there is a divergent case found, the assumption that the system is stable can be overturned. This cannot be established with the ordinary Lyapunov criterion.

7.2 Special behaviors of differential equations

In the stability discussions studied earlier, the solution of a stable nonlinear differential equation may settle down in a stable region, when t is large. While if a system is unstable, the solution may diverge. In real nonlinear differential equations, some strange behavior of differential equations may be witnessed. For instance, there can be limit cycles or chaotic behaviors. In this section, limit cycles, periodic solutions, chaos, and attractors are demonstrated with examples. Then the concepts of Poincaré map and sections are illustrated.

7.2.1 Limit cycles

In Chapter 4, the phase plane trajectory of van der Pol equation was demonstrated, where the trajectory would start from a certain point, and then settle done on a closed-path. Periodic motion then formed the path. The closed-path is known as the limit cycle of the differential equation. Limit cycle phenomena were first discovered by a French mathematician Jules Henri Poincaré (1854–1912) and a Swedish mathematician

Ivar Otto Bendixson (1861–1935). Limit cycles stimulated research in many fields such as physics, chemistry, and biology.

Theorem 7.4 (Poincaré–Bendixson theorem). *Each bounded solution of a two-dimensional autonomous system*

$$\begin{cases} y_1'(t) = f_1(y_1(t), y_2(t)), \\ y_2'(t) = f_2(y_1(t), y_2(t)) \end{cases} \tag{7.2.1}$$

either approaches $f_1(y_1(t), y_2(t)) = f_2(y_1(t), y_2(t)) = 0$, or a periodic solution, or a limit cycle, when t tends to infinity.

In this section, examples are used to demonstrate limit cycles through numerical solutions of differential equations.

Example 7.13. Consider a given differential equation model[32]

$$\begin{cases} x_1'(t) = A + x_1^2(t)x_2(t) - (B+1)x_1(t), \\ x_2'(t) = Bx_1(t) - x_1^2(t)x_2(t) \end{cases}$$

with $A = 1$ and $B = 3$, and initial values $x_1(0) = x_2(0) = 0$. Draw the phase plane trajectory. Selecting different initial values study the limit cycles.

Solutions. The equation was used to describe the dynamic process of a chemical reaction, known as Brusselator. It can be described by an anonymous function. Then regular commands can be used to solve the equations directly.

```
>> ff=odeset; ff.RelTol=100*eps; ff.AbsTol=100*eps;
   A=1; B=3; x0=[0; 0];
   f=@(t,x)[A+x(1)^2*x(2)-(B+1)*x(1); B*x(1)-x(1)^2*x(2)];
   tic, [t,x]=ode45(f,[0,10],x0,ff); toc %solve differential equation
```

The elapsed time of the solution process is about 0.56 seconds, and the number of points is 193 241. Since the relative error tolerance is set to a small number, the solution obtained is of high accuracy.

If the values are set to $x_1(0) = 4$ and $x_2(0) = 0$, the numerical solution must be found again. The phase plane trajectories from the two initial values are obtained as shown in Figure 7.6. It can be seen that no matter how the initial values are selected, the final phase plane curves settle down on the same closed-path, that is, the limit cycle.

```
>> tic, [t1,x1]=ode45(f,[0,200],[4; 0.5],ff); toc
   plot(x(:,1),x(:,2),x1(:,1),x1(:,2)) %draw the phase plane plots
```

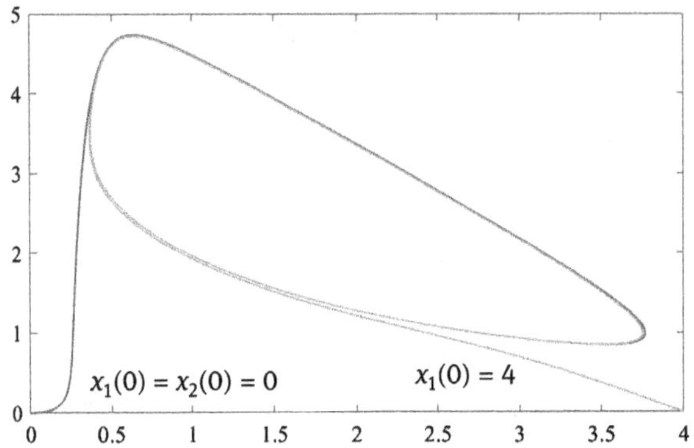

Figure 7.6: Limit cycle of the differential equation.

Example 7.14. Assume that in Example 7.13 the parameters such as A and B are not constants. They are dynamic functions of t, satisfying the differential equations[32] below. It was noticed that for the differential equations provided in the reference, the limit cycles are not observed. Therefore some modifications were made, and the new model is

$$\begin{cases} A'(t) = -k_1 A(t), \\ B'(t) = -k_2 B(t) X(t), \\ D'(t) = k_2 B(t) X(t), \\ E'(t) = k_4 X(t), \\ X'(t) = k_1(A(t)+1) - k_2(B(t)+4)X(t) + k_3 X^2(t) Y(t) - k_4 X(t), \\ Y'(t) = k_2(B(t)+4)X(t) - k_3 X^2(t) Y(t). \end{cases}$$

Assuming that $k_i = 1$, $i = 1, 2, 3, 4$, and $A(0) = 1$, $B(0) = 4$, the initial values of all the other state variables are zero. Solve again the new differential equations, and observe the limit cycles.

Solutions. Selecting state variables $x_1(t) = A(t)$, $x_2(t) = B(t)$, $x_3(t) = D(t)$, $x_4(t) = E(t)$, $x_5(t) = X(t)$, and $x_6(t) = Y(t)$, the original differential equations can be manually rewritten into the standard form:

$$\begin{cases} x_1'(t) = -k_1 x_1(t), \\ x_2'(t) = -k_2 x_2(t) x_5(t), \\ x_3'(t) = k_2 x_2(t) x_5(t), \\ x_4'(t) = k_4 x_5(t), \\ x_5'(t) = k_1(x_1(t)+1) - k_2(x_2(t)+4)x_5(t) + k_3 x_5^2(t) x_6(t) - k_4 x_5(t), \\ x_6'(t) = k_2(x_2(t)+4)x_5(t) - k_3 x_5^2(t) x_6(t). \end{cases}$$

The following commands can be used to describe the first-order explicit differential equations. Then the equations are solved, in 2.98 seconds, with the number of points being 1 325 665. The new phase plane trajectory is obtained as shown in Figure 7.7. If the solver is changed to `ode15s()`, the time is increased to 34.4 seconds.

```
>> k=[1,1,1,1]; x0=[1;4; zeros(4,1)];
   ff=odeset; ff.RelTol=100*eps; ff.AbsTol=100*eps;
   f=@(t,x)[-k(1)*x(1); -k(2)*x(2)*x(5);
       k(2)*x(2)*x(5); k(4)*x(5);
       k(1)*(x(1)+1)-k(2)*(x(2)+4)*x(5)+k(3)*x(5)^2*x(6)-k(4)*x(5);
       k(2)*(x(2)+4)*x(5)-k(3)*x(5)^2*x(6)];
   tic, [t,x]=ode45(f,[0,1000],x0,ff); toc
   plot(x(:,5),x(:,6)), length(t)  % draw phase plane plot
```

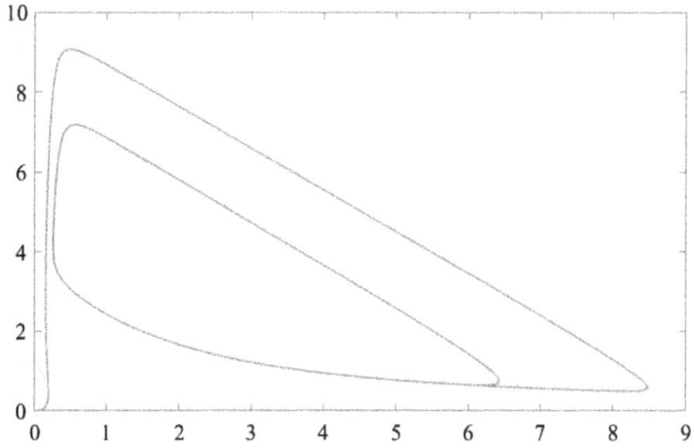

Figure 7.7: Limit cycle of the new differential equation.

Example 7.15. In fact, by observing the differential equations in Example 7.14, we notice that the signals $D(t)$ and $E(t)$ do not appear on the right-hand of the equations, indicating that they have no impact on the other state signals. Therefore they can be removed from the differential equations. Solve again the simplified differential equations.

Solutions. Letting $x_1(t) = A(t)$, $x_2(t) = B(t)$, $x_3(t) = X(t)$, and $x_4(t) = Y(t)$, the new differential equations can be manually written as

$$\begin{cases} x_1'(t) = -k_1x_1(t), \\ x_2'(t) = -k_2x_2(t)x_3(t), \\ x_3'(t) = k_1(x_1(t)+1) - k_2(x_2(t)+4)x_3(t) + k_3x_3^2(t)x_4(t) - k_4x_3(t), \\ x_4'(t) = k_2(x_2(t)+4)x_3(t) - k_3x_3^2(t)x_4(t). \end{cases}$$

Solving the simplified equations again, it can be seen that the solutions are exactly the same as those obtained earlier. The number of points is still the same. Since two redundant states were removed, the time needed was reduced to 2.83 seconds.

```
>> k=[1,1,1,1]; x0=[1; 4; 0; 0];
   ff=odeset; ff.RelTol=100*eps; ff.AbsTol=100*eps;
   f=@(t,x)[-k(1)*x(1); -k(2)*x(2)*x(3);
       k(1)*(x(1)+1)-k(2)*(x(2)+4)*x(3)+k(3)*x(3)^2*x(4)-k(4)*x(3);
       k(2)*(x(2)+4)*x(3)-k(3)*x(3)^2*x(4)];
   tic, [t,x]=ode45(f,[0,1000],x0,ff); toc
   plot(x(:,3),x(:,4)), length(t)  %draw the phase plane plot
```

Example 7.16 (Multiple limit cycle problems). Consider the following differential equations:[22]

$$\begin{cases} x'(t) = -y(t) + x(t)f\left(\sqrt{x^2(t)+y^2(t)}\right), \\ y'(t) = x(t) + y(t)f\left(\sqrt{x^2(t)+y^2(t)}\right) \end{cases}$$

where $f(r) = r^2\sin(1/r)$. Observe the limit cycles under different initial values.

Solutions. Letting $x_1(t) = x(t)$ and $x_2(t) = y(t)$, and substituting the $f(\cdot)$ function directly into the equation, the standard form the first-order explicit differential equations can be written as

$$\boldsymbol{x}'(t) = \begin{bmatrix} -x_2(t) + x_1(t)(x_1^2(t)+x_2^2(t))\sin 1/\sqrt{x_1^2(t)+x_2^2(t)} \\ x_1(t) + x_2(t)(x_1^2(t)+x_2^2(t))\sin 1/\sqrt{x_1^2(t)+x_2^2(t)} \end{bmatrix}.$$

The following anonymous function can be used to describe the first-order explicit differential equations. If two different initial values are selected as (0.1,0.2) and (0.1,0.01), we can solve the differential equations and then draw the phase plane trajectories in the same plot, as shown in Figure 7.8. It can be found that the two curves settle down as two different limit cycles.

```
>> D=@(x)x(1)^2+x(2)^2;
   f=@(t,x)[-x(2)+x(1)*D(x)*sin(1/sqrt(D(x)));
            x(1)+x(2)*D(x)*sin(1/sqrt(D(x)))];
   ff=odeset; ff.RelTol=100*eps; ff.AbsTol=100*eps;
   [t,x]=ode45(f,[0,100],[0.1; 0.2],ff); plot(x(:,1),x(:,2))
   [t,x]=ode45(f,[0,100],[0.1; 0.01],ff); line(x(:,1),x(:,2))
```

In fact, if a smaller initial value is tried, more limit cycles can be found. This indicates that the same differential equation may have different limit cycles. For this particular differential equation, there may be infinitely many limit cycles.

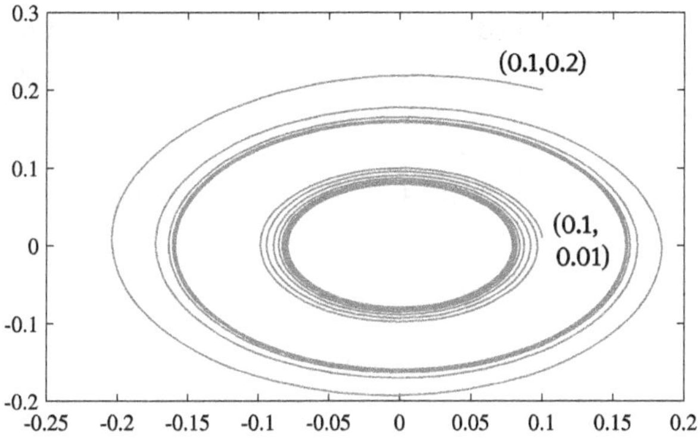

Figure 7.8: Different initial values may yield different limit cycles.

7.2.2 Periodic solutions

It has been indicated that if the real parts of a pair of complex conjugate eigenvalues of a linear differential equation with constant coefficients are zeros, when the transient response vanishes and the states of the system may have oscillations with equal magnitude. The system responses then can be regarded as periodic. Besides, the limit cycles can also be understood as periodic solutions of nonlinear differential equations. In fact, many differential equation solutions are periodic. For instance, such is the multibody system discussed earlier. In this section, examples are proposed to explore the periodicity of differential equation solutions, and how the period can be extracted from the numerical solutions.

Example 7.17. Solve the following linear differential equations with constant coefficients:

$$y^{(4)}(t) + 2y'''(t) + 3y''(t) + 4y'(t) + 2y(t) = u(t)$$

where $u(t) = 1$, $y(0) = 2$, $y'(0) = 1$, $y''(0) = y'''(0) = 0$, and $t \in (0, 30)$.

Solutions. It can be seen that this is a linear differential equation with constant coefficients. The analytical solution can be found with function `dsolve()`.

```
>> syms t y(t); d1y=diff(y);
   d1y=diff(y); d2y=diff(y,2); d3y=diff(y,3); d4y=diff(y,4);
   Y=dsolve(d4y+2*d3y+3*d2y+4*d1y+2*y==1,...
         y(0)==2, d1y(0)==1, d2y(0)==0, d3y(0)==0)
```

The analytical solution of the equation is

$$Y(t) = \frac{19}{9}e^{-t} + \frac{5}{3}te^{-t} - \frac{11}{18}\cos\sqrt{2}t + \frac{13\sqrt{2}}{18}\sin\sqrt{2}t + \frac{1}{2}.$$

It can be seen that the e^{-t} and the related terms are transient in the system responses. When t increases, these terms vanish gradually. The remaining ones are oscillating terms, with an additional constant 1/2. These terms may retain oscillation with equal magnitudes. The periodicity is $T_0 = 2\pi/\sqrt{2}$.

The analytical solution can be used to directly draw the time domain responses of the solution $y(t)$, as seen in Figure 7.9. It is seen that when the transient response vanishes, the output signal is oscillatory with equal magnitudes.

```
>> fplot(Y,[0,30])  % draw the exact curve of the output
```

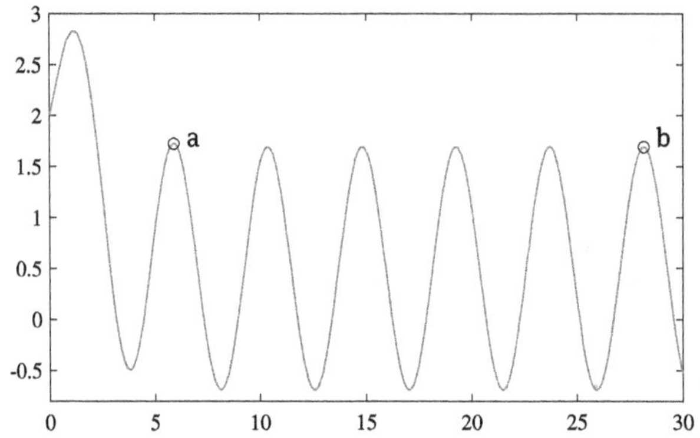

Figure 7.9: Time domain solution of signal $y(t)$.

How can we find the period of the signal? Two important subintervals $t \in (5, 10)$ and $t \in (25, 30)$ can be observed. The maximum value and time instance can be measured in these subintervals. In this period there are five cycles visible. The time difference can be computed and divided by 5, so that the period can be estimated. This may be more accurate than finding the period of a single cycle.

```
>> t=0:0.00001:30;
   y=(19*exp(-t))/9-(11*cos(2^(1/2)*t))/18+(5*t.*exp(-t))/3+...
     (13*2^(1/2)*sin(2^(1/2)*t))/18+1/2;
   i01=find(t>=5&t<10); [xm1,i1]=max(y(i01(1):end));
   i02=find(t>=25); [xm2,i2]=max(y(i02(1):end));
   t1=t(i01(1)+i1), t2=t(i02(1)+i2), T=(t2-t1)/5
   hold on, plot(t1,xm1,'ko',t2,xm2,'ko')
```

It can be seen that the time instances at a and b points can be found as $t_1 = 5.9233$ and $t_2 = 28.1493$, the average period is found as $T = 4.4452$, which is quite close to the theoretical value of $T_0 = 2\pi/\sqrt{2} = 4.4429$.

Linear differential equations may also have their limit cycles. The phase plane trajectory between $y(t)$ and $y'(t)$ can be obtained as shown in Figure 7.10, which is the limit cycle curve.

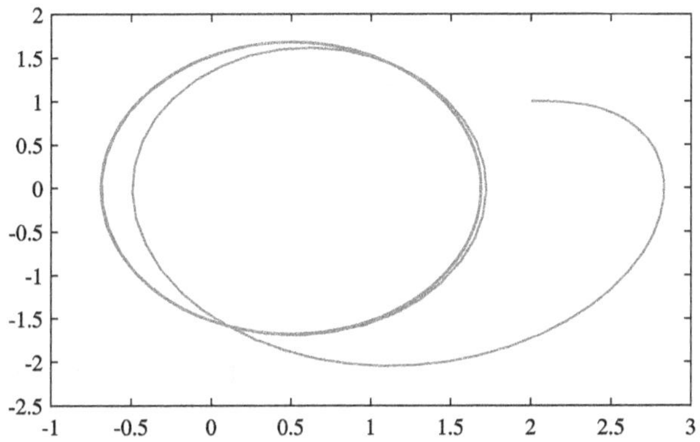

Figure 7.10: Limit cycles of linear differential equations.

Example 7.18. Solve again the differential equations in Example 7.13. Draw the time responses of the state variables and observe their periodicity.

Solutions. An anonymous function can be used to describe the differential equation model, and then regular commands can be used to solve the equations directly. The time responses of the equation can be found as shown in Figure 7.11. It can be seen that the time responses exhibit clear periodicity. However, the difference with a linear system is that the curves are no longer sine or cosine curves. They are distorted curves.

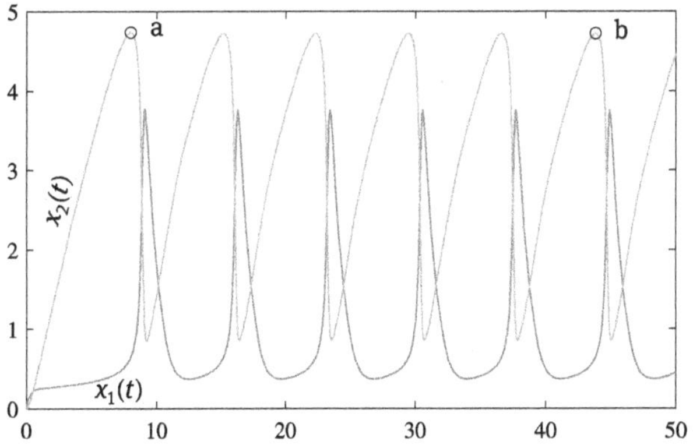

Figure 7.11: Time responses of the states.

```
>> ff=odeset; ff.RelTol=100*eps; ff.AbsTol=100*eps;
   A=1; B=3; x0=[0; 0];
   f=@(t,x)[A+x(1)^2*x(2)-(B+1)*x(1); B*x(1)-x(1)^2*x(2)];
   [t,x]=ode45(f,[0,10],x0,ff); plot(t,x) %differential equation solutions
```

Two circles are labeled in the figure, to indicate the first maximum points of $x_2(t)$ values. Now the subintervals $t \in (0, 10)$ and $t \in (40, 50)$ are used to extract the peak values and measure the time instances. They are respectively $t_1 = 7.9917$ and $t_2 = 43.7779$, therefore the period found is $T = 7.1573$.

```
>> i0=find(t>=10); [xm1,i1]=max(x(1:i0(1),2)); t1=t(i1)
   i0=find(t>=40); [xm2,i2]=max(x(i0(1):end,2));
   t2=t(i0(1)+i2), T=(t2-t1)/5    %compute the average period
   hold on, plot(t1,xm1,'ko',t2,xm2,'ko')
```

7.2.3 Chaos and attractors

The so-called chaos means that, for a determined system, if there is only a small difference in the initial values, the output signal changes significantly in a form which looks like a random one. This phenomenon is also known as chaotic behavior, which is not a random but deterministic one.

Although the observation of this phenomenon may be traced back to the Henri Poincaré's works, it first received attention in a paper by an American meteorologist Edward Lorenz [49]. In 1961, when Lorenz was solving differential equations, and simply used an initial value of 0.506127 instead of 0.506, completely different results were obtained.[28] He began studying the phenomenon and achieved a series of results. He coined a new term, "butterfly effect", to describe the phenomenon, since the phase space trajectory of his equation looks like a butterfly. The equation was solved in Chapter 3, and the butterfly curve can be seen in Figure 3.6.

What really became a sensation and widely spread was the title of Edward Lorenz's plenary talk in 1972 "Does the flap of a butterfly's wings in Brazil set off a tornado in Texas?".[50] Therefore, the butterfly effect became the precursor of chaos.

Several chaotic differential equations like Lorenz and Chua equations are simulated in this section, and the concept of attractors is presented.

Example 7.19. In [49], Edward Lorenz proposed a system of simplified differential equations:

$$\begin{cases} x'(t) = -\sigma x(t) + \sigma y(t), \\ y'(t) = -x(t)z(t) + rx(t) - y(t), \\ z'(t) = x(t)y(t) - bz(t) \end{cases}$$

where $\sigma = 10$, $b = 8/3$, and $r = 28$. The initial values are $x(0) = z(0) = 0$ and $y(0) = 0.01$. This example is close to that in Example 3.7. Find the state value at $t = 100$. If $y(0) = 0.02$, solve the differential equation again and observe the impact of the initial value to the results.

Solutions. If we let $x_1(t) = x(t)$, $x_2(t) = y(t)$ and $x_3(t) = z(t)$, the differential equations can be described directly and solved.

```
>> s=10; b=8/3; r=28; x0=[0; 0.01; 0];          % set parameters
   ff=odeset; ff.RelTol=100*eps; ff.AbsTol=100*eps;
   f=@(t,x)[-s*x(1)+s*x(2);
      -x(1)*x(3)+r*x(1)-x(2); x(1)*x(2)-b*x(3)];
   [t,x]=ode45(f,[0,100],x0,ff); x1=x(end,:) % final state value
   subplot(211), plot(t,x(:,2))
   x0(2)=0.02; [t,x]=ode45(f,[0,100],x0,ff); % another initial value
   x2=x(end,:), subplot(212), plot(t,x(:,2))
```

Although the phase space trajectories of the sets of solutions seem very close, the butterfly shaped trajectories are found in Figure 3.6, the progress has significant differences, as shown in Figure 7.12. The shapes of the curves of $y(t)$ are very close in the first 30 seconds. Afterward they are totally different. The terminal values are $\boldsymbol{x}_1(100) = [-14.2366, -20.8535, 26.7349]^{\mathrm{T}}$ and $\boldsymbol{x}_2(100) = [8.8711, 13.3504, 20.2361]^{\mathrm{T}}$, which implies that they have significant differences. It can be seen that a very small difference in initial values yields significant results.

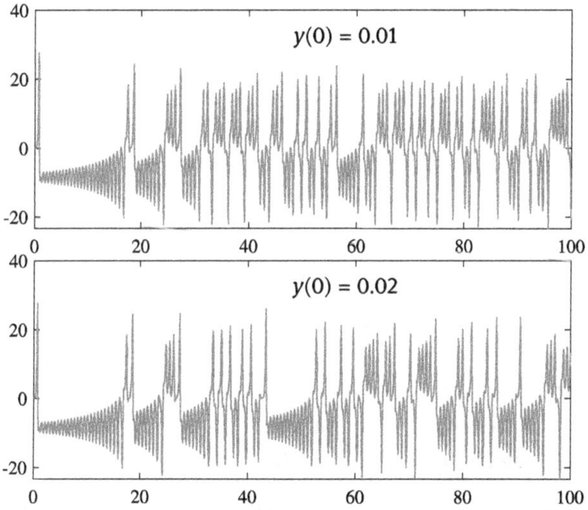

Figure 7.12: The curves of $y(t)$ signals in Lorenz equations.

An American scholar Leon Ong Chua (1936–) built a hardware circuit, known as Chua circuit, to reproduce chaotic behavior. The basic form of the circuit is illustrated in Figure 7.13(a), with the mathematical model being:[22]

$$\begin{cases} i'(t) = v_1(t)/L, \\ v_1'(t) = (v_2(t) - v_1(t))/(C_1R) - i(t)/C_1, \\ v_2'(t) = (v_1(t) - v_2(t))/(C_2R) - v_2(t)/(C_2r) \end{cases}$$

where r is the nonlinear element shown in Figure 7.13(b). The nonlinear element is referred to as Chua diode. Later, nonlinear elements of various forms appeared to model chaotic behavior of different kinds.

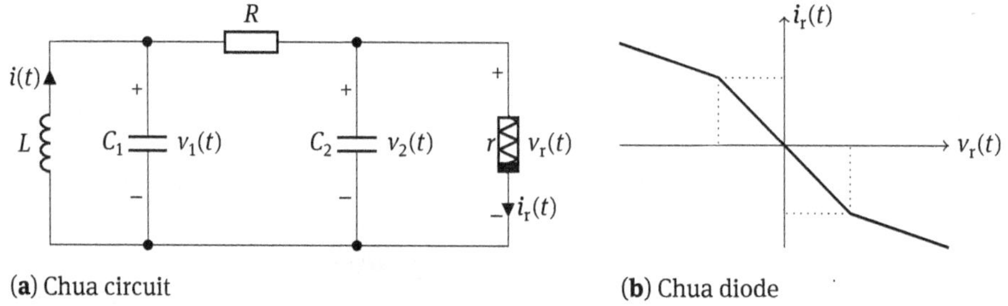

Figure 7.13: Chua circuit and diode curve.

Chua circuit opened up a new world for simulating chaotic behaviors with hardware circuits. Later similar circuits were proposed to model different chaotic behaviors,[7] including solid circuits and simulation models.

Example 7.20. According to Chua circuit, the dimensionless differential equations are

$$\mathbf{x}'(t) = \begin{bmatrix} \alpha(x_2(t) - h(x_1(t))) \\ x_1(t) - x_2(t) + x_3(t) \\ -\beta x_2(t) \end{bmatrix}$$

where the nonlinear function is

$$h(a) = m_1 a + \frac{1}{2}(m_0 - m_1)(|a+1| - |a-1|).$$

The Chua circuit has 4 control parameters α, β, m_0, and m_1. If the control parameters are selected as $\alpha = 9$, $\beta = 14.2886$, $m_0 = -1/7$, and $m_1 = 2/7$, simulate the Chua circuit and draw the phase plane trajectory of $x_1(t)$ and $x_2(t)$.

Solutions. The control parameters can be input into MATLAB workspace, and anonymous functions can be used to describe the Chua nonlinear element and state space equations, so that they can be solved. The phase plane trajectory is obtained as shown in Figure 7.14. Note that since the simulation time interval is large, the selection of the initial values are not thus important.

```
>> alfa=9; bet=14.2886; m0=-1/7; m1=2/7; x0=[0; 0; 0.2];
   h=@(a)m1*a+0.5*(m0-m1)*(abs(a+1)-abs(a-1));
   f=@(t,x)[alfa*(x(2)-h(x(1))); x(1)-x(2)+x(3); -bet*x(2)];
   ff=odeset; ff.RelTol=100*eps; ff.AbsTol=100*eps;
   [t,x]=ode45(f,[0,1000],x0,ff); plot(x(:,1),x(:,2))
```

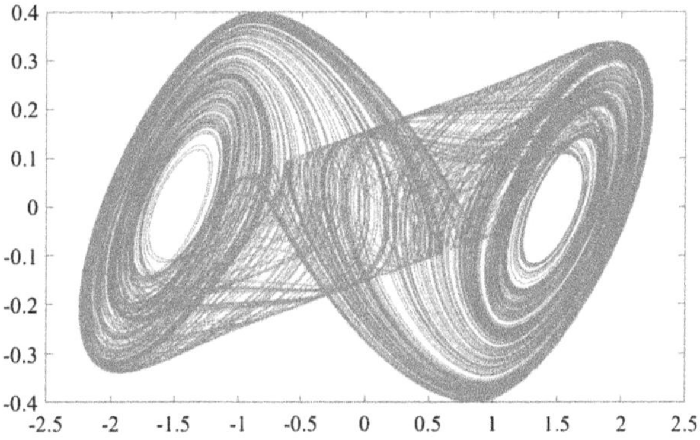

Figure 7.14: Phase plane trajectory of Chua circuit.

It can be seen from the chaotic behavior shown in Figure 7.14 that the trajectory is formed around two central points. The two central points are referred to as attractors. The curve in this example is known as a double-scroll attractor curve. The butterfly effect in Lorenz equations also has two attractors.

If the following nonlinear function is introduced, the chaotic behavior of an n-scroll attractor can be witnessed in theory. Note that q and n are not in one-to-one correspondence:

$$h(a) = m_{2q-1} + \frac{1}{2}\sum_{i=1}^{2q-1}(m_{i-1} - m_i)(|a + c_i| - |a - c_i|). \tag{7.2.2}$$

Example 7.21. Selecting the parameters $\alpha = 9$, $\beta = 14.2886$, $m_0 = 0.9/7$, $m_1 = -3/7$, $m_2 = 3.5/7$, $m_3 = -2.4/7$, $c_1 = 1$, $c_2 = 2.15$, and $c_3 = 4$, the Chua circuit can be simulated again. Observe the phase plane trajectory.

Solutions. The following statements can be used to directly construct function $h(a)$. Note that the function is suitable for any number of attractors. The vectors $\boldsymbol{m}$ and $\boldsymbol{c}$ should be written as row and column vectors, respectively, and the length of $\boldsymbol{m}$ is 1 more than that of $\boldsymbol{c}$. With such a nonlinear function, the 3-scroll attractors in phase plane are drawn, as seen in Figure 7.15. It can be seen from the curve that the phase plane trajectory indeed looks like a 3-scroll attractor model.

```
>> alfa=9; bet=14.2886; x0=[0; 0; 0.2];
   m=[0.9,-3,3.5,-2.4]/7; c=[1, 2.15, 4]';
   h=@(a)m(end)*a+0.5*(m(1:end-1)-m(2:end))*(abs(a+c)-abs(a-c));
   f=@(t,x)[alfa*(x(2)-h(x(1)));
          x(1)-x(2)+x(3); -bet*x(2)];
   ff=odeset; ff.RelTol=100*eps; ff.AbsTol=100*eps;
   [t,x]=ode45(f,[0,1000],x0,ff); plot(x(:,1),x(:,2))
```

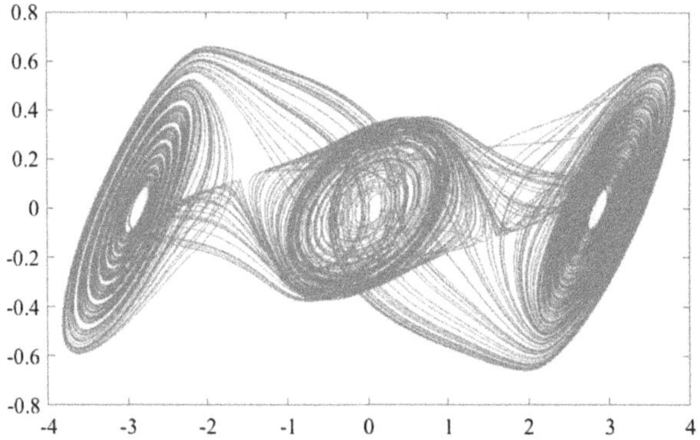

Figure 7.15: Phase plane trajectory of a 3-scroll attractor.

If the parameters are changed to $m_0 = -1/7$, $m_1 = 2/7$, $m_2 = -4/7$, $m_3 = 2/7$, $c_1 = 1$, $c_2 = 2.15$, and $c_3 = 3.6$, and again, $q = 2$, the following statements can be used to draw the phase plane trajectory of the Chua circuit, as shown in Figure 7.16. It is seen that this is a chaotic curve of the 4-scroll attractor model.

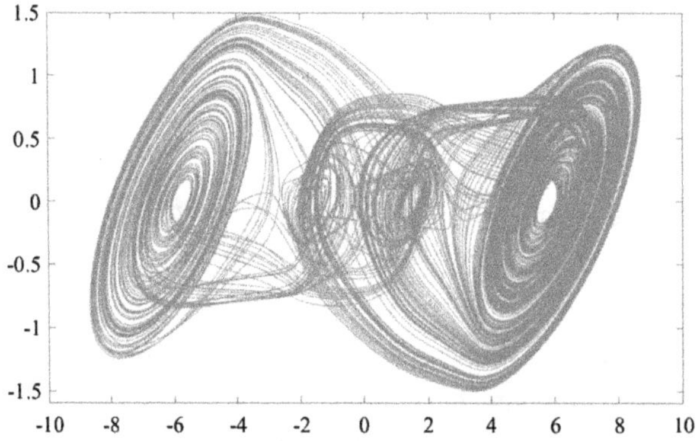

Figure 7.16: Phase plane portrait of a 4-scroll attractor.

```
>> m=[-1/7,2/7,-4/7,2/7]; c=[1;2.15;3.6];
   h=@(a)m(end)*a+0.5*(m(1:end-1)-m(2:end))*(abs(a+c)-abs(a-c));
   f=@(t,x)[alfa*(x(2)-h(x(1)));
           x(1)-x(2)+x(3); -bet*x(2)];
   [t,x]=ode45(f,[0,1000],x0,ff); plot(x(:,1),x(:,2))
```

It can be seen from the trajectory that the central point in-between the two attractors seems to be covered with the trajectories. In fact, this is not the case, since the plot shown in the figure is the projection on the $x_1(t)$–$x_2(t)$ plane. In a real three-dimensional display, as show in Figure 7.17, the attractors are located at different values of $x_3(t)$, such that the attractors are not covered.

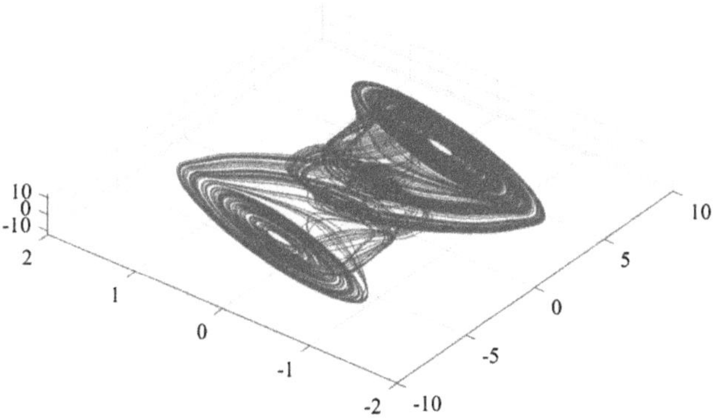

Figure 7.17: Phase space trajectory of 4-scroll attractors.

7.2.4 Poincaré mapping

Suppose in the differential equation $\boldsymbol{x}'(t) = \boldsymbol{f}(t, \boldsymbol{x}(t))$, the input $\boldsymbol{f}(t, \boldsymbol{x}(t))$ is a periodic function with period T. Imagine that there is a plane. In each cycle, the trajectory penetrates the plane once, and leaves a mark on it. A series of such marks can be generated in this way, to form the Poincaré section. The map is known as Poincaré map. Poincaré map is named after a French mathematician Jules Henri Poincaré (1854–1912). An example is introduced to demonstrate Poincaré map.

Example 7.22. Consider the following Duffing equation:

$$x''(t) + \delta x'(t) + \alpha x(t) + \beta x^3(t) = \gamma \cos \omega t$$

where the parameters are $\alpha = 1$, $\beta = 5$, $\delta = 0.02$, $\gamma = 8$, and $\omega = 0.5$. Draw the Poincaré section for the differential equation.

Solutions. Duffing equation is named after a German engineer Georg Duffing (1861–1944), who used it in describing nonlinear oscillation problems. It can be seen that the period of the input signal is $T = 2\pi/\omega$. Suppose 100 000 cycles of simulation are made and, with each cycle, a point in Poincaré section is computed, then the Poincaré section is obtained as shown in Figure 7.18. Note that the results should be marked by dots, rather than segments. Otherwise Poincaré section cannot be drawn.

```
>> a=1; b=5; d=0.02; gam=8; w=0.5; T=2*pi/w; %given parameters
   f=@(t,x)[x(2);
              -d*x(2)-a*x(1)-b*x(1)^3+gam*cos(w*t)];
   [t,x]=ode45(f,0:T:100000*T,[1; 0]);           %each cycle returns a point
   plot(x(:,1),x(:,2),'.','MarkerSize',2)        %draw the result in dots
```

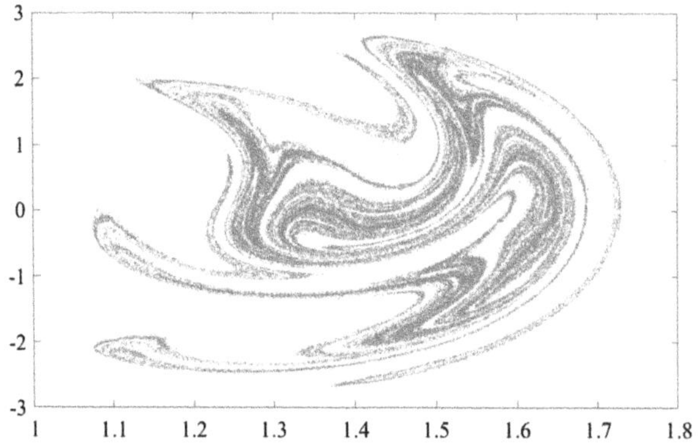

Figure 7.18: Poincaré section of Duffing equation.

Example 7.23. In fact, the beautiful Poincaré section shown in Figure 7.18 is questionable. In the solution process, the function `ode45()` is called with no control options. That is, the default relative error tolerance `RelTol`, or 10^{-3}, is used. This value is too large, and there may be large errors in this problem. Select a reasonable error tolerance and draw Poincaré section again.

Solutions. If a stricter, but reliable error tolerance 10^{-10} is used, the Poincaré section can be drawn again. After 593.3 seconds of waiting, the genuine Poincaré section can be redrawn, as shown in Figure 7.19. Compared with the plot in Figure 7.18, the accurate one looks not beautiful at all, and seems disordered. But it is the actual one.

```
>> a=1; b=5; d=0.02; gam=8; w=0.5; T=2*pi/w; %know parameters
   ff=odeset; ff.RelTol=1e-10; ff.AbsTol=1e-10;
   f=@(t,x)[x(2); -d*x(2)-a*x(1)-b*x(1)^3+gam*cos(w*t)];
   tic, [t,x]=ode45(f,0:T:100000*T,[1; 0],ff); toc
   plot(x(:,1),x(:,2),'.','MarkerSize',4)       %show the dots
```

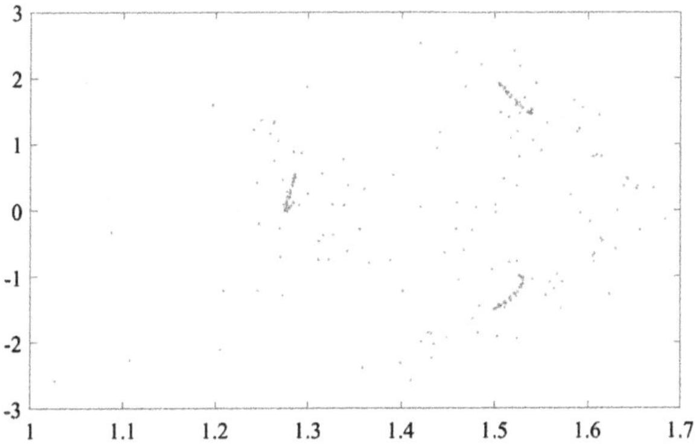

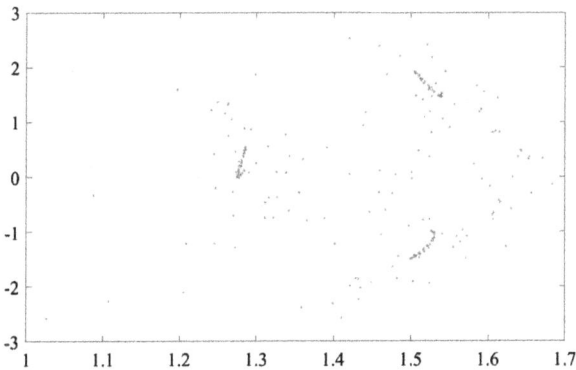

Figure 7.19: Genuine Poincaré section obtained by an accurate solver.

7.3 Linearization approximation of differential equations

Compared with linear differential equations, nonlinear ones are more complicated to handle. For the numerical solutions, the solution process is no more complicated for nonlinear differential equations than for linear ones, however, for property analysis, nonlinear differential equations are far more complicated. Therefore, sometimes, nonlinear differential equations must be approximated as linear ones. The approximation method is known as linearization of nonlinear differential equations.

It is not practical to carry out linearization for nonlinear differential equations over the entire time span, and it is also meaningless. Linearization must be carried out in a certain neighborhood of a point of interest. How can we select a meaningful neighborhood? The local range of the equilibrium point is usually a good choice. Therefore the concept of computation of equilibrium point is first presented in this section, followed by the linearization method. The properties of equilibrium points are also addressed in the section.

7.3.1 Equilibrium points

Imagine a particle moving in the space. What conditions must be satisfied such that the particle stops moving? When the projections of speed at all the axes are zero, the particle stops moving. The position the particle where it stops moving is referred to as the equilibrium point. Normally speaking, a differential equation may have several equilibria.

How can we find all the equilibria of a given differential equation? Suppose the first-order explicit autonomous differential equation is known as

$$\boldsymbol{x}'(t) = \boldsymbol{f}(\boldsymbol{x}(t)) \tag{7.3.1}$$

where $\boldsymbol{x}'(t)$ is the speed of the state variables, or understood as the projection of the speed of the particle at all the axes. If all the components of the speed equal to zero, the particle stops moving, so that an equilibrium point of the particle can be found. Unfortunately, this method applies only to autonomous differential equations. If a differential equation contains t explicitly, the number of equations is smaller than the number of unknowns, the equilibrium points cannot be found.

Theorem 7.5. *The equilibria of an autonomous state equation* $\boldsymbol{x}'(t) = \boldsymbol{f}(\boldsymbol{x}(t))$ *can be found from the following algebraic equation:*

$$\boldsymbol{f}(\boldsymbol{x}(t)) = \boldsymbol{0}. \tag{7.3.2}$$

The following examples are used to demonstrate the positions of the equilibria, and their properties are also illustrated.

Example 7.24. Find the equilibria for the following nonlinear differential equations with two states:

$$\begin{cases} x'(t) = x(t)(1 - x(t) + y(t)), \\ y'(t) = y(t)(1 + x(t) - 2y(t)). \end{cases}$$

Solutions. Since the expressions of $x'(t)$ and $y'(t)$ are known, the algebraic equations can be solved directly to find the equilibria:

$$\begin{cases} x'(t) = x(t)(1 - x(t) + y(t)) = 0, \\ y'(t) = y(t)(1 + x(t) - 2y(t)) = 0. \end{cases}$$

A symbolic function `solve()` in MATLAB can be used to analytically find the solution of the algebraic equations.

```
>> syms x y;
   [x0,y0]=solve(x*(1-x+y)==0, y*(1+x-2*y)==0) %solve algebraic equations
```

The solutions obtained are $\boldsymbol{x}_0 = [0,1,0,3]^{\mathrm{T}}$, $\boldsymbol{y}_0 = [0,0,1/2,2]^{\mathrm{T}}$, which implies that there are four equilibria $(0,0)$, $(1,0)$, $(0,1/2)$, and $(3,2)$.

What is the relationship between the equilibria and solutions? A simulation experiment can be made. In the square $0 \leqslant x, y \leqslant 1$, 100 initial points are generated randomly, and the differential equations are solved such that the phase plane trajectories can be drawn, as shown in Figure 7.20. It can be seen that all the solutions flow to the equilibrium point $(3,2)$.

```
>> f=@(t,x)[x(1)*(1-x(1)+x(2)); x(2)*(1+x(1)-2*x(2))];
   for i=1:100, x0=rand(2,1); %select 100 random initial values
      [t,x]=ode45(f,[0,100],x0,ff); line(x(:,1),x(:,2))
   end
```

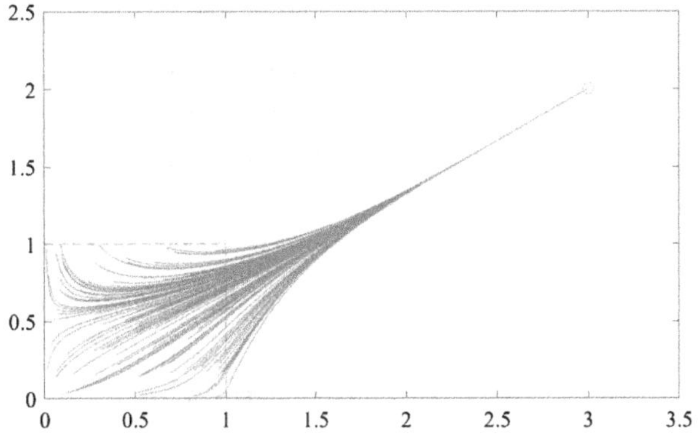

Figure 7.20: Trajectories from different initial values.

Which initial values may yield solutions flowing to other equilibria? If an initial point is located on the x axis, the solution may flow to the equilibrium point $(1,0)$; if the initial point is on the y axis, the solution may settle down at $(0,1/2)$. The origin $(0,0)$ is an isolated point. Unless the initial point is selected at the origin, all other positions make the point move away and flow to other equilibria.

It can be seen from the example that if the algebraic equation is analytically solvable, function `solve()` can be used to find all the equilibria in one function call. Unfortunately, in real applications, it is not the case. Function `vpasolve()` can be used, however, only one solution can be found at a time. If the equation set has multiple solutions, a dedicated solver provided in Volume IV, `more_sols()`, can be used to find all the solutions in one function call.

Example 7.25. Find all the equilibria for the Chua circuit studied in Example 7.20.

Solutions. Chua circuit is itself a first-order explicit differential equation. Therefore there is no need for standard form conversion. The corresponding algebraic equations can be solved directly. Since the algebraic equations are nonlinear, and may have multiple solutions, ordinary solvers are not suitable to solve equations like that. Function `more_sols()` can be executed directly, with the initial solution set to `zeros(3,1,0)`, indicating the equation is a 3×1 matrix equation, and 0 means there is no known solution at the beginning. After the solution process, the solution can be returned to MATLAB workspace in a three-dimensional array $\boldsymbol{X}$, whose component $\boldsymbol{X}(:,:,i)$ is the ith solution.

```
>> alfa=9; bet=14.2886; m0=-1/7; m1=2/7; x0=[0; 0; 0.2];
   h=@(a)m1*a+0.5*(m0-m1)*(abs(a+1)-abs(a-1));
   f=@(x)[alfa*(x(2)-h(x(1))); x(1)-x(2)+x(3); -bet*x(2)];
   more_sols(f,zeros(3,1,0)), X
```

With the above statements, three equilibria can be found at (0, 0, 0), (−1.5, 0, 1.5), and (1.5, 0, −1.5). Compared with Figure 7.13, it can be seen that the latter two equilibria are the double-scroll attractors. The origin is an isolated equilibrium point.

In fact, if the phase space trajectory of Chua circuit is drawn, and rotated properly, the effect in Figure 7.21 can be found. In fact, besides the two equilibria on the two sides, all trajectories avoid the origin. Therefore, the origin is also one of the equilibria.

```
>> F=@(t,x)f(x); x1=X(1,1,:); x2=X(2,1,:); x3=X(3,1,:);
   ff=odeset; ff.RelTol=100*eps; ff.AbsTol=100*eps;
   [t,x]=ode45(F,[0,1000],x0,ff);  % solve the differential equation
   plot3(x(:,1),x(:,2),x(:,3),x1(:),x2(:),x3(:),'o')
```

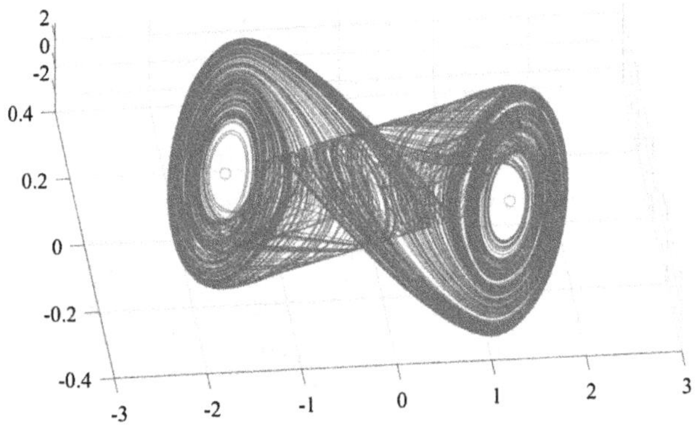

Figure 7.21: Equilibria and attractors in Chua circuit.

Example 7.26. Find the equilibria of the 3-scroll attractors circuit in Example 7.21 and locate the positions of the attractors.

Solutions. The initial algebraic equations can be expressed by an anonymous function, and the equilibria can be found.

```
>> alfa=9; bet=14.2886; x0=[0; 0; 0.2];
   m=[0.9,-3,3.5,-2.4]/7; c=[1, 2.15, 4]';
   h=@(a)m(end)*a+0.5*(m(1:end-1)-m(2:end))*(abs(a+c)-abs(a-c));
   f=@(x)[alfa*(x(2)-h(x(1))); x(1)-x(2)+x(3); -bet*x(2)];
   more_sols(f,zeros(3,1,0)), X % find all the solutions
```

It is somewhat surprising to see that there are altogether seven equilibria found: at (0, 0, 0), (−5.6354, 0, 5.6354), (5.6354, 0, −5.6354), (−1.3, 0, 1.3), (2.8786, 0, −2.8786), (−2.8786, 0, 2.8786), and (1.3, 0, −1.3). The following statements can be used to solve again the differential equations, and draw the phase space trajectories. The equilibria are all superimposed on the chaotic trajectories, as shown in Figure 7.22. It can be seen

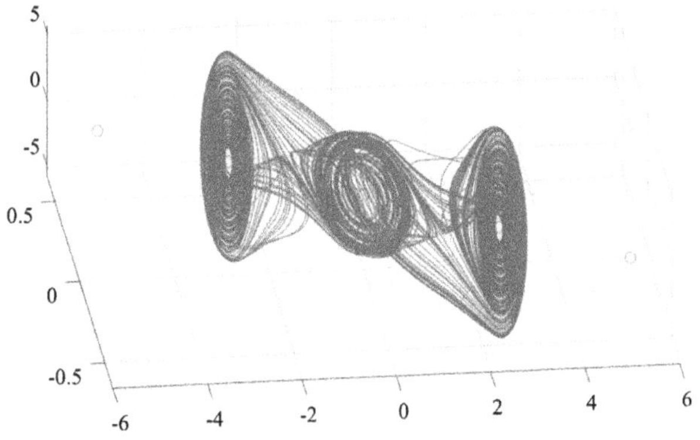

Figure 7.22: Equilibria and attractors.

the leftmost and rightmost ones are unstable equilibria. The second, fourth, and sixth from left are the 3-scroll attractors, while the third and fifth are isolated equilibria.

```
>> F=@(t,x)f(x); x0=[0; 0; 0.2];
   ff=odeset; ff.RelTol=100*eps; ff.AbsTol=100*eps;
   [t,x]=ode45(F,[0,1000],x0,ff);               % solve differential equation
   x1=X(1,1,:); x2=X(2,1,:); x3=X(3,1,:); % draw equilibrium points
   plot3(x(:,1),x(:,2),x(:,3),x1(:),x2(:),x3(:),'o')
```

The equilibria can further be classified into stable, unstable, and saddle points. They can be further judged according to the linearized models.

7.3.2 Linearization of nonlinear differential equations

For an equilibrium point $\boldsymbol{x}_0$, the linearized approximation to the original differential equations can be obtained. In this section, linearization is explained and MATLAB based solutions are illustrated.

First let us consider a simple autonomous differential equation

$$\begin{cases} x'(t) = F(x(t), y(t)), \\ y'(t) = G(x(t), y(t)) \end{cases} \tag{7.3.3}$$

for which an equilibrium point (x_0, y_0) is known. Selecting a nearby point $(x_0+\Delta x, y_0+\Delta y)$, such that $|\Delta x| \ll 1$ and $|\Delta y| \ll 1$, it can be found that

$$\begin{cases} F(x_0+\Delta x, y_0+\Delta y) = \dfrac{\partial}{\partial x}F(x_0, y_0)\Delta x + \dfrac{\partial}{\partial y}F(x_0, y_0)\Delta y + \cdots, \\ G(x_0+\Delta x, y_0+\Delta y) = \dfrac{\partial}{\partial x}G(x_0, y_0)\Delta x + \dfrac{\partial}{\partial y}G(x_0, y_0)\Delta y + \cdots \end{cases} \tag{7.3.4}$$

where the higher-order terms of Δx and Δy can be omitted, and the terms $F(x_0, y_0)$ and $G(x_0, y_0)$ can also be omitted, since at the equilibrium point, they are both zeros. It is not hard to find from (7.3.4) that

$$\begin{cases} \Delta x'(t) \approx \dfrac{\partial}{\partial x} F(x_0, y_0)\Delta x(t) + \dfrac{\partial}{\partial y} F(x_0, y_0)\Delta y(t), \\ \Delta y'(t) \approx \dfrac{\partial}{\partial x} G(x_0, y_0)\Delta x(t) + \dfrac{\partial}{\partial y} G(x_0, y_0)\Delta y(t). \end{cases} \tag{7.3.5}$$

Letting $z_1(t) = \Delta x(t)$ and $z_2(t) = \Delta y(t)$, the linear state space equation can be written for $\boldsymbol{z}(t)$ as

$$\boldsymbol{z}'(t) = \boldsymbol{J}\boldsymbol{z}(t) \tag{7.3.6}$$

where $\boldsymbol{J}$ is the Jacobian matrix having the mathematical form

$$\boldsymbol{J} = \begin{bmatrix} \dfrac{\partial}{\partial x} F(x_0, y_0) & \dfrac{\partial}{\partial y} F(x_0, y_0) \\ \dfrac{\partial}{\partial x} G(x_0, y_0) & \dfrac{\partial}{\partial y} G(x_0, y_0) \end{bmatrix}. \tag{7.3.7}$$

Theorem 7.6. *More generally, the nonlinear first-order explicit differential equations in* (3.1.1) *can be linearized into the form of* (7.3.6), *where*

$$\boldsymbol{J} = \begin{bmatrix} \partial f_1/\partial x_1 & \partial f_1/\partial x_2 & \dots & \partial f_1/\partial x_n \\ \partial f_2/\partial x_1 & \partial f_2/\partial x_2 & \dots & \partial f_2/\partial x_n \\ \vdots & \vdots & \ddots & \vdots \\ \partial f_n/\partial x_1 & \partial f_n/\partial x_2 & \dots & \partial f_n/\partial x_n \end{bmatrix}_{(x_1(0), x_2(0), \dots, x_n(0))}. \tag{7.3.8}$$

Linearization of differential equations can be carried out through symbolic computation, since the function `jacobian()` provided in Symbolic Math Toolbox can be used in computing the Jacobian matrix. In this section, linearization method is demonstrated through examples.

Example 7.27. Find the linearized models of the following differential equations:

$$\begin{cases} x_1'(t) = -2x_1(t) + x_2(t) - x_3(t) - x_1^2(t)\mathrm{e}^{x_1(t)}, \\ x_2'(t) = x_1(t) - x_2(t) - 4x_1^3(t)x_2(t) - x_3^2(t), \\ x_3'(t) = x_1(t) + x_2(t) - x_3(t) + 8\mathrm{e}^{x_1(t)}(x_2^2(t) + x_3^2(t)). \end{cases}$$

Solutions. It can be seen from the model that the equilibrium point of the original system is $(0, 0, 0)$, so the Jacobian matrix of the linearized model from symbolic computation can be obtained. Also the linearized matrix and the eigenvalues at the equilibrium point are

```
>> syms x1 x2 x3
   F=[-2*x1+x2-x3-x1^2*exp(x1);
      x1-x2-4*x1^3*x2-x3^2;
      x1+x2-x3+8*exp(x1)*(x2^2+x3^2)];
   A=jacobian(F,[x1;x2;x3])          %Jacobi matrix
   A0=subs(A,{x1,x2,x3},{0,0,0})  %linearized model at equilibrium point
   double(eig(A0))                    %eigenvalues of the linearized model
```

The Jacobian matrix $\boldsymbol{A}$ of the differential equation and the linearization matrix $\boldsymbol{A}_0$ at the equilibrium point can be found. The eigenvalues of matrix $\boldsymbol{A}_0$ are −2.4656 and −0.7672 ± 0.7926j. Since all of them have negative real parts, the linearized model is stable:

$$\boldsymbol{A}=\begin{bmatrix} -x_1^2\mathrm{e}^{x_1}-2x_1\mathrm{e}^{x_1}-2 & 1 & -1 \\ -12x_2x_1^2+1 & -4x_1^3-1 & -2x_3 \\ 8\mathrm{e}^{x_1}(x_2^2+x_3^2)+1 & 16x_2\mathrm{e}^{x_1}+1 & 16x_3\mathrm{e}^{x_1}-1 \end{bmatrix},\quad \boldsymbol{A}_0=\begin{bmatrix} -2 & 1 & -1 \\ 1 & -1 & 0 \\ 1 & 1 & -1 \end{bmatrix}.$$

Now observe the numerical solutions of the original nonlinear differential equations. If an anonymous function is used to describe them, and very small initial values are selected randomly, numerical solutions can be found. If 100 simulations are made, the solutions are shown in Figure 7.23. It can be seen that the solutions from some of the initial values are divergent, therefore the original system of differential equations is unstable.

```
>> f=@(t,x)[-2*x(1)+x(2)-x(3)+x(1)^2*exp(x(1));
            x(1)-x(2)-4*x(1)^3*x(2)-x(3)^2;
            x(1)+x(2)-x(3)+8*exp(x(1))*(x(2)^2+x(3)^2)];
   ff=odeset; ff.RelTol=100*eps; ff.AbsTol=100*eps;
```

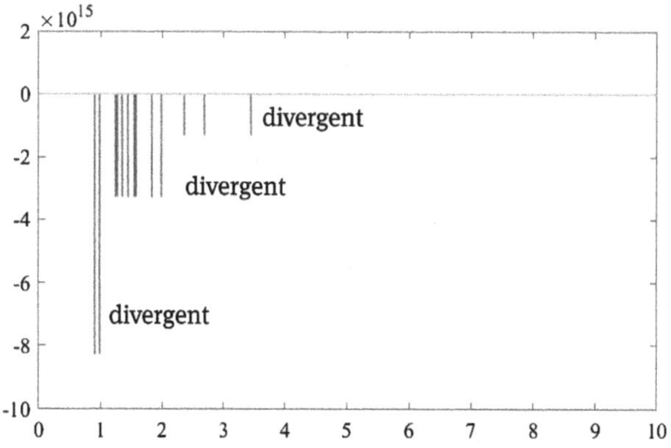

Figure 7.23: Solutions of the nonlinear differential equations from 100 initial values.

```
for i=1:100, i, x0=-0.15+0.3*rand(3,1);
   [t,x]=ode45(f,[0,10],x0,ff); line(t,x); % solve differential equations
end
```

Example 7.28. Find the linearized model of the Chua circuit in Example 7.20, and find the eigenvalues of the matrix at the equilibria.

Solutions. The following commands can be used to compute the linearized model:

```
>> alfa=9; bet=14.2886; m0=-1/7; m1=2/7;
   syms x1 x2 x3 a;
   h(a)=m1*a+0.5*(m0-m1)*(abs(a+1)-abs(a-1));
   f=[alfa*(x2-h(x1)); x1-x2+x3; -bet*x2];
   J=jacobian(f,[x1,x2,x3])        % compute Jacobian matrix
```

The Jacobian matrix can be found as

$$J = \begin{bmatrix} 27\mathrm{sgn}(x_1+1)/14 - 27\mathrm{sgn}(x_1-1)/14 - 18/7 & 9 & 0 \\ 1 & -1 & 1 \\ 0 & -71\,443/5\,000 & 0 \end{bmatrix}.$$

Substituting the three equilibria $(0,0,0)$, $(-1.5,0,1.5)$, and $(1.5,0,-1.5)$ into the Jacobian matrix, the corresponding matrices can be found.

```
>> J1=subs(J,{x1,x2,x3},{0,0,0})
   J2=subs(J,{x1,x2,x3},{-1.5,0,1.5})
   J3=subs(J,{x1,x2,x3},{1.5,0,-1.5}) % the three equilibrium points
   double(eig(J1)), double(eig(J2)), double(eig(J3))
```

Therefore the three Jacobian matrices are obtained, and their eigenvalues have positive real parts, such that the three linearized models are all unstable.

$$J_1 = \begin{bmatrix} 9/7 & 9 & 0 \\ 1 & -1 & 1 \\ 0 & -71\,443/5\,000 & 0 \end{bmatrix}, \quad J_2 = J_3 = \begin{bmatrix} -18/7 & 9 & 0 \\ 1 & -1 & 1 \\ 0 & -71\,443/5\,000 & 0 \end{bmatrix}.$$

For this specific problem, the unstable linearized model is not of any value, since the approximation to the original nonlinear model is not satisfactory, and one is not likely to see chaotic behavior in the linearized model.

It can be seen from the two examples that the stability judgements from linearized models yield totally different results as for the original nonlinear differential equations. This sufficiently indicates that linearized models should not be used to judge the stability of the original nonlinear differential equations. If a nonlinear differential equation has nonnegligible nonlinearities, no useful results may be concluded from its linearized models. Sometimes erroneous or misleading results can be obtained. Therefore the behavior analysis should be made based on the original differential equations.

7.3.3 Properties of equilibria

The computation of equilibria is discussed in this section. The Jacobian matrix computation is also introduced. If one substitutes the information of a certain equilibrium point into the Jacobian matrix, the system can be approximated around this equilibrium point. The eigenvalues of the Jacobian matrix $\boldsymbol{A}$ determine the properties of this equilibrium point.

If all the eigenvalues have negative real parts, the equilibrium point is stable. The trajectory converges to the equilibrium point. If some of the eigenvalues have a positive real part, while others have negative ones, the point is referred to as a saddle point. If all the eigenvalues have positive real parts, a trajectory may diverge.

In fact, for the three equilibria in Example 7.28, some of the eigenvalues have a positive real part, while some have negative. Therefore they are all saddle points.

7.4 Bifurcation of differential equations

Bifurcation is a special behavior of some nonlinear differential equations subject to parameter variations. The so-called bifurcation means that under certain parameters, the stability of the differential equation may change. Here only a simple example of bifurcation is given.

Example 7.29. Consider the differential equation in Example 7.13. If $A = 1$, observe the impact of parameter B to the behavior of the differential equation.

Solutions. The Jacobian matrix can be obtained under the symbolic framework, and the equilibria are found. It is found that $x_{10} = A$, $x_{20} = A/B$. Substituting these values into the Jacobian matrix, the eigenvalues of the matrix can be found.

```
>> syms A B x1 x2
   F=[A+x1^2*x2-(B+1)*x1; B*x1-x1^2*x2]; % two given functions
   J=jacobian(F,[x1 x2])              % compute Jacobian matrix
   [x10,x20]=solve(F,[x1,x2])         % compute the equilibrium
   J0=subs(J,{x1,x2},{x10,x20})       % substitute them into Jacobian matrix
   e1=simplify(eig(J0))               % compute eigenvalues of Jacobian
```

The Jacobian matrix $\boldsymbol{J}$ and $\boldsymbol{J}_0$ at the equilibrium point are

$$\boldsymbol{J} = \begin{bmatrix} 2x_1x_2 - B - 1 & {x_1}^2 \\ B - 2x_1x_2 & -{x_1}^2 \end{bmatrix}, \quad \boldsymbol{J}_0 = \begin{bmatrix} B-1 & A^2 \\ -B & -A^2 \end{bmatrix}.$$

The eigenvalues of the Jacobian matrix at the equilibrium point are

$$\lambda_{1,2} = \frac{1}{2}(B - A^2 - 1) \pm \frac{1}{2}\sqrt{(A^2 - 2A - B + 1)(A^2 + 2A - B + 1)}.$$

It can be seen that $(B-A^2-1)/2 = 0$ is a branching point. If $B > A^2+1$, then at least one eigenvalue is positive, so the differential equation is unstable. When $B < A^2 + 1$, the differential equation is stable. The point $B = A^2+1$ can be regarded as a bifurcation point.

The conclusion “when $B > A^2+1$, the differential equation is unstable” is reached based on the Jacobian matrix and the linearized model. In fact, when $A = 1$ and $B = 3$, the limit cycle in Figure 7.6 was found such that the differential equation is still stable. Therefore for the original differential equation, there is no impact at the “bifurcation point”.

In other words, although a bifurcation point was found, the point was obtained from the linearized model, whereas the original differential equation does not have these bifurcation phenomena. The so-called bifurcation in this example may be the feature brought by the linearized model. Bifurcation issues are not further discussed in this book.

7.5 Exercises

7.1 Judge the stability of $s^6 + 6s^5 + 16s^4 + 25s^3 + 24s^2 + 14s + 4$. Compose a linear differential equation with constant coefficients from the polynomial, and validate the stability by the solution curve.

7.2 Assess the stability of the following linear differential equation, and validate the solution with the analytical solution and the curve of the solution if

$$\boldsymbol{x}'(t) = \begin{bmatrix} -2 & -1 & -1 & 1 & 0 \\ 1 & -3 & -2 & 0 & 1 \\ -2 & 2 & -2 & -1 & 0 \\ -1 & 2 & -1 & -4 & 1 \\ -4 & 3 & 1 & -1 & -3 \end{bmatrix} \boldsymbol{x}(t) + \begin{bmatrix} 1 \\ 1 \\ -1 \\ 0 \\ -2 \end{bmatrix} u(t)$$

where $\boldsymbol{x}(0) = [-1, 2, 1, 0, 1]^{\mathrm{T}}$ and $u(t) = 1$.

7.3 The mathematical form of Duffing equation is

$$\begin{cases} x'(t) = y(t), \\ y'(t) = x(t) - x^3(t) - \epsilon y(t) + \gamma \cos \omega t \end{cases}$$

where $y = 0.3$ and $\omega = 1$. If the parameter ϵ is selected as 0.22 and 0.15, respectively, solve the differential equation, and observe the phase trajectory.

7.4 In [7], a set of Chua circuit parameters with 10-scroll attractors are given, where $\alpha = 9.35$ and $\beta = 11.4$. When i is even, $m_i = -1.4$, while when is odd, $m_i = -0.6$. Besides, $c_1 = 1$, $c_2 = 1.9$, $c_3 = 2.6$, $c_4 = 3.75$, $c_5 = 4.75$, $c_6 = 5.85$, $c_7 = 6.46$, $c_8 = 7.5$, and $c_9 = 8.55$. Observe the phase space trajectory with MATLAB, and also note the attractors.

7.5 Find the equilibria and linearized models of the following differential equation. Assess the behavior of each equilibrium point if

$$\begin{cases} x'(t) = 2x(t)(1 - x(t)/2) - x(t)y(t), \\ y'(t) = 3y(t)(1 - y(t)/3) - 2x(t)y(t). \end{cases}$$

7.6 Select an initial value $\boldsymbol{x}_0 = [\texttt{eps}, 0]^{\mathrm{T}}$ and solve the differential equation in Example 7.24. Observe to which of the equilibria the differential equation trajectory may finally converge to.

7.7 Find all the equilibria of the 4-scroll attractor in Example 7.21, and observe the behaviors of each of them.

8 Fractional-order differential equations

It is known that $\mathrm{d}^n y/\mathrm{d}x^n$ is used to represent the nth order derivative of y with respect to x. What if $n = 1/2$? This was the question asked by a French mathematician Guillaume François Antoine L'Hôpital (1661–1704) to one of the inventors of calculus, Gottfried Wilhelm Leibniz, 300 years ago.[70] This marked the beginning of fractional calculus. Strictly speaking, the term "fractional-order" is misused, proper names should be "noninteger-order" or "arbitrary-order". The irrational number $\sqrt{2}$ can be used as the order, but it is not a fraction. Since the term "fractional-order" is already widely used in the research community, it is also used in this book, while it really means arbitrary-order.

Although the fractional calculus research has more than 300 years of history, the earlier research was concentrated only on theoretical issues. In recent decades, this research has been introduced into many fields. For instance, in automatic control, fractional-order control systems is a relatively new and active research topic. In this chapter, we concentrate on introducing the definitions and various computing methods. The solutions of linear and nonlinear fractional-order differential equations are fully discussed.

A dedicated MATLAB toolbox for fractional calculus and fractional-order control has been designed[74, 76] by the author of this book. It is named FOTF Toolbox, downloadable for free from the following website:[75]

```
http://cn.mathworks.com/matlabcentral/fileexchange/60874-fotf-toolbox
```

The toolbox is also included in the resources of this book.

The numerical issues of fractional calculus are addressed in this chapter. In Section 8.1, the definitions and properties of fractional calculus are introduced. Numerical computation problems in fractional calculus are introduced. In Section 8.2, Mittag-Leffler functions are introduced, and, based on them, the analytical solutions of linear commensurate-order fractional-order differential equations are studied with partial fraction expansion techniques. Of course, the number of fractional-order equations, which can be solved in this way, is quite small. The focus of this chapter is on numerical solutions. In Section 8.3, the numerical solutions of linear fractional-order differential equations are addressed, and in Section 8.4, the numerical solutions of nonlinear fractional-order differential equations are presented. We concentrate on finding the $o(h^p)$ high-precision solutions.

8.1 Definitions and numerical computation in fractional calculus

Fractional calculus is a direct extension of traditional calculus, which is referred to as integer-order calculus in this book. In this section, some widely used definitions of fractional calculus are introduced. The relationships between different definitions

https://doi.org/10.1515/9783110675252-008

are also presented. The numerical algorithms and solvers, especially those with high precision, are presented, along with MATLAB implementations.

8.1.1 Definitions of fractional calculus

During the development of fractional calculus theory, many definitions of fractional calculus appeared. For instance, the Cauchy integral formula is directly extended from integer-order calculus; there also Grünwald–Letnikov, Riemann–Liouville, and Caputo definitions. In this section, some of the definitions are presented, followed by the properties of fractional calculus.

Definition 8.1 (Grünwald–Letnikov definition). The definition of a fractional derivative of order α is

$$ {}_{t_0}^{\mathrm{GL}}\mathscr{D}_t^{\alpha} f(t) = \lim_{h\to 0} \frac{1}{h^{\alpha}} \sum_{j=0}^{[(t-t_0)/h]} (-1)^j \binom{\alpha}{j} f(t-jh) \tag{8.1.1} $$

where t_0 is the initial time instance, and we assume that, when $t < t_0$, $f(t) \equiv 0$. The symbol $\binom{\alpha}{j}$ denotes the binomial coefficient. The computation methods will be presented later.

Grünwald–Letnikov definition is named after a Czech mathematician Anton Karl Grünwald (1838–1920) and a Russian mathematician Aleksey Vasilievich Letnikov (1837–1888). They proposed the definition of a fractional-order derivative in 1867 and 1868, respectively.

For integer-order derivatives, the finite difference method can be used to evaluate them numerically from the information at several points, while when computing fractional-order derivatives, all the points in the past have to be considered. Therefore fractional-order derivatives are regarded as the derivatives with memory.

Definition 8.2 (Riemann–Liouville definition). The fractional-order integral is defined as

$$ {}_{t_0}^{\mathrm{RL}}\mathscr{D}_t^{-\alpha} f(t) = \frac{1}{\Gamma(\alpha)} \int_{t_0}^{t} \frac{f(\tau)}{(t-\tau)^{1-\alpha}} \mathrm{d}\tau \tag{8.1.2} $$

where $\alpha > 0$ and t_0 is the initial time instance. If $t_0 = 0$, a simple notation is used, ${}^{\mathrm{RL}}\mathscr{D}_t^{-\alpha} f(t)$. If there is no conflict in the definitions, the mark RL can be omitted. Riemann–Liouville definition is one of the commonly used definitions in fractional calculus. Especially, the subscripts on the two sides of $\mathscr{D}$ are the lower and upper bounds in the integral expression.[37]

Riemann–Liouville definition is named after a German mathematician Georg Friedrich Bernhard Riemann (1826–1866) and a French mathematician Joseph Liouville (1809–1882). They proposed this definition in 1847 and 1834, respectively.

From the integral definition, the derivative definition can be derived. Assume that the order is $m = \lceil\alpha\rceil$, then the fractional-order derivative is defined as

$$\begin{aligned}{}_{t_0}^{\mathrm{RL}}\mathscr{D}_t^{\alpha} f(t) &= \frac{\mathrm{d}^m}{\mathrm{d}t^m}\left[{}_{t_0}^{\mathrm{RL}}\mathscr{D}_t^{-(m-\alpha)} f(t)\right] \\ &= \frac{1}{\Gamma(m-\alpha)}\frac{\mathrm{d}^m}{\mathrm{d}t^m}\left[\int_{t_0}^{t}\frac{f(\tau)}{(t-\tau)^{1+\alpha-m}}\mathrm{d}\tau\right].\end{aligned} \tag{8.1.3}$$

Definition 8.3 (Caputo definition). Caputo definition of a fractional derivative is

$${}_{t_0}^{\mathrm{C}}\mathscr{D}_t^{\alpha} f(t) = \frac{1}{\Gamma(m-\alpha)}\int_{t_0}^{t}\frac{f^{(m)}(\tau)}{(t-\tau)^{1+\alpha-m}}\mathrm{d}\tau \tag{8.1.4}$$

where $m = \lceil\alpha\rceil$ is an integer.

Caputo definition was proposed by an Italian mathematician and geophysicist Michele Caputo in 1971,[14] who aimed at solving problems of fractional calculus with nonzero initial values.

Caputo and Riemann–Liouville fractional-order integrals are exactly the same in definition:

$${}_{t_0}^{\mathrm{C}}\mathscr{D}_t^{-\alpha} f(t) = \frac{1}{\Gamma(\alpha)}\int_{t_0}^{t}\frac{f(\tau)}{(t-\tau)^{1-\alpha}}\mathrm{d}\tau = {}_{t_0}^{\mathrm{RL}}\mathscr{D}_t^{-\alpha} f(t), \quad \alpha > 0. \tag{8.1.5}$$

8.1.2 Relationships and properties of different fractional calculus definitions

It can be shown that[56] for a very wide category of practical functions, Grünwald–Letnikov and Riemann–Liouville definitions are completely equivalent. The two definitions are not distinguished in this book. Caputo and Riemann–Liouville definitions are different in the order of integral and derivative symbols. The difference and relationship will be further illustrated next.

If $y(t)$ has nonzero initial value, then, when $\alpha \in (0,1)$, it is seen by comparing the Caputo and Riemann–Liouville definitions that

$${}_{t_0}^{\mathrm{C}}\mathscr{D}_t^{\alpha} f(t) = {}_{t_0}^{\mathrm{RL}}\mathscr{D}_t^{\alpha}(f(t) - f(t_0)) \tag{8.1.6}$$

where the derivative of the constant $f(t_0)$ is ${}_{t_0}^{\mathrm{RL}}\mathscr{D}_t^{\alpha} f(t_0) = f(t_0)(t-t_0)^{-\alpha}/\Gamma(1-\alpha)$, such that the relationship between Caputo and Riemann–Liouville definitions can derived as

$${}_{t_0}^{\mathrm{C}}\mathscr{D}_t^{\alpha} f(t) = {}_{t_0}^{\mathrm{RL}}\mathscr{D}_t^{\alpha} f(t) - \frac{f(t_0)(t-t_0)^{-\alpha}}{\Gamma(1-\alpha)}. \tag{8.1.7}$$

Theorem 8.1. *If the order $\alpha > 1$, denoting $m = \lceil \alpha \rceil$, then*

$$ {}_{t_0}^{\mathrm{C}}\mathscr{D}_t^{\alpha} f(t) = {}_{t_0}^{\mathrm{RL}}\mathscr{D}_t^{\alpha} f(t) - \sum_{k=0}^{m-1} \frac{f^{(k)}(t_0)}{\Gamma(k-\alpha+1)}(t-t_0)^{k-\alpha}, \tag{8.1.8} $$

and, for $0 \leqslant \alpha \leqslant 1$, (8.1.7) *is just a special case of the above formula.*

If $\alpha < 0$, it has been indicated that the Riemann–Liouville and Caputo definitions are exactly the same, so that either of the two can be used.

Some of the properties in fractional calculus are summarized below, without proofs:[55]

(1) The fractional-order derivatives ${}_{t_0}\mathscr{D}_t^{\alpha} f(t)$ of an analytic function $f(t)$ are analytic in t and α.
(2) If $\alpha = n$ is an integer, the fractional-order and integer-order derivatives are identical, and ${}_{t_0}\mathscr{D}_t^{0} f(t) = f(t)$.
(3) Fractional-order operator is linear, that is, for any constants a and b,

$$ {}_{t_0}\mathscr{D}_t^{\alpha}\left[af(t) + bg(t)\right] = a\,{}_{t_0}\mathscr{D}_t^{\alpha} f(t) + b\,{}_{t_0}\mathscr{D}_t^{\alpha} g(t). \tag{8.1.9} $$

The Laplace transform of a fractional-order integral can be written as

$$ \mathscr{L}[\mathscr{D}_t^{-\gamma} f(t)] = s^{-\gamma}\mathscr{L}[f(t)]. \tag{8.1.10} $$

Theorem 8.2. *Under Riemann–Liouville definition, the Laplace transform of the fractional-order derivative satisfies*

$$ \mathscr{L}[{}_{t_0}^{\mathrm{RL}}\mathscr{D}_t^{\alpha} f(t)] = s^{\alpha}\mathscr{L}[f(t)] - \sum_{k=1}^{n-1} s^k\, {}_{t_0}^{\mathrm{RL}}\mathscr{D}_t^{\alpha-k-1} f(t)\Big|_{t=t_0}. \tag{8.1.11} $$

Laplace transforms of fractional-order integrals under Caputo definition are exactly the same as for Riemann–Liouville definition, and are not further discussed.

Theorem 8.3. *The Laplace transform of the fractional-order derivative under Caputo definition satisfies*

$$ \mathscr{L}\left[{}_{t_0}^{\mathrm{C}}\mathscr{D}_t^{\gamma} f(t)\right] = s^{\gamma}F(s) - \sum_{k=0}^{n-1} s^{\gamma-k-1} f^{(k)}(t_0). \tag{8.1.12} $$

Especially, if the initial values of function $f(t)$ and its derivatives are all 0, it is found that $\mathscr{L}[{}_{t_0}\mathscr{D}_t^{\alpha} f(t)] = s^{\alpha}\mathscr{L}[f(t)]$.

It can be seen that, in Caputo definition, the initial values of the function and its integer-order derivatives are involved. This is the case expected in real applications. While in Riemann–Liouville definition, the initial values of fractional-order derivatives are involved, which are not provided in real applications. Therefore Caputo equations are more suitable for dynamic system description for systems with nonzero initial values.

8.1.3 Numerical computation for Grünwald–Letnikov derivatives

The direct method for evaluating fractional-order derivatives is to use Grünwald–Letnikov definition

$$
\begin{aligned}
{}_{t_0}\mathscr{D}_t^{\alpha} f(t) &= \lim_{h\to 0} \frac{1}{h^{\alpha}} \sum_{j=0}^{[(t-t_0)/h]} (-1)^j \binom{\alpha}{j} f(t-jh) \\
&\approx \frac{1}{h^{\alpha}} \sum_{j=0}^{[(t-t_0)/h]} w_j^{(\alpha)} f(t-jh)
\end{aligned}
\tag{8.1.13}
$$

where $w_j^{(\alpha)} = (-1)^j \binom{\alpha}{j}$ are the polynomial expansion coefficients of $(1-z)^{\alpha}$. These coefficients can be recursively from

$$
w_0^{(\alpha)} = 1, \quad w_j^{(\alpha)} = \left(1 - \frac{\alpha+1}{j}\right) w_{j-1}^{(\alpha)}, \quad j = 1, 2, \ldots \tag{8.1.14}
$$

If the step-size h is small enough, (8.1.13) can be used to directly compute the numerical values. It can be shown that[56] the precision is $o(h)$. Therefore with Grünwald–Letnikov definition, the following solver can be written to compute the fractional-order derivatives, based on Grünwald–Letnikov definition, for given functions:

```
function dy=glfdiff(y,t,gam)
if strcmp(class(y),'function_handle'), y=y(t); end %function handle
h=t(2)-t(1); w=1; y=y(:); t=t(:);               %data stored in column vector
for j=2:length(t), w(j)=w(j-1)*(1-(gam+1)/(j-1)); end  %binomial coeffs
for i=1:length(t), dy(i)=w(1:i)*[y(i:-1:1)]/h^gam; end %derivative
```

The syntax of the function is $\boldsymbol{y}_1$=`glfdiff(`$\boldsymbol{y}$,$\boldsymbol{t}$,γ`)`, where $\boldsymbol{t}$ is an equally-spaced time vector, $\boldsymbol{y}$ contains the samples of the signal, or the function handle of the signal, while γ is the fractional order. The returned argument $\boldsymbol{y}_1$ contains the samples of the fractional-order derivatives. If γ is a negative number, integrals are evaluated.

For the expected precision of $o(h^p)$, a high-precision function algorithm and its implementation function `glfdiff9()` is proposed in [76]. The syntax of the function is $\boldsymbol{y}_1$=`glfdiff9(`$\boldsymbol{y}$,$\boldsymbol{t}$,γ,p`)`, where the details in the algorithm and its MATLAB implementation are not provided. The interested readers may refer to [76].

8.1.4 Numerical computation for Caputo derivatives

It can be seen from the presentation that the integrals in Caputo and Grünwald–Letnikov definitions are exactly the same. Function `glfdiff9()` can be used to evaluate the fractional-order integrals directly. If $\alpha > 0$, the compensations in (8.1.8) can be used to compute the Caputo fractional-order derivatives. High-precision function `caputo9()` can also be used to evaluate Caputo derivatives, with the syntax y_1=`caputo9(`$\boldsymbol{y}$,$\boldsymbol{t}$,α,p`)`, where $\alpha \leqslant 0$. The Caputo derivatives can be found directly, with precision $o(h^p)$.

Example 8.1. For the given function $f(t) = \mathrm{e}^{-t}$, find its 0.6th order Caputo derivative. Select different step-sizes and values of p and assess the precision. The analytical expression of the Caputo derivative is $y_0(t) = -t^{0.4}\mathrm{E}_{1,1.4}(-t)$.

Solutions. $\mathrm{E}_{\alpha,\beta}(\cdot)$ is referred to as a Mittag-Leffler function with two parameters. It will be further explained in the next section. Mittag-Leffler function can be numerically evaluated with function `ml_func()` provided in the FOTF Toolbox.

Selecting a step-size of $h = 0.01$, different p can be tried to evaluate fractional-order derivatives. Compared with analytical solution, the errors are listed in Table 8.1. It can be seen that, when $p = 6$, the maximum error is as low as 10^{-13}, many orders of magnitude higher than for the other existing algorithms. If the order p is further increased, it is not likely to increase the accuracy due to the limitations in the double precision data structure. The quality may even deteriorate.

Table 8.1: The maximum errors for the step-size $h = 0.01$.

order p	1	2	3	4	5	6	7
maximum error	0.0018	1.19×10^{-5}	8.89×10^{-8}	7.07×10^{-10}	5.85×10^{-12}	3.14×10^{-13}	7.33×10^{-13}

```
>> t=0:0.01:5; y=exp(-t);      % generate the samples
   y0=-t.^0.4.*ml_func([1,1.4],-t,0,eps); T=[]; % exact values
   for p=1:7, y1=caputo9(y,t,0.6,p); T=[T [y1-y0']]; end
   max(abs(T))        % find the maximum value in each column |T|
```

If a larger step-size $h = 0.1$ is selected, for different values of p, the numerical Caputo derivatives can also be evaluated with maximum errors listed in Table 8.2. It can be seen that even if a large step-size like this is chosen, the error is still at the 10^{-10} level, if $p = 8$.

Table 8.2: Maximum errors when $h = 0.1$.

order p	3	4	5	6	7	8	9
max error	7.82×10^{-5}	5.98×10^{-6}	4.73×10^{-7}	3.74×10^{-8}	3.12×10^{-9}	4.94×10^{-10}	1.14×10^{-8}

```
>> t=0:0.1:5; y=exp(-t); T=[];                %regenerate samples
   y0=-t.^0.4.*ml_func([1,1.4],-t,0,eps); %exact values
   for p=3:9, y1=caputo9(y,t,0.6,p); T=[T [y1-y0']]; end
   max(abs(T))
```

If the mathematical form of $y(t)$ or its samples is not known, block diagram method should be used to find its fractional-order derivatives. The related presentations will be given in Chapter 9. Besides, in Volume II, the numerical computation and implementation of high-order fractional-order derivatives are presented. Interested readers may refer to the related materials, and they are not discussed further here.

8.2 Analytical solution of linear commensurate-order differential equations

Exponential functions play an important role in integer-order linear differential equations. Like the exponential functions, the basic functions for the linear fractional-order differential equations are Mittag-Leffler functions. The definition and numerical computation methods for such functions are given in this section, then an important Laplace transform formula is given, from which the analytical solutions of commensurate-order linear differential equations are formed and studied.

8.2.1 Mittag-Leffler functions

Integer-order differential equations are represented by exponential functions $e^{\lambda t}$. Taylor series expansion is expressed as

$$e^z = 1 + \frac{z}{1!} + \frac{z^2}{2!} + \cdots = \sum_{k=0}^{\infty} \frac{z^k}{k!} = \sum_{k=0}^{\infty} \frac{z^k}{\Gamma(k+1)}. \tag{8.2.1}$$

A Swedish mathematician Magnus Gustaf (Gösta) Mittag-Leffler (1846–1927) extended the function in 1903, by changing the k term by αk in the Gamma function, so a special function was defined. The function is referred to as a Mittag-Leffler function.

Definition 8.4. Mittag-Leffler function with one parameter is defined as

$$\mathrm{E}_\alpha(z) = \sum_{k=0}^{\infty} \frac{z^k}{\Gamma(\alpha k+1)} \tag{8.2.2}$$

where $\alpha \in \mathscr{C}$, with $\mathscr{C}$ being the set of complex numbers. The convergence condition of the infinite series is $\mathrm{Re}(\alpha) > 0$.

If 1 in the above gamma function is replaced by another free constant β, the series then becomes the Mittag-Leffler function with two parameters.

Definition 8.5. The mathematical form of the Mittag-Leffler function with two parameters is

$$\mathrm{E}_{\alpha,\beta}(z)=\sum_{k=0}^{\infty}\frac{z^k}{\Gamma(\alpha k+\beta)} \tag{8.2.3}$$

where $\alpha,\beta\in\mathscr{C}$, and the convergence conditions for the infinite series of $z\in\mathscr{C}$ are $\mathrm{Re}(\alpha)>0$ and $\mathrm{Re}(\beta)>0$.

Definition 8.6. More generally, Mittag-Leffler functions with three and four parameters are respectively defined as[64]

$$\mathrm{E}_{\alpha,\beta}^{\gamma}(z)=\sum_{k=0}^{\infty}\frac{(\gamma)_k}{\Gamma(\alpha k+\beta)}\frac{z^k}{k!},\quad \mathrm{E}_{\alpha,\beta}^{\gamma,q}(z)=\sum_{k=0}^{\infty}\frac{(\gamma)_{kq}}{\Gamma(\alpha k+\beta)}\frac{z^k}{k!} \tag{8.2.4}$$

where $\alpha,\beta,\gamma\in\mathscr{C}$. For any $z\in\mathscr{C}$, the convergence conditions for the infinite series are $\mathrm{Re}(\alpha)>0$, $\mathrm{Re}(\beta)>0$, and $\mathrm{Re}(\gamma)>0$. It is noted that $q\in\mathscr{N}$, where $\mathscr{N}$ is the set of integers. The symbol $(\gamma)_k$ is also known as Pochhammer symbol.

Definition 8.7. Pochhammer symbol is defined as

$$(\gamma)_k=\gamma(\gamma+1)(\gamma+2)\cdots(\gamma+k-1)=\frac{\Gamma(k+\gamma)}{\Gamma(\gamma)}. \tag{8.2.5}$$

Pochhammer symbol is also known as the rising factorial.

There are many properties for Mittag-Leffler functions [74, 76]. They are not presented here.

A MATLAB function `ml_func()` for computing Mittag-Leffler functions is written by the author. It can be used to numerically evaluate Mittag-Leffler functions with up to four parameters, with the syntax

`y=ml_func(v,t,n,ϵ)`

where for Mittag-Leffler function with one parameter, $\boldsymbol{v}=\alpha$; while for that having two parameters, $\boldsymbol{v}=[\alpha,\beta]$. The input argument $\boldsymbol{v}$ can also be selected as $[\alpha,\beta,\gamma]$ and $[\alpha,\beta,\gamma,q]$ for Mittag-Leffler functions with three and four parameters. The argument n is the integer order of the Mittag-Leffler function. For Mittag-Leffler function computation, $n=0$. The argument ϵ is the error tolerance. The returned argument $\boldsymbol{y}$ is the nth order derivative of the Mittag-Leffler function at the time vector $\boldsymbol{t}$.

8.2.2 Commensurate-order linear differential equations

In this section, the mathematical form of linear fractional-order differential equations is presented. The concept of commensurate-order differential equations is given, which is the basis of analytical solutions to be presented in the next section.

Definition 8.8. A class of linear time-invariant (LTI) fractional-order differential equations is[56]

$$a_1\mathscr{D}^{\eta_1}y(t)+\cdots+a_n\mathscr{D}^{\eta_n}y(t)=b_1\mathscr{D}^{\gamma_1}u(t)+\cdots+b_m\mathscr{D}^{\gamma_m}u(t) \tag{8.2.6}$$

where b_i and a_i are real coefficients, while γ_i and η_i are the orders. The signal $u(t)$ can be regarded as the input to the system, while $y(t)$ is the output signal.

Definition 8.9. As a special case, if there exists an order α such that the above differential equation can be written as

$$\begin{aligned}c_1\mathscr{D}^{n\alpha}y(t)+c_2\mathscr{D}^{(n-1)\alpha}y(t)+\cdots+c_n\mathscr{D}^{\alpha}y(t)+c_{n+1}y(t)\\ =d_1\mathscr{D}^{m\alpha}u(t)+\cdots+b_m\mathscr{D}^{\alpha}u(t)+b_{m+1}u(t),\end{aligned} \tag{8.2.7}$$

the equation is referred to as a commensurate-order differential equation, and the order α is the base order.

Definition 8.10. If the commensurate-order differential equation has zero initial values, and denoting $\lambda = s^\alpha$, the integer-order transfer function of the operator λ can be established as

$$G(\lambda)=\frac{d_1\lambda^m+d_2\lambda^{m-1}+\cdots+d_m\lambda+d_{m+1}}{c_1\lambda^n+c_2\lambda^{n-1}+\cdots+c_n\lambda+c_{n+1}}. \tag{8.2.8}$$

8.2.3 An important Laplace transform formula

An important Laplace transform formula is presented first. Then several special cases are considered, from which the analytical solutions of certain fractional-order differential equations can be formulated.

Theorem 8.4. *An very important Laplace transform formula is*[15]

$$\mathscr{L}^{-1}\left[\frac{s^{\alpha\gamma-\beta}}{(s^\alpha+a)^\gamma}\right]=t^{\beta-1}\mathrm{E}_{\alpha,\beta}^{\gamma}\left(-at^\alpha\right) \tag{8.2.9}$$

where $\mathrm{E}_{\alpha,\beta}^{\gamma}(\cdot)$ is the Mittag-Leffler function with three parameters.

For different parameter combinations, the following formulas can be derived.

Theorem 8.5. *If $\gamma = 1$ and $\alpha\gamma = \beta$, then (8.2.9) can be written as*

$$\mathscr{L}^{-1}\left[\frac{1}{s^\alpha+a}\right]=t^{\alpha-1}\mathrm{E}_{\alpha,\alpha}(-at^\alpha). \tag{8.2.10}$$

If consider from the control systems viewpoint, since the Laplace transform of an impulsive signal is 1, the above formula can be regarded as the analytical solution of the fractional-order system $1/(s^\alpha + a)$ driven by an impulsive signal. It can be seen that the behavior of Mittag-Leffler functions is similar to that of exponential functions $\mathrm{e}^{-\lambda t}$ in integer-order systems.

Theorem 8.6. *If $\gamma = 1$ and $\alpha\gamma - \beta = -1$, that is, if $\beta = \alpha + 1$, then (8.2.9) can be written as*

$$\mathscr{L}^{-1}\left[\frac{1}{s(s^\alpha + a)}\right] = t^\alpha \mathrm{E}_{\alpha,\alpha+1}(-at^\alpha). \tag{8.2.11}$$

This formula can be regarded as the analytical solution of the fractional-order system $1/(s^\alpha + a)$ driven by a step input signal.

Theorem 8.7. *If $\gamma = k$ is an integer, and $\alpha\gamma = \beta$, that is, if $\beta = \alpha k$, then (8.2.9) can be written as*

$$\mathscr{L}^{-1}\left[\frac{1}{(s^\alpha + a)^k}\right] = t^{\alpha k-1} \mathrm{E}^k_{\alpha,\alpha k}\left(-at^\alpha\right), \tag{8.2.12}$$

and it can be regarded as the analytical solution of the fractional-order transfer function $1/(s^\alpha + a)^k$ driven by an impulsive signal

Theorem 8.8. *If $\gamma = k$ is an integer, and $\alpha\gamma - \beta = -1$, that is, if $\beta = \alpha k + 1$, then (8.2.9) can be written as*

$$\mathscr{L}^{-1}\left[\frac{1}{s(s^\alpha + a)^k}\right] = t^{\alpha k} \mathrm{E}^k_{\alpha,\alpha k+1}(-at^\alpha). \tag{8.2.13}$$

It can be regarded as the analytical solution of the fractional-order transfer function $1/(s^\alpha + a)^k$, driven by a step input.

8.2.4 Partial fraction expansion-based analytical approach

In the analytical analysis of integer-order linear differential equations, the partial fraction expansion technique plays a very important part. This idea can also be extended to the analytical solution of commensurate-order systems. In this section, the partial fraction expansions of commensurate-order systems are presented, and then step and impulse responses based methods are presented.

Theorem 8.9. *If there is a set of distinct poles $-p_i$ for λ in the commensurate-order transfer function, the integer-order transfer function of λ can be expressed by a partial fraction expansion in the form of*

$$G(\lambda) = \sum_{i=1}^{n} \frac{r_i}{\lambda + p_i} = \sum_{i=1}^{n} \frac{r_i}{s^\alpha + p_i}. \tag{8.2.14}$$

If there exist repeated poles at $-p_i$, with multiplicity m, the partial fraction expansion of the relevant part is written as

$$\frac{r_{i1}}{s^\alpha + p_i} + \frac{r_{i2}}{(s^\alpha + p_i)^2} + \cdots + \frac{r_{im}}{(s^\alpha + p_i)^m} = \sum_{j=1}^{m} \frac{r_{ij}}{(s^\alpha + p_i)^j}. \tag{8.2.15}$$

Definition 8.11. For commensurate-order systems with base order α, the partial fraction expansion can be written as

$$G(s)=\sum_{i=1}^{N}\sum_{j=1}^{m_i}\frac{r_{ij}}{(s^\alpha+p_i)^j} \tag{8.2.16}$$

where p_i and r_{ij} are complex numbers, the multiplicity m_i of the ith pole p_i is an integer, and $m_1+m_2+\cdots+m_N=n$.

If function `residue()` in MATLAB is used, an integer-order transfer function can be converted into the partial fraction expansion form. Therefore analytical solutions of certain commensurate-order differential equations can be found.

If the partial fraction expansion of the fractional-order transfer function is obtained, the Laplace transforms of the impulse and step responses can be obtained as

$$\mathscr{L}^{-1}\left[\sum_{i=1}^{n}\frac{r_i}{s^\alpha+p_i}\right]=\sum_{i=1}^{n}r_i t^{\alpha-1}\mathrm{E}_{\alpha,\alpha}(-p_i t^\alpha), \tag{8.2.17}$$

$$\mathscr{L}^{-1}\left[\sum_{i=1}^{n}\frac{r_i}{s(s^\alpha+p_i)}\right]=\sum_{i=1}^{n}r_i t^{\alpha}\mathrm{E}_{\alpha,\alpha+1}(-p_i t^\alpha). \tag{8.2.18}$$

Theorem 8.10. *More generally, since commensurate-order transfer functions are factorized in the form defined in Definition* 8.11, *the analytical solutions for impulsive input signals can be written as*

$$y_\delta(t)=\mathscr{L}^{-1}\left[\sum_{i=1}^{N}\sum_{j=1}^{m_i}\frac{r_{ij}}{(s^\alpha+p_i)^j}\right]=\sum_{i=1}^{N}\sum_{j=1}^{m_i}r_{ij}\,t^{\alpha j-1}\mathrm{E}_{\alpha,\alpha j}^{j}(-at^\alpha), \tag{8.2.19}$$

while for step responses, the analytical solution is

$$y_u(t)=\mathscr{L}^{-1}\left[\sum_{i=1}^{N}\sum_{j=1}^{m_i}\frac{r_{ij}}{s(s^\alpha+p_i)^j}\right]=\sum_{i=1}^{N}\sum_{j=1}^{m_i}r_{ij}\,t^{\alpha j}\mathrm{E}_{\alpha,\alpha j+1}^{j}(-at^\alpha). \tag{8.2.20}$$

Example 8.2. Find the analytical solution of the following fractional-order differential equations for step inputs:

$${}_0\mathscr{D}^{0.8}y(t)+0.75\,{}_0\mathscr{D}^{0.4}y(t)+0.9y(t)=5u(t).$$

Solutions. It can be seen that the base order is $\alpha=0.4$. Denote $\lambda=s^{0.4}$, then the commensurate-order transfer function model can be derived as follows, which is an integer-order transfer function of λ:

$$G(s)=\frac{5}{s^{0.8}+0.75s^{0.4}+0.9}\quad\Longrightarrow\quad G(\lambda)=\frac{5}{\lambda^2+0.75\lambda+0.9}.$$

With the following statements:

```
>> [r,p,k]=residue(5,[1 0.75 0.9]) %partial fraction expansion
```

the partial fraction expansion of the transfer function $G(\lambda)$ is found as

$$G(s) = \frac{-2.8689\mathrm{j}}{s^{0.4} + 0.3750 - 0.8714\mathrm{j}} + \frac{2.8689\mathrm{j}}{s^{0.4} + 0.3750 + 0.8714\mathrm{j}}$$

From the above expansion, it is immediately seen that the analytical solution of the step response can be written as

$$\begin{aligned} y(t) = &-2.87\mathrm{j}\, t^{0.4}\mathrm{E}_{0.4,1.4}\left((-0.38 + 0.87\mathrm{j})t^{0.4}\right) \\ &+ 2.87\mathrm{j}\, t^{0.4}\mathrm{E}_{0.4,1.4}\left((-0.38 - 0.87\mathrm{j})t^{0.4}\right). \end{aligned}$$

Example 8.3. Find the analytical and numerical solutions for the impulse response of the following fractional-order system:

$$G(s) = \frac{s^{1.2} + 3s^{0.4} + 5}{s^{1.6} + 10s^{1.2} + 35s^{0.8} + 50s^{0.4} + 24}.$$

Solutions. Letting $\lambda = s^{0.4}$, the original system can be written as

$$G(\lambda) = \frac{\lambda^3 + 3\lambda + 5}{\lambda^4 + 10\lambda^3 + 35\lambda^2 + 50\lambda + 24}.$$

With `residue()` function in MATLAB

```
>> n=[1 0 3 5]; d=[1 10 35 50 24]; [r,p,K]=residue(n,d)
```

the partial fraction expansion of the system can be written as

$$G(s) = \frac{71/6}{s^{0.4} + 4} + \frac{-31/2}{s^{0.4} + 3} + \frac{9/2}{s^{0.4} + 2} + \frac{1/6}{s^{0.4} + 1}.$$

With the properties in (8.2.10), the analytical solution of the impulse response can be written as

$$\begin{aligned} y(t) = &\,71t^{-0.6}\mathrm{E}_{0.4,0.4}(-4t^{0.4})/6 - 31t^{-0.6}\mathrm{E}_{0.4,0.4}(-3t^{0.4})/2 \\ &+ 9t^{-0.6}\mathrm{E}_{0.4,0.4}(-2t^{0.4})/2 + t^{-0.6}\mathrm{E}_{0.4,0.4}(-t^{0.4})/6. \end{aligned}$$

Based on the analytical solution formula, the following MATLAB commands can also be used to evaluate the numerical solutions, as shown in Figure 8.1.

```
>> t=0:0.001:0.2; t1=t.^(0.4);
   y=71/6*t.^(-0.6).*ml_func([0.4,0.4],-4*t1)-...
```

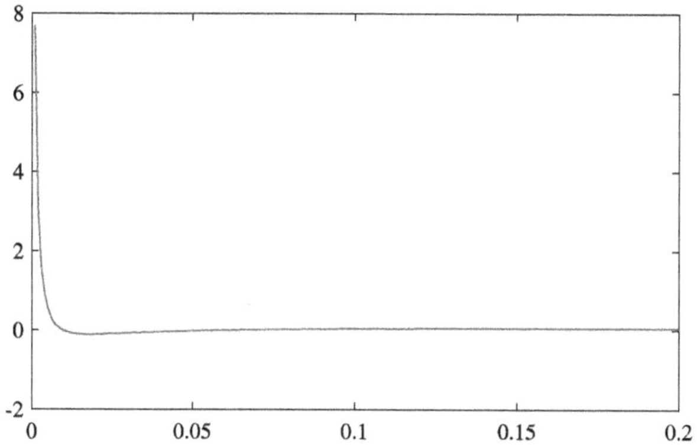

Figure 8.1: Numerical solution of the impulse response.

```
   31/2*t.^(-0.6).*ml_func([0.4,0.4],-3*t1)+...
    9/2*t.^(-0.6).*ml_func([0.4,0.4],-2*t1)+...
   1/6*t.^(-0.6).*ml_func([0.4,0.4],-t1);
plot(t,y)    % solution curves of the differential equations
```

Example 8.4. Solve the following fractional-order differential equation with zero initial values:

$$\mathscr{D}^{1.2}y(t) + 5\mathscr{D}^{0.9}y(t) + 9\mathscr{D}^{0.6}y(t) + 7\mathscr{D}^{0.3}y(t) + 2y(t) = u(t)$$

where $u(t)$ is a unit step or impulsive input.

Solutions. Selecting the base order 0.3 and letting $\lambda = s^{0.3}$, the integer-order transfer function of λ can be found as

$$G(\lambda) = \frac{1}{\lambda^4 + 5\lambda^3 + 9\lambda^2 + 7\lambda + 2}.$$

With the following commands, the partial fraction expansion can be found:

```
>> num=1; den=[1 5 9 7 2]; [r,p]=residue(num,den)
```

whose mathematical form is

$$G(\lambda) = -\frac{1}{\lambda+2} + \frac{1}{\lambda+1} - \frac{1}{(\lambda+1)^2} + \frac{1}{(\lambda+1)^3}.$$

If the input signal $u(t)$ is a unit impulsive signal, the Laplace transform of the output signal is

$$Y(s) = G(s) = -\frac{1}{s^{0.3}+2} + \frac{1}{s^{0.3}+1} - \frac{1}{(s^{0.3}+1)^2} + \frac{1}{(s^{0.3}+1)^3}.$$

It can be seen from (8.2.10) and (8.2.12) that the analytical solution of the impulse response is

$$y_1(t) = -t^{-0.7}\mathrm{E}_{0.3,0.3}\left(-2t^{0.3}\right) + t^{-0.7}\mathrm{E}_{0.3,0.3}\left(-t^{0.3}\right) - t^{-0.4}\mathrm{E}^2_{0.3,0.6}\left(-t^{0.3}\right) + t^{-0.1}\mathrm{E}^3_{0.3,0.9}\left(-t^{0.3}\right).$$

If the input $u(t)$ is a unit step signal, the Laplace transform of the output signal is

$$Y(s) = \frac{1}{s}G(s) = -\frac{1}{s(s^{0.3}+2)} + \frac{1}{s(s^{0.3}+1)} - \frac{1}{s(s^{0.3}+1)^2} + \frac{1}{s(s^{0.3}+1)^3}.$$

The analytical solution of the system is then

$$y_2(t) = -t^{0.3}\mathrm{E}_{0.3,1.3}\left(-2t^{0.3}\right) + t^{0.3}\mathrm{E}_{0.3,1.3}\left(-t^{0.3}\right) - t^{0.6}\mathrm{E}^2_{0.3,1.6}\left(-t^{0.3}\right) + t^{0.9}\mathrm{E}^3_{0.3,1.9}\left(-t^{0.3}\right).$$

The curves of the step and impulse responses of the output signal can be obtained as shown in Figure 8.2.

```
>> t=0:0.002:0.5;
   y1=-t.^-0.7.*ml_func([0.3,0.3],-2*t.^0.3)...
      +t.^-0.7.*ml_func([0.3,0.3],-t.^0.3)...
      -t.^-0.4.*ml_func([0.3,0.6,2],-t.^0.3)...
      +t.^-0.1.*ml_func([0.3,0.9,3],-t.^0.3);
   y2=-t.^0.3.*ml_func([0.3,1.3],-2*t.^0.3)...
      +t.^0.3.*ml_func([0.3,1.3],-t.^0.3)...
      -t.^0.6.*ml_func([0.3,1.6,2],-t.^0.3)...
      +t.^0.9.*ml_func([0.3,1.9,3],-t.^0.3);
   plot(t,y1,t,y2) % impulse and step responses
```

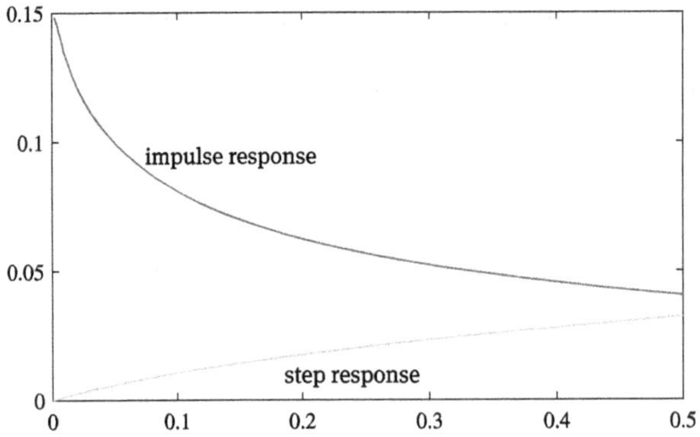

Figure 8.2: The step and impulse responses.

It is worth mentioning that the analytical solutions have too many restrictions. On the one hand, the differential equations must be linear and have commensurate orders. On the other hand, only special input signals such as step and impulse are allowed. Although the theory can be expanded for signals such as ramp input, the signal types are too much restricted. For ordinary input signals, the method here cannot be used. If the analytical solutions cannot be found, numerical solutions become the only choice. In the latter presentation of this chapter, numerical methods are discussed for fractional-order differential equations.

8.3 Numerical solutions of linear fractional-order differential equations with constant coefficients

Consider the linear fractional-order differential equation with constant coefficients in (8.2.6). If the initial values of the output $y(t)$ and its derivatives are all zero, the equation is referred to as an equation with zero initial values. In this case, the differential equations under different definitions are equivalent. There is no constraint such as commensurate order. The input signal can be of any form, in mathematical formula or in sampled data. The target of this section is to find numerical solutions of linear fractional-order differential equations with constant coefficients.

8.3.1 A closed-form solution

If the initial values of $u(t)$, $y(t)$, and their derivatives are all zero, the right-hand side of the expression in (8.2.6) is equivalently denoted as $\hat{u}(t)$, so the original differential equation can be simplified as

$$a_1 \mathscr{D}_t^{\gamma_1} y(t) + a_2 \mathscr{D}_t^{\gamma_2} y(t) + \cdots + a_{n-1} \mathscr{D}_t^{\gamma_{n-1}} y(t) + a_n \mathscr{D}_t^{\gamma_n} y(t) = \hat{u}(t) \tag{8.3.1}$$

where $\hat{u}(t)$ is a linear combination of $u(t)$ and its fractional-order derivatives, which can be evaluated independently as

$$\hat{u}(t) = b_1 \mathscr{D}_t^{\eta_1} u(t) + b_2 \mathscr{D}_t^{\eta_2} u(t) + \cdots + b_m \mathscr{D}_t^{\eta_m} u(t). \tag{8.3.2}$$

For simplicity, assume that $\gamma_1 > \gamma_2 > \cdots > \gamma_{n-1} > \gamma_n > 0$. If the following two special cases appear, transformations must be carried out first.

(1) If the orders of the above equations are not the same as those above, sorting must be made first.

(2) If there exists a negative order γ_i, integro-differential equations are involved. They are not easy to solve. A new variable $z(t) = \mathscr{D}_t^{\gamma_n} y(t)$ should be introduced. The original equation can be transformed into an equation of $z(t)$, and for the results, numerical integrals should be evaluated to find the output signal $y(t)$.

Theorem 8.11. *The closed-form solution of linear fractional-order differential equation with zero initial values is*[80]

$$y_t = \frac{1}{\sum\limits_{i=1}^{n} \dfrac{a_i}{h^{\gamma_i}}} \left[\hat{u}_t - \sum_{i=1}^{n} \frac{a_i}{h^{\gamma_i}} \sum_{j=1}^{[(t-t_0)/h]} w_j^{(\gamma_i)} y_{t-jh} \right]. \tag{8.3.3}$$

Now consider the general form of the linear fractional-order differential equation in (8.2.6). The algorithm introduced here seeks to find first the equivalent signal $\hat{u}(t)$ on the right, and then solve the differential equation in (8.3.1). The idea is illustrated in Figure 8.3(a). Unfortunately, this idea is not feasible. For instance, if the input signal is a step one, and there happen to be integer-order derivatives, the contributions of the terms may be neglected. Wrong results can be found. Therefore new alternative ideas must be employed.

A different idea is adopted in real programming. For instance, the original linear problem can be equivalently tackled by using $u(t)$ to compute directly the output $\hat{y}(t)$, then to compose $\hat{y}(t)$ as a weighted sum of fractional-order derivatives. The idea is illustrated in Figure 8.3(b). Since the original system is linear, it can be divided into two parts, $N(s)$ and $1/D(s)$, where $N(s)$ and $D(s)$ are pseudopolynomials. In a linear system framework, they can be swapped, such that the obtained $y(t)$ are the same. The algorithm is implemented in a MATLAB solver given later.

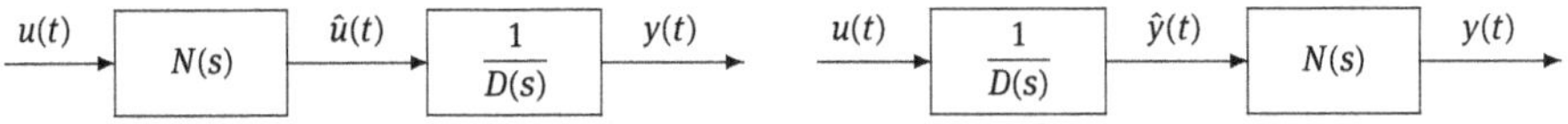

(a) regular order **(b)** swapped order

Figure 8.3: The illustration of the computation.

Based on the above algorithm, the following MATLAB function `fode_sol()` can be written, for solving linear fractional-order differential equations with zero initial values. In the function, $\boldsymbol{W}$ is a matrix whose jth row stores the w_j vector of the jth order.

```
function y=fode_sol(a,na,b,nb,u,t)
h=t(2)-t(1); D=sum(a./[h.^na]); nT=length(t);
D1=b(:)./h.^nb(:); nA=length(a); vec=[na nb];
y1=zeros(nT,1); W=ones(nT,length(vec));
for j=2:nT, W(j,:)=W(j-1,:).*(1-(vec+1)/(j-1)); end
for i=2:nT
   A=[y1(i-1:-1:1)]'*W(2:i,1:nA);
   y1(i)=(u(i)-sum(A.*a./[h.^na]))/D;
end
for i=2:nT, y(i)=(W(1:i,nA+1:end)*D1)'*[y1(i:-1:1)]; end
```

The syntax of the function is `y=fode_sol(`$\boldsymbol{a}$`,`$\boldsymbol{n}_a$`,`$\boldsymbol{b}$`,`$\boldsymbol{n}_b$`,`$\boldsymbol{u}$`,`$\boldsymbol{t}$`)`, where the time and input vectors are provided in $\boldsymbol{t}$ and $\boldsymbol{u}$.

Example 8.5. If the input signal is $u(t) = \sin t^2$, find the numerical solution of the following linear fractional-order differential equation:

$$\mathscr{D}_t^{3.5}y(t) + 8\mathscr{D}_t^{3.1}y(t) + 26\mathscr{D}_t^{2.3}y(t) + 73\mathscr{D}_t^{1.2}y(t) + 90\mathscr{D}_t^{0.5}y(t) = 30u'(t) + 90\mathscr{D}^{0.3}u(t).$$

Solutions. This fractional-order differential equation cannot be solved with the analytical methods discussed earlier, since the input is not a step or impulsive signal. Therefore the method presented in this section can be used to find the numerical solutions directly.

It is found from the equations that the vectors $\boldsymbol{a}$, $\boldsymbol{n}_a$, $\boldsymbol{b}$, and $\boldsymbol{n}_b$ can immediately be found. The equally-spaced time and input vectors can then be computed such that function `fode_sol()` can be called to solve the original equation. The results obtained are shown in Figure 8.4. The two step-sizes, 0.002 and 0.001, are used to cross-validate the results. The two results are the same, meaning that the solutions are correct.

```
>> a=[1,8,26,73,90]; na=[3.5,3.1,2.3,1.2,0.5];
   nb=[1,0.3]; b=[30,90];
   t=0:0.002:10; u=sin(t.^2); y=fode_sol(a,na,b,nb,u,t);
   t1=0:0.001:10; u1=sin(t1.^2); y1=fode_sol(a,na,b,nb,u1,t1);
   plot(t,y,t1,y1)  %solutions of the fractional-order differential equation
```

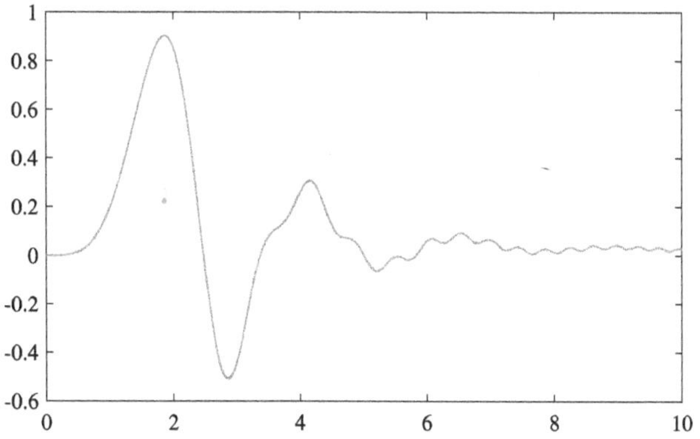

Figure 8.4: Numerical solutions under different step-sizes.

8.3.2 Riemann–Liouville differential equations

The above illustrated closed-form algorithm only applies for Riemann–Liouville linear differential equation solutions. Unfortunately, the precision is only $o(h)$. Therefore the algorithm has certain limitations in practical use. In [74], an $o(h^p)$ closed-form algorithm is given.

If the differential equation in (8.3.1) has initial values, another closed-form formula similar to (8.3.3) can be constructed from Grünwald–Letnikov definition. If the coefficients w_j in (8.3.3) are obtained by a high precision algorithm, then the following closed-form solution can be formulated:

(1) Replace the operator in (8.3.1) with Grünwald–Letnikov operator.
(2) For each order, use a high precision recursive formula to compute w_j.
(3) Find the numerical solution y_k from (8.3.3).

Based on such an algorithm, the MATLAB function `fode_sol9()` can be written to solve fractional-order differential equations with zero initial values. The syntax of the function is very close to that of `fode_sol()`, where an extra p parameter is appended.

```
function y=fode_sol9(a,na,b,nb,u,t,p)
h=t(2)-t(1); n=length(t); vec=[na nb]; u=u(:);
g=double(genfunc(p)); t=t(:); W=[];
for i=1:length(vec), W=[W; get_vecw(vec(i),n,g)]; end
D1=b(:)./h.^nb(:); nA=length(a); y1=zeros(n,1);
W=W.'; D=sum((a.*W(1,1:nA))./[h.^na]);
for i=2:n
   A=[y1(i-1:-1:1)]'*W(2:i,1:nA);
   y1(i)=(u(i)-sum(A.*a./[h.^na]))/D;
end
for i=2:n, y(i)=(W(1:i,nA+1:end)*D1)'*[y1(i:-1:1)]; end
```

The necessary condition of using the pth order algorithm is that the first p values in $y(t)$ must be zero or very close to zero, to avoid the impact to the results, if the first few terms w_j are missing. For certain differential equations, where the necessary conditions are not satisfied, high precision algorithms to be presented next can be used instead.

Example 8.6. Solve the fractional-order differential equation

$$y'''(t) + {}_0^{\mathrm{RL}}\mathscr{D}_t^{2.5}y(t) + y(t) = -1 + t - t^2/2 - t^{0.5}\mathrm{E}_{1,1.5}(-t).$$

It is known that the analytical solution of the equation is $y(t) = -1 + t - t^2/2 + \mathrm{e}^{-t}$. Solve the differential equations for different values of p, and find the errors.

Solutions. Selecting a slightly larger step-size $h = 0.1$, the following commands can be used to solve the differential equation for different p. It can be seen that the numerical solutions are as shown in Figure 8.5. It is immediately seen that the errors are rather large in the solutions.

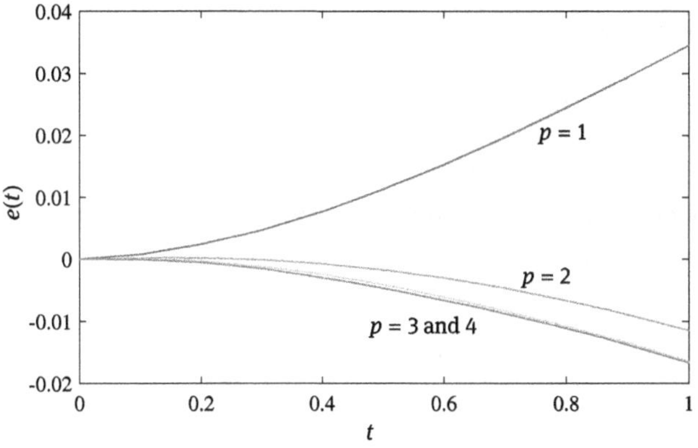

Figure 8.5: Computational error for different p values.

```
>> a=[1 1 1]; na=[3 2.5 0]; b=1; nb=0;
   t=0:0.1:1; y=-1+t-t.^2/2+exp(-t);
   u=-1+t-t.^2/2-t.^0.5.*ml_func([1,1.5],-t);
   y1=fode_sol9(a,na,b,nb,u,t,1); e1=y-y1;
   y2=fode_sol9(a,na,b,nb,u,t,2); e2=y-y2;
   y3=fode_sol9(a,na,b,nb,u,t,3); e3=y-y3;
   y4=fode_sol9(a,na,b,nb,u,t,4); e4=y-y4;
   plot(t,e1,t,e2,t,e3,t,e4) %draw the error curves
```

For the four selected values of p, the error curves can be obtained with the above statements, as shown in Figure 8.5. It is seen that the precision for $p = 2$ is clearly better than that for $p = 1$, obtained by `fode_sol()`. Further increasing the value to $p = 3$ and $p = 4$, the precision is reduced, since the first four initial values of $y(t)$ are respectively 0, −0.0002, −0.0013, and −0.0042. The first two are close to zero, but the latter two are large. Therefore the approximation in the solution is not good.

Note that the condition "the initial values of $y(t)$ and their derivatives are sufficiently small" is too strict. Therefore this algorithm is not suitable for large values of p. The high-precision Caputo equation solver to be presented next is recommended.

8.3.3 Caputo differential equations

Definition 8.12. The general form of a linear Caputo differential equation is

$$a_1\,{}_{t_0}^{\mathrm{C}}\mathscr{D}_t^{\gamma_1}y(t) + a_2\,{}_{t_0}^{\mathrm{C}}\mathscr{D}_t^{\gamma_2}y(t) + \cdots + a_n\,{}_{t_0}^{\mathrm{C}}\mathscr{D}_t^{\gamma_n}y(t) = \hat{u}(t), \tag{8.3.4}$$

and it is safe to assume that $\gamma_1 > \gamma_2 > \cdots > \gamma_n$.

Definition 8.13. The initial values of a Caputo equation are

$$y(t_0) = c_0, \quad y'(0) = c_1, \dots, y^{(q-1)}(0) = c_{q-1}. \tag{8.3.5}$$

If $c_i = 0, 0 \leqslant i \leqslant q-1$, the differential equation is reduced to a zero initial value one. Otherwise, it is a differential equation with nonzero initial values. For the differential equation to have a unique solution, $q = \lceil \gamma_1 \rceil$.

If the right-hand side of the equation is a dynamic expression $\hat{u}(t)$, the equivalent signal $\hat{u}(t)$ should be evaluated first. The signal should be evaluated directly with high-precision function `caputo9()`,

$$\hat{u}(t) = b_1 \mathscr{D}_t^{\eta_1} u(t) + b_2 \mathscr{D}_t^{\eta_2} u(t) + \cdots + b_m \mathscr{D}_t^{\eta_m} u(t). \tag{8.3.6}$$

It is worth mentioning that, in the Caputo definition, the initial values are the values of the signal $y(t)$ and its integer-order derivatives. Therefore this definition is more suitable for describing practical systems. Further explorations are needed for the numerical solutions of Caputo equations.

If the differential equation in (8.3.4) has nonzero initial values, an auxiliary function $T(t)$ should be introduced such that the original equation can be mapped into a differential equation of $z(t)$ with zero initial values.

Definition 8.14. Build up the Taylor auxiliary function[57] such that

$$T(t) = \sum_{k=0}^{q-1} \frac{y^{(k)}(t_0)}{k!} (t - t_0)^k \tag{8.3.7}$$

and

$$y(t) = z(t) + T(t) \tag{8.3.8}$$

where the initial values of the auxiliary function $T(t)$ are the same as those of signal $y(t)$, while $z(t)$ has zero initial values.

Substituting the signal $y(t)$ by $z(t) + T(t)$ in (8.3.4), the differential equation of $y(t)$ can be mapped into an equation of signal $z(t)$, which contains zero initial values:

$$a_1 {}_{t_0}^{\rm C}\mathscr{D}_t^{\gamma_1} z(t) + a_2 {}_{t_0}^{\rm C}\mathscr{D}_t^{\gamma_2} z(t) + \cdots + a_n {}_{t_0}^{\rm C}\mathscr{D}_t^{\gamma_n} z(t) = \hat{u}(t) - P(t) \tag{8.3.9}$$

and is expressed as

$$P(t) = (a_1 {}_{t_0}^{\rm C}\mathscr{D}_t^{\gamma_1} + a_2 {}_{t_0}^{\rm C}\mathscr{D}_t^{\gamma_2} + \cdots + a_n {}_{t_0}^{\rm C}\mathscr{D}_t^{\gamma_n}) T(t). \tag{8.3.10}$$

Since the $z(t)$ signal has zero initial values, ${}_{t_0}^{\rm C}\mathscr{D}^{\alpha} z(t) = {}_{t_0}^{\rm GL}\mathscr{D}^{\alpha} z(t)$. Therefore, the equivalent signal on the right-hand side is $\hat{u}(t) - P(t)$. The numerical solutions z_m of this equation can be obtained from (8.3.3), used for the zero initial value problems.

Theorem 8.12. *The closed-form solution of a linear Caputo equation is*

$$y_m = \frac{1}{\sum\limits_{i=1}^{n} \frac{a_i}{h^{\gamma_i}}} \left(\hat{u}_m - P_m - \sum_{i=1}^{n} \frac{a_i}{h^{\gamma_i}} \sum_{j=1}^{m} w_j y_{m-j} \right) + T_m. \tag{8.3.11}$$

Example 8.7. Solve the Bagley–Torwik differential equation[20]

$$Ay''(t) + B\mathscr{D}^{3/2}y(t) + Cy(t) = C(t+1), \quad \text{and} \quad y(0) = y'(0) = 1.$$

Show that the solutions are independent of constants A, B, and C.

Solutions. From the given initial values, an auxiliary function can be written as $T(t) = t+1$. Therefore the original signal can be decomposed as $y(t) = z(t)+t+1$, where $z(t)$ is the signal with zero initial values. The original differential equation can be rewritten as the Grünwald–Letnikov solution in terms of signal $z(t)$ with zero initial values.

Let us observe again the Caputo definition, since the 1.*th order derivative of $y(t)$ is taken, the second-order derivative must be evaluated, then we need to evaluate fractional-order integrals. This means that the compensating term $t + 1$ will vanish. The original equation can be rewritten as

$$Az''(t) + B\mathscr{D}^{3/2}z(t) + Cz(t) + C(t+1) = C(t+1).$$

It can be seen that the $C(t + 1)$ terms on both sides can be canceled. The original differential equation is rewritten as

$$Az''(t) + B\mathscr{D}^{3/2}z(t) + Cz(t) = 0.$$

It can be seen that since $z(t)$ has zero initial values, and there is no external excitation in the above equation, it implies that $z(t) \equiv 0$. Therefore the solution of the original equation is $y(t) = t + 1$. That is, the solution of the equation is independent of the constants A, B, and C.

The auxiliary function $T(t)$ is the Taylor series expansion of the output signal $y(t)$. The difference between $T(t)$ and $y(t)$ is very small in the very beginning of the simulation process. Since $y(t)$ is a bounded function, the value of $|y(t) - T(t)|$ may increase when t increases. When t is sufficiently large, $|T(t)|$ is an increasing function such that $|z(t)| = |y(t) - T(t)|$ is also an increasing function. Since the two terms are evaluated separately, they cannot be perfectly canceled when t is very large. This may lead to huge computational error when t is large. It is obvious that the computation error cannot be maintained small. Examples will be shown next to illustrate this kind of phenomenon.

8.3.4 Computing equivalent initial values

It has been indicated that, if the pth order algorithm is used, the necessary condition is that the first p values of $z(t) = y(t) - T(t)$ are zero or close to zero. This has been demonstrated in Example 8.6. Since $z(t)$ here is a zero initial value function, the condition is satisfied.

Since the order p can be selected independently, there are two possibilities for p and the actual highest order q in the Caputo equations. One is that $p \leqslant q$, so the necessary conditions are satisfied, and high precision solutions can be found. The other one is when $p > q$, then p equivalent initial values are needed. Only then the high-precision solution can be found. In the latter case, two-step method is used.

No matter what the value of q in Caputo equation, the Taylor auxiliary function $T(t)$ is constructed such that

$$T(t) = \sum_{k=0}^{p-1} c_k \frac{t^k}{k!}. \tag{8.3.12}$$

If $p \leqslant q$, then, when $k = 0, 1, \dots, p-1$, letting $c_k = y^{(k)}(0)$, the first step is completed. If $p > q$, then, when $k = 0, 1, \dots, q-1$, letting $c_k = y^{(k)}(0)$, the remaining $p - q$ initial values c_k can be computed. To let $T(t)$ and $y(t)$ have the same initial values, the undetermined constants should still be denoted by c_k, when $q \leqslant k \leqslant p-1$. Therefore the $z(t)$ signal has zero initial values.

For simplicity, function $T(t)$ is rewritten as

$$T(t) = \sum_{k=0}^{p-1} c_k T_k \tag{8.3.13}$$

where $T_k = t^k/k!$. Since $z(t)$ has zero initial values, the first p values in the interpolation polynomial are zero, so that the first p initial values of $T(t)$ and $y(t)$ are the same. In other words, the first p initial values of $T(t)$ satisfy the original Caputo differential equation. Equation (8.3.13) can be substituted into the original equation such that

$$\sum_{k=0}^{p} c_k x_k(t) = \hat{u}(t) \tag{8.3.14}$$

where

$$x_k = \left(a_1 {}_{t_0}^{\mathrm{C}}\mathscr{D}_t^{\eta_1} + a_2 {}_{t_0}^{\mathrm{C}}\mathscr{D}_t^{\eta_2} + \cdots + a_n {}_{t_0}^{\mathrm{C}}\mathscr{D}_t^{\eta_n}\right) T_k(t). \tag{8.3.15}$$

It can be seen that the first p points can be evaluated directly. In (8.3.14), letting $t = h, 2h, \dots, Kh$, where $K = p - q$, the following linear algebraic equation can be established:

$$\begin{bmatrix} x_0(h) & x_1(h) & \dots & x_p(h) \\ x_0(2h) & x_1(2h) & \dots & x_p(2h) \\ \vdots & \vdots & \ddots & \vdots \\ x_0(Kh) & x_1(Kh) & \dots & x_p(Kh) \end{bmatrix} \begin{bmatrix} c_0 \\ c_1 \\ \vdots \\ c_{p-1} \end{bmatrix} = \begin{bmatrix} \hat{u}(h) \\ \hat{u}(2h) \\ \vdots \\ \hat{u}(Kh) \end{bmatrix}. \tag{8.3.16}$$

From this equation it can be seen that c_k, $0 \leqslant k \leqslant q-1$ are the initial values of the equation. The number of the unknowns is K, and the quantities can be evaluated directly from the linear algebraic equation in (8.3.16). Therefore the constants c_i, $0 \leqslant i \leqslant p-1$ can be regarded as the new equivalent initial values. With these initial values, (8.3.13) can be used to build up the Taylor auxiliary function $T(t)$, such that the first p terms in $y(t)$ can be computed directly from $T(t)$. The above ideas are included in the following algorithm:

(1) For $0 \leqslant k \leqslant p-1$, construct $T_k = t^k/k!$, and compute x_k from (8.3.15).
(2) Letting $K = p - q$, the coefficient matrix in (8.3.16) can be established with x_k.
(3) For the given $\hat{u}(h), \hat{u}(2h), \dots, \hat{u}(Kh)$, the values are obtained from the equivalent input on the right-hand side of (8.3.4).
(4) The first q coefficients c_k equal to the given initial values. The rest can be found from the linear algebraic equation in (8.3.16), such that all equivalent initial values are found.

Based on the above considerations, the MATLAB function `caputo_ics()` can be written for finding the first p equivalent initial values, aiming at obtaining high precision solutions of Caputo differential equations:

```
function [c,y]=caputo_ics(a,na,b,nb,y0,u,t)
na1=ceil(na); q=max(na1); K=length(t);
p=K+q-1; y0=y0(:); u=u(:); t=t(:); d1=y0./gamma(1:q)';
I1=1:q; I2=(q+1):p; X=zeros(K,p);
for i=1:p, for k=1:length(a)
   if i>na1(k)
      X(:,i)=X(:,i)+a(k)*t.^(i-1-na(k))*gamma(i)/gamma(i-na(k));
end, end, end
u1=0; for i=1:length(b), u1=u1+b(i)*caputo9(u,t,nb(i),K-1); end
X(1,:)=[]; u2=u1(2:end)-X(:,I1)*d1; d2=inv(X(:,I2))*u2;
c=[d1;d2]; y=0; for i=1:p, y=y+c(i)*t.^(i-1); end
```

The syntax of the function is

`[c,y]=caputo_ics(`$\boldsymbol{a}$,$\boldsymbol{n}_a$,$\boldsymbol{b}$,$\boldsymbol{n}_b$,$\boldsymbol{y}_0$,$\boldsymbol{u}$,$\boldsymbol{t}$`)`

where vectors $\boldsymbol{a}$, $\boldsymbol{n}_a$, $\boldsymbol{b}$, and $\boldsymbol{n}_b$ are respectively the coefficients and orders on the two sides of the differential equation, while $\boldsymbol{y}_0$ is the initial value vector of length q. The vectors $\boldsymbol{u}$ and $\boldsymbol{t}$ are the input and time vectors, having length $p+1$, from which the value p is loaded into the function. The returned argument $\boldsymbol{c}$ contains the equivalent initial values, and vector $\boldsymbol{y}$ returns the first p values of the solution.

If the precision requirement is not very high, that is, $p \leqslant q$, the given initial values $\boldsymbol{y}_0$ are sufficient. If the precision requirements are high, that is, $p > q$, the given vector $\boldsymbol{y}_0$ is not sufficient. The equivalent initial values are, in fact, the high order terms of Taylor series of $y(t)$. Since the $y(t)$ signal is not known in real applications, high order terms in its Taylor series can only be found by solving linear algebraic equations.

Example 8.8. Solve the following Caputo differential equation:

$$y'''(t) + \frac{1}{16}{}_0^{\rm C}\mathscr{D}_t^{2.5}y(t) + \frac{4}{5}y''(t) + \frac{3}{2}y'(t) + \frac{1}{25}{}_0^{\rm C}\mathscr{D}_t^{0.5}y(t) + \frac{6}{5}y(t) = \frac{172}{125}\cos\frac{4t}{5}$$

with initial values of $y(0) = 1$, $y'(0) = 4/5$, and $y''(0) = -16/25$. The analytical solution is $y(t) = \sqrt{2}\sin(4t/5 + \pi/4)$. If $0 \leqslant t \leqslant 30$, for different step-sizes h try to reconstruct the Taylor series coefficients and validate the results.

Solutions. The following commands can be used to compute the Taylor series coefficients from the analytical solutions:

```
>> syms t; y=sqrt(2)*sin(4*t/5+pi/4) % analytical solution
   F=taylor(y,t,'Order',7), y0a=sym2poly(F)
```

It can be seen that the first seven terms in Taylor series are

$$y(t) = 1 + \frac{4}{5}t - \frac{8}{25}t^2 - \frac{32}{375}t^3 + \frac{32}{1\,875}t^4 + \frac{128}{46\,875}t^5 - \frac{256}{703\,125}t^6 + o(h^6).$$

If the first few terms in the Taylor series are to be rebuilt, different values of p and step-sizes h are tried. The equivalent initial values can be found, and the errors are given in Table 8.3.

```
>> a=[1 1/16 4/5 3/2 1/25 6/5]; na=[3 2.5 2 1 0.5 0];
   y0=[1 4/5 -16/25]; b=1; nb=0;
   h=[0.1,0.05,0.02,0.01,0.005,0.002,0.001];
   for i=1:7, for p=1:6
      t=[0:h(i):p*h(i)]; u=172/125*cos(4/5*t);
      y=sqrt(2)*sin(4*t/5+pi/4);  % compute the analytical solution
      [ee,yy]=caputo_ics(a,na,b,nb,y0,u,t); err(p,i)=norm(yy-y');
   end, end
```

It is seen that the errors obtained are all very small. Therefore the equivalent initial values can be used, where the necessary conditions are satisfied. They can be used in finding high precision solutions of Caputo differential equations.

Table 8.3: Computation errors for different orders and step-sizes.

p	$h = 0.1$	$h = 0.05$	$h = 0.02$	$h = 0.01$	$h = 0.005$	$h = 0.002$	$h = 0.001$
1	5.0×10^{-6}	3.2×10^{-7}	8.2×10^{-9}	5.1×10^{-10}	3.2×10^{-11}	8.2×10^{-13}	5.1×10^{-14}
2	1.6×10^{-6}	5.7×10^{-8}	6.1×10^{-10}	2.0×10^{-11}	6.2×10^{-13}	6.0×10^{-15}	2.0×10^{-16}
3	3.74×10^{-7}	5.9×10^{-9}	2.4×10^{-11}	3.8×10^{-13}	6.0×10^{-15}	3.0×10^{-16}	2.0×10^{-16}
4	4.00×10^{-8}	3.77×10^{-10}	6.84×10^{-13}	5.3×10^{-15}	3.0×10^{-16}	2.0×10^{-16}	3.0×10^{-16}
5	7.22×10^{-9}	2.89×10^{-11}	1.90×10^{-14}	2.22×10^{-16}	3.8×10^{-16}	3.1×10^{-16}	3.8×10^{-16}
6	5.16×10^{-10}	1.35×10^{-12}	4.96×10^{-16}	5.43×10^{-16}	4.4×10^{-16}	3.1×10^{-16}	3.8×10^{-16}

8.3.5 High precision algorithms

The equivalent initial values are, in fact, the first p values of the numerical solution $y(t)$. Taylor auxiliary function $T(t)$ can be established directly. With the above methods, the high precision solutions of Caputo differential equations can then be found. Summarizing the above ideas, the following algorithm is proposed:
(1) Compute the equivalent initial values.
(2) Establish the auxiliary function $T(t)$ from the equivalent initial values. Signal $y(t)$ is decomposed as $T(t) + z(t)$.
(3) Find the high precision solutions $z(t)$ of the differential equations with zero initial values.
(4) From $y(t) = T(t) + z(t)$, the high precision solution can be constructed.

Based on the above algorithm, the following MATLAB function `fode_caputo9()` can be written, where two functions are embedded, that is, `fode_sol9()` and `caputo_ics()`:

```
function y=fode_caputo9(a,na,b,nb,y0,u,t,p)
T=0; dT=0; t=t(:); u=u(:);
if p>length(y0)
   yy0=caputo_ics(a,na,b,nb,y0,u(1:p),t(1:p));
   y0=yy0(1:p).*gamma(1:p)';
elseif p==length(y0)
   yy0=caputo_ics(a,na,b,nb,y0,u(1:p+1),t(1:p+1));
   y0=yy0(1:p+1).*gamma(1:p+1)';
end
for i=1:length(y0), T=T+y0(i)/gamma(i)*t.^(i-1); end
for i=1:length(na), dT=dT+a(i)*caputo9(T,t,na(i),p); end
u=u-dT; y=fode_sol9(a,na,b,nb,u,t,p)+T';
```

The syntax of the function is `y=fode_caputo9(`$\boldsymbol{a}$,$\boldsymbol{n}_a$,$\boldsymbol{b}$,$\boldsymbol{n}_b$,$\boldsymbol{y}_0$,$\boldsymbol{u}$,$\boldsymbol{t}$,p`)`. This function can be used to directly solve linear Caputo differential equations with nonzero initial values.

Example 8.9. Solve again the equation in Example 8.8, and assess the precision of the solutions for different orders and step-sizes.

Solutions. Selecting a large step-size $h = 0.1$ and different orders p, the numerical solutions of the differential equations can be found. Compared with the analytical solutions, the errors at different time instances are measured, as shown in Table 8.4. It can be seen that, when p increases, the errors are significantly reduced. The case of $p = 6$ is a counterexample. When t is large, the error is also large. It is still significantly better than when using the $o(h)$ algorithm. In this example, $p = 5$ is the best choice.

```
>> h=0.1; t=0:h:30; y=sqrt(2)*sin(4*t/5+pi/4);
   u=172/125*cos(4/5*t); ii=31:30:301; y=y(ii);
   a=[1 1/16 4/5 3/2 1/25 6/5]; na=[3 2.5 2 1 0.5 0];
   b=1; nb=0; y0=[1 4/5 -16/25]; T=[];
   for p=1:6
      y1=fode_caputo9(a,na,b,nb,y0,u,t,p);
      err=y-y1(ii); T=[T abs(err.')];
   end
```

Table 8.4: Solution errors for step-size $h = 0.1$.

time t	$p = 1$	$p = 2$	$p = 3$	$p = 4$	$p = 5$	$p = 6$
3	0.00103	0.01469	0.002614	0.000146	5.150×10^{-7}	1.911×10^{-7}
6	0.000353	0.00974	0.000599	4.576×10^{-5}	3.034×10^{-7}	2.068×10^{-7}
9	0.023482	0.00347	0.001267	4.883×10^{-5}	1.111×10^{-9}	6.854×10^{-7}
12	0.04216	0.004681	0.000448	0.000137	3.720×10^{-7}	4.163×10^{-6}
15	0.03881	0.01212	0.000228	0.0001925	1.016×10^{-6}	1.816×10^{-5}
18	0.00876	0.016612	0.001434	0.0002022	1.313×10^{-6}	5.091×10^{-5}
21	0.03357	0.016534	0.000876	0.0001621	2.251×10^{-6}	0.0001259
24	0.066352	0.011885	0.001411	8.124×10^{-5}	1.074×10^{-6}	0.0002638
27	0.069885	0.004088	0.000479	1.812×10^{-5}	1.767×10^{-6}	0.00050
30	0.040484	0.004329	0.0006258	0.000112	3.506×10^{-6}	0.000895

If smaller step-size $h = 0.01$ is selected, the computational errors are found for different selections of p, as shown in Table 8.5. It can be seen that the accuracy are not in general as good as in the case of step-size $h = 0.1$. This is because that when the step-size is reduced, the total number of points increases, and the accumulative error also increases. In this example, the total number of points is 3 001, and $p = 3$ is a good choice.

```
>> h=0.01; t=0:h:30; y=sqrt(2)*sin(4*t/5+pi/4);
   u=172/125*cos(4/5*t); ii=301:300:3001; y=y(ii); T=[];
   for p=1:6, y1=fode_caputo9(a,na,b,nb,y0,u,t,p);
```

Table 8.5: Computational errors for $h = 0.01$.

time t	$p = 1$	$p = 2$	$p = 3$	$p = 4$	$p = 5$	$p = 6$
3	0.0002661	0.0008371	0.0002531	9.896×10^{-7}	6.242×10^{-9}	1.622×10^{-7}
6	0.0004601	8.888×10^{-5}	4.260×10^{-5}	2.256×10^{-6}	9.681×10^{-8}	6.299×10^{-5}
9	0.0003474	0.0005399	0.0001029	1.245×10^{-5}	1.145×10^{-7}	0.001254
12	0.000589	0.0010615	5.969×10^{-5}	2.683×10^{-5}	1.155×10^{-5}	0.0075575
15	0.0007764	0.0013316	5.806×10^{-5}	6.094×10^{-5}	4.532×10^{-5}	0.026356
18	0.0038395	0.001292	0.0001429	0.00010372	0.0001179	0.069312
21	0.0071045	0.00094829	0.00014851	0.00016832	0.0001865	0.15579
24	0.0086094	0.00038365	9.830×10^{-5}	0.00026625	0.0003858	0.30641
27	0.0067604	0.00026481	0.0001268	0.00034755	0.0009031	0.55351
30	0.001628	0.00084186	7.436×10^{-5}	0.00052638	0.0019239	0.94305

```
    err=y-y1(ii); T=[T abs(err.')]; % compute the errors
end
```

It is seen that to avoid the increase of cumulative errors, the step-size should be suitably chosen, such that the total number of points does not exceed 1 500, otherwise the cumulative errors may affect the final solutions.

Example 8.10. Consider the zero initial value problem in Example 8.6. In the previous example, the behavior of the high precision algorithm was not satisfactory. Solve again the problem with the high precision Caputo equation solver.

Solutions. It was pointed out in Example 8.6 that the high precision algorithm failed because the necessary conditions of the algorithm were not satisfied. With the new algorithm, the equivalent initial values are computed, such that the differential equation can be solved with the new solver. The error curve is shown in Figure 8.6, from which it is seen that, if $p = 4$, there is no apparent error.

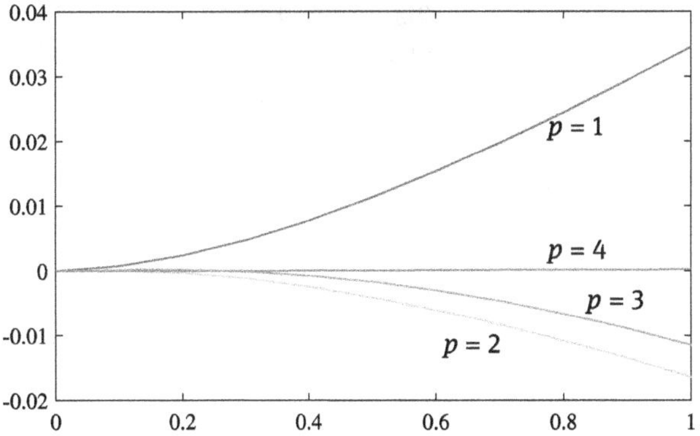

Figure 8.6: Computational errors for different values of p.

```
>> a=[1 1 1]; na=[3 2.5 0]; b=1; nb=0;
   t=0:0.1:1; y=-1+t-t.^2/2+exp(-t); y0=0;
   u=-1+t-t.^2/2-t.^0.5.*ml_func([1,1.5],-t);
   y1=fode_caputo9(a,na,b,nb,y0,u,t,1); e1=y-y1;
   y2=fode_caputo9(a,na,b,nb,y0,u,t,2); e2=y-y2;
   y3=fode_caputo9(a,na,b,nb,y0,u,t,3); e3=y-y3;
   y4=fode_caputo9(a,na,b,nb,y0,u,t,4); e4=y-y4;
   plot(t,e1,t,e2,t,e3,t,e4) % draw the error curves
```

Therefore it is recommended here to solve various fractional-order differential equations with the solver `fode_caputo9()`, no matter what the initial values. Note that the number of points should not be too large.

8.4 Solution of nonlinear fractional-order differential equations

Linear Caputo differential equations were studied, and a high precision algorithm was proposed for solving linear Caputo differential equations, together with its MATLAB solvers. In this section, general numerical algorithms for nonlinear Caputo differential equations are studied.

Definition 8.15. The general for of a nonlinear Caputo differential equation is

$$F(t, y(t), {}_{t_0}^{\mathrm{C}}\mathscr{D}_t^{\alpha_1} y(t), \dots, {}_{t_0}^{\mathrm{C}}\mathscr{D}_t^{\alpha_n} y(t)) = 0 \tag{8.4.1}$$

where $F(\cdot)$ is a function of time t, unknown $y(t)$, and its fractional-order derivatives. It is safe to assume that $\alpha_n > \alpha_{n-1} > \cdots > \alpha_2 > \alpha_1 > 0$.

Definition 8.16. Denoting $q = \lceil \alpha_n \rceil$ in Definition 8.15, the necessary initial values for the fractional-order differential equation satisfy (8.3.5).

In this section, a special predictor–corrector algorithm is presented. The algorithms are divided into two independent, prediction and correction parts. A suitable predictor solution is found first. Based on this solution, a correction algorithm is used to find the solutions of the differential equations. Further a high-precision matrix algorithm for implicit Caputo equations is proposed.

8.4.1 Predictor method

Consider again a nonlinear explicit Caputo differential equation:

$${}_0^{\mathrm{C}}\mathscr{D}_t^{\alpha} y(t) = f(t, y(t), {}_0^{\mathrm{C}}\mathscr{D}_t^{\alpha_1} y(t), \dots, {}_0^{\mathrm{C}}\mathscr{D}_t^{\alpha_{n-1}} y(t)). \tag{8.4.2}$$

Let us review the algorithm in Chapter 5. The key point is to introduce a Taylor auxiliary function $T(t)$, which ensures that the output signal can be decomposed into $y(t) = z(t) + T(t)$, where $z(t)$ has zero initial values, and $T(t)$ is the Taylor auxiliary function defined in (8.3.7). The definition is given again here:

$$T(t) = \sum_{k=0}^{q-1} \frac{y^{(k)}(0)}{k!} t^k = \sum_{k=0}^{q-1} \frac{y_k}{k!} t^k. \tag{8.4.3}$$

It can be seen from (8.4.2) that

$${}_0^{\mathrm{C}}\mathscr{D}_t^{\alpha} z(t) = f(t, z(t), {}_0^{\mathrm{C}}\mathscr{D}_t^{\alpha_1} z(t), \dots, {}_0^{\mathrm{C}}\mathscr{D}_t^{\alpha_{n-1}} z(t)). \tag{8.4.4}$$

Besides, since ${}_0^{\mathrm{C}}\mathscr{D}_t^{\gamma} z(t) = {}_0^{\mathrm{RL}}\mathscr{D}_t^{\gamma} z(t)$, it is found that

$${}_0^{\mathrm{RL}}\mathscr{D}_t^{\alpha} z(t) = f(t, z(t), {}_0^{\mathrm{RL}}\mathscr{D}_t^{\alpha_1} z(t), \dots, {}_0^{\mathrm{RL}}\mathscr{D}_t^{\alpha_{n-1}} z(t)). \tag{8.4.5}$$

Selecting a step-size h, the first q points of function z_i are zero, and $y_i = T_i$, $i = 1, 2, \dots, q$. Therefore, for each value of k, a loop structure can be used to solve the equation. Assuming that z_k are known, to find the numerical solution of T_{k+1}, the predict value of z_{k+1} is denoted as $z_{k+1}^{\mathrm{p}} = z_k$. Substituting it into (8.4.5), it is found that

$${}_0^{\mathrm{RL}}\mathscr{D}_t^{\alpha} z(t) = \hat{f}. \tag{8.4.6}$$

If $\hat{f}$ is regarded as a known function, (8.4.6) is a single-term differential equation. The solution $\hat{z}_{k+1}$ can be found. If $\|\hat{z}_{k+1} - z_{k+1}^{\mathrm{p}}\| < \epsilon$, where ϵ is a preselected error tolerance, the quantity $\hat{z}_{k+1}$ can be regarded as the numerical solution of the original equation. Otherwise, let $z_{k+1}^{\mathrm{p}} = \hat{z}_{k+1}$ and continue the iteration process. The prediction algorithm is summarized below:

(1) Set up the Taylor auxiliary function $T(t)$ such that the signal is decomposed as $y(t) = z(t) + T(t)$. Set up (8.4.5).
(2) Select a step-size h, and for $i = 1, 2, \dots, q$, let $y_i = T_i$, $z_i = 0$.
(3) Let $k = q + 1$, and initiate the loop structure.
(4) Denote $z_{k+1}^{\mathrm{p}} = z_k$ and compute ${}_0^{\mathrm{RL}}\mathscr{D}_t^{\alpha_i} z_{k+1}$. Substituting it into (8.4.5), from (8.4.6), $\hat{z}_{k+1}$ can be solved for.
(5) If $\|\hat{z}_{k+1} - z_{k+1}^{\mathrm{p}}\| < \epsilon$, the solution $z_{k+1} = \hat{z}_{k+1}$ is accepted, otherwise, denote $z_{k+1}^{\mathrm{p}} = \hat{z}_{k+1}$. Continue the iteration process to find the solutions of the equations.

Based on the above algorithms, the following MATLAB function is written:

```
function [y,t]=nlfep(fun,alpha,y0,tn,h,p,err)
m=ceil(tn/h)+1; t=(0:(m-1))'*h; ha=h.^alpha; z=0;
[T,dT,w,d2]=aux_func(t,y0,alpha,p);
```

```
y=z+T(1); dy=zeros(1,d2-1);
for k=1:m-1
   zp=z(end); yp=zp+T(k+1); y=[y; yp]; z=[z; zp];
   [zc,yc]=repeat_funs(fun,t,y,d2,w,k,z,ha,dT,T);
   while abs(zc-zp)>err
      yp=yc; zp=zc; y(end)=yp; z(end)=zp;
      [zc,yc]=repeat_funs(fun,t,y,d2,w,k,z,ha,dT,T);
end, end
% make the repeatable function as a subfunction
function [zc,yc]=repeat_funs(fun,t,y,d2,w,k,z,ha,dT,T)
for j=1:(d2-1)
   dy(j)=w(1:k+1,j+1)'*z((k+1):-1:1)/ha(j+1)+dT(k,j+1);
end, f=fun(t(k+1),y(k+1),dy);
zc=((f-dT(k+1,1))*ha(1)-w(2:k+1,1)'*z(k:-1:1))/w(1,1);
yc=zc+T(k+1);
```

where the supporting function `aux_func()` is given below. The target is to reduce the repeatable code in the correction algorithm to be presented later. The code is as follows:

```
function [T,dT,w,d2]=aux_func(t,y0,alpha,p)
an=ceil(alpha); y0=y0(:); q=length(y0); d2=length(alpha);
m=length(t); g=double(genfunc(p));
for i=1:d2, w(:,i)=get_vecw(alpha(i),m,g)'; end
b=y0./gamma(1:q)'; T=0; dT=zeros(m,d2);
for i=1:q, T=T+b(i)*t.^(i-1); end
for i=1:d2
   if an(i)==0, dT(:,i)=T;
   elseif an(i)<q
      for j=(an(i)+1):q
         dT(:,i)=dT(:,i)+(t.^(j-1-alpha(i)))*...
                 b(j)*gamma(j)/gamma(j-alpha(i));
end, end, end
```

The syntax of the prediction function is $[\boldsymbol{y},\boldsymbol{t}]$=`nlfep(fun,`$\alpha,\boldsymbol{y}_0,t_n,h,p,\epsilon$`)`, where `fun` is the function handle of the Caputo differential equation, whose format is to be illustrated in full details through examples later. The initial value vector is provided in $\boldsymbol{y}_0$. The argument t_n is the stoping time, h is the step-size, and p and ϵ are respectively the order of the algorithm and error tolerance. It is worth mentioning that in step (4) it is better select $p = 1$. A larger value of p may yield even larger errors.

It can be seen that the solutions obtained are not the solutions of the original differential equations. They can be regarded as the initial values for the corrector solvers to be introduced next.

Example 8.11. Find the numerical solution of the following nonlinear fractional-order differential equation:[20]

$$ {}_0^{\mathrm{C}}\mathscr{D}_t^{1.455} y(t) = -t^{0.1} \frac{\mathrm{E}_{1,1.545}(-t)}{\mathrm{E}_{1,1.445}(-t)} \mathrm{e}^t y(t) {}_0^{\mathrm{C}}\mathscr{D}_t^{0.555} y(t) + \mathrm{e}^{-2t} - [y'(t)]^2 $$

where $y(0) = 1$ and $y'(0) = -1$.

Solutions. The analytical solution is e^{-t}. In [20], the original form of the equation is given where the two Mittag-Leffler functions are described wrongly as $\mathrm{E}_{1.545}(-t)$ and $\mathrm{E}_{1.445}(-t)$. If the analytical solution $y(t) = \mathrm{e}^{-t}$ is substituted back to the original equation, it can be seen that the two sides are different. To ensure the analytical solutions, the Mittag-Leffler functions should be modified into two parameter functions.

It can be seen that the order vector is $\boldsymbol{\alpha} = [1.455, 0.555, 1]$, $\boldsymbol{y}_0 = [1, -1]$. Introducing the signal as $d_1(t) = {}_0^{\mathrm{C}}\mathscr{D}_t^{0.555} y(t)$ and $d_2(t) = y'(t)$, the standard form of the original equation can be written as follows:

$$ {}_0^{\mathrm{C}}\mathscr{D}_t^{1.455} y(t) = -t^{0.1} \frac{\mathrm{E}_{1,1.545}(-t)}{\mathrm{E}_{1,1.445}(-t)} \mathrm{e}^t y(t) d_1(t) + \mathrm{e}^{-2t} - d_2^2(t). $$

Note that the vectorized Caputo differential equations can be described by anonymous functions

```
>> f=@(t,y,Dy)-t.^0.1.*ml_func([1,1.545],-t).*exp(t)./...
     ml_func([1,1.445],-t).*y.*Dy(:,1)+exp(-2*t)-Dy(:,2).^2;
```

where variables $\boldsymbol{t}$ and $\boldsymbol{y}$ are time and output column vectors. The argument $\boldsymbol{D}_y$ is a matrix, whose columns correspond the fractional-order derivatives of the state space expressions.

The following commands can be used to solve Caputo differential equations. The prediction results are shown in Figure 8.7. It can be seen that the solution is not satisfactory in the predictions. The accuracy when $p = 2$ is rather worse than the case of $p = 1$. The elapsed time is about 0.1 seconds. The maximum errors are respectively 0.0264 and 0.0364.

```
>> alpha=[1.455,0.555,1]; y0=[1,-1];
   tn=1; h=0.01; err=1e-8; p=1; %input parameters
   tic, [yp1,t]=nlfep(f,alpha,y0,tn,h,p,err); toc
   tic, [yp2,t]=nlfep(f,alpha,y0,tn,h,2,err); toc
   plot(t,yp1,t,yp2,t,exp(-t))  %predictor result and comparisons
   max(abs(yp1-exp(-t))), max(abs(yp2-exp(-t)))
```

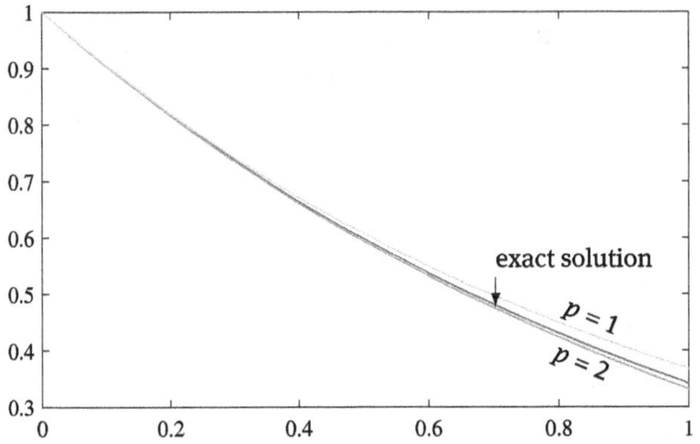

Figure 8.7: The prediction solution.

8.4.2 Corrector method

A better solver is presented in this section. The method is the corrector method. For the same step-size h, the predictor solution $\boldsymbol{y}_\mathrm{p}$ can be employed. If it is substituted into the right-hand side of (8.4.2), a single-term differential equation can be established. With iteration methods, the solutions of the differential equations can be found. The ideas are implemented in the following vectorized algorithm:

(1) Assume that the predictor solution is $\boldsymbol{y}_\mathrm{p}$.
(2) Substituting $\boldsymbol{y}_\mathrm{p}$ into (8.4.2), a corrector solution can be found as $\hat{\boldsymbol{y}}$.
(3) If $\|\hat{\boldsymbol{y}}-\boldsymbol{y}_\mathrm{p}\| < \epsilon$, the solution $\hat{\boldsymbol{y}}$ is accepted, otherwise, let $\boldsymbol{y}_\mathrm{p} = \hat{\boldsymbol{y}}$ and go back to step (2) to continue the iteration process, until the solutions are found.

Based on the above algorithm, the following MATLAB function can be written. It can be used to directly solve explicit Caputo differential equations of any complexity:

```
function y=nlfec(fun,alpha,y0,yp,t,p,err)
yp=yp(:); t=t(:); h=t(2)-t(1); m=length(t); ha=h.^alpha;
[T,dT,w,d2]=aux_func(t,y0,alpha,p);
[z,y]=repeat_funs(fun,t,yp,T,d2,alpha,dT,ha,w,m,p);
while norm(z)>err, yp=y; z=zeros(m,1);
   [z,y]=repeat_funs(fun,t,yp,T,d2,alpha,dT,ha,w,m,p);
end
% the repetitive subfunction
function [z,y]=repeat_funs(fun,t,yp,T,d2,alpha,dT,ha,w,m,p)
for i=1:d2, dyp(:,i)=glfdiff9(yp-T,t,alpha(i),p)'+dT(:,i); end
f=fun(t,yp,dyp(:,2:d2))-dyp(:,1); y=yp; z=zeros(m,1);
for i=2:m, ii=(i-1):-1:1;
   z(i)=(f(i)*(ha(1))-w(2:i,1)'*z(ii))/w(1,1); y(i)=z(i)+yp(i);
end
```

The syntax of the function is $\boldsymbol{y}$=`nlfec(fun,`α`,`$\boldsymbol{y}_0$`,`$\boldsymbol{y}_\mathrm{p}$`,`$\boldsymbol{t}$`,`$p$`,`$\epsilon$`)`, where the input arguments are almost the same as those in `nlfep()`. Compared with the above algorithm, it can be seen that the predictor algorithm is used to provide the initial values, or the solution, for iterative process. They can be replaced by others, e. g., $\boldsymbol{y}_\mathrm{p}$=`ones(size(`$\boldsymbol{t}$`))`.

Example 8.12. Solve again the Caputo differential equation in Example 8.11.

Solutions. From the predictor results, selecting $p = 2$, the following commands are used to find the corrector solutions. The maximum error is 3.9337×10^{-5}, and the elapsed time is 2.098 seconds. It can be seen that the efficiency of the algorithm is rather high.

```
>> f=@(t,y,Dy)-t.^0.1.*ml_func([1,1.545],-t).*exp(t)./...
      ml_func([1,1.445],-t).*y.*Dy(:,1)+exp(-2*t)-Dy(:,2).^2;
   alpha=[1.455,0.555,1]; y0=[1,-1];
   tn=1; h=0.01; err=1e-8; p=1;
   tic, [yp1,t]=nlfep(f,alpha,y0,tn,h,p,err); toc
   tic, [y2,t]=nlfec(f,alpha,y0,yp1,t,2,err); toc
   max(abs(y2-exp(-t)))   % compute the maximum error
```

If a smaller step-size $h = 0.001$ is used, the following statements can be used to solve the differential equation again. The maximum error is reduced to 3.9716×10^{-7}, and the elapsed time is 3.39 seconds. Further reducing the step-size to $h = 0.0001$, the maximum error is 6.8851×10^{-9}, with the elapsed time of 80.11 seconds. If one selects $p = 3$, the maximum error is reduced to 3.8361×10^{-9}, and the elapsed time is 367.06 seconds. Although with $p = 3$ the precision was improved, the theoretical level of $o(h^3)$ was not really reached.

```
>> h=0.001; tic, [yp,t]=nlfep(f,alpha,y0,tn,h,1,err); toc
   err=1e-10; tic, [y2,t]=nlfec(f,alpha,y0,yp,t,2,err); toc
   max(abs(y2-exp(-t)))   % compute maximum error
```

In fact, a larger step-size can still be tried, for instance, $h = 0.01$. If $p = 4$, the maximum error obtained is 7.0546×10^{-9}, and the elapsed time is 65.37 seconds. It can be seen that the efficiency is higher than using smaller step-sizes.

```
>> h=0.01; [yp,t]=nlfep(f,alpha,y0,tn,h,1,err); tic
   [y2,t]=nlfec(f,alpha,y0,yp,t,4,err); toc
   max(abs(y2-exp(-t)))    % compute maximum error
```

8.4.3 Matrix method for implicit Caputo differential equations

Up to now, the nonlinear Caputo differential equation algorithms were all for explicit differential equations. If implicit equations in Definition 8.15 are involved, implicit solvers are expected. The general form of nonlinear implicit fractional-order differential equations is given by

$$F(t, y(t), {}_{t_0}^{\mathrm{C}}\mathscr{D}_t^{\alpha_1} y(t), \dots, {}_{t_0}^{\mathrm{C}}\mathscr{D}_t^{\alpha_n} y(t)) = 0 \tag{8.4.7}$$

and the initial values are y_i, $i = 0, 1, \dots, \lceil \max(\alpha_i) \rceil - 1$. Signal $y(t)$ can still be decomposed as $y(t) = z(t) + T(t)$, where $T(t)$ is the same as defined in (8.4.3). The $z(t)$ signal has zero initial values. If the differential equation is revised into Riemann–Liouville definition, the following equations are still satisfied:

$$F(t, y(t), {}_{t_0}^{\mathrm{RL}}\mathscr{D}_t^{\alpha_1} z(t), \dots, {}_{t_0}^{\mathrm{RL}}\mathscr{D}_t^{\alpha_n} z(t)) = 0. \tag{8.4.8}$$

Recalling the matrix method of the pth order algorithm, Riemann–Liouville equations can be found. The original signal $y(t)$ can be decomposed into $u(t) + v(t)$, where

$$u(t) = \sum_{k=0}^{p} c_k (t - t_0)^k \tag{8.4.9}$$

and $c_k = y_k/k!$. Then the α_ith order differential equation can be written directly as

$${}_{t_0}^{\mathrm{RL}}\mathscr{D}_t^{\alpha_i} y(t) = \frac{1}{h^{\alpha_i}} \boldsymbol{W}^{\alpha_i} \boldsymbol{v} + \sum_{k=0}^{p} c_k \frac{\Gamma(k+1)}{\Gamma(k+1-\alpha_i)} (t - t_0)^{k-\alpha_i} \tag{8.4.10}$$

where $\boldsymbol{W}^{\alpha_i}$ is a lower-triangular matrix with the coefficients w_i. For simplicity, denote $\boldsymbol{B}_i = \boldsymbol{W}^{\alpha_i}/h^{\alpha_i}$.

Note that for finding the α_ith order Caputo derivative, the $(\lceil \alpha_i \rceil + 1)$th order integer-order derivative should be evaluated first. Then integration can be performed. The terms with orders lower than $\lceil \alpha_i \rceil$ are all eliminated. Then, (8.4.10) can be rewritten as

$${}_{t_0}^{\mathrm{RL}}\mathscr{D}_t^{\alpha_i} y(t) = \frac{1}{h^{\alpha_i}} \boldsymbol{W}^{\alpha_i} \boldsymbol{v} + \sum_{k=\lceil \alpha_i \rceil + 1}^{q} \frac{c_k \Gamma(k+1)}{\Gamma(k+1-\alpha_i)} (t - t_0)^{k-\alpha_i}. \tag{8.4.11}$$

For the time vector $\boldsymbol{t} = [0, h, 2h, \dots, mh]$, the following nonlinear algebraic equation can be composed:

$$\boldsymbol{f}(\boldsymbol{t}, \boldsymbol{B}_1 \boldsymbol{v}, \boldsymbol{B}_2 \boldsymbol{v}, \dots, \boldsymbol{B}_n \boldsymbol{v}) = \boldsymbol{0}. \tag{8.4.12}$$

The equation can be solved with the MATLAB solver `fsolve()`. The following MATLAB code is written to implement the algorithm, where $\boldsymbol{B}_i$ and matrix $\boldsymbol{d}_u$ can be passed to the equation involving f as additional parameters; $\boldsymbol{B}_i$ are described as three-dimensional arrays:

```
function [y,t]=nlfode_mat(f,alpha,y0,tn,h,p,yp)
y0=y0(:); alfn=ceil(alpha); m=ceil(tn/h)+1;
```

```
t=(0:(m-1))'*h; d1=length(y0); d2=length(alpha);
B=zeros(m,m,d2); g=double(genfunc(p));
for i=1:d2, w=get_vecw(alpha(i),m,g);
   B(:,:,i)=rot90(hankel(w(end:-1:1)))/h^alpha(i);
end
c=y0./gamma(1:d1)'; du=zeros(m,d2);
u=0; for i=1:d1, u=u+c(i)*t.^(i-1); end
for i=1:d2
   if alfn(i)==0, du(:,i)=u;
   elseif alfn(i)<d1
      for k=(alfn(i)+1):d1
         du(:,i)=du(:,i)+(t.^(k-1-alpha(i)))*c(k)...
                    *gamma(k)/gamma(k-alpha(i));
end, end, end
if nargin==6, yp=zeros(m,1); end
v=fsolve(f,yp,[],t,u,B,du); y=u+v;
```

The syntax of the function is `[`$\boldsymbol{y}$`,`$\boldsymbol{t}$`]=nlfode_mat(fun,`$\boldsymbol{\alpha}$`,`$\boldsymbol{y}_0$`,`t_n`,`h`,`p`,`$\boldsymbol{y}_\mathrm{p}$`)`, where `fun` is the function handle of the implicit differential equations; $\boldsymbol{\alpha}$ is the order vector for the differential equation. The argument $\boldsymbol{y}_0$ contains the initial values; h and t_n are respectively step-size and stoping time. The argument p is the order of the algorithm. The optional vector $\boldsymbol{y}_\mathrm{p}$ stores the predictor solution. If it is not known, a zero vector can be used instead.

Example 8.13. Solve the fractional-order differential equation in Example 8.11 with matrix methods.

Solutions. The implicit differential equation can be written from the given explicit one

$${}_0^{\mathrm{C}}\mathscr{D}_t^{1.455}y(t)+t^{0.1}\frac{\mathrm{E}_{1,1.545}(-t)}{\mathrm{E}_{1,1.445}(-t)}\,\mathrm{e}^t y(t)\,{}_0^{\mathrm{C}}\mathscr{D}_t^{0.555}y(t)-\mathrm{e}^{-2t}+\left[y'(t)\right]^2=0.$$

The order vector is $\boldsymbol{\alpha}=[1.455,0.555,1]$, and we denote ${}_0^{\mathrm{C}}\mathscr{D}_t^{\alpha_i}y(t)$ in matrix form as $\boldsymbol{d}_u$`(:,`i`)` $+\boldsymbol{B}$`(:,:,`i`)*`$\boldsymbol{v}$. The signal $y(t)$ is denoted as $u+v$. Selecting $p=1$ and $p=2$, the following statements can be used to directly solve the differential equation. The maximum errors are 0.0090 and 0.0010, and the elapsed time is 13 seconds. If a smaller step-size $h=0.001$ is selected, the maximum errors are respectively 9.7603×10^{-4} and 3.5270×10^{-5}. The elapsed times are respectively 300 and 200 seconds.

```
>> alpha=[1.455,0.555,1]; y0=[1,-1]; tn=1; h=0.01;
   f=@(v,t,u,B,du)(du(:,1)+B(:,:,1)*v)+t.^0.1.*exp(t)...
         .*ml_func([1,1.545],-t)./ml_func([1,1.445],-t)...
         .*(v+u).*(du(:,2)+B(:,:,2)*v)-exp(-2*t)+...
```

```
   (du(:,3)+B(:,:,3)*v).^2;
tic, [y1,t]=nlfode_mat(f,alpha,y0,tn,h,1); toc
tic, [y2,t]=nlfode_mat(f,alpha,y0,tn,h,2); toc
max(abs(y1-exp(-t))), max(abs(y2-exp(-t)))
```

It is worth mentioning that the algorithm is meant for implicit fractional-order differential equations. The benefit of it cannot be fully demonstrated when tackling explicit equations.

Although significant improvement in accuracy and speed are made to the nonlinear fractional-order differential equation solvers, the algorithms have certain limitations: (1) the solution process is rather time consuming; (2) it cannot be extended to handle any differential equations, such as implicit ones, differential-algebraic equations, and delay differential equations. In Chapter 9, block diagram based methods are presented to solve fractional-order differential equations with higher efficiency and higher usability.

8.5 Exercises

8.1 In Section 8.2, the analytical solutions of step and impulse responses for commensurate-order systems are provided. If the input signal is a ramp function, $u(t) = t$, derive the analytical solutions of the responses.

8.2 Consider the single-term differential equation

$$ {}_0^{\mathrm{C}}\mathscr{D}_t^{0.7} y(t) = \begin{cases} \dfrac{1}{\Gamma(1.3)} t^{0.3}, & 0 \leqslant t \leqslant 1, \\ \dfrac{1}{\Gamma(1.3)} t^{0.3} - \dfrac{2}{\Gamma(2.3)} (t-1)^{1.3}, & t > 1, \end{cases} $$

where the initial value is $y(0) = 0$, and find the numerical solution of the fractional-order differential equation for $t \in (0, 10)$. The analytical solution of the fractional-order differential equation is given below:

$$ y(t) = \begin{cases} t, & 0 \leqslant t \leqslant 1, \\ t - (t-1)^2, & t > 1. \end{cases} $$

Assess the precision of the numerical solution.

8.3 Solve the linear fractional-order differential equation

$$ y'''(t) + {}_0^{\mathrm{C}}\mathscr{D}_t^{2.5} y(t) + y''(t) + 4y'(t) + {}_0^{\mathrm{C}}\mathscr{D}_t^{0.5} y(t) + 4y(t) = 6\cos t, $$

where the initial values are $y(0) = 1$, $y'(0) = 1$, and $y''(0) = -1$. Evaluate the accuracy and speed of the numerical solution. The analytical solution of the problem is $y(t) = \sqrt{2}\sin(t + \pi/4)$.

8.4 For the given linear fractional-order differential equation[56]

$$0.8\mathscr{D}_t^{2.2}y(t) + 0.5\mathscr{D}_t^{0.9}y(t) + y(t) = 1, \quad y(0) = y'(0) = y''(0) = 0,$$

find its numerical solution. If the orders are approximated, for instance, 2.2 is regarded as 2, while 0.9 is approximated by 1, solve the integer-order differential equation and observe the differences.

8.5 Solve the linear fractional-order differential equation with zero initial values:

$$\mathscr{D}^{1.2}y(t) + 5\mathscr{D}^{0.9}y(t) + 9\mathscr{D}^{0.6}y(t) + 7\mathscr{D}^{0.3}y(t) + 2y(t) = u(t)$$

where $u(t)$ is a unit impulsive or step signal.

8.6 Solve the following single-term Caputo equation[20]

$$\mathscr{D}^{\alpha}y(t) = \frac{40\,320}{\Gamma(9-\alpha)}t^{8-\alpha} - 3\frac{\Gamma(5+\alpha/2)}{\Gamma(5-\alpha/2)}t^{4-\alpha/2} + \frac{9}{4}\Gamma(\alpha+1) + \left(\frac{3}{2}t^{\alpha/2} - t^4\right)^3 - y^{3/2}(t)$$

where the time interval is $t \in (0,1)$, with initial values $y(0) = 0$ and $y'(0) = 0$. It is known that the analytical solution of the equation is $y(t) = t^8 - 3t^{4+\alpha/2} + 9t^{\alpha}/4$. If $\alpha = 1.25$, assess the speed and accuracy of the MATLAB solver.

8.7 Consider the following nonlinear Caputo equation:

$${}_0^{\mathrm{C}}\mathscr{D}_t^{\sqrt{2}}y(t) = -t^{1.5-\sqrt{2}}\frac{\mathrm{E}_{1,3-\sqrt{2}}(-t)}{\mathrm{E}_{1,1.5}(-t)}\,\mathrm{e}^t\,y(t)\,{}_0^{\mathrm{C}}\mathscr{D}_t^{0.5}y(t) + \mathrm{e}^{-2t} - \left[y'(t)\right]^2,$$

with given initial values $y(0) = 1$ and $y'(0) = -1$. If the analytical solution is $y(t) = \mathrm{e}^{-t}$, solve the numerical solution of the Caputo equation for $t \in (0,2)$, and assess the precision and elapsed time.

8.8 Solve the following nonlinear fractional-order differential equations with zero initial values, where $f(t) = 2t + 2t^{1.545}/\Gamma(2.545)$. If the equation is a Caputo one, and $y(0) = -1$ and $y'(0) = 1$, solve again the differential equation:

$$\mathscr{D}^2x(t) + \mathscr{D}^{1.455}x(t) + \left[\mathscr{D}^{0.555}x(t)\right]^2 + x^3(t) = f(t).$$

9 Block diagram solutions of ordinary differential equations

Initial value problems of various differential equations were systematically introduced in the previous chapters. There the solution methods were globally referred to as command-driven methods. The procedures of the command-driven methods were that the entire differential equation should be expressed by anonymous or MATLAB functions, then appropriate solvers could be called to find the numerical solutions of the differential equations.

From the numerical analysis viewpoint, the solution pattern like this is perfect. A direct command can be used to solve the whole differential equation. From the application viewpoint, there are limitations in the command-driven methods. Since for large-scale systems the entire system is composed of many independent parts, each can be by itself a differential equation or implicit differential equation, or composed of other models. Therefore, to convert such a complicated system into a single standard model is a very tedious, if not impossible, job. For instance, in an aeroplane control problem, if a command-driven method is to be used, the internal system may have mechanical, electrical, electronic components, and may have other nonlinearities and discrete controllers. A set of state variables must be selected to form a state vector, and one has to write a set of independent first-order explicit or implicit differential equations so that the solution process can be invoked. If in the solution process, a component is replaced by another one, the state variables should be selected again, and differential equations must be rewritten. This is an impossible task. A slight carelessness may lead to wrong simulation results.

A more practical way to solve this problem is the use of a block diagram based solution pattern. A large-scale system can be divided into several independent subsystems, and these subsystems are interconnected in certain ways. Therefore when replacing components, a simulation subsystem can be created, without affecting any other subsystems.

Simulink is an ideal tool to implement diagram based simulation modeling and simulation. Details on Simulink will be further presented in Volume VI. In this section, Section 9.1 introduces Simulink and provides some essential knowledge about it. This can be regarded as a fundamental background for latter materials in this chapter. In Section 9.2, the Simulink modeling methodology for differential equations is proposed. Examples are used to demonstrate how to draw differential equations in Simulink using a graphical approach. It is also demonstrated on how to set simulation control parameters. In Section 9.3, examples of a variety of differential equations, such as differential-algebraic equations, delay differential equations, stochastic differential equations, switched differential equations, and so on, are illustrated. In Section 9.4, modeling and simulation of fractional-order differential equations are introduced.

https://doi.org/10.1515/9783110675252-009

9.1 Simulink essentials

Simulink is a block diagram based modeling and simulation environment in MATLAB. To learn block diagram based modeling and simulation methods, essential knowledge on Simulink should be studied first. In this section, Simulink is briefly introduced first, and some of the commonly used blocks are presented. This will establish some foundation for modeling and simulation studies of differential equations.

9.1.1 An introduction to Simulink

Simulink environment was first released by MathWorks in 1990. The original name was SimuLAB. In 1992, it was renamed Simulink, whose two key words – simu and link – are respectively for simulation and modeling. The environment can be used to represent complicated systems in a block diagram modeling format. Various differential equations can be described graphically with such a powerful tool.

Of course, differential equation modeling and solution is a very small part in Simulink modeling. It provides also a significant number of blocks used for control and many other systems. The multidomain physical modeling facilities enable the researchers to build mechanical, motor, power electronic, and communication systems to create simulation models using building block patterns. The Simulink environment is very powerful. With Simulink and its blocksets, differential equations of arbitrary complexity can be solved. For the relevant materials the readers are referred to [68] or [77]. In the latter book, only differential equation related systems are discussed.

In this section, a brief introduction to the commonly used blocks in differential equation modeling are introduced, and then an example is used to demonstrate how to create graphical models and solve differential equations with Simulink.

9.1.2 Relevant blocks in Simulink

Typing `open_system('simulink')` in MATLAB command window, a model window shown in Figure 9.1 is opened. It can be seen that there are many group icons. They are composed of lower-level blocks or groups. For instance, the input signal block group named Sources, the continuous block group called Continuous, the user-defined blocks and program group titled User-defined Functions, and so on. The blocks in each group are plentiful, and, theoretically speaking, dynamic systems of any complexity can be modeled with Simulink.

There are a huge number of blocks supplied in Simulink. It is not possible to present them in a book, or even in a chapter like this. For modeling and solution of differential equations, a user-specified group is created. The user can simply type `odegroup` to open the user-defined blockset, as shown in Figure 9.2.

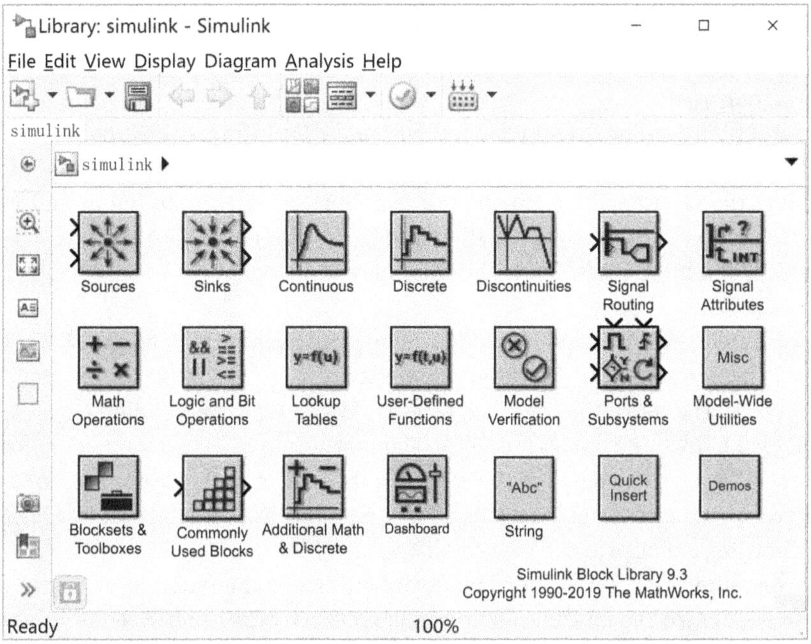

Figure 9.1: Main window of Simulink environment.

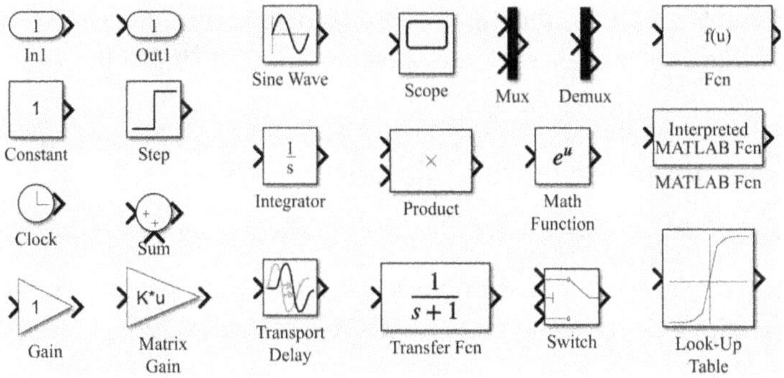

Figure 9.2: User-defined blockset for differential equations.

The commonly used blocks are summarized below:

(1) In1 and out1. Normally, the out1 block is used to generate the output signal, which may generate a variable `yout` in MATLAB workspace. Besides the signals of interest can also connected to Scope block.

(2) Clock block generates signal t, and it can also be used to establish time-varying models.

(3) **Commonly used input blocks** include a **Constant** block to generate a constant signal, **Sine** block to generate a sinusoidal signal, and **Step** block to generate a step input signal.
(4) **Integrator** block can be used to define key signals, for instance, its input can be regarded as the first-order derivative of the output signal. If in differential equation modeling the input of the ith integrator is defined as $x_i'(t)$, then the output is $x_i(t)$. If a first-order differential equation is to be described, the key signals can be defined by the integrators. For high-order linear differential equations, **Transfer Fcn** block can also be used.
(5) **Transport Delay** block can be used to find the signal value at $t - \tau$. The block is useful in modeling delay differential equations.
(6) **Gain blocks.** The related blocks such as **Gain**, **Matrix Gain**, and **Sliding Gain** are useful gain blocks, the names are self-explanatory. **Gain** block is used to amplify signals. If the input signal is u, the output is then Ku. A dialog box can also be used to change to matrix gain matrix. **Sliding Gain** block allows the user to adjust the gain using a mouse to drag the scroll bar.
(7) **Math computing blocks.** Dedicated blocks for arithmetic computations on signals are provided. Also logical and comparison operation blocks can be used. An algebraic equation solver block is provided, and it is useful in modeling differential-algebraic equations or implicit differential equations.
(8) **Math function blocks.** Nonlinear function evaluations of signals are supported, such as trigonometric functions and logarithmic functions.
(9) **Signal vectorization blocks. Mux** block is used to compose several signals into a vectorized one, while **DeMux** block separates a vector signal into individual ones.

9.2 Block diagram modeling methodologies in differential equations

How to represent differential equations with block diagrams? In fact, representing differential equations with a block diagram is not difficult. A new methodology should be used. In this section, key signal defining is the first step. Based on the key signals, the differential equations are drawn in a step-by-step manner. Simulation methods can then adopted to find the numerical solutions of the differential equations.

9.2.1 Integrator chains and key signal generation

If one observes closely the ordinary differential equation models, it is easily seen that differential equations are used to describe the relationship among time t, unknown signals $x_i(t)$, and their derivatives. Sometimes, even high-order derivatives are involved. These signals are key ones in differential equation modeling.

Signal t can be provided by the **Clock** block in the **Sources** group or on the commonly used group in Figure 9.2.

There are two ways in defining unknown signals and their arbitrary derivatives. One way is to use n **Derivative** block connected serially, as shown in Figure 9.3. Assuming that the signal flowing into the first block is $y(t)$, the signals $y'(t)$, $y''(t)$, ..., $y^{(n)}(t)$ can then be defined. This method seems simple and straightforward, yet there are two problems. The first is that numerical differentiation algorithms are not quite reliable. Therefore high-order differentiation may bring in numerical troubles, and the numerical stability is questionable; the other is that there is no place to specify the initial values. Therefore this method is not further considered.

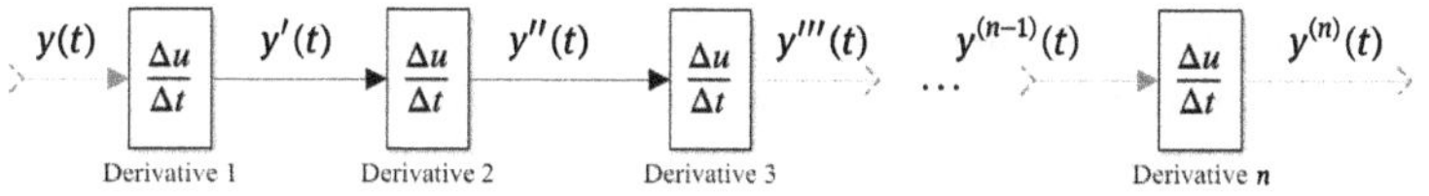

Figure 9.3: Key signal defining with a differentiator chain.

A practical way to define key signals is to compose an integrator chain, as shown in Figure 9.4. If the signal flowing out of the last integrator is defined as $y(t)$, then the signals in other integrators can then be defined as $y'(t), y''(t), \ldots, y^{(n)}(t)$. Besides, double clicking the integrators, the corresponding initial values can be filled into the dialog boxes. Therefore, a set of key signals with initial values can be defined in this way. The key signal construction method is used throughout this book.

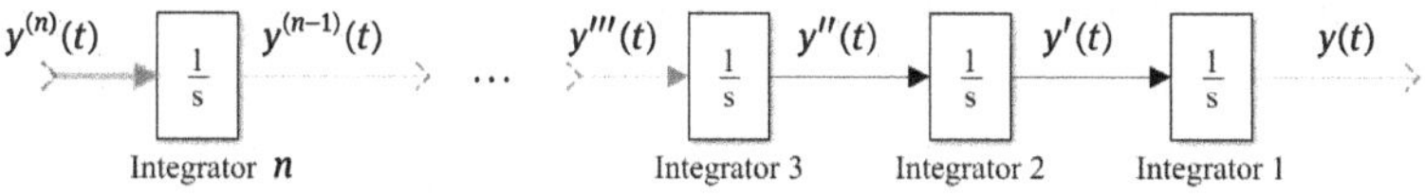

Figure 9.4: Defining key signals with an integrator chain.

With the key signals, other blocks can be used to manipulate these signals, and finally, the differential equation model can be established.

9.2.2 Differential equation description methods

In practical differential equation modeling process, the following actions are usually taken to draw block diagrams:

(1) If two signal are equal, they can be joined. This is usually used to close the simulation loop.

(2) Assuming that signal $u(t)$ is known and a **Gain** block is connected to it, by double-clicking the block to enter the value of k, the signal $ku(t)$ can be constructed.

(3) If a nonlinear action on signal $v(t)$ is expected, for instance, the nonlinear signal is $(v(t) + v^3(t)/3)\sin v(t)$, then signal $v(t)$ can be fed into block Fcn. Double-clicking the block, one can fill `(u+u^3/3)*sin(u)`. Note that the input signal to the Fcn block is denoted as `u`. Through this modeling method, the output of the block is the expected nonlinear function.
(4) Several signals can be fed into the Mux block, and the output of this block is the vector composed of these inputs. If the vector signal is fed into Demux block, it is decomposed into scalar signals.
(5) If the $u(t)$ signal is known, and the $u(t-d)$ delayed signal is expected, $u(t)$ can be fed into Transport Delay block. Double-clicking the block, parameter d can be filled to the block. The output signal is the expected delayed signal.
(6) If a signal is to be observed, it can be connected to a Scope block or Out block, so that it can be displayed or returned.

An example is given next to demonstrate the Simulink modeling method of differential equations.

Example 9.1. Consider Lorenz equations in Example 3.7. For convenience, the differential equations are given again:

$$\begin{cases} x_1'(t) = -\beta x_1(t) + x_2(t)x_3(t), \\ x_2'(t) = -\rho x_2(t) + \rho x_3(t), \\ x_3'(t) = -x_1(t)x_2(t) + \sigma x_2(t) - x_3(t) \end{cases}$$

where we let $\beta = 8/3$, $\rho = 10$ and $\sigma = 28$. The initial values are $x_1(0) = x_2(0) = 0$, and $x_3(0) = \epsilon$. Establish a Simulink model.

Solutions. It can be seen from the differential equations that there are 6 (3 pairs) key signals to be defined first, $x_1(t)$ and $x_1'(t)$, $x_2(t)$ and $x_2'(t)$, as well as $x_3(t)$ and $x_3'(t)$. Three integrators are used to define these key signals, as shown in Figure 9.5. Since only first-order derivatives are expected, there is no need to build integrator chains. One integrator can be used to define a signal and its derivative. Besides, double-clicking the integrator, the initial value of the signal can be specified in the dialog box.

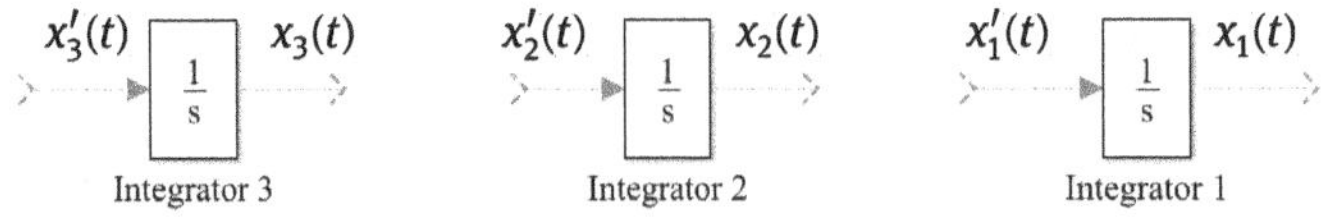

Figure 9.5: Key signals by integrators.

Let us see the first differential equation. There is a term $3x_1(t)/8$, which can be expressed by the $x_1(t)$ signal followed by a Gain block. Double-clicking the block, the parameter 3/8 can be specified in the dialog box. The output of the block is the expected $3x_1(t)/8$. If $x_2(t)x_3(t)$ signal is also expected, the two signals $x_2(t)$ and $x_3(t)$ can be fed into a Product block. Then the output of the block is the expected $x_2(t)x_3(t)$. The two signals constructed above can be added up by feeding them into a Sum block. Note that before adding them, the sign of $3x_1(t)/8$ should be altered. Double-clicking the Sum bock, the signs of the two signals can be edited. In this case, the signs should be changed to −+. The output of the Sum signal is $x_1'(t)$. Since it is $x_1'(t)$, and the input terminal of the first integrator is also $x_1'(t)$, they can be connected directly to form a closed loop. The first equation can be described graphically in this manner, as shown in Figure 9.6.

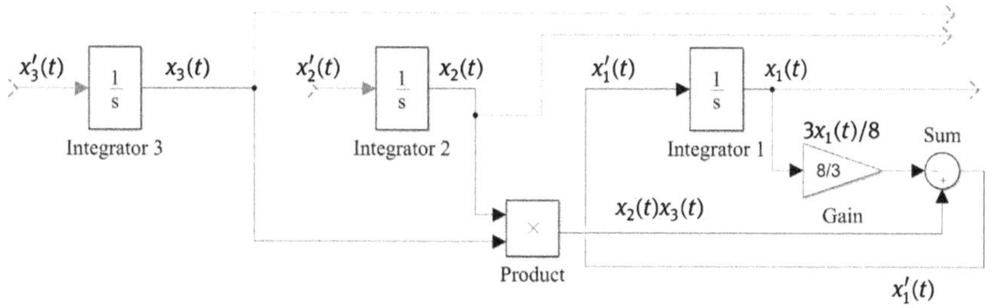

Figure 9.6: Establishing the first equation.

Similarly, the second and third differential equations can be described graphically, so that the entire system of differential equations is represented in the model. The facilities in Simulink can be used to find the numerical solution of the differential equations.

The modeling process presented here is far too complicated for differential equations. It is not recommended, and it is only used to illustrate the ideas in Simulink modeling. For differential equations, a better modeling method will be illustrated next.

Example 9.2. Compared with Example 3.7, it can be seen that the modeling method is trivial and error-prone. Is there a better way for differential equations? Establish a Simulink model.

Solutions. A vector integrator is needed in the modeling process. Just using the integrator block, double-clicking it, and specifying a vector as its initial values, the block is automatically changed into a vector integrator. For this particular example, the initial vector `[0,0,1e-10]` is specified, so that the integrator is a 3×1 vector, whose output is defined as $\boldsymbol{x}(t) = [x_1(t), x_2(t), x_3(t)]^{\mathrm{T}}$, and then the input becomes $\boldsymbol{x}'(t)$. The relationship between them can be described by a MATLAB Function block. A MATLAB function as follows can be written:

```
function dy=lorenz_mod(x)
dy=[-x(1)+x(2)*x(3);
   -10*x(2)+10*x(3);  -x(1)*x(2)+28*x(2)-x(3)];
```

Filling-in the function name of the block, the Simulink model for the given differential equations can be established as shown in Figure 9.7. Modeling in this way is as simple as using the `ode45()` solver.

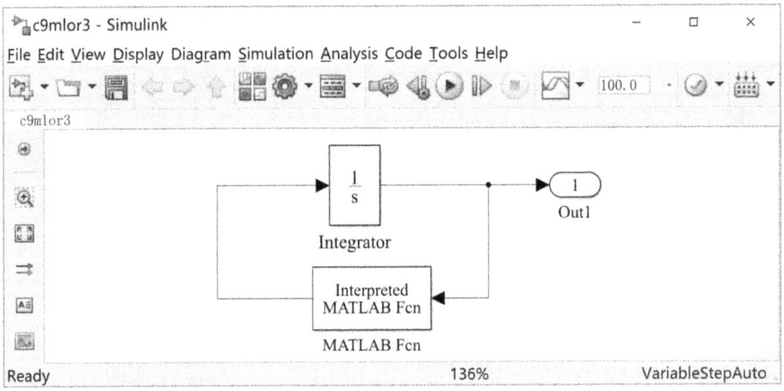

Figure 9.7: Simulink modeling of Lorenz equations (model: c9mlor3.slx).

In fact, if a differential equation can be written in the form of first-order explicit differential equations, this framework can be used to build the Simulink model directly.

9.2.3 Solutions of differential equations

With the Simulink model, clicking the ▶ icon in the toolbar, a simulation can be started, and a numerical solution of the equations can be found directly. An alternative solution for the differential equations is to call function `sim()`, with the syntax

[$\boldsymbol{t}$,$\boldsymbol{x}$,$\boldsymbol{y}$]=`sim(mod_name,tspan,options)`

where the format of $\boldsymbol{t}$ and $\boldsymbol{y}$ is the same as that in function `ode45()`. The argument $\boldsymbol{x}$ returns the time response data of the state variables. The actual states $\boldsymbol{x}$ are assigned internally in Simulink model. The output vector $\boldsymbol{y}$ returns the signals connected into the outport Out in the model. If the user is interested in a particular signal, the signal can be connected to an Out block.

In the input arguments, `mod_name` is the Simulink model name. The suffix of the model name is `slx`, and in the earlier versions, the suffix was `mdl`. The difference is that the latter is in an ASCII file. The other two input arguments are the same as in `ode45()`.

In normal cases, it is not necessary to provide `options`, since the control options can be assigned in a visual manner in Simulink modeling. The argument `tspan` can also be omitted, since it can be assigned in the Simulink model.

9.2.4 Algorithms and parameter settings

Although a simple simulation method was demonstrated for the Simulink model, it is not recommended in this book, since it was pointed out that the validation process is a very important issue in differential equation solutions. Control parameters should be set by the user to ensure that the solution is correct.

The control parameters can be specified with dialog boxes. The user may click the Simulation → Model Configuration Parameters menu item, or click the button in the toolbar, such that a dialog box in Figure 9.8 is opened. It can be seen that at Start time and Stop time edit boxes, the start and stop time of simulation can be specified. Besides, Type listbox can be used to set fixed- or variable-step algorithms. Normally, Variable-step option is recommended, so as to find high-efficiency simulation results.

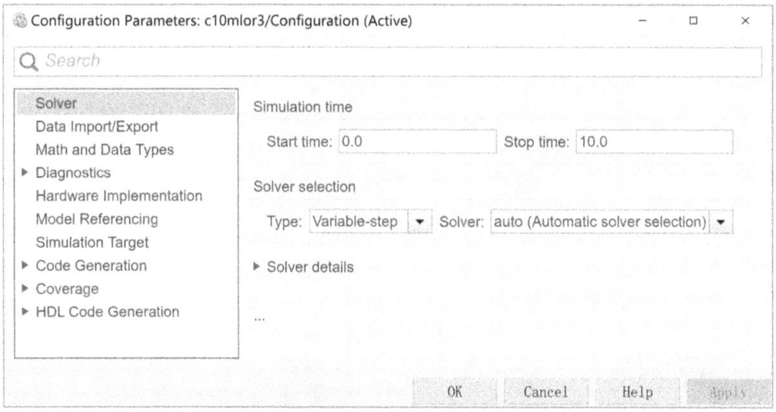

Figure 9.8: Simulink parameter setting dialog box.

If the Solvers listbox is opened, the simulation algorithms can be selected from it. For average users, the automatic selection method is recommended. Simulink will select an algorithm automatically based on the model. If the listbox is opened, the options ode45, ode15s, as well as others, can be selected.

If the item Solver details is selected, the dialog box is expanded as shown in Figure 9.9. Options like Relative tolerance, Absolute tolerance, and others are allowed.

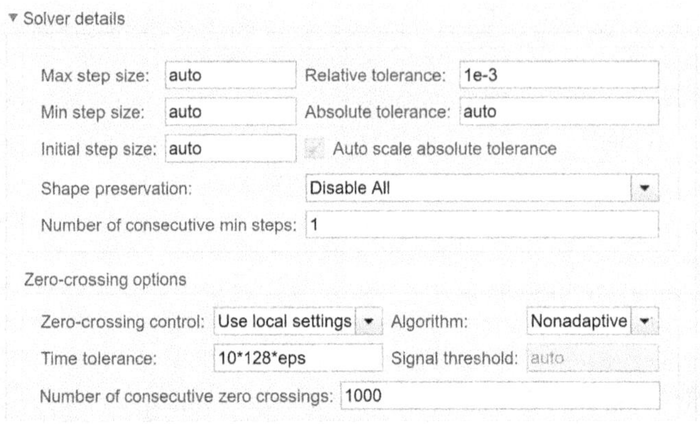

Figure 9.9: Detailed control options.

To ensure accurate simulation, it is recommended to assign them to the minimum allowed values, such as `100*eps`.

Since output blocks are used in the modeling, there is another setting to be made. Clicking the **Data Import/Export** on the left, the dialog box is changed to the form in Figure 9.10. It is recommended to change the **Format** listbox from the default **Dataset** to **Array**, so as to better use the simulation results.

Now observe the **Save to workspace or file** item in the dialog box. The first three are **Time**, **States**, and **Output**, with the first and third checked. If the button is used to start the simulation process, two variables `tout` and `yout` are automatically

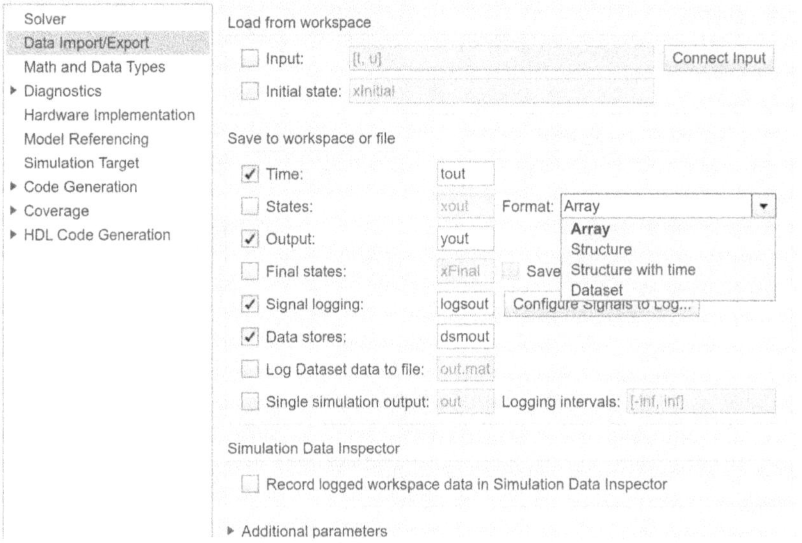

Figure 9.10: Data import and export setting.

returned to MATLAB workspace. The former is time and the latter contains the output signals. If there is more than one output block used, the output signals are returned in each column of `yout` matrix. Of course, this implies that Format is set to Array. If it is set to the default Dataset, the handling becomes complicated, and is not presented here.

Since by default States is not checked, the variable `xout` is not returned after the simulation process.

Example 9.3. Establish a Simulink model to describe and solve the Lorenz equation in Example 3.7.

Solutions. For the Simulink model `c9mlor3.slx` of Lorenz equation in Example 9.2, two ways can be used to invoke the solution process. For instance, clicking the ▶ button, two variables are established in MATLAB workspace, under the names `tout` and `yout`. The following commands can be used to draw the simulation results. The results are the same as in Figure 3.5.

```
>> plot(tout,yout) % find the numerical solution
```

Another simulation method is to employ the following statements. The same results can be found. Note that function curves of $\boldsymbol{y}$ are drawn, not $\boldsymbol{x}$. The arguments $\boldsymbol{x}$ are the states internally assigned in Simulink, they may not be the same, as those expected.

```
>> [t,x,y]=sim('c9mlor3',[0,100]); plot(t,y) % solve and draw
```

9.3 Modeling examples of differential equations

The modeling and simulation methods for first-order explicit differential equations were illustrated earlier, and a framework was constructed. It can be seen that first-order explicit differential equations of any complexity can be modeled with the unified framework. Examples are given next to demonstrate the modeling of ordinary differential equation sets, differential-algebraic equations, and delay differential equations.

9.3.1 Simple differential equations

For normal ordinary differential equation sets, there is usually no need to convert them manually into first-order explicit differential equations, since several sets of integrator chains are adequate in constructing the unknown functions and their derivatives. Several examples are given here to illustrate the modeling process of differential equation sets.

Example 9.4. Use Simulink to solve the high-order differential equation in Example 4.7. For convenience of presentation, the differential equation is given again

$$y(t)y^{(6)}(t) + 6y'(t)y^{(5)}(t) + 15y''(t)y^{(4)}(t) + 10(y'''(t))^2 = 2\sin^3 2t$$

where $y(0) = 1$ and $y^{(i)}(0) = 0$, $i = 1, 2, \ldots, 5$.

Solutions. To set up the simulation model, the explicit form of the highest-order derivative of $y(t)$ is written. In fact, Example 4.7 was used to write the explicit form

$$y^{(6)}(t) = \frac{1}{y(t)}\left[-6y'(t)y^{(5)}(t) - 15y''(t)y^{(4)}(t) - 10(y'''(t))^2 + 2\sin^3 2t\right].$$

Of course, with the method in Example 4.7, the first-order explicit differential equations is a modeling method. A MATLAB function is written. Compared with the anonymous function in Example 4.7, the input argument is $\boldsymbol{x}$, rather than the t and $\boldsymbol{x}$ in the anonymous function.

```
function dx=c9model a(x)
dx=[x(2:6);
    (-6*x(2)*x(6)-15*x(3)*x(5)-10*x(4)^2+2*sin(2*t)^3)/x(1)];
```

With this function, the differential equation can be modeled in the framework in Figure 9.7.

A direct modeling method is demonstrated here. Since the highest-order derivative is 6, therefore 6 integrators connected serially can be used to define the key signals. With them, the expressions on the left-hand side of the equal sign can be constructed then divided by $y(t)$. The result is the same as the $y^{(6)}(t)$ signal, so both can be connected, and the closed-loop simulation model can be established as shown in Figure 9.11.

From the given model, some modeling details are not presented. For this particular system, only $y(0)$ is nonzero, while the others are zero. How can these be set? We should double-click the integrators and fill in the values to the parameter dialog box. The stoping time and error tolerance are assigned using an appropriate dialog box. If a simulation process is initiated, the same result is obtained as in Example 4.7.

Example 9.5. Draw a concise simulation model for Example 9.4.

Solutions. It can be seen from the model constructed above that the display is in a mess. There are too many crossing lines. In fact, in Simulink modeling, if there is no solid dot in the display, although there are crossing lines, one should not be worried. There are no faulty connections.

If neat and concise models are expected, Mux blocks should be used to construct vectorized signals. Based on the vectorized operations, the new model in Figure 9.12 is constructed. Note that when constructing vectorized signals, the time signal should

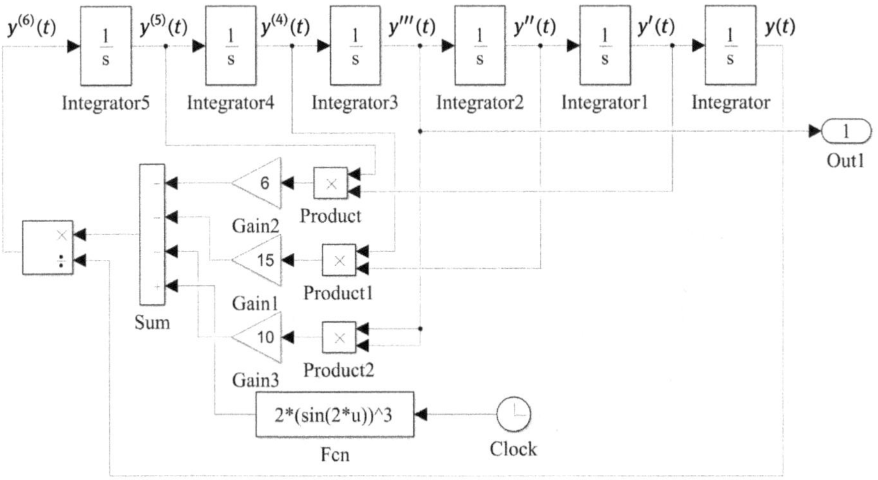

Figure 9.11: Simulation model I (file: c9mode1.slx).

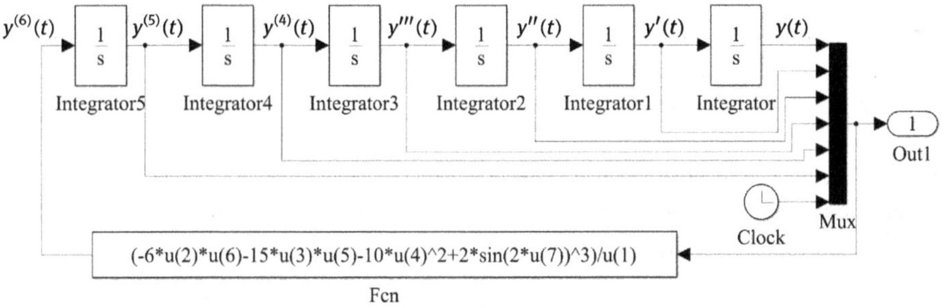

Figure 9.12: Simulation model II (file c9mode2.slx).

be included. For instance, in this example, the time signal is used as the seventh input. If the simulation process is invoked, the result obtained is exactly the same as in Example 4.7.

Example 9.6. Construct a Simulink model for the Apollo equations in Example 4.12.

Solutions. If Simulink modeling is expected, normally there is no need to convert the equation into first-order explicit differential equations first. Two integrator chains can be used to define the key signals, and convert them into a vectorized signal. The intermediate signals $r_1(t)$ and $r_2(t)$ can be composed. With Mux1, another vectorized signal can be composed, whose former first four signals are taken from Mux block, the latter two are respectively $r_1(t)$ and $r_2(t)$. Through block Fcn, the two differential equation models are established as shown in Figure 9.13. The simulation result is the same as in Figure 4.12.

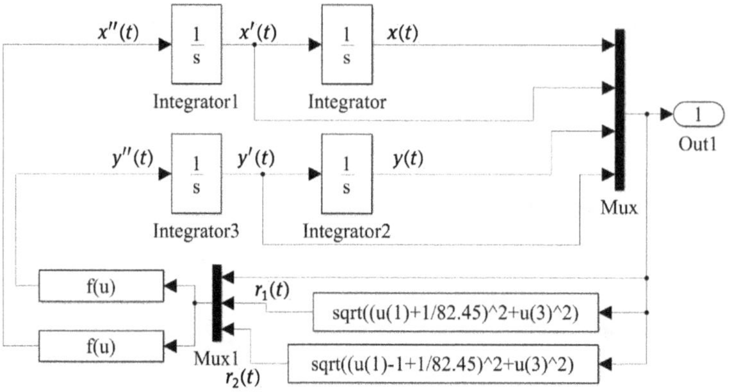

Figure 9.13: Simulation model of Apollo differential equation (model c9mode3.slx).

Here four Fcn blocks are used. The two on the right are already displayed, while for the two on the left, due to the space restrictions, their contents are listed below:

```
-2*u(2)+u(3)-(1-1/82.45)*u(3)/u(5)^3-1/82.45*u(3)/u(6)^3
2*u(4)+u(1)-(1-1/82.45)*(u(1)+1/82.45)/u(5)^3-...
    1/82.45*(x(1)-1+1/82.45)/u(6)^3
```

For complicated differential equation modeling, the methods introduced thus far are rather complicated. State space equations should be used instead. Then the framework in Example 9.2 should be used.

Since block Clock can be used in generating the time signal t, time-varying differential equations can also be modeled with Simulink. Examples are given next to show the modeling and simulation process of time-varying differential equations.

Example 9.7. Model the time-varying differential equations in Example 4.5 with Simulink. For convenience, the mathematical model is given again here, where the independent variable is changed from x to t:

$$t^5 y'''(t) = 2(ty'(t) - 2y(t)), \quad y(1) = 1, \quad y'(1) = 0.5, \quad y''(1) = -1, \quad t \in (1, \pi).$$

Solutions. From the given differential equation model, the explicit form can be written directly as follows:

$$y'''(t) = \frac{2}{t^5}(ty'(t) - 2y(t)).$$

Since the highest order here is three, three integrators in a chain can be used to define the key signals. The vectorized signal is defined composed of the key signals. For convenience, the fourth channel is the time signal t. The simulation system model

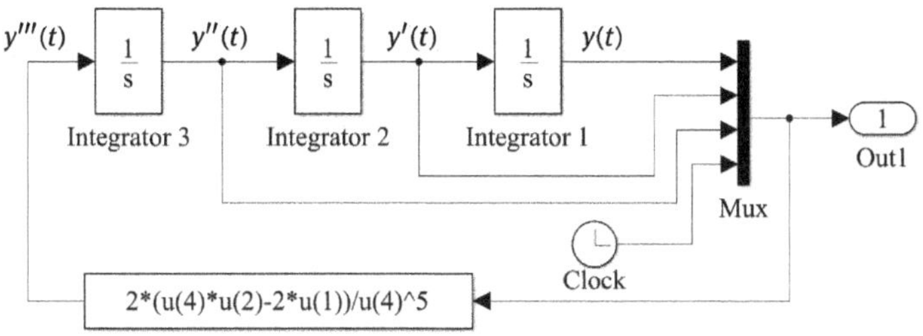

Figure 9.14: Mathematical time-varying differential equations (file c9mode4.slx).

shown in Figure 9.14 is constructed. Performing simulation to this model, we may find the numerical solution of the original differential equations.

In Example 4.5, an attempt was made to perform simulation from starting time $t_0 = 1$ to stoping time $t_n = 0.2$. Unfortunately, in the current Simulink mechanism, the stoping time is not allowed to be smaller than the initial time. The solution in this interval cannot be obtained in this way.

In the solution process, if there is no solution found for a long period of time, the differential equations are probably stiff. Stiff solvers such as ode15s should be assigned in the dialog box in Figure 9.8. The stiff solvers can then be used for solving the differential equations.

9.3.2 Differential-algebraic equations

Compared with ordinary differential equations, differential-algebraic equations are more complicated to solve. In Simulink modeling, the difficulty is how to describe the algebraic constraints. In the solution process, the difficulty is that, in each simulation step, algebraic equations must be solved. This may make the solution process extremely slow.

If algebraic constraints are to be described, the **Algebraic Constraints** block in the **Math** group can be used to directly describe them. The method involving this block is tricky. Examples are used next to demonstrate the modeling and simulation process for differential-algebraic equations.

Example 9.8. Establish the differential-algebraic equations of Example 5.14 in Simulink. For convenience, the mathematical model is given below:

$$\begin{cases} x_1'(t) = -0.2x_1(t) + x_2(t)x_3(t) + 0.3x_1(t)x_2(t), \\ x_2'(t) = 2x_1(t)x_2(t) - 5x_2(t)x_3(t) - 2x_2^2(t), \\ 0 = x_1(t) + x_2(t) + x_3(t) - 1. \end{cases}$$

The initial values are $x_1(0) = 0.8$ and $x_2(0) = x_3(0) = 0.1$.

Solutions. Although the differential-algebraic equations have three state variables, only two of them describe dynamic relationships, $x_1'(t)$ and $x_2'(t)$. The derivative of state $x_3(t)$ does not appear in the differential-algebraic equations. Therefore two integrators are used in describing the key signals. Now, the block Mux is used to construct the three states in a vector, and we assume that the third signal $x_3(t)$ is idle for the time being. With the vectorized signal, the first two differential equations can be set up, as shown in Figure 9.15.

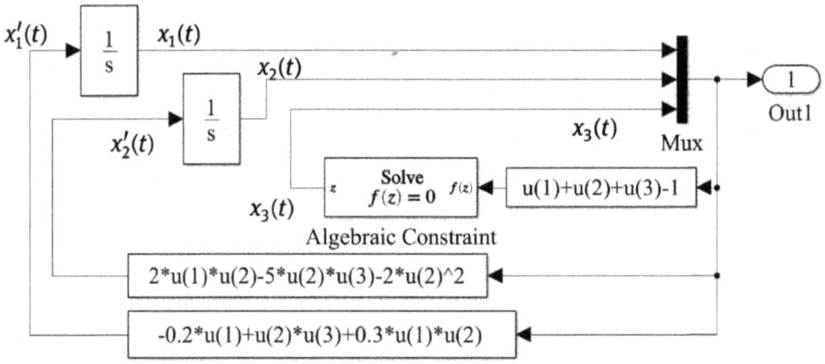

Figure 9.15: Simulink model for differential-algebraic equations (file: `c9mdae1.slx`).

How can we describe algebraic equations? A block function Fcn can be used to describe one side of the equation, for this particular problem, $x_1(t) + x_2(t) + x_3(t) - 1$, then we can feed it into the **Algebraic Constraint** block. Its output can be any signal. If the signal is connected to a given port, then the port receives the signal. It happens that Mux block has one idle port, therefore the signal can be connected to it, and the signal is then $x_3(t)$. In this way, the simulation model of the entire differential-algebraic equations is established. Performing simulation of the model, the numerical solution is exactly the same as that obtained in Example 5.14.

With the following statements, a warning message “Warning: Block diagram ‘c9mdae1’ contains 1 algebraic loop(s)” is received. This is a normal phenomenon. The differential-algebraic equations also contain an algebraic loop. The message can be neglected. The elapsed time is 0.24 seconds, with the number of computed points being 1 544. The efficiency is higher than in Example 5.14.

```
>> tic, [t,x,y]=sim('c9mdae1'); toc, length(t)
```

Example 9.9. Solve the differential-algebraic equations in Example 9.8 again, without using the constraint block.

Solutions. For this particular problem, it can be seen that $x_3(t) = 1 - x_1(t) - x_2(t)$. Therefore, from the given signals $x_1(t)$ and $x_2(t)$, the signal $x_3(t)$ can be directly constructed. The Mux block can be used to describe the $\boldsymbol{x}(t)$ vector, and the simulation model is shown in Figure 9.16. The simulation results obtained are the same.

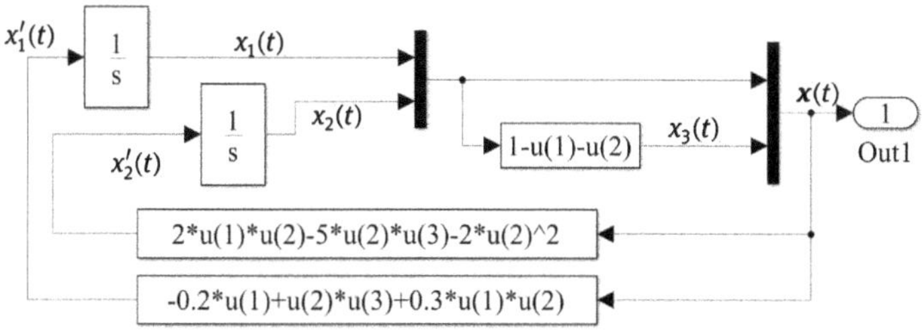

Figure 9.16: Simulink model of differential-algebraic equations (file:`c9mdae1a.slx`).

It should be noted that the algebraic loop in the simulation model is eliminated. It can be seen that the elapsed time and number of points are almost the same as previously.

Implicit differential equations and some differential-algebraic equations are normally difficult to express in Simulink models. For instance, it is hard to describe the differential-algebraic equations in Examples 5.16 and 5.17 in Simulink. It is also difficult to express the implicit differential equations like the one in Example 5.10 in Simulink. Therefore it can be seen that Simulink modeling has limitations in implicit differential equations modeling.

9.3.3 Switched differential equations

The key in switched differential equations modeling in Simulink is the used of the switch blocks. Switch blocks are located in the Signal Routing group, where Switch and Multiport Switch blocks can be used directly. Here examples are used to demonstrate the modeling and simulation of switched differential equations.

Example 9.10. Build the switched differential equations in Example 5.25 using Simulink.

Solutions. A vectorized integrator can be used to define the key signals $\boldsymbol{x}(t)$ and $\boldsymbol{x}'(t)$. The Fcn block is used to generate the $x_1(t)x_2(t)$ signal, and feed it into the control port of the switching signal. If the first signal is positive, the signal $\boldsymbol{A}_2\boldsymbol{x}(t)$ is connected, while when the control signal is negative, the other branch is connected. The switched system is constructed as shown in Figure 9.17. Since zero-crossings are detected automatically in Simulink models, the model can be normally executed, such that correct numerical solutions are found. The results are the same as those obtained in Example 5.25.

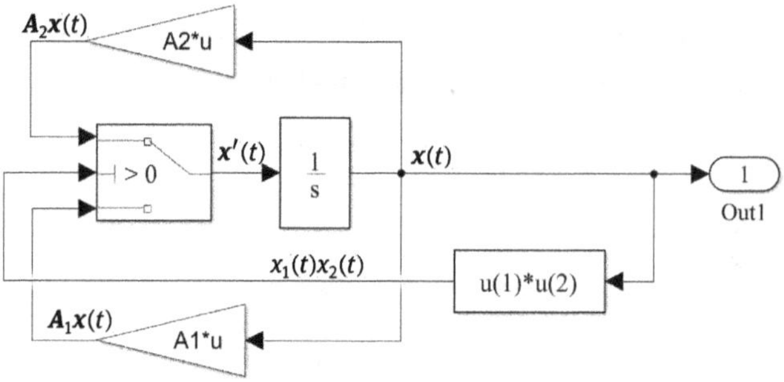

Figure 9.17: Simulation model of switched differential equations (file: `c9mswi1.slx`).

In the modeling process, a trick is used. That is, matrices $\boldsymbol{A}_1$ and $\boldsymbol{A}_2$ are automatically loaded into MATLAB workspace with the model. Specifically, the menu item File → Model Properties → Model Properties is chosen. Selecting Callback, the dialog box shown in Figure 9.18 is displayed. In the PreLoadFcn option, the matrices $\boldsymbol{A}_1$ and $\boldsymbol{A}_2$ are assigned. In this case, every time the Simulink model is opened, the parameters are loaded automatically.

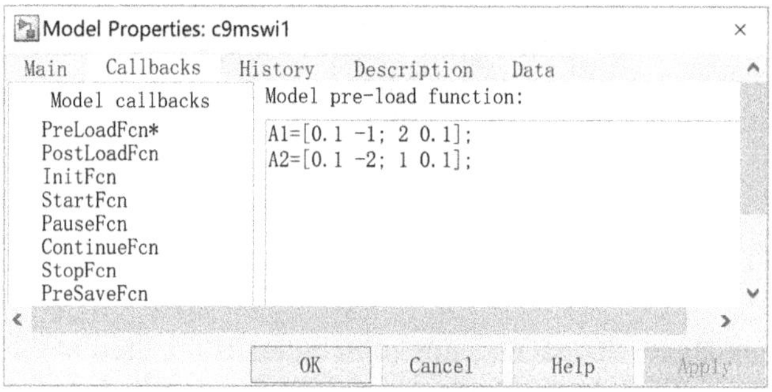

Figure 9.18: Preset function specifications.

Example 9.11. Consider the nonlinear switched differential equations in Example 5.28. Use Simulink to simulate the system and draw the control curve.

Solutions. Low-level modeling method is used directly with Simulink for the switched differential equations, as shown in Figure 9.19. In the model, three Interpreted MATLAB Function blocks are written to compute respectively $\boldsymbol{y}(t)$, the dynamical model of the wheeled robot and the switching laws. The contents of the three blocks are:

```
function y=c9mswi2a(x)
c=cos(x(3)); s=sin(x(3));
y=[x(3); x(1)*c+x(2)*s; x(1)*s-x(2)*c];
```

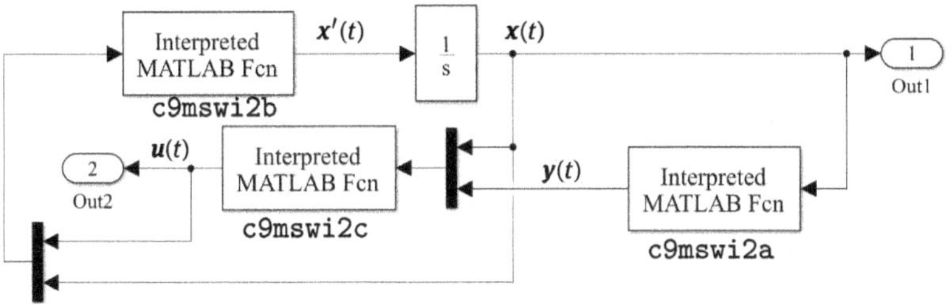

Figure 9.19: Nonlinear switched differential equations model (file: c9mswi2.slx).

```
function dx=c9mswi2b(x)
u=x(1:2); x=x(3:5);
dx=[u(1)*cos(x(3)); u(1)*sin(x(3)); u(2)];
function u=c9mswi2c(x)
y=x(4:6); x=x(1:3);
if abs(x(3))>norm(x)/2
   u=[-4*y(2)-6*y(3)/y(1)-y(3)*y(1); -y(1)];
else
   sgn=-1; if y(2)*y(3)>=0, sgn=1; end
   u=[-y(2)-y(3)*sgn; -sgn];
end
```

Executing the model, the numerical solution of the switched differential equations is obtained. The phase plane trajectory is the same as in Figure 5.22.

```
>> [t,~,y]=sim('c9mswi2'); plot(y(:,1),y(:,2))
```

Besides, the two control signals are obtained as shown in Figure 9.20. It can be seen that, at a certain point, the control law is switching all the time. Therefore it is not a

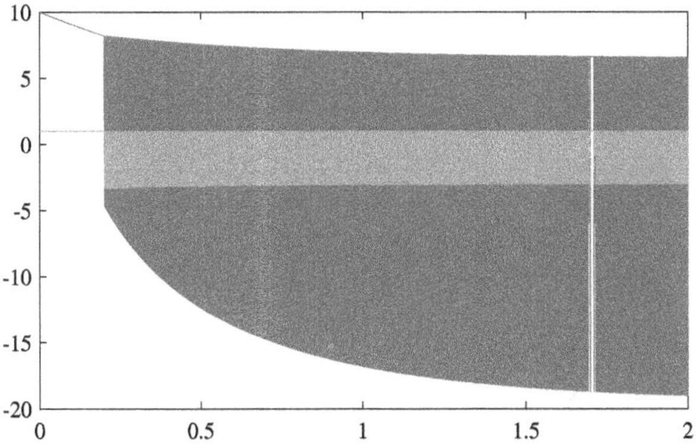

Figure 9.20: Control signal curves.

control strategy, and should be avoided in real applications.

```
>> plot(t,y(:,[4,5]))
```

9.3.4 Discontinuous differential equations

Simulink modeling and simulation of discontinuous differential equations can also be tried through an example. It can be seen from the example that the efficiency and accuracy are higher than for the command-driven methods.

Example 9.12. Solve again the discontinuous differential equations in Example 5.29 in Simulink.

Solutions. The Simulink modeling of the example is comparatively simple and straightforward. The Simulink model in Figure 9.21 can be established. If the absolute and relative error tolerances are both set to 10^{-11}, the solution can be found in 40.01 seconds, the same as in Example 5.29. Note that the error tolerance here is much smaller than for the command-driven method. The number of points is 5 835 447, which is similar to that of the command-driven method.

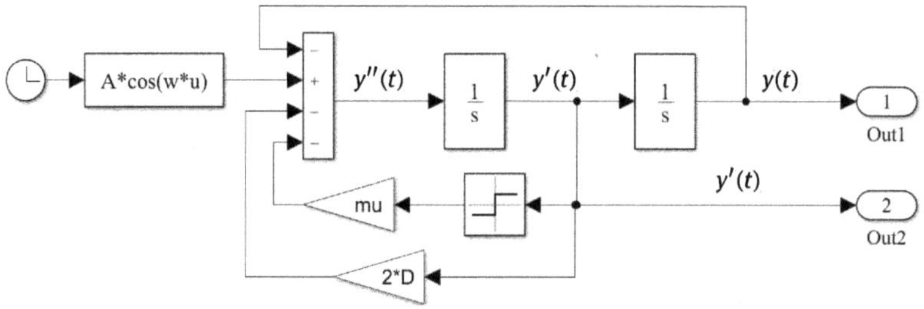

Figure 9.21: Simulink model of discontinuous differential equations (file: `c9mswi3.slx`).

9.3.5 Delay differential equations

In the Continuous group in Simulink, various delay blocks are provided, including Transport Delay block, Variable Transport Delay block, and Variable Time Delay block. If a signal is fed into a delay block, the output of the block can be regarded as the delay signal. In this example, examples are used to demonstrate various delay differential equations.

Example 9.13. Set up the simulation model for the delay differential equations in Example 6.2:

$$\begin{cases} x'(t) = 1 - 3x(t) - y(t-1) - 0.2x^3(t-0.5) - x(t-0.5), \\ y''(t) + 3y'(t) + 2y(t) = 4x(t) \end{cases}$$

where, when $t \leqslant 0$, it is known that $x(t) = y(t) = y'(t) = 0$.

Solutions. It can be seen from the second equation that the transfer function $4/(s^2 + 3s + 2)$ can be used to describe the relationship between $x(t)$ and $y(t)$. An integrator can be used to define the key signals $x(t)$ and $x'(t)$. The Simulink model in Figure 9.22 can be constructed. Executing the model, the results are the same as in Example 6.2.

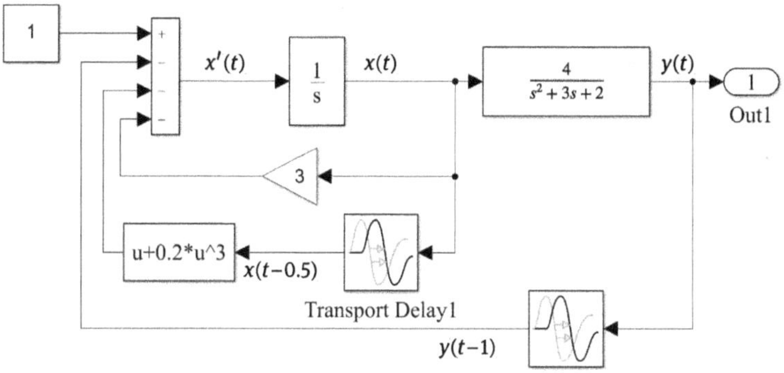

Figure 9.22: Simulation model of delay differential equation (file: `c9mdde1.slx`).

Example 9.14. Construct the variable delay differential equation in Example 6.7, with zero initial values:

$$\begin{cases} x_1'(t) = -2x_2(t) - 3x_1(t - 0.2|\sin t|), \\ x_2'(t) = -0.05x_1(t)x_3(t) - 2x_2(t-0.8) + 2, \\ x_3'(t) = 0.3x_1(t)x_2(t)x_3(t) + \cos(x_1(t)x_2(t)) + 2\sin 0.1t^2. \end{cases}$$

Solutions. Variable delay signal can be implemented by using Variable Transport Delay block. This block has two input channels. The first is the original signal, while the other describes the variable delay. The signal is driven by t, and the output is the variable delay signal. In this example, vectorized signals are still used, including three state signals, with the 4th, 5th, and 6th, respectively, being the constant delay signal $x_2(t-0.8)$, variable delay signal $x_1(t - 0.2|\sin t|)$, and time t. Therefore the following MATLAB function can be written to describe the right-hand side of the equations:

```
function dx=c9mdde2a(x)
dx=[-2*x(2)-3*x(5);
    -0.05*x(1)*x(3)-2*x(4)+2;
    0.3*x(1)*x(2)*x(3)+cos(x(1)*x(2))+2*sin(0.1*x(6)^2)];
```

The output signal is then $\boldsymbol{x}'(t)$, which can be linked to the input terminal of the vectorized integrator, to form the closed loop. The simulation model in Figure 9.23 can be constructed. The simulation results are the same as in Example 6.7.

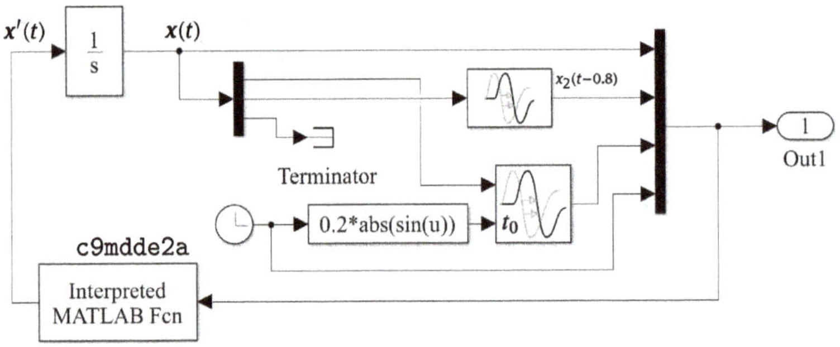

Figure 9.23: Simulation model of delayed differential equations (file: `c9mdde2.slx`).

Example 9.15. Reconstruct the neutral-type delay differential equations in Example 6.15. The model is given as

$$\boldsymbol{x}'(t) = \boldsymbol{A}_1\boldsymbol{x}(t-0.15) + \boldsymbol{A}_2\boldsymbol{x}'(t-0.5) + \boldsymbol{B}u(t)$$

where the input is $u(t) \equiv 1$, and the matrices are

$$\boldsymbol{A}_1 = \begin{bmatrix} -13 & 3 & -3 \\ 106 & -116 & 62 \\ 207 & -207 & 113 \end{bmatrix}, \quad \boldsymbol{A}_2 = \begin{bmatrix} 0.02 & 0 & 0 \\ 0 & 0.03 & 0 \\ 0 & 0 & 0.04 \end{bmatrix}, \quad \boldsymbol{B} = \begin{bmatrix} 0 \\ 1 \\ 2 \end{bmatrix}.$$

Solutions. It is not easy to solve neutral-type delay differential equations, even if the solver `ddensd()` is used. Simulink modeling is relatively simple. If a vectorized integrator block is used, the key signals are $\boldsymbol{x}(t)$ and $\boldsymbol{x}'(t)$. These signals can be used to build the delay signals, then construct the simulation model as shown in Figure 9.24. The solutions obtained for this model are the same as those in Example 6.15.

It is a pity that compared with the delay differential equation solver `ddesd()`, the handling of delay terms $x_i(\alpha t)$ is not allowed in Simulink.

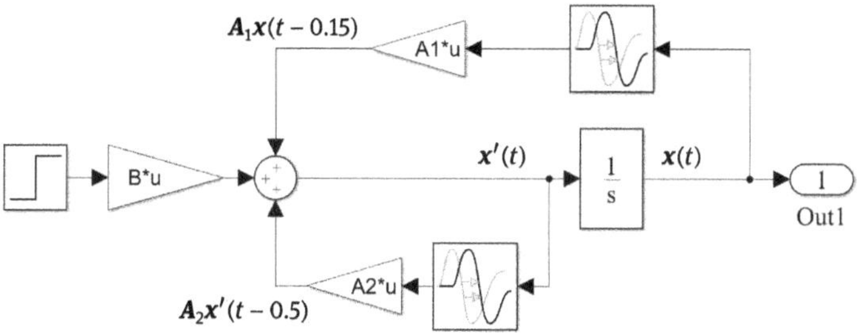

Figure 9.24: Simulink model of delay differential equations (file: `c9mdde3.slx`).

9.3.6 Delay differential equations with nonzero history functions

If delay differential equations are with nonzero initial values, a similar modeling technique can be used to handle the modeling problem directly. If $t \leqslant t_0$, the state variables are given functions, and the previous method cannot be used in the modeling. Switch blocks should be used for setting the functions of $t \leqslant 0$. In this section, examples are used to demonstrate these problems.

Example 9.16. Consider the delay differential equations in Example 6.8. Since the history function is nonzero when $t \leqslant 0$, direct modeling in Simulink is not simple. Use Simulink to represent the history function, and perform simulation again.

Solutions. The Simulink model in Figure 9.23 is used as the fundamental model. Based on it, the nonzero history function can be expressed by a switch block, controlled by time t. If time is larger than 0, the output of the integrator is fed into the first input port of the switch, such that the equation for $t > 0$ is described. The description is the same as in the case studied earlier. If $t \leqslant 0$, the clock bock is used to drive an Interpreted MATLAB Function block to generate nonzero history function, defined as

```
function x=c9mdde2b(u)
x=[sin(u+1); cos(u); exp(3*u)];
```

Therefore, when $t \leqslant 0$, this channel is connected to set the states $\boldsymbol{x}(t)$ as the history function. With these ideas in mind, the Simulink model is constructed directly as shown in Figure 9.25. Simulating the system model, the results obtained are the same as in Figure 6.7.

Example 9.17. Solve in Simulink the neutral-type delay differential equations of Example 6.14.

Solutions. It was pointed out in Example 6.14 that function `ddensd()` can be used in solving neutral-type delay differential equations, however, the error tolerance should not be selected too small. Therefore Simulink is used again to solve the

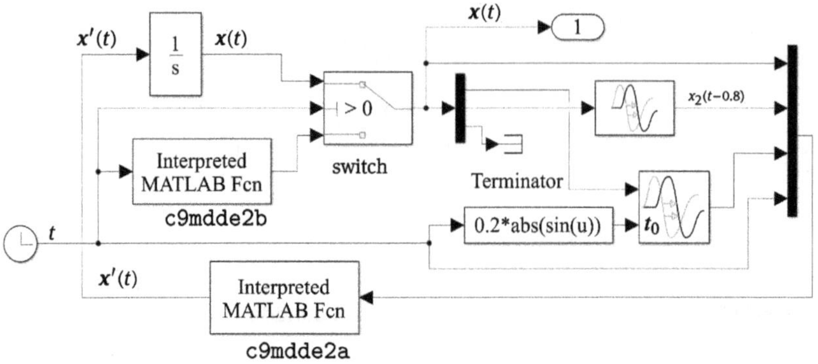

Figure 9.25: Neutral-type delay differential equation with nonzero history functions (file: `c9mdde2x.slx`).

differential equations, and see whether more accurate results can be found. From the given neutral-type differential equations, the simulation model in Figure 9.26 can be constructed. The following commands can be used to simulate the system, and the error can also be computed:

```
>> [t,~,y]=sim('c9mdde4');
   z=exp(t)+(t-1).*exp(t-1).*(t>=1 & t<2)+...
      (exp(t-1)+(t-2).*(t+2*exp(1)).*exp(t-2)/2).*(t>=2);
   norm(z-y), plot(t,y,t,z) %compare with exact values
```

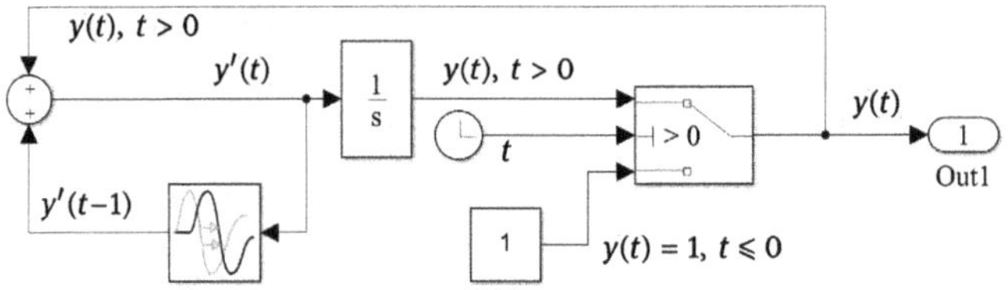

Figure 9.26: Neutral-type delay differential equations with nonzero history functions (file: `c9mdde4.slx`).

Although very tough error tolerance is set in the Simulink model, the norm of the error found is 0.0014, which is much larger than that in Example 6.14. The error can be distinguished from the plot.

9.3.7 Stochastic differential equations

Numerical solutions of linear stochastic differential equations were studied in Section 5.5. For linear stochastic differential equations, a discretization method can be used to find the discrete model, so that pseudorandom numbers can be used to drive the discrete system to find the simulation results. Statistical analysis can be performed for the simulation results. Unfortunately, this method cannot be used arbitrarily in the case of nonlinear stochastic differential equations. Generalized simulation methods for nonlinear stochastic differential equations are expected.

A block Band-Limited White Noise is provided in MATLAB. With such a block, pseudorandom numbers can be generated to drive the stochastic differential equations. Fixed-step simulation algorithm should be selected to solve them. The simulation method like this is no longer restricted to linear stochastic differential equations. An example is given here to solve linear stochastic differential equations, also nonlinear stochastic differential equations are handled.

Example 9.18. Solve again the linear stochastic differential equations in Example 5.31 using Simulink:

$$\mathrm{d}^4y(t) + 10\mathrm{d}^3y(t) + 35\mathrm{d}^2y(t) + 50\mathrm{d}y(t) + 24y(t) = \mathrm{d}^3\xi(t) + 7\mathrm{d}^2\xi(t) + 24\mathrm{d}\xi(t) + 24\xi(t).$$

Solutions. From the given differential equations, the transfer function model can be established as

$$G(s) = \frac{s^3 + 7s^2 + 24s + 24}{s^4 + 10s^3 + 35s^2 + 50s + 24}.$$

With the transfer function model, the simulation model in Figure 9.27 can be established. The step-size can be set to $T = 0.1$, with fixed-step simulation algorithm. The following commands can be used to solve the differential equations, and the results are virtually the same as in Example 5.31.

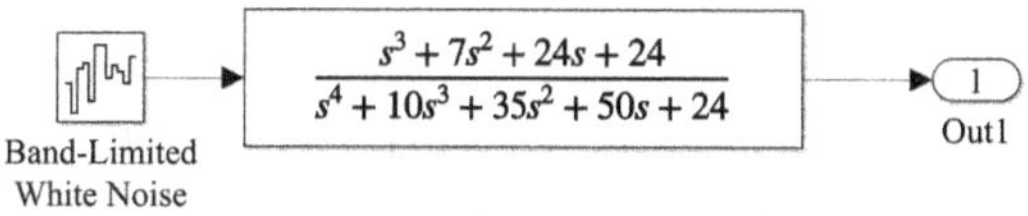

Figure 9.27: Linear stochastic differential equations (file: `c9mrand1.slx`).

```
>> G=tf([1,7,24,24],[1,10,35,50,24]); v=norm(G,2);
   xx=linspace(-2.5,2.5,30); yy=hist(yout,xx);
   yy=yy/(length(yout)*(xx(2)-xx(1)));   % compute PDF
   yp=exp(-xx.^2/(2*v^2))/sqrt(2*pi)/v;  % exact one
   bar(xx,yy,'c'), line(xx,yp)            % draw the solution
```

Example 9.19. Consider the block diagram of a nonlinear system shown in Figure 9.28, where the linear part of the transfer function $G(s)$ is

$$G(s) = \frac{s^3 + 7s^2 + 24s + 24}{s^4 + 10s^3 + 35s^2 + 50s + 24}.$$

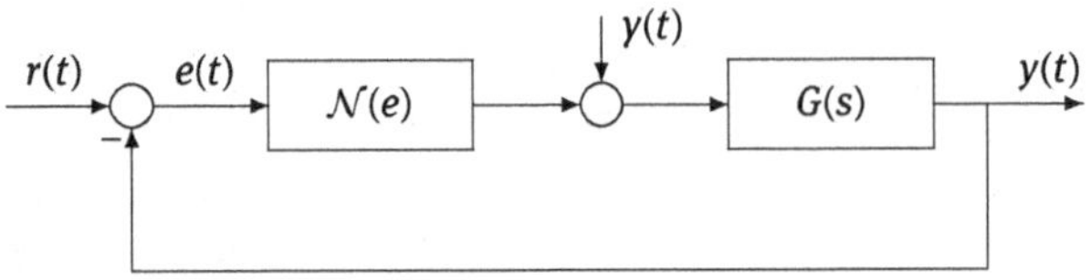

Figure 9.28: Nonlinear system with a stochastic input.

The saturation element is given by

$$\mathcal{N}(e) = \begin{cases} 2\,\mathrm{sign}(e), & |e| > 1, \\ 2e, & |e| \leqslant 1. \end{cases}$$

If the deterministic signal is given by $d(t) = 0$, draw the probability density function of signal $e(t)$.

Solutions. With the **Band-Limited White Noise** block to simulate the white noise input signal, and selecting fixed-step simulation methods, it is not hard to construct the simulation model, as shown in Figure 9.29.

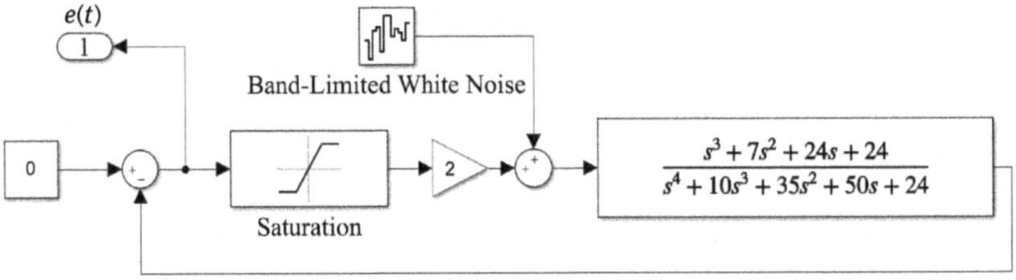

Figure 9.29: Simulation model of nonlinear stochastic differential equations (file: `c9mrand2.slx`).

Simulating the stochastic differential equations, the error signal $e(t)$ can be found. The probability density function can be obtained in histograms as shown in Figure 9.30.

```
>> c=linspace(-2,2,20); y1=hist(yout,c);
   bar(c,y1/(length(tout)*(c(2)-c(1)))) %draw PDF
```

9.4 Simulink modeling of fractional-order differential equations

The Riemann–Liouville and Caputo definitions were introduced for fractional calculus in Chapter 8. Command-driven methods were presented for solving fractional-order differential equations. If a fractional-order differential equation is too complicated, the command-driven format is not suitable for solving. Simulink can be used to draw the differential equations, and find the numerical solutions.

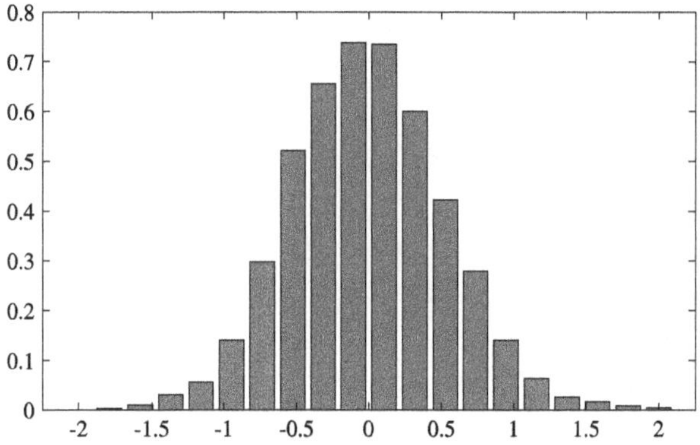

Figure 9.30: The probability density function of signal $e(t)$.

The approximation block for a fractional-order operator is introduced first in this section. Then the modeling techniques under Riemann–Liouville and Caputo definitions are introduced. Examples are also used to demonstrate fractional-order implicit differential equations and delay differential equations.

9.4.1 Fractional-order operator block

The numerical analysis of fractional-order derivatives of a given signal was studied in Section 8.1. The precondition of computing the fractional-order derivative is that the original signal is known, either in mathematical expression or in samples. If the signal comes from another part of the whole system, and it is not known, the methods in Section 8.1 cannot be used directly. An online computing method or algorithm must be introduced. Online computing methods under different fractional-order definitions are introduced in this section. For convenience, the online computing blocks are referred to as filters.

In [74, 55], various filters are introduced to approximate the behaviors of fractional-order derivatives using high-order integer-order filters. A good representative is the Oustaloup filter.[54]

Theorem 9.1. *Assuming that the frequency range of interest is $(\omega_\mathrm{b}, \omega_\mathrm{h})$, an Nth order continuous filter can be designed as*

$$G_\mathrm{f}(s) = K \prod_{k=1}^{N} \frac{s+\omega_k'}{s+\omega_k} \tag{9.4.1}$$

where the poles, zeros, and gain of the filter are

$$\omega_k' = \omega_\mathrm{b}\omega_\mathrm{u}^{(2k-1-\gamma)/N}, \quad \omega_k = \omega_\mathrm{b}\omega_\mathrm{u}^{(2k-1+\gamma)/N}, \quad K = \omega_\mathrm{h}^{\gamma}, \tag{9.4.2}$$

with $\omega_\mathrm{u} = \sqrt{\omega_\mathrm{h}/\omega_\mathrm{b}}$.

Based on the above algorithm, the following function is written to design the continuous filters directly. If signal $f(t)$ is fed into such a filter, the output signal can be regarded as the ${}^{\mathrm{RL}}\mathscr{D}_t^{\gamma} f(t)$ signal.

```
function G=ousta_fod(gam,N,wb,wh)
if round(gam)==gam, G=tf('s')^gam;   % for order is integer
else, k=1:N; wu=sqrt(wh/wb);         % find intermediate frequency
   wkp=wb*wu.^((2*k-1-gam)/N); wk=wb*wu.^((2*k-1+gam)/N); % zero/pole
   G=zpk(-wkp,-wk,wh^gam); G=tf(G); % construct Oustaloup filter
end
```

The syntax of the function is G_1=`ousta_fod`$(\gamma, N, \omega_{\mathrm{b}}, \omega_{\mathrm{h}})$, where γ is the fractional order, which can be positive or negative; N is the filter order; ω_{b} and ω_{h} are respectively the user selected lower and upper bounds of the frequency range. The fitting outside this interval is always neglected. Due to the powerful and effective facilities in MATLAB, a large frequency range such as $(10^{-5}, 10^5)$ can be selected, and the order can be set large such as $N = 25$. Accurate computation is implemented in solving fractional-order differential equations.

The author of this book has constructed a fractional-order operator block in Simulink. The block implements many filters, including Oustaloup filter, to approximate fractional-order differentiator or integrator. The block is contained in FOTF blockset, with many other blocks,[76, 74] and can be used directly. Double-clicking the block, a dialog box shown in Figure 9.31 is displayed. The parameters of the filter can be assigned in the block.

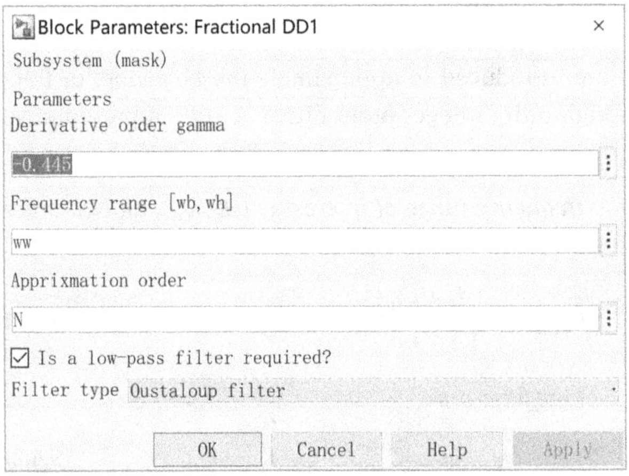

Figure 9.31: Dialog box of Oustaloup filter parameters.

With such a block, if a signal is fed into the block, the output of the block is the Riemann–Liouville derivative or integral of the input signal. The block cannot be used directly for Caputo derivatives. The computation of Caputo derivatives will be presented later.

9.4.2 Modeling and solution of Riemann–Liouville differential equations

With the modeling strategy of integer-order differential equations, the unknown function and its derivatives can be constructed with an integrator chain. Oustaloup filters can be used to compute fractional-order derivatives. Finally, Simulink models can be established for fractional-order differential equations. In this section, the modeling of linear and nonlinear fractional-order differential equations are demonstrated.

Example 9.20. Find the numerical solution of the linear fractional-order differential equation in Example 8.5. For convenience of presentation, the differential equation is recalled below:

$$\mathscr{D}_t^{3.5}y(t) + 8\mathscr{D}_t^{3.1}y(t) + 26\mathscr{D}_t^{2.3}y(t) + 73\mathscr{D}_t^{1.2}y(t) + 90\mathscr{D}_t^{0.5}y(t) = 90\sin t^2.$$

Solutions. The original fractional-order differential equation can be converted into the explicit form as follows:

$$\mathscr{D}_t^{3.5}y(t) = -8\mathscr{D}_t^{3.1}y(t) - 26\mathscr{D}_t^{2.3}y(t) - 73\mathscr{D}_t^{1.2}y(t) - 90\mathscr{D}_t^{0.5}y(t) + 30u'(t) + 90\mathscr{D}^{0.3}u(t).$$

The differential equation can be described by the Simulink model shown in Figure 9.32. Simulating the model, the numerical solution of the fractional-order differential equation can be found.

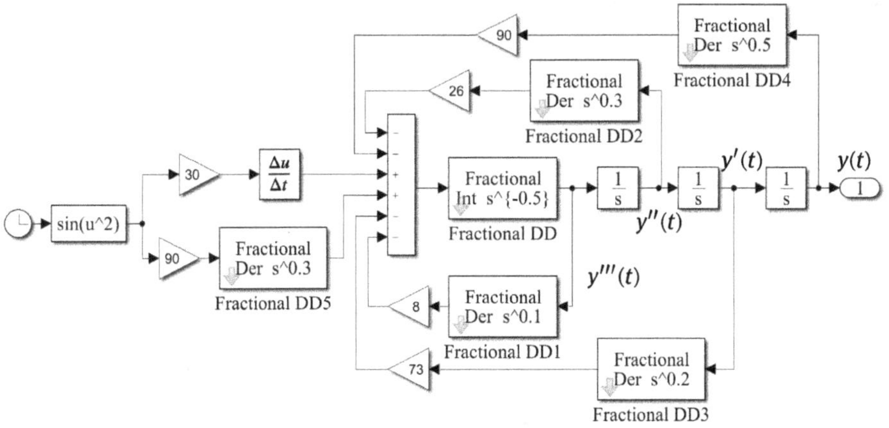

Figure 9.32: Simulink model of fractional-order differential equation (file: `c9mfode1.slx`).

Example 9.21. Solve the following nonlinear fractional-order differential equation:

$$\frac{3\mathscr{D}^{0.9}y(t)}{3+0.2\mathscr{D}^{0.8}y(t)+0.9\mathscr{D}^{0.2}y(t)}+\left|2\mathscr{D}^{0.7}y(t)\right|^{1.5}+\frac{4}{3}y(t)=5\sin 10t.$$

Solutions. Based on the equation, the explicit form of function $y(t)$ can be written as

$$y(t)=\frac{3}{4}\left[5\sin 10t-\frac{3\mathscr{D}^{0.9}y(t)}{3+0.2\mathscr{D}^{0.8}y(t)+0.9\mathscr{D}^{0.2}y(t)}-\left|2\mathscr{D}^{0.7}y(t)\right|^{1.5}\right].$$

The simulation model for describing $y(t)$ is described in Figure 9.33. It can be seen from the model that the fractional-order derivative signals can be constructed directly with the filters. The precision of the simulation results mainly depends upon the fitting of the filters to fractional-order derivatives. The selection of frequency range and order of the filter has a certain impact on the solution accuracy. In Figure 9.34, comparisons

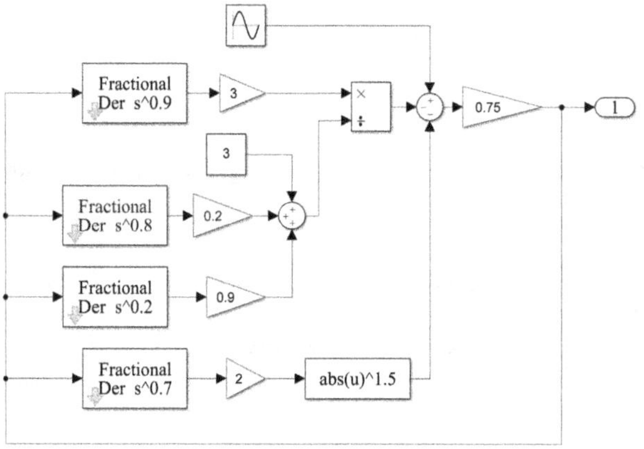

Figure 9.33: Simulink model (file: c9mfode2.slx).

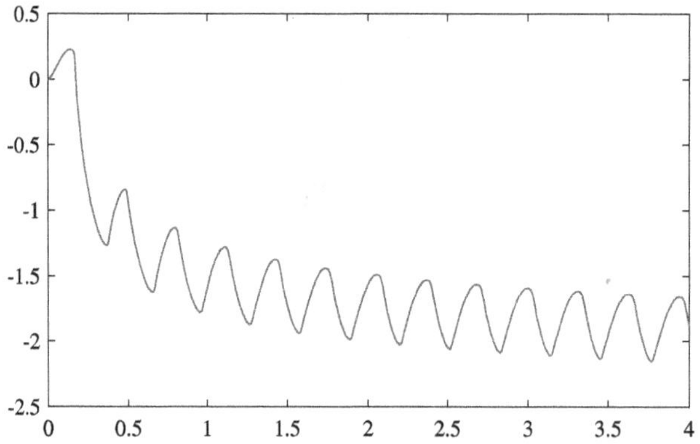

Figure 9.34: Simulation result of the nonlinear fractional-order differential equations.

are made to different frequency ranges and orders, and the results are virtually the same. The one with a slightly larger error is from the fitting with $\omega_\text{b} = 0.001$, $\omega_\text{h} = 1000$ and order $n = 5$. Therefore for this example, selecting $n = 9$ and a larger frequency range may give identical results.

9.4.3 Block for Caputo derivatives

Oustaloup and other filters are normally designed for handling zero initial value Riemann–Liouvile problems. They cannot be used to generate directly Caputo derivatives of certain signals. Alternative methods should be explored to compute Caputo derivatives. Two important properties are presented first to describe the relationship between Caputo and Riemann–Liouville derivatives.

Theorem 9.2. *Caputo derivative can be generated from Oustaloup filter*

$$ {}_{t_0}^{\text{C}}\mathscr{D}_t^{\gamma} y(t) = {}_{t_0}^{\text{RL}}\mathscr{D}_t^{-(\lceil\gamma\rceil-\gamma)} \left[y^{(\lceil\gamma\rceil)}(t) \right]. \tag{9.4.3} $$

The physical interpretation of the theorem is that the γth order Caputo derivative of $y(t)$ can be constructed by feeding the integer-order derivative $y^{(\lceil\gamma\rceil)}(t)$ through the $(\lceil\gamma\rceil - \gamma)$th order Oustaloup integrator.

It can be seen from the idea of key signals in the integrator chain that the integer-order derivatives of $y(t)$ are usually constructed by the chain. If one wants to compute the 2.4th order Caputo derivative, the signal $y'''(t)$ is used, and fed into the 0.6th order Oustaloup integrator. The output of the integrator is the expected ${}^{\text{C}}\mathscr{D}^{2.4}y(t)$.

Theorem 9.3. *Taking the $(\lceil\gamma\rceil - \gamma)$th order Riemann–Liouville derivatives of both sides of (9.4.3), the following relationship can be established:*

$$ {}_{t_0}^{\text{RL}}\mathscr{D}_t^{\lceil\gamma\rceil-\gamma} \left[{}_{t_0}^{\text{C}}\mathscr{D}_t^{\gamma} y(t) \right] = y^{(\lceil\gamma\rceil)}(t). \tag{9.4.4} $$

The physical interpretation of the theorem is that, for the γth order Caputo derivative signal ${}_{t_0}^{\text{C}}\mathscr{D}_t^{\gamma} y(t)$, taking $(\lceil\gamma\rceil - \gamma)$th order Riemann–Liouville derivative, the integer-order derivative $y^{(\lceil\gamma\rceil)}(t)$ can be found. That is, an integer-order derivative can be constructed by feeding Caputo derivatives to an Oustaloup filter.

With the two theorems and suitable processing with the Oustaloup filters, Caputo derivatives can be constructed so that fractional-order Caputo differential equations can be established.

9.4.4 Modeling and solving of Caputo differential equations

The construction of Caputo derivatives is introduced first. With the method and theoretic foundations, the integrator chain can be constructed to define the unknown signal and its integer-order derivatives, then, with appropriate properties, the Caputo derivatives can also be defined, such that Caputo differential equations can be expressed in Simulink. Examples are shown here to model and solve two classes of Caputo differential equations.

Example 9.22. Use Simulink to solve the nonlinear Caputo differential equation in Example 8.11. For convenience, the differential equation is recalled:

$$ {}_0^{\mathrm{C}}\mathscr{D}_t^{1.455}y(t) = -t^{0.1}\frac{\mathrm{E}_{1,1.545}(-t)}{\mathrm{E}_{1,1.445}(-t)}\mathrm{e}^t y(t){}_0^{\mathrm{C}}\mathscr{D}_t^{0.555}y(t) + \mathrm{e}^{-2t} - [y'(t)]^2 $$

where $y(0) = 1$ and $y'(0) = -1$.

Solutions. Since the highest order in the equation is 1.455, $q = 2$, so two integrators are needed to define the key signals $y(t)$, $y'(t)$, and $y''(t)$. The initial values are assigned accordingly to the two integrators. Now if the key signal ${}_0^{\mathrm{C}}\mathscr{D}_t^{0.555}y(t)$ is expected, it can be seen from (9.4.3) that signal $y'(t)$ should be fed into the 0.445th order Oustaloup integrator. The output of the integrator is the expected ${}_0^{\mathrm{C}}\mathscr{D}_t^{0.555}y(t)$ signal. With these key signals, the right-hand side of the equation can be expressed with low-level blocks, that is, we construct the ${}_0^{\mathrm{C}}\mathscr{D}_t^{1.455}y(t)$ signal. Now, the simulation loop should be closed. It can be seen from (9.4.4) that, if the above signal is fed into the 0.445th order Oustaloup derivative block, the output of the block is $y''(t)$. This signal happens to be the starting signal in the integrator chain. Therefore the two signals defining $y''(t)$ can be joined together to construct the closed loop, as shown in Figure 9.35. For simplicity, the static nonlinear function evaluation regarding t is described in an Interpreted MATLAB Fcn block, defined as

```
function y=c9mmlfs(u) % describe the static function in MATLAB
y=u^0.1*exp(u)*ml_func([1,1.545],-u)./ml_func([1,1.445],-u);
```

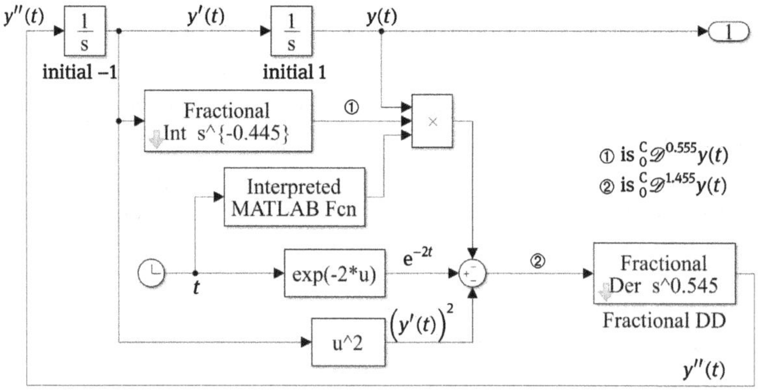

Figure 9.35: Simulink model of the nonlinear fractional-order differential equation (file: c9mfode3.slx).

The following parameters for the Oustaloup filters can be selected as follows. Simulation results can then be obtained. Compared with the given analytical solution, the maximum error is 5.5179×10^{-7}, and the elapsed time is 0.63 seconds.

```
>> N=30; ww=[1e-6 1e6]; % select Oustaloup filter parameters
   tic, [t,x,y]=sim('c9mfode3'); toc
   max(abs(y-exp(-t))) % solve equation and validate the results
```

Example 9.23. Solve the following implicit fractional-order differential equation:

$$
\begin{aligned}
&{}_0^{\mathrm{C}}\mathscr{D}_t^{0.2}y(t)\,{}_0^{\mathrm{C}}\mathscr{D}_t^{1.8}y(t) + {}_0^{\mathrm{C}}\mathscr{D}_t^{0.3}y(t)\,{}_0^{\mathrm{C}}\mathscr{D}_t^{1.7}y(t) \\
&\quad = -\frac{t}{8}\left[\mathrm{E}_{1,1.8}\left(-\frac{t}{2}\right)\mathrm{E}_{1,1.2}\left(-\frac{t}{2}\right) + \mathrm{E}_{1,1.7}\left(-\frac{t}{2}\right)\mathrm{E}_{1,1.3}\left(-\frac{t}{2}\right)\right]
\end{aligned}
$$

where $y(0) = 1$ and $y'(0) = -1/2$. The exact solution is $y(t) = \mathrm{e}^{-t/2}$.

Solutions. The implicit Caputo equation can be converted into the standard form as

$$
\begin{aligned}
&{}_0^{\mathrm{C}}\mathscr{D}_t^{0.2}y(t)\,{}_0^{\mathrm{C}}\mathscr{D}_t^{1.8}y(t) + {}_0^{\mathrm{C}}\mathscr{D}_t^{0.3}y(t)\,{}_0^{\mathrm{C}}\mathscr{D}_t^{1.7}y(t) \\
&\quad + \frac{t}{8}\left[\mathrm{E}_{1,1.8}\left(-\frac{t}{2}\right)\mathrm{E}_{1,1.2}\left(-\frac{t}{2}\right) + \mathrm{E}_{1,1.7}\left(-\frac{t}{2}\right)\mathrm{E}_{1,1.3}\left(-\frac{t}{2}\right)\right] = 0.
\end{aligned}
$$

Based on the above modeling method, the key signals $y(t)$, $y'(t)$, and $y''(t)$ are defined first. The Caputo derivatives $\mathscr{D}^{0.2}y(t)$, $\mathscr{D}^{0.3}y(t)$, $\mathscr{D}^{1.7}y(t)$, and $\mathscr{D}^{1.8}y(t)$ are then defined. With these key signals, the left-hand side of the standard form given above can be constructed and fed into the Algebraic Constraint block. The target is to have the output signal of the block equal to $\mathscr{D}^{1.8}y(t)$. To ensure this relationship, a further 0.2th order derivative is taken to yield $y''(t)$. The signal can be joined with signal $y''(t)$ in the integrator chain, to close the loop for the implicit Caputo differential equation. The model is as shown in Figure 9.36. Since the initial values are defined in the integrator chain, it is not necessary to consider them elsewhere. The Oustaloup filter can be used to express fractional-order derivatives. The contents of Interpreted MATLAB Fcn block is

```
function y=c9mimpfs(u) % describe the implicit equation in MATLAB
y=1/8*u*(ml_func([1,1.8],-u/2)*ml_func([1,1.2],-u/2)+...
    ml_func([1,1.7],-u/2)*ml_func([1,1.3],-u/2)); % implicit equation
```

With the following Oustaloup filter parameters, the numerical solution of the implicit differential equation can be found, with the maximum error of 3.8182×10^{-5}, and elapsed time of 151.2 seconds. Compared with other models, the simulation took too long. This is because of the algebraic loop in the model. In each simulation step, an algebraic equation is invoked once. In the solution process, a warning message "Block diagram contains 1 algebraic loop" is displayed. This is unavoidable and should be neglected.

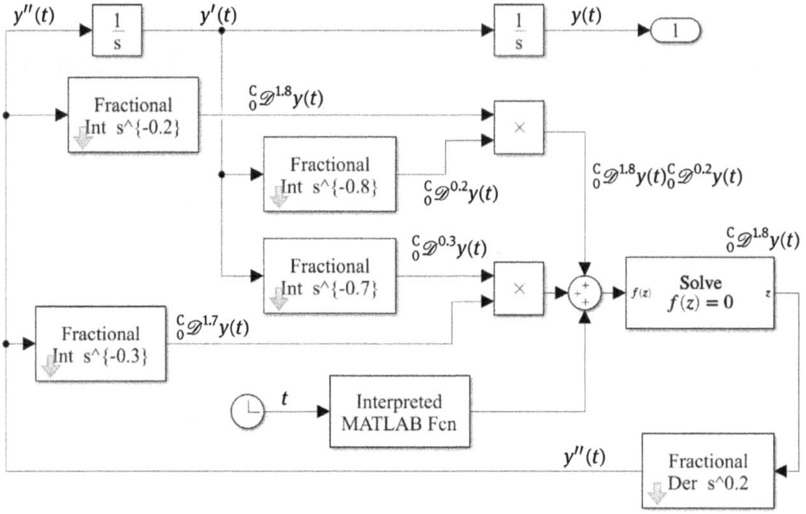

Figure 9.36: Simulink model of implicit differential equation (file: `c9mimps.slx`).

```
>> ww=[1e-5 1e5]; n=30; tic, [t,x,y]=sim('c9mimps'); toc
   max(abs(y-exp(-t/2)))
```

9.4.5 Fractional-order delay differential equations

It is rather difficult to find an example for fractional-order delay differential equations with analytical solutions in the literature. And it is difficult to find a fractional-order delay differential equation. Simulink can still be used to construct fractional-order delay differential equation models, and then find their numerical solutions. There is no better method to check whether the numerical solutions are correct than by selecting different parameters of Oustaloup filters or simulation algorithms to see whether consistent results are obtained. In this section, an example is used to demonstrate the solution process of fractional-order delay differential equations.

Example 9.24. Modifying slightly the nonlinear fractional-order differential equation in Example 9.22, the following delay Caputo differential equation can be created:

$$ {}_0^{\rm C}\mathscr{D}_t^{1.455}y(t) = -t^{0.1}\frac{{\rm E}_{1,1.545}(-t)}{{\rm E}_{1,1.445}(-t)}{\rm e}^t y(t){}_0^{\rm C}\mathscr{D}_t^{0.555}y(t-0.3) + {\rm e}^{-2t} - [y'(t-0.4)]^2 $$

where $y(0) = 1$ and $y'(0) = -1$. If $t < 0$, $y(t) = y'(t) = 0$. Find the numerical solution of the differential equation.

Solutions. For the `c9mfode3.slx` model given in Figure 9.35, a slight modification is made such that delay constant can be appended to the key signals, and the model in

Figure 9.37 can be constructed. Simulating the model and trying different parameters for the Oustaloup filter, the same output curves are found as seen in Figure 9.38.

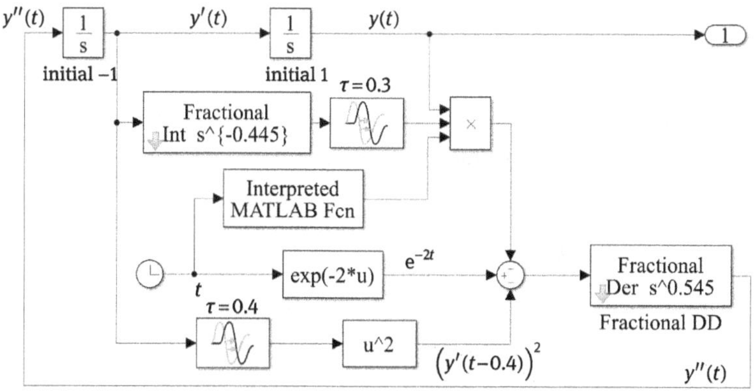

Figure 9.37: Delay differential equation in Simulink (file: `c9mfdde1.slx`).

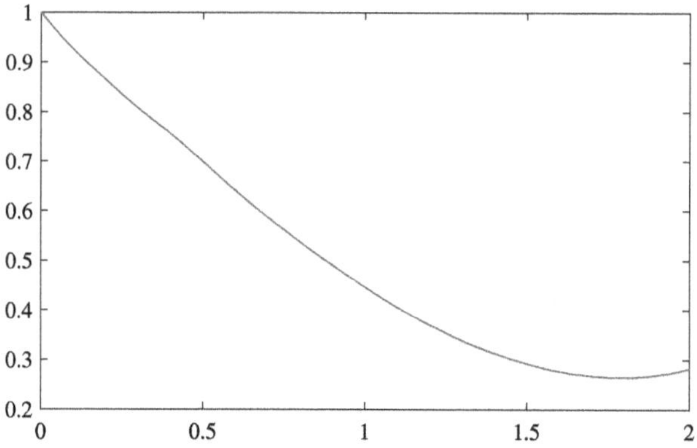

Figure 9.38: Simulation results of fractional-order delay differential equation.

It is worth pointing out that the fractional-order delay differential equation studied here has zero initial values. For nonzero constant history functions, the method in Example 9.16 can be used to model and simulate the system again.

In fact, it is seen from the previous examples that, in theory, the modeling technique can be used to describe differential equations of any complexity. If the parameters in Oustaloup filter are chosen well, the numerical solutions of any fractional-order differential equations can be found effectively.

9.5 Exercises

9.1 Consider a simple linear differential equation

$$y^{(4)} + 3y''' + 3y'' + 4y' + 5y = \mathrm{e}^{-3t} + \mathrm{e}^{-5t}\sin(4t + \pi/3)$$

with the initial values $y(0) = 1$, $y'(0) = y''(0) = 1/2$, $y'''(0) = 0.2$. Establish the simulation model with Simulink, and draw the simulation results.

9.2 Consider the above model. Assume that the given differential equation is changed into a time-varying one:

$$y^{(4)} + 3ty''' + 3t^2y'' + 4y' + 5y = \mathrm{e}^{-3t} + \mathrm{e}^{-5t}\sin(4t + \pi/3).$$

The initial values are $y(0) = 1$, $y'(0) = y''(0) = 1/2$, and $y'''(0) = 0.2$. Establish the simulation model with Simulink, and draw the simulation results.

9.3 In Example 9.6, a Simulink model was created for the Apollo differential equations. Establish a Simulink model with the method in Example 9.2. Compare the modeling efficiency and results.

9.4 Consider the differential-algebraic equations in Example 5.14:

$$\begin{cases} x_1'(t) = -0.2x_1(t) + x_2(t)x_3(t) + 0.3x_1(t)x_2(t), \\ x_2'(t) = 2x_1(t)x_2(t) - 5x_2(t)x_3(t) - 2x_2^2(t), \\ 0 = x_1(t) + x_2(t) + x_3(t) - 1 \end{cases}$$

with known initial values $x_1(0) = 0.8$, $x_2(0) = x_3(0) = 0.1$. In this example, an equivalent transform was introduced to convert the system into explicit differential equations to find its solution. If such a transform is not used, find the solution of the differential-algebraic equations with Algebraic Constraints block. Compare the simulation results.

9.5 If, in Example 9.13, the transfer function model is not used, draw the Simulink model and compare the simulation results.

9.6 Consider the delay differential equation

$$y^{(4)}(t) + 4y'''(t - 0.2) + 6y''(t - 0.1) + 6y''(t) + 4y'(t - 0.2) + y(t - 0.5) = \mathrm{e}^{-t^2}.$$

Assume that the equation has zero initial values when $t \leqslant 0$. Establish a Simulink model to describe this equation. Also solve it with function `dde23()` and compare the results. Draw the curve of signal $y(t)$.

9.7 Establish a Simulink model for the differential equations in Example 4.18 and compare the simulation results.

9.8 Assume that a Simulink model is as shown in Figure 9.39. Write down its mathematical model.

9.9 Construct a Simulink model for the single-term Caputo differential equations in Problem 8.6.

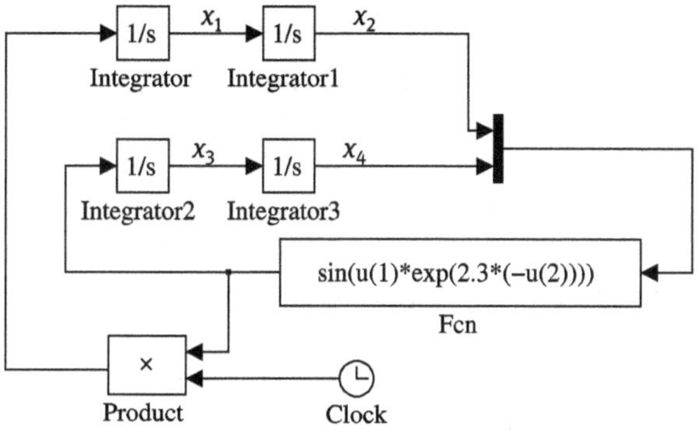

Figure 9.39: Simulink model for Problem 9.8.

9.10 Solve the following nonlinear fractional-order differential equations with zero initial values, where $f(t) = 2t + 2t^{1.545}/\Gamma(2.545)$. If the equation is a Caputo one, and $y(0) = -1$, $y'(0) = 1$, solve the equation again:

$$\mathscr{D}^2 x(t) + \mathscr{D}^{1.455} x(t) + \left[\mathscr{D}^{0.555} x(t)\right]^2 + x^3(t) = f(t).$$

9.11 Assuming that a nonlinear fractional-order differential equation is described in a Simulink model shown in Figure 9.40, write down the mathematical model of this differential equation, and draw the output signal $y(t)$.

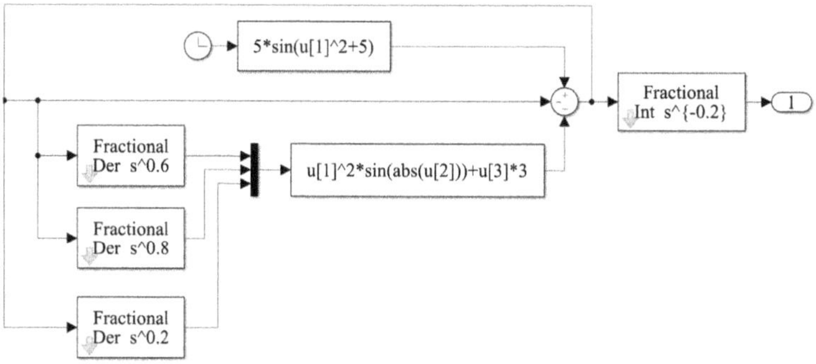

Figure 9.40: Simulink model of a nonlinear fractional-order differential equation (file: `c9mfode4.mdl`).

9.12 Solve Bagley–Torwik differential equation[20]

$$Ay''(t) + B\mathscr{D}^{3/2} y(t) + Cy(t) = C(t+1)$$

with $y(0) = y'(0) = 1$. Show that the solution of the equation is independent to the values of the constants A, B, and C.

9.13 Three benchmark problems on fractional-order differential equations are proposed in [74]. Two of them were presented earlier. Use Simulink to solve directly the three problems. Find the numerical solutions and assess the errors and efficiency:

(1) ${}_0^{\mathrm{C}}\mathscr{D}_t^{1.6}y(t) = t^{0.4}\mathrm{E}_{1,1.4}(-t)$, $0 \leqslant t \leqslant 100$, $x(0) = 1$, $x'(0) = -1$. The solution is $y(t) = \mathrm{e}^{-t}$.

(2) The Caputo linear differential equation with nonzero initial values

$$y'''(t) + {}_0^{\mathrm{C}}\mathscr{D}_t^{2.5}y(t) + y''(t) + 4y'(t) + {}_0^{\mathrm{C}}\mathscr{D}_t^{0.5}y(t) + 4y(t) = 6\cos t.$$

If the initial values are $y(0) = 1$, $y'(0) = 1$, $y''(0) = -1$, and $0 \leqslant t \leqslant 10$, the analytical solution is $y(t) = \sqrt{2}\sin(t + \pi/4)$.

(3) Fractional-order nonlinear state space model

$$\begin{cases} {}_0^{\mathrm{C}}\mathscr{D}_t^{0.5}x(t) = \dfrac{1}{2\Gamma(1.5)}\left(\left[(y(t)-2)(z(t)-3)\right]^{1/6} + \sqrt{t}\right), \\ {}_0^{\mathrm{C}}\mathscr{D}_t^{0.2}y(t) = \Gamma(2.2)\left[x(t)-1\right], \\ {}_0^{\mathrm{C}}\mathscr{D}_t^{0.6}z(t) = \dfrac{\Gamma(2.8)}{\Gamma(2.2)}\left[y(t)-2\right] \end{cases} \tag{9.5.1}$$

where $x(0) = 1$, $y(0) = 2$, and $z(0) = 3$. The analytical solution of the state space equations is $x(t) = t+1$, $y(t) = t^{1.2}+2$, $z(t) = t^{1.8}+3$, and $0 \leqslant t \leqslant 5\,000$.

10 Boundary value problems for ordinary differential equations

The ordinary differential equation studied so far were based on the assumption that the initial values are known. That is, the $\boldsymbol{x}_0$ vector was known. Solutions were then found for the state variables at other time instances. These differential equations are known as initial value problems. In practice, some of the state variables are known at time $t = t_0$, and some at $t = t_n$. Such problems are the so-called boundary value problems. Solvers such as `ode45()` cannot be used to solve boundary value problems directly. In this chapter, we explore how to solve boundary value problems for differential equations.

In Section 10.1, the general form of boundary value problems is proposed. Necessary interpretations and comments are made on the mathematical forms. In Section 10.2, three low-level algorithms are presented to solve two-point problems of second-order differential equations. These may be considered as the basis for general purpose solvers for the description and solution of the boundary value problems. If the readers are interested only in finding the numerical solutions to boundary value problems using MATLAB, this section can be skipped. The subsequent materials can be visited directly.

In Section 10.3, a general purpose solver for two-point boundary value problems is discussed. Various differential equations are solved with this solver. Complicated boundary value problems are also studied, including semi-infinite interval boundary value problems.

For some specific problems, the solver provided in Section 10.3 is limited. Some problems cannot be solved with it. In Section 10.4, an optimization based solution pattern is presented. Some of the problems which cannot be solved with other methods are considered, including the boundary value problems of implicit differential equations, delay differential equations, and fractional-order differential equations. Simulink based methods are often adopted, such that the boundary value problems of differential equations of any complexity can be solved. These solution methods do not exist in any other literature. With such an idea, boundary value problems in even more complicated cases can be solved.

10.1 Standard boundary value problems

The boundary value problems are widely encountered in real applications. In the beginning of [2], 25 practical applications of boundary value problems are illustrated. In this section, the mathematical form of boundary value problems is presented, and boundary conditions of different forms are presented.

https://doi.org/10.1515/9783110675252-010

Definition 10.1. The first-order explicit differential equations with undetermined coefficients are given by

$$\boldsymbol{y}'(t) = \boldsymbol{f}(t, \boldsymbol{y}(t), \boldsymbol{\theta}) \tag{10.1.1}$$

where $\boldsymbol{y}(t)$ is the state variable vector while $\boldsymbol{\theta}$ is a vector consisting of all the undetermined coefficients. The boundary conditions of the equation are

$$\boldsymbol{\phi}(\boldsymbol{y}(a), \boldsymbol{y}(b), \boldsymbol{\theta}) = \boldsymbol{0}. \tag{10.1.2}$$

It can be seen from the formulas that compared with initial value problems, some of the initial values of the states and some terminal values satisfy the algebraic equations in (10.1.2). These equations can be given in simple as well as complicated forms. The problem to be studied in this chapter is of finding the numerical solutions from the given differential equations and boundary value conditions.

Although the mathematical form in (10.1.1) is introduced, it does not imply the differential equations to be studied can only be described by explicit forms. In this chapter, boundary value problems of other forms, such implicit differential equations, delay differential equations, and fractional-order differential equations, are also considered. The differential equations which cannot be explained with (10.1.1) will also be discussed.

10.2 Shooting methods in two-point boundary value problems for second-order differential equations

In the linear differential equations studied earlier, the analytical solution of a second-order differential equation has two undetermined constants, C_1 and C_2. If $y(a)$ and $y'(a)$ are known, C_1 and C_2 can be determined uniquely. The problem like this is the initial value problem, discussed in the earlier chapters.

The given quantities in boundary value problems are no longer $y(a)$ and $y'(a)$, they are $y(a)$ and $y(b)$:

$$y(a) = y_\mathrm{a}, \quad y(b) = y_\mathrm{b}. \tag{10.2.1}$$

From the given values at these two points, the undetermined constants C_1 and C_2 can still be determined. The differential equation problem in this case is changed into a boundary value problem. For second-order nonlinear differential equations, if the boundary conditions are known, the numerical solution of the differential equations may or may not be uniquely determined.

Note that (10.2.1) is a special form of (10.1.2). This type of boundary value problems is considered in the section.

The idea of solving boundary value problem is that, if we assume that the value of $\hat{y}'(a)$ is also known, functions like `ode45()` can be used to solve the problem. The solution $\hat{y}(b)$ must have errors compared with the given $y(b)$. Using the error information, the initial value $\hat{y}'(a)$ can be repeatedly modified, until a consistent solution is found, such that $|\hat{y}(b) - y(b)| < \epsilon$. When the consistent initial value $\hat{y}'(a)$ is found, it can be used to directly find the solutions of the original boundary value problems. This type of methods is also known as shooting methods. In this section, different shooting methods are introduced so as to illustrate the idea and methods in solving boundary value problems for second-order differential equations.

10.2.1 Shooting algorithms for linear time-varying differential equations

First consider a simple problem of finding the numerical solutions of linear time-varying second-order differential equations described as boundary value problems. Generally speaking, second-order linear time-varying differential equations have no analytical solutions. Numerical solutions are the only way to study such systems.

Definition 10.2. The mathematical form of second-order linear time-varying differential equations can be expressed as

$$y''(t) + p(t)y'(t) + q(t)y(t) = f(t) \tag{10.2.2}$$

where $p(t)$, $q(t)$, and $f(t)$ are all given functions. The boundary conditions studied in this section are given by (10.2.1).

The basic idea of a shooting method is to find the corresponding initial values of $y(a)$ and $y'(a)$, such that the boundary values in (10.1.2) are satisfied. Then the initial value problem solvers can be used to solve the boundary value problems.

The shooting method for linear differential equations can be summarized by the following procedures:

(1) Find the initial value $y_1(b)$ for the following homogenous differential equation:

$$y_1''(t) + p(t)y_1'(t) + q(t)y_1(t) = 0, \quad y_1(a) = 1, \quad y_1'(a) = 0. \tag{10.2.3}$$

(2) Find the final value $y_2(b)$ from the following initial value problem of a homogeneous differential equation:

$$y_2''(t) + p(t)y_2'(t) + q(t)y_2(t) = 0, \quad y_2(a) = 0, \quad y_2'(a) = 1. \tag{10.2.4}$$

(3) Solve the following initial value problem for the inhomogeneous differential equations and find $y_\mathrm{p}(b)$:

$$y_\mathrm{p}''(t) + p(t)y_\mathrm{p}'(t) + q(t)y_\mathrm{p}(t) = f(t), \quad y_\mathrm{p}(a) = 0, \quad y_\mathrm{p}'(a) = 1. \tag{10.2.5}$$

(4) If $y_2(b) \neq 0$, then compute

$$m = \frac{y_\mathrm{b} - y_\mathrm{a}y_1(b) - y_\mathrm{p}(b)}{y_2(b)}. \tag{10.2.6}$$

(5) Solve the initial value problem, and let $y(t)$ be the numerical solution of the original boundary value problem:

$$y''(t) + p(t)y'(t) + q(t)y(t) = f(t), \quad y(a) = y_a, \quad y'(a) = m. \tag{10.2.7}$$

To find the numerical solutions, the first-order explicit differential equations should be found. Selecting the state variables $x_1(t) = y(t)$ and $x_2(t) = y'(t)$, the following standard form can be found, and denoted as $f_2(t, \boldsymbol{x}(t))$:

$$\begin{cases} x_1'(t) = x_2(t), \\ x_2'(t) = -q(t)x_1(t) - p(t)x_2(t) + f(t). \end{cases} \tag{10.2.8}$$

Another auxiliary differential equation for $f_1(t, \boldsymbol{x}(t))$ should be constructed, which is used to describe the homogeneous part:

$$\begin{cases} x_1'(t) = x_2(t), \\ x_2'(t) = -q(t)x_1(t) - p(t)x_2(t). \end{cases} \tag{10.2.9}$$

With the above equations and algorithm, the following MATLAB function can be written:

```
function [t,y]=shooting(p,q,f,tspan,x0f,varargin)
ga=x0f(1); gb=x0f(2);                       % extract the boundaries
f1=@(t,x)[x(2); -q(t)*x(1)-p(t)*x(2)];      % homogeneous equations
f2=@(t,x)f1(t,x)+[0; f(t)];                 % inhomogeneous equations
[t,y1]=ode45(f1,tspan,[1;0],varargin{:});   % (10.2.3)
[t,y2]=ode45(f1,tspan,[0;1],varargin{:});   % (10.2.4)
[t,yp]=ode45(f2,tspan,[0;0],varargin{:});   % (10.2.5)
m=(gb-ga*y1(end,1)-yp(end,1))/y2(end,1);    % (10.2.6), find initial values
[t,y]=ode45(f2,tspan,[ga;m],varargin{:});   % (10.2.7)
```

It can be seen that each statement in the function faithfully implements the mathematical description. It is immediately seen that MATLAB has advantages in solving scientific computing problems.

The syntax of the function is

[$\boldsymbol{t}$,$\boldsymbol{y}$]=shooting(p,q,f,tspan,$\boldsymbol{x}_0$,controls)

where p, q and f are respectively the function handles of $p(t)$, $q(t)$, and $f(t)$; tspan is the vector of the start and stop times; $\boldsymbol{x}_0$=[y_a;y_b] contains the boundary values, where $\boldsymbol{x}_0$ is just a notation, not initial values. Besides it can be seen that in solving initial value problems, the solver ode45() is used directly, and the "controls" are allowed to pass varargin to the ode45() solver. If the user thinks that ode45() solver is not suitable for certain problems, other solvers can be used instead. Simple examples are given next to demonstrate the use of the new solver.

Example 10.1. Find the solutions of the following linear differential equations in the interval $[0, \pi]$:

$$(\sin(t+1)+2)^2 y''(t) - 2(\cos(t+1)+2)^2 y'(t) - 16\sin(t+1)y(t) = 4e^{-2t}(4\cos(t+1)+9)$$

where $y(0) = 1$ and $y(\pi) = e^{-2\pi}$. It is known that the analytical solution of the differential equation is $y(t) = e^{-2t}$.

Solutions. Since $(\sin(t+1)+2)^2 \neq 0$, the original differential equation can be converted into the following second-order standard form:

$$y''(t) - \frac{2(\cos(t+1)+2)^2}{(\sin(t+1)+2)^2} y'(t) - \frac{16\sin(t+1)}{(\sin(t+1)+2)^2} y(t) = \frac{4e^{-2t}(4\cos(t+1)+9)}{(\sin(t+1)+2)^2}$$

where the three time-varying functions are found as

$$p(t) = -\frac{2(\cos(t+1)+2)^2}{(\sin(t+1)+2)^2}, \quad q(t) = -\frac{16\sin(t+1)}{(\sin(t+1)+2)^2}, \quad f(t) = \frac{4e^{-2t}(4\cos(t+1)+9)}{(\sin(t+1)+2)^2}.$$

The denominator can be denoted by an anonymous function $D(t)$. Then other anonymous functions can be set up to describe the time-varying functions $p(t)$, $q(t)$, and $f(t)$.

```
>> D=@(t)(sin(t+1)+2)^2; % the common denominator
   p=@(t)-2*(cos(t+1)+2)^2/D(t); q=@(t)-16*sin(t+1)/D(t);
   f=@(t)4*exp(-2*t)*(4*cos(t+1)+9)/D(t);
```

With the following commands, the solution to the original problem can be found, as shown in Figure 10.1. It can be seen that the solution $y_1(x)$ satisfies both boundary value conditions. Besides, in the solution process, the error tolerance is set to be very strict, so the solution satisfies the original differential equations.

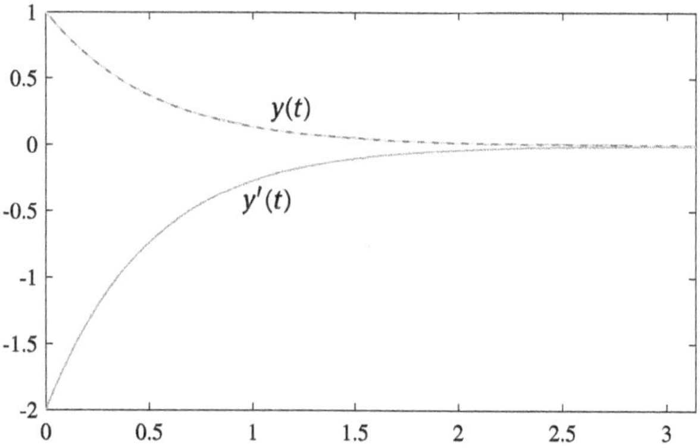

Figure 10.1: Solution of the two-point boundary value problem.

```
>> ff=odeset; ff.AbsTol=100*eps; ff.RelTol=100*eps;
   x0=[1; exp(-2*pi)];                    % given boundary values
   [t,y]=shooting(p,q,f,[0,pi],x0,ff); % solve the problem
   y0=exp(-2*t); plot(t,y,t,y0,'--'), e=norm(y(:,1)-y0)
```

It is seen from validating the result that the error norm between the numerical and analytical solutions is $e = 2.0422 \times 10^{-12}$. The precision obtained is rather high.

10.2.2 Finite difference algorithm for linear differential equations

The numerical solution approach for the boundary value problems of linear time-varying differential equations was presented earlier, and it can be used to effectively solve certain boundary value problems. In this section, a matrix based numerical method is presented. We also consider how to increase the effectiveness of the method.

Theorem 10.1. *For a set of equally-spaced samples $y(x_1), y(x_2), \dots, y(x_n)$, the $o(h^2)$ central-point difference method can be used to approximate the terms involving the first- and second-order derivatives of $y(t)$:*

$$y'(x_i) = \frac{y(x_{i+1}) - y(x_{i-1})}{2h}, \quad y''(x_i) = \frac{y(x_{i+1}) - 2y(x_i) + y(x_{i-1})}{h^2} \tag{10.2.10}$$

where $i = 2, 3, \dots, n-1$, and h is the step-size.

The first- and second-order difference formulas can be substituted into (10.2.2) so that

$$\frac{y(x_{i+1}) - 2y(x_i) + y(x_{i-1})}{h^2} + p(x_i)\frac{y(x_{i+1}) - y(x_{i-1})}{2h} + q(x_i)y(x_i) = f(x_i) \tag{10.2.11}$$

where $i = 1, 2, \dots, n$ and n is the number of intermediate points. We have $h = (b-a)/(n-1)$ and $x_i = a + (i-1)h$.

Summarizing the above, it is found that

$$\left(1 + \frac{h}{2}p(x_i)\right)y(x_{i+1}) + (-2 + h^2 q(x_i))y(x_i) + \left(1 - \frac{h}{2}p(x_i)\right)y(x_{i-1}) = h^2 f(x_i). \tag{10.2.12}$$

Introducing the notations

$$t_i = -2 + h^2 q(x_i), \quad v_i = 1 + \frac{h}{2}p(x_i), \quad w_i = 1 - \frac{h}{2}p(x_i), \quad b_i = h^2 f(x_i) \tag{10.2.13}$$

where $i = 1, \dots, n-1$, the parameters at the boundaries are modified as

$$b_1 = b_1 - w_1 y_a, \quad b_{n-1} = b_{n-1} - v_{n-1} y_b. \tag{10.2.14}$$

The following linear algebraic equation can be set up:

$$v_i y(x_{i+1}) - t_i y(x_i) + w_i y(x_{i-1}) = b_i \tag{10.2.15}$$

whose matrix form is

$$\begin{bmatrix} t_1 & v_1 & & & \\ w_2 & t_2 & v_2 & & \\ & \ddots & \ddots & \ddots & \\ & & w_{n-2} & t_{n-2} & v_{n-2} \\ & & & w_{n-1} & t_{n-1} \end{bmatrix} \begin{bmatrix} y(x_2) \\ y(x_3) \\ \vdots \\ y(x_{n-1}) \\ y(x_n) \end{bmatrix} = \begin{bmatrix} b_1 \\ b_2 \\ \vdots \\ b_{n-2} \\ b_{n-1} \end{bmatrix} \tag{10.2.16}$$

and $y(x_1) = y_a$, $y(x_{n+1}) = y_b$. Solving this matrix equation, the numerical solution of the differential equations $y(x_i)$ can be found.

The mathematical form of the algorithm is simple, however, a low-precision central-point finite difference method is adopted, so the accuracy of the function may not be very high. The theoretical precision is $o(h^2)$.

The MATLAB implementation of the algorithm is

```
function [x,y]=fdiff0(p,q,f,tspan,x0f,n)
t0=tspan(1); tn=tspan(2); ga=x0f(1); gb=x0f(2);
x=linspace(t0,tn,n); h=x(2)-x(1); x0=x(1:n-1);
t=-2+h^2*q(x0); b=h^2*f(x0);
p0=p(x0); v=1+h*p0/2; w=1-h*p0/2;   % compute the 4 vectors in (10.2.13)
b(1)=b(1)-w(1)*ga; b(n-1)=b(n-1)-v(n-1)*gb; % modified parameters (10.2.14)
A=diag(t)+diag(v(1:end-1),1)+diag(w(2:end),-1); % the tri-diagonal matrix
y=inv(A)*b(:);  y=[ga; y; gb]';      % solve the linear algebraic equation
```

The syntax of the function is

$[\boldsymbol{t},\boldsymbol{y}]$=fdiff0($p$,$q$,$f$,tspan,$\boldsymbol{x}_0$,$n$)

The syntax of the function is very close to that of the `shooting()` function. The differences are that the functions p, q, and f are expressed in dot operations. Besides, n is the number of points to compute. It can be seen that in the algorithm, the points are equally-spaced, therefore the algorithm may not be practical. Examples are used here to demonstrate the use of the function.

Example 10.2. Solve again the two-point linear time-varying differential equations in Example 10.1 with the new algorithm.

Solutions. Try now $n = 5\,000$ to solve the problem. The values of y can be found, and function `fdiff0()` can be used to draw the $y(t)$ curve. After 2.72 seconds of waiting, the maximum error is found as 4.7675×10^{-4}. It can be seen that the algorithm is of very low efficiency, and the error is too large.

```
>> p=@(t)-2*(cos(t+1)+2).^2./(sin(t+1)+2).^2;
   q=@(t)-16*sin(t+1)./(sin(t+1)+2).^2;
   f=@(t)4*exp(-2*t).*(4*cos(t+1)+9)./(sin(t+1)+2).^2;
   tic, [t,y]=fdiff0(p,q,f,[0,pi],[1,exp(-2*pi)],5000); toc
   y0=exp(-2*t); plot(t,y,t,y0,'--'), norm(y-y0,inf)
```

Since in the solution process a huge matrix of size $5\,000\times 5\,000$ is used, the efficiency is not satisfactory. To increase the accuracy, the value of n should also be increased. When $n = 10\,000$, the elapsed time increases to 27.16 seconds while the error is only reduced to 2.3836×10^{-4}. Compared with the precision in Example 10.1, the error is too large. Further increase of n may result in memory problems, and the speed is significantly reduced, such that the solutions cannot be achieved on the current computers. Therefore this algorithm is not recommended for solving boundary value problems of linear time-varying differential equations.

Of course, with the properties of the tri-diagonal matrix in (10.2.16), sparse matrices can be used to solve the problem. A sparse matrix based method is implemented. The modifications are made on sparse matrix representation of $\boldsymbol{A}$. Repeating the commands in Example 10.2, even though n is increased to 2 000 000, the solution can be found in 1.246 seconds. The error norm is 1.1340×10^{-6}. It can be seen that the solver is improved significantly in efficiency, but the error is still too large. The solver is not of any use.

```
function [x,y]=fdiff(p,q,f,tspan,x0f,n)
t0=tspan(1); tn=tspan(2); ga=x0f(1); gb=x0f(2);
x=linspace(t0,tn,n); h=x(2)-x(1); x0=x(1:n-1);
t=-2+h^2*q(x0); b=h^2*f(x0); i=1:n-2;
p0=p(x0); v=1+h*p0/2; w=1-h*p0/2;  %compute the 4 vectors in (10.2.13)
b(1)=b(1)-w(1)*ga; b(n-1)=b(n-1)-v(n-1)*gb; %modified parameters (10.2.14)
A=sparse([i i+1 i n-1],[i+1 i i n-1],[v(1:end-1),w(2:end),t]);
y=A\b; y=[ga; y(1:end-1); gb]';  %solve the linear algebraic equation
```

Since an $o(h^2)$ algorithm is used in the solver, the solution method is not of high accuracy. The $o(h^p)$ high precision algorithm from Volume II can be used. High accuracy results may be achieved, however, the programming may become more complicated.

It can be seen from the previous examples that in solving the boundary value problems of second-order linear time-varying differential equations, the first method is recommended.

10.2.3 Shooting algorithms for nonlinear differential equations

In the previous sections, only linear time-varying differential equations were considered. However, the two methods cannot be used in handling the general cases in (10.1.1) and (10.2.1). In this section, numerical solutions of simple boundary value problems of second-order nonlinear differential equations are explored.

Definition 10.3. The mathematical form of a second-order nonlinear explicit differential equation is given by

$$y''(t) = F(t, y(t), y'(t)), \quad y(a) = y_a, \quad y(b) = y_b. \tag{10.2.17}$$

Assuming that the original problem can be converted into the following initial value problem:

$$y''(t) = F(t, y(t), y'(t)), \quad y(a) = y_a, \quad y'(a) = m(t), \tag{10.2.18}$$

the problem is reduced to solving $y(b, m(t)) = y_b$. The following Newton–Raphson iterative method can be used in finding the value of m:

$$m_{i+1} = m_i - \frac{y(b, m_i) - y_b}{(\partial y(t)/\partial m)(b, m_i)} = m_i - \frac{v_1(b) - y_b}{v_3(b)} \tag{10.2.19}$$

where the new state variables are $v_1(t) = y(t, m_i)$, $v_2(t) = y'(t, m_i)$, $v_3(t) = \partial y(t, m_i)/\partial m$, and $v_4(t) = \partial y'(t, m_i)/\partial m$. The initial value problem of the following auxiliary differential equations can be posed:

$$\begin{cases} v_1'(t) = v_2(t), \\ v_2'(t) = F(t, v_1(t), v_2(t)), \\ v_3'(t) = v_4(t), \\ v_4'(t) = \dfrac{\partial}{\partial y} F(t, v_1(t), v_2(t)) v_3(t) + \dfrac{\partial}{\partial y'} F(t, v_1(t), v_2(t)) v_4(t) \end{cases} \tag{10.2.20}$$

where $v_1(a) = y_a$, $v_2(a) = m$, $v_3(a) = 0$, and $v_4(a) = 1$. It is required that $\partial F/\partial y$ and $\partial F/\partial y'$ are found explicitly. In the computation, a value m is assigned, and the initial value problem in (10.2.20) can be solved. Substituting the result into (10.2.19) and iterating once, the new value of m is obtained and substituted into (10.2.20), and we continue computing. If the difference of the values of m obtained in two steps lies in an allowed region, then either of these m values can be used to solve the initial value problem in (10.2.18). The first two terms in the auxiliary equations are the explicit form of the differential equations. The latter two can be computed with symbolic manipulation.

Based on the above algorithm, the following MATLAB function can be implemented. It can be called when directly solving the boundary value problems of second-order nonlinear differential equations.

```
function [t,y]=nlbound(funcn,funcv,tspan,x0f,tol)
if nargin==4, tol=100*eps; end
ga=x0f(1); gb=x0f(2); m=1; m0=0;
ff=odeset; ff.RelTol=tol;        % set the solution precision
while (norm(m-m0)>tol), m0=m; % shooting with a loop
   [t,v]=ode45(funcv,tspan,[ga;m;0;1],ff);
   m=m0-(v(end,1)-gb)/(v(end,3));  % implement (10.2.19)
end
[t,y]=ode45(funcn,tspan,[ga;m],ff);
```

The syntax of the function is

[$\boldsymbol{t}$,$\boldsymbol{y}$]=nlbound(f_1,f_2,tspan,$\boldsymbol{y}_0$,tol)

where f_1 and f_2 are respectively the function handles of the explicit and the auxiliary differential equations. The argument tol is the relative error tolerance, with default value of 100eps. Examples are used next to demonstrate the solution process.

Example 10.3. Solve the boundary value problem of the nonlinear differential equation

$$y''(t) = F(t, y(t), y'(t)) = 2y(t)y'(t), \quad y(0) = -1,\ y(\pi/2) = 1.$$

Solutions. The partial derivatives $\partial F/\partial y = 2y'(t)$ and $\partial F/\partial y' = 2y(t)$ can easily be found. Substituting them into (10.2.20), the fourth equation $v_4'(t) = 2x_2(t)x_3(t) + 2x_1(t)x_4(t)$ can be constructed. Therefore anonymous functions can be used to describe the original and auxiliary differential equations.

```
>> f1=@(t,x)[x(2); 2*x(1)*x(2)];
   f2=@(t,v)[f1(t,v(1:2)); v(4); 2*v(2)*v(3)+2*v(1)*v(4)];
```

MATLAB commands can be used to solve the original problem, and the solution curves are as shown in Figure 10.2. It can be seen that the boundary values of $x_1(t)$ satisfy the given conditions.

```
>> [t,y]=nlbound(f1,f2,[0,pi/2],[-1,1]); plot(t,y)
```

For this simple differential equation, the following commands can be used to find the analytical solution:

```
>> syms t y(t)
   y=dsolve(diff(y,2)==2*y*diff(y),y(0)==-1,y(pi/2)==1)
```

The analytical solution is $y(t) = \tan(t - \pi/4)$. Therefore, the following commands can be used to assess whether the solutions are accurate or not. The norm of the error is

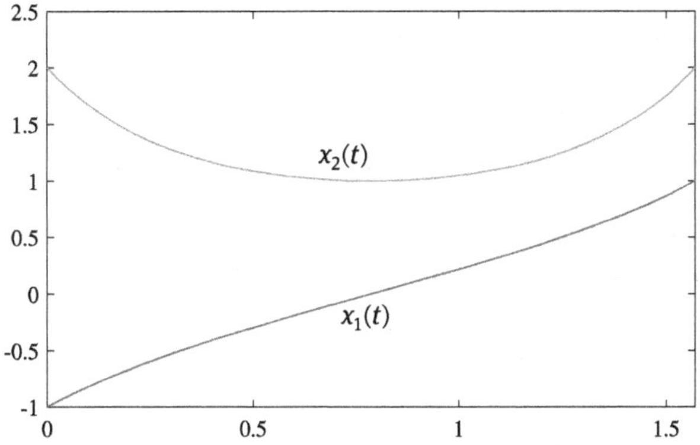

Figure 10.2: Solution of a two-point boundary value problem.

found to be 1.3068×10^{-13}, meaning that high precision solutions can be found.

```
>> y0=tan(t-pi/4); e1=norm(y(:,1)-y0)
```

Example 10.4. Solve the boundary value problem for the linear time-varying differential equation in Example 10.1.

Solutions. The explicit standard form is

$$y''(t) = F(t, y(t), y'(t)) = -p(t)y'(t) - q(t)y(t) + f(t).$$

Selecting the state variables $y_1(t) = y(t)$ and $y_2(t) = y'(t)$, the corresponding state equation can be written as

$$\boldsymbol{y}'(t) = \begin{bmatrix} y_2(t) \\ -p(t)y_2(t) - q(t)y_1(t) + f(t) \end{bmatrix}. \tag{10.2.21}$$

The partial derivatives $\partial F/\partial y$ and $\partial F/\partial y'$ can be written respectively as

$$\frac{\partial F}{\partial y} = -q(t), \quad \frac{\partial F}{\partial y'} = -p(t).$$

Therefore the auxiliary equation can be written as

$$\boldsymbol{v}'(t) = \begin{bmatrix} v_2(t) \\ -p(t)v_2(t) - q(t)v_1(t) + f(t) \\ v_4(t) \\ -q(t)v_3(t) - p(t)v_4(t) \end{bmatrix}.$$

With the following anonymous functions, the two equations can be described, and then the boundary value problem can be solved. The norm of the error is 1.3011 ×

10^{-12}, and the elapsed time is 0.14 seconds. It can be seen that the method here is the most efficient.

```
>> D=@(t)(sin(t+1)+2)^2;
   p=@(t)-2*(cos(t+1)+2)^2/D(t); q=@(t)-16*sin(t+1)/D(t);
   f=@(t)4*exp(-2*t)*(4*cos(t+1)+9)/D(t);
   F1=@(t,y)[y(2); -p(t)*y(2)-q(t)*y(1)+f(t)]; %original equation
   F2=@(t,v)[F1(t,v(1:2));
             v(4); -q(t)*v(3)-p(t)*v(4)];        %auxiliary equation
   ff=odeset; ff.AbsTol=100*eps; ff.RelTol=100*eps;
   tic, [t,y]=nlbound(F1,F2,[0,pi],[1,exp(-2*pi)],ff); toc
   y0=exp(-2*t); plot(t,y,t,y0,'--'), norm(y(:,1)-y0)
```

It should be noted that for complicated differential equations, evaluating the derivatives with respect to $y(t)$ and $y'(t)$ may become too tedious. Therefore the auxiliary equations cannot be easily constructed. Besides, the method given here has limitations. It cannot be extended to solve high-order differential equations. Complicated boundary value conditions are not easily described.

In real applications, the numerical solutions of complicated differential equations with difficult boundary conditions are expected. On the one hand, high-order differential equations are to be solved; on the other hand, more complicated boundary conditions are to be tackled. The idea given here is to convert the original problem into an initial value problem. Several algebraic equations are to be solved. If the number of unknowns equals the number of equations, algebraic equations must be solved, then the differential equations can be solved. Compared with typical boundary value problems, the differential equations studied here will be more general. Apart from traditional boundary value problems, the equations with undetermined coefficients are also considered.

In the next section, the general purpose solvers provided in MATLAB are discussed. The objective is to solve the two-point boundary value problems of high-order differential equations. The solution patterns are made more straightforward and more general purpose.

10.3 Two-point boundary value problems for high-order differential equations

The above mentioned low-level algorithms have certain limitations, since only second-order differential equations can be handled, and also the formulas of the boundary conditions are restricted. High-order differential equations with the general form of boundary conditions in (10.1.2) cannot be solved. Besides, differential equations with undetermined coefficients cannot be solved.

Definition 10.4. The so-called two-point boundary value problem is that if the solution interval is (a, b), the given conditions should be the function values at terminal points a and b. The forms of boundary conditions can be different. They can be generally described as in (10.1.2).

The methods discussed so far contained the solvers from low-level algorithms. In this section, a more general differential equations solver is presented to deal with various boundary value problems.

10.3.1 Direct solver in MATLAB

The solver `bvp5c()` provided in MATLAB is a very good one. It can be used to solve well many boundary value problems. To solve a complicated boundary value problem of differential equations, the following procedures are used:

(1) Parameter interpolation. Function `bvpinit()` can be used to input information. Of course, it is not limited to describing the boundary values, other undetermined coefficients can also be handled.

The syntax of the function is

sinit=bvpinit($\boldsymbol{v}$,$\boldsymbol{x}_0$,$\boldsymbol{\theta}_0$)

where $\boldsymbol{v}$ is the vector of testing time, created by $\boldsymbol{v}$=linspace(a,b,n), or a colon expression. Note that to ensure fast speed, n should not be set too large. Normally, $n = 5$ is fine. Apart from vector $\boldsymbol{v}$, the initial values of the states $\boldsymbol{x}_0$ and the interpolation of the undetermined coefficients $\boldsymbol{\theta}_0$ should also be provided.

The undetermined coefficient problems will be presented in detail later. If there are no undetermined coefficients, $\boldsymbol{\theta}_0$ can be omitted.

(2) MATLAB description of differential equations and boundary conditions. The descriptions of differential equations are the same as those in initial value problems. Boundary conditions in (10.1.2) will be demonstrated later through examples.

(3) Boundary value problem solutions. Function `bvp5c()` can be called directly to solve the boundary value problems:

```
sol=bvp5c(fun1,fun2,sinit,options)
```

where `fun1` and `fun2` are respectively the function handles of differential equations and boundary conditions, which can be anonymous or MATLAB functions. The `options` are virtually the same as used in `ode45()` solver. The returned `sol` is a structured variable, with sol.$\boldsymbol{x}$ member representing $\boldsymbol{t}$ row vector, sol.$\boldsymbol{y}$ is a matrix whose rows contain the information of each state variable. The member `sol.parameters` returns the undetermined coefficients $\boldsymbol{\theta}$, if any.

Examples are introduced later to demonstrate the syntaxes of the functions, and also the solution process. If differential equations are to be solved, two MATLAB functions must be written to describe respectively the first-order explicit differential equations and boundary conditions. The former is exactly the same as for initial value problems. Another MATLAB function is still needed to describe the boundary conditions. Details are presented later in examples.

10.3.2 Solution of simple boundary value problems

The second-order differential equations studied in Section 10.2 can also be solved with `bvp5c()` function. Of course, this function can also be used in handling high-order differential equations. In this section, the earlier examples are solved again, and the accuracy and efficiency are assessed.

Example 10.5. Use the `bvp5c()` solver to solve again the boundary value problem

$$y''(t) = F(t, y(t), y'(t)) = 2y(t)y'(t), \quad y(0) = -1, \ y(\pi/2) = 1.$$

Solutions. Letting $x_1(t) = y(t)$ and $x_2(t) = y'(t)$, the first-order explicit differential equations can be obtained as follows. Anonymous functions can be used to describe the differential equations

$$\boldsymbol{x}'(t) = \begin{bmatrix} x_2(t) \\ 2x_1(t)x_2(t) \end{bmatrix}.$$

We concentrate next on observing the boundary conditions, and show how to describe them. The two terminal points of interest are denoted by $a = 0$ and $b = \pi/2$. The boundary conditions can be written as

$$x_1(a) + 1 = 0, \quad x_1(b) - 1 = 0.$$

Simpler notations are defined in MATLAB, where the two boundary values can be written as

$$\boldsymbol{x}_a(1) + 1 = 0, \quad \boldsymbol{x}_b(1) - 1 = 0.$$

Here $\boldsymbol{x}_a$ is the state variable vector at time a, while $\boldsymbol{x}_b$ is the state at time b. The notation "(1)" here means the first component in the state vector. Note that the expression here is only a notation.

The MATLAB descriptions of the differential equations and the boundary conditions are presented as

```
>> f1=@(t,x)[x(2); 2*x(1)*x(2)];
   f2=@(xa,xb)[xa(1)+1; xb(1)-1]; %equations and boundary values
```

With the descriptions on the differential equations and boundary conditions, the following statements can be used to solve the boundary value problem directly. Five interpolation points can be selected, and the results are identical to those in Figure 10.2. If the analytical solution is used for comparison, it can be seen the norm of the error is 1.2331×10^{-14}. The solution is indeed the most accurate solution under the double precision data type.

```
>> ff=odeset; ff.RelTol=100*eps; ff.AbsTol=100*eps;
   N=5; S1=bvpinit(linspace(0,pi/2,N),rand(2,1));
   s1=bvp5c(f1,f2,S1,ff);  %select 5 interpolation points
   t=s1.x; y1=s1.y; y0=tan(t-pi/4);
   plot(s1.x,s1.y,t,y0); norm(y1(1,:)-y0)%compare the solutions
```

Example 10.6. Solve again the problem in Example 10.1 with the provided solver.

Solutions. The first-order explicit differential equations can be found from the original differential equations as

$$\boldsymbol{x}'(t) = \begin{bmatrix} x_2(t) \\ -p(t)x_2(t) - q(t)x_1(t) + f(t) \end{bmatrix}.$$

The boundary conditions can be written as

$$\boldsymbol{x}_a(1) - 1 = 0, \quad \boldsymbol{x}_b(1) - \mathrm{e}^{-2\pi} = 0.$$

Therefore anonymous functions can be used to describe the differential equations and boundary conditions, and the solver `bvp5c()` can be called to find the solutions directly:

```
>> D=@(t)(sin(t+1)+2)^2; p=@(t)-2*(cos(t+1)+2)^2/D(t);
   q=@(t)-16*sin(t+1)/D(t); f=@(t)4*exp(-2*t)*(4*cos(t+1)+9)/D(t);
   F=@(t,x)[x(2); -p(t)*x(2)-q(t)*x(1)+f(t)];
   G=@(xa,xb)[xa(1)-1; xb(1)-exp(-2*pi)];
   N=5; S=bvpinit(linspace(0,pi,N),rand(2,1));
   tic, sol=bvp5c(F,G,S,ff); toc               %boundary values
   t=sol.x; y1=sol.y; y0=exp(-2*t);            %analytical solution
   plot(t,y1,t,y0,'--'); norm(y1(1,:)-y0)  %compare solutions
```

The elapsed time is 0.76 seconds and the norm of the error is 3.4325×10^{-15}, which is more accurate than for the other methods. The efficiency is also higher than for the solvers in Example 10.1.

Example 10.7. Solve the high-order differential equation problem:[35]

$$\begin{cases} y_1''(t) = \lambda^2 y_1(t) + y_2(t) + t + (1-\lambda^2)\mathrm{e}^{-t}, \\ y_2''(t) = -y_1(t) + \mathrm{e}^{y_2(t)} + \mathrm{e}^{-\lambda t} \end{cases}$$

where the boundary conditions are $y_1(0) = 2$, $y_2(0) = 0$, $y_1(1) = \mathrm{e}^{-\lambda}+\mathrm{e}^{-1}$, and $y_2(1) = -1$. The analytical solution is $y_1(t) = \mathrm{e}^{-\lambda t} + \mathrm{e}^{-t}$ and $y_2(t) = -t$. If $\lambda = 20$, find the solution for $t \in (0, 1)$.

Solutions. As in the case of other methods discussed so far, if high-order differential equations are solved, the first thing to do is to convert the original differential equations into first-order explicit differential equations. Selecting a set of state variables as $x_1(t) = y_1(t)$, $x_2(t) = y_1'(t)$, $x_3(t) = y_2(t)$, and $x_4(t) = y_2'(t)$, the standard form can be composed:

$$\boldsymbol{x}'(t) = \begin{bmatrix} x_2(t) \\ \lambda^2 x_1(t) + x_3(t) + t + (1-\lambda^2)\mathrm{e}^{-t} \\ x_4(t) \\ -x_1(t) + \mathrm{e}^{x_3(t)} + \mathrm{e}^{-\lambda t} \end{bmatrix}.$$

Besides, assuming that a and b are the terminal points, the boundary conditions can be written as

$$\boldsymbol{x}_a(1) - 2 = 0, \quad \boldsymbol{x}_a(3) = 0, \quad \boldsymbol{x}_b(1) - \mathrm{e}^{-\lambda} + \mathrm{e}^{-1} = 0, \quad \boldsymbol{x}_b(3) + 1 = 0.$$

Expressing the differential equations and boundary conditions with anonymous functions, the following commands can be written to solve the problem. The solutions are shown in Figure 10.3, with analytical solutions superimposed. It can be seen that the two solutions are very close, and the norm of the error matrix is 1.3573×10^{-14}.

```
>> lam=20;
   f=@(t,x)[x(2); lam^2*x(1)+x(3)+t+(1-lam^2)*exp(-t);
              x(4); -x(1)+exp(x(3))+exp(-lam*t)];
   g=@(xa,xb)[xa(1)-2; xa(3); xb(1)-exp(-lam)-exp(-1); xb(3)+1];
```

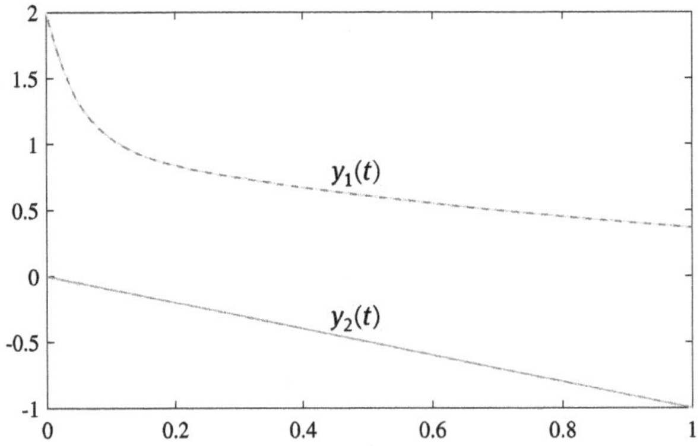

Figure 10.3: Solution of the boundary value problem.

```
ff=odeset; ff.RelTol=100*eps; ff.AbsTol=100*eps;
N=10; S1=bvpinit(linspace(0,1,N),rand(4,1));
s1=bvp5c(f,g,S1,ff);                 % 10 interpolations are selected
t=s1.x; y=s1.y; y1=exp(-lam*t)+exp(-t); y2=-t;
plot(t,y([1,3],:),t,y1,'--',t,y2,'--');
norm([y(1,:)-y1; y(3,:)-y2])  % find the norm of the error matrix
```

In the solution process, a warning message "Warning: Unable to meet the tolerance without using more than 2500 mesh points" is displayed, claiming that the mesh grid number n is not sufficient. The value of n can be set to 3 000, yet, when solving the problem again, the warning message persists. The norm of error matrix is increased to 0.0013. It is obvious that the result is incorrect. Therefore the warning like this can be neglected.

```
>> N=3000; S1=bvpinit(linspace(0,1,N),rand(4,1));
   s1=bvp5c(f,g,S1,ff);               % select 3000 interpolation points
   t=s1.x; y=s1.y; y1=exp(-lam*t)+exp(-t); y2=-t;
   norm([y(1,:)-y1; y(3,:)-y2]) % find the norm of the error matrix
```

Example 10.8. Solve the following boundary value problem:[2]

$$(x^3u''(x))'' = 1, \quad 1 \leqslant x \leqslant 2$$

where $u(1) = u''(1) = u(2) = u''(2) = 0$, and the analytical solution is

$$u(x) = \frac{1}{4}(10\ln 2 - 3)(1-x) + \frac{1}{2}\left[\frac{1}{x} + (3+x)\ln x - x\right].$$

Solutions. It is obvious that the differential equations like this cannot be solved. They should be converted first into the standard form. One way, of course, is to derive the equation manually, but a better way is to derive the left-hand side of the equation with symbolic commands.

```
>> syms c u(x)
   diff(x^3*diff(u,2),2), latex(ans) % derive the left-hand side expression
```

With these commands, the mathematical form is found as

$$x^3u^{(4)}(x) + 6x^2u'''(x) + 6xu''(x) = 1.$$

Since the interval is $x \in (1,2)$, $x \neq 0$, the explicit form can be found immediately as

$$u^{(4)}(x) = -\frac{6}{x}u'''(x) - \frac{6}{x^2}u''(x) + \frac{1}{x^3}.$$

Selecting $y_1(x) = u(x), y_2(x) = u'(x), y_3(x) = u''(x)$, and $y_4(x) = u'''(x)$, the standard form can be derived as

$$y'(x) = \begin{bmatrix} y_2(x) \\ y_3(x) \\ y_4(x) \\ -6y_4(x)/x - 6y_3(x)/x^2 + 1/x^3 \end{bmatrix}.$$

The standard form of the boundary conditions is

$$y_a(1) = 0, \quad y_a(3) = 0, \quad y_b(1) = 0, \quad y_b(3) = 0.$$

Calling the following statements, the equations can be solved directly. The elapsed time is 0.33 seconds, and the error is 2.4773×10^{-15}. The curves of $u(x)$ and $u''(x)$ are shown in Figure 10.4. The $u(x)$ curve is exactly the same as the theoretical one.

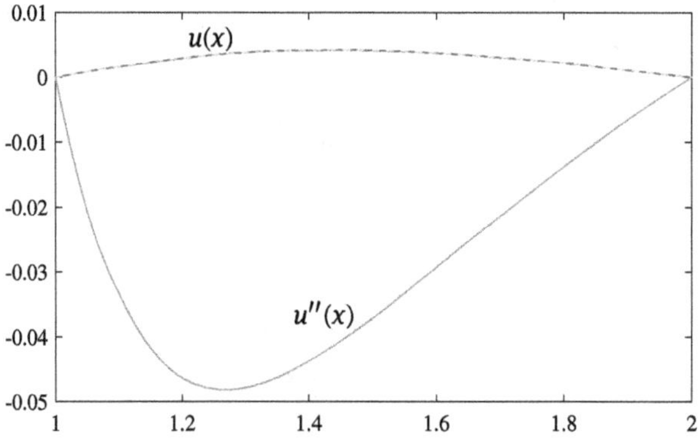

Figure 10.4: Solution of the differential equations.

```
>> f=@(x,y)[y(2:4); -6*y(4)/x-6*y(3)/x^2+1/x^3];
   g=@(ya,yb)[ya(1); ya(3); yb(1); yb(3)];
   ff=odeset; ff.RelTol=100*eps; ff.AbsTol=100*eps;
   tic, N=10; x0=rand(4,1); S1=bvpinit(linspace(1,2,N),x0);
   s1=bvp5c(f,g,S1,ff); x=s1.x; y=s1.y; toc
   y0=(10*log(2)-3)*(1-x)/4+(1./x+(3+x).*log(x)-x)/2;
   plot(x,y([1,3],:),x,y0,'--'); norm(y(1,:)-y0)
```

10.3.3 Descriptions of complicated boundary conditions

Solutions with simple boundary conditions were found in the previous presentation. If the boundary conditions are complicated, such as the one in (10.1.2), the boundary conditions should be modified carefully into the standard forms, such that the solution process can be performed. For differential equations to have unique solutions, the number of boundary conditions must equal the number of unknowns plus the number of undetermined coefficients. Examples are shown next to demonstrate the description of complicated boundary conditions.

Example 10.9. Consider again the problem in Example 10.2. If the boundary conditions are changed into the following form:

$$y'(\pi/2) = 1, \quad y(\pi/2) - y(0) = 1,$$

function `bvp5c()` can still be used to solve the boundary value problem. Solve it again.

Solutions. If the symbols a and b are used to indicate the terminal points, the boundary conditions under MATLAB notation can be written as

$$\boldsymbol{x}_b(2) - 1 = 0, \quad \boldsymbol{x}_b(1) - \boldsymbol{x}_a(1) - 1 = 0.$$

Therefore the boundary condition function `f2` can also be described, and the following statements can be used to solve the differential equations. The results are as shown in Figure 10.5:

```
>> f1=@(t,x)[x(2); 2*x(1)*x(2)];
   f2=@(xa,xb)[xb(2)-1; xb(1)-xa(1)-1];
   ff=odeset; ff.RelTol=100*eps; ff.AbsTol=100*eps;
   N=5; S1=bvpinit(linspace(0,pi/2,N),rand(2,1));
   s3=bvp5c(f1,f2,S1,ff); plot(s3.x,s3.y)
```

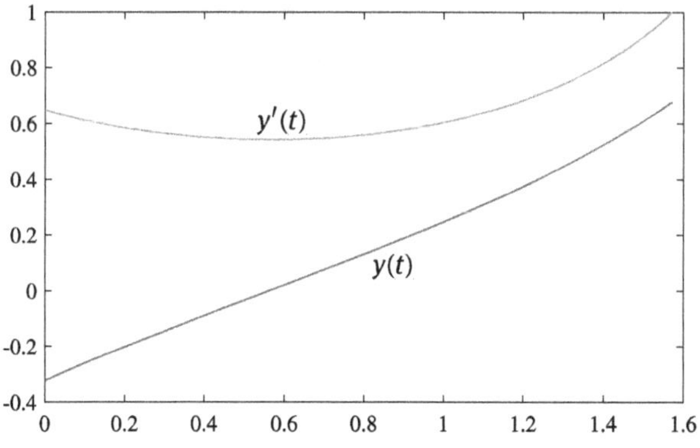

Figure 10.5: Solution of the differential equations under new conditions.

10.3.4 Boundary value problems with undetermined coefficients

If a boundary value problem has undetermined coefficients, for instance, the undetermined coefficient λ in Example 10.7, the solver `bvp5c()` can still be used in solving the differential equations. The solution of the equations as well as the undetermined coefficients are found. An extra boundary condition is needed for each undetermined coefficient. If the number of boundary conditions is not adequate, the differential equations cannot be solved. In this section, the differential equations with undetermined parameters are studied through examples.

Example 10.10. Consider the problem in Example 10.7. If λ is unknown, five boundary conditions are needed to solve the problem. Assume that an extra condition $y_1'(0) = -6$ is introduced, and the $y_1(1)$ condition is changed to $y_1(1) = \mathrm{e}^{-5} + \mathrm{e}^{-1}$. Solve again the boundary value problem.

Solutions. If λ is unknown, then λ is set to the undetermined parameter v. The differential equations are then rewritten as

$$\boldsymbol{x}'(t) = \begin{bmatrix} x_2(t) \\ v^2 x_1(t) + x_3(t) + t + (1 - v^2)\mathrm{e}^{-t} \\ x_4(t) \\ -x_1(t) + \mathrm{e}^{x_3(t)} + \mathrm{e}^{-vt} \end{bmatrix}.$$

The standard form of the five boundary conditions is

$$\boldsymbol{x}_a(1) - 2 = 0, \quad \boldsymbol{x}_a(3) = 0, \quad \boldsymbol{x}_b(1) - \mathrm{e}^{-5} - \mathrm{e}^{-1} = 0, \quad \boldsymbol{x}_b(3) + 1 = 0, \quad \boldsymbol{x}_a(2) + 6 = 0.$$

With the information, the problem is described by anonymous functions, and the solutions of the differential equations can be found. For better solving the original problem, the number of interpolation points, n, is slightly increased. Note that when describing the differential equations and boundary conditions, apart from the fixed arguments, an extra parameter $\boldsymbol{v}$ is used.

```
>> f=@(t,x,v)[x(2); v^2*x(1)+x(3)+t+(1-v^2)*exp(-t);
              x(4); -x(1)+exp(x(3))+exp(-v*t)];
   g=@(xa,xb,v)[xa(1)-2; xa(3); xb(1)-exp(-5)-exp(-1); ...
                xb(3)+1; xa(2)+6];
   ff=odeset; ff.RelTol=100*eps; ff.AbsTol=100*eps;
   N=30; x0=rand(4,1); v0=rand(1);
   S1=bvpinit(linspace(0,1,N),x0,v0); s1=bvp5c(f,g,S1,ff);
   t=s1.x; y=s1.y; y1=exp(-5*t)+exp(-t); y2=-t;
   plot(t,y([1,3],:),t,y1,'--',t,y2,'--');
   norm([y(1,:)-y1; y(3,:)-y2]), s1.parameters %compute error
```

For this particular problem, there is an about 50 % chance that the exact solution of the equations can be found. That is, $\lambda = 5$. The error found is 1.0819×10^{-14}. This is the best solution. Sometimes unsatisfactory results are found. For instance, with negative values of λ. Sometimes an error message "Unable to solve the collocation equations – a singular Jacobian encountered". If this happens, the solver can be called again. The commands should be called repeatedly, to find the correct results.

In real applications, when such problems are solved, while the analytical solution is not known, the code can be executed repeatedly. If in several executions, the results are the same, they can probably be the genuine solution of the problem.

Example 10.11. For the differential equations given below, find the constants α and β, and solve the differential equations:

$$\begin{cases} x'(t) = 4x(t) - \alpha x(t)y(t), \\ y'(t) = -2y(t) + \beta x(t)y(t) \end{cases}$$

with $x(0) = 2$, $y(0) = 1$, $x(3) = 4$, and $y(3) = 2$.

Solutions. The state variables $x_1(t) = x(t)$ and $x_2(t) = y(t)$ can be selected. Besides, letting $v_1 = \alpha$ and $v_2 = \beta$, the original problem is converted into the following standard form:

$$\boldsymbol{x}'(t) = \begin{bmatrix} 4x_1(t) - v_1x_1(t)x_2(t) \\ -2x_2(t) + v_2x_1(t)x_2(t) \end{bmatrix}.$$

Denoting $a = 0$ and $b = 3$, the boundary conditions are

$$\boldsymbol{x}_a(1) - 2 = 0, \quad \boldsymbol{x}_a(2) - 1 = 0, \quad \boldsymbol{x}_b(1) - 4 = 0, \quad \boldsymbol{x}_b(2) - 2 = 0.$$

Therefore the following commands can be used to describe the differential equations and boundary conditions:

```
>> f=@(t,x,v)[4*x(1)+v(1)*x(1)*x(2);
              -2*x(2)+v(2)*x(1)*x(2)];
   g=@(xa,xb,v)[xa(1)-2; xa(2)-1; xb(1)-4; xb(2)-2];
```

The function `bvpinit()` can be called for the initialization process first, for the time grids, and parameters α and β. Since there are two initial states and two undetermined coefficients, they can both be set to `rand(2,1)`. With these parameters, function `bvp5c()` can be called to determine the parameters α and β and solve the boundary value problem. The results of the two states are obtained as shown in Figure 10.6.

```
>> x1=[1;1]; v1=[-1;1];
   sinit=bvpinit(linspace(0,3,20),x1,v1); %interpolation
```

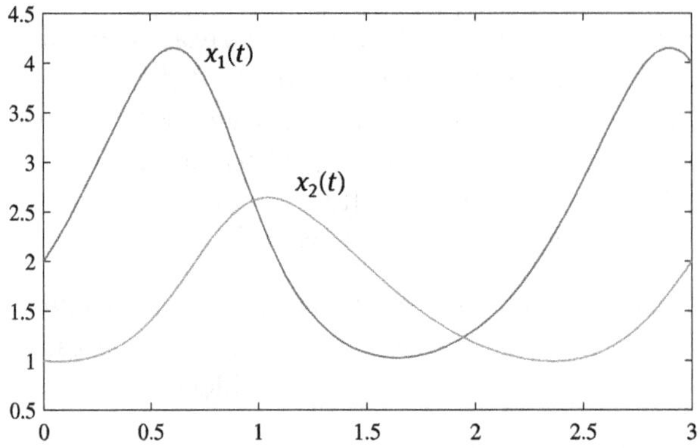

Figure 10.6: Solution of the differential equations.

```
ff=odeset; ff.RelTol=100*eps; ff.AbsTol=100*eps;
sol=bvp5c(f,g,sinit,ff); sol.parameters %undetermined coefficients
plot(sol.x,sol.y);                      %time domain responses
```

Meanwhile, the undetermined coefficients found are $\alpha = -2.3721$ and $\beta = 0.8934$. It can be seen from the simulation curve that the boundary conditions are also satisfied. Therefore the solution obtained is correct. In the selection of $\boldsymbol{x}_1$ and $\boldsymbol{v}_1$, it should be noted that if they are not chosen properly, singular Jacobian matrices may be found, and the solution process is unsuccessful. If that happens, another set of initial values should be chosen.

10.3.5 Boundary value problems with semi-infinite intervals

The boundary value problems discussed so far were defined on the interval $[a, b]$, with a, b being given numbers. In this section, semi-infinite interval boundary value problems are explored.

Definition 10.5. If the boundary value problem is defined on an infinite interval, that is, $a = -\infty$ or $b = \infty$, the boundary value problem is referred to as semi-infinite interval problem.

For ordinary semi-infinite boundary value problems, it seems that there are no good solution methods in mathematics. The usual way is to select a large value of L to replace infinity terms. Of course, when the solutions are found, their curves should be examined to see whether at L the curves are flat. If they are, this means that the solution is valid, otherwise it means that the selected L may be too small. The value of L should be increased and one should try again. In this section, semi-infinite boundary value problems are explored.

Example 10.12. Consider the differential equations in Example 10.1. If the boundary conditions are $y(0) = 1$ and $y(\infty) = 0$, solve the differential equations.

Solutions. The state variables $y_1(t) = y(t)$ and $y_2(t) = y'(t)$ can be selected first. The state space equation in (10.2.21) can be constructed. Besides, selecting a stoping time of $t = L$, the boundary condition can be written as

$$y_a(1) - 1 = 0, \quad y_b(1) = 0.$$

Letting $L = 20$, the following statements can be used to approximately solve the problem. The norm of the error between analytical and numerical solution can be found as 3.2678×10^{-14}. The solutions are shown in Figure 10.7. It can be seen that the selection of the stoping time is successful, since when $L = 20$, the $y(t)$ curve becomes flat.

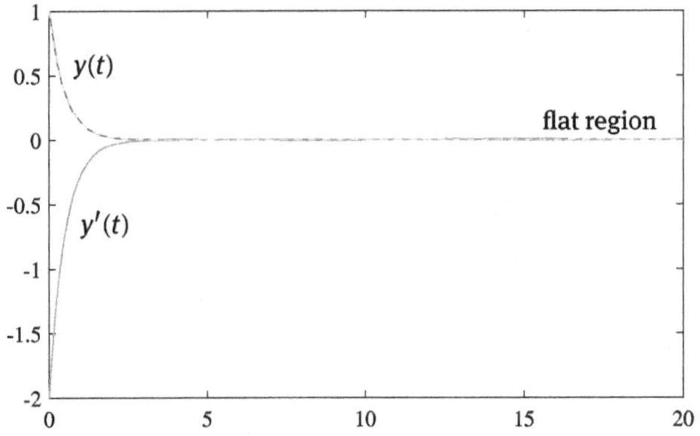

Figure 10.7: Numerical solution.

```
>> D=@(t)(sin(t+1)+2)^2;
   p=@(t)-2*(cos(t+1)+2)^2/D(t); q=@(t)-16*sin(t+1)/D(t);
   f=@(t)4*exp(-2*t)*(4*cos(t+1)+9)/D(t);
   F=@(t,y)[y(2); -p(t)*y(2)-q(t)*y(1)+f(t)];
   G=@(ya,yb)[ya(1)-1; yb(1)]; L=20; x0=rand(2,1);
   sinit=bvpinit(linspace(0,L,10),x0);
   ff=odeset; ff.RelTol=100*eps; ff.AbsTol=100*eps;
   sol=bvp5c(F,G,sinit,ff); t=sol.x; y=sol.y;
   y0=exp(-2*t); plot(t,y,t,y0,'--'), norm(y(1,:)-y0)
```

In fact, if the analytical solution is not known, different values of L can be tried, and one can see whether consistent solutions are found. If they are not consistent, a larger value of L can be tried, and solutions should be validated again.

For this particular example, if $L = 5$ is selected, the norm of the error is increased to 1.2695×10^{-4}. The error is large, indicating that L is not chosen properly. It should be increased.

```
>> L=5; x0=rand(2,1);
   sinit=bvpinit(linspace(0,L,10),x0);
   ff=odeset; ff.RelTol=100*eps; ff.AbsTol=100*eps;
   sol=bvp5c(F,G,sinit,ff); t=sol.x; y=sol.y;
   y0=exp(-2*t); plot(t,y,t,y0,'--'), norm(y(1,:)-y0)
```

It should be noted that when $L = 20$, the initial value problem solution from the theoretical initial value $\boldsymbol{x}_0 = [1, -2]^{\mathrm{T}}$ may cause problems. The solution is divergent. If $L = 10$, the forward simulation is successful. Therefore the problem is sensitive to initial values. In numerical solution studies, care must be taken.

```
>> L=20;
   [t,x]=ode45(F,[0,L],[1;-2],ff); plot(t,x(:,1))
```

10.3.6 Differential equations with floating boundary values

It has been indicated earlier that for linear differential equations with m unknown functions and undetermined coefficients, there should be m boundary conditions to uniquely determine the solutions. In theory, the m boundary conditions must be linearly independent. In real problems, it is not possible to judge whether the conditions are linearly independent or not. Besides, if the m equations are nonlinear algebraic equations, it is hard to ensure the uniqueness of the solutions. Sometimes the differential equations have more than one solution.

In this section, a special example is explored. In the example, the initial and terminal values are not given as fixed numbers. The relationship between them is known. The problem is referred to as a boundary value problem with floating bounds. An example is used to show that the solver `bvp5c()` fails to find solutions. In later sections, effective methods are introduced.

Example 10.13. Solve the following boundary value problem:

$$u''(t) = \min(u(t) + 2, \max(-u(t), u(t) - 2))$$

where $u(0) = u(2\pi)$ and $u'(0) = u'(2\pi)$. In [19], the analytical solution of the equation is given as $u(t) = \sin t$ and $u = -\sin t$.

Solutions. Selecting the state variables as $x_1(t) = u(t)$ and $x_2(t) = u'(t)$, the standard form of first-order explicit differential equations is written as

$$\boldsymbol{x}'(t) = \begin{bmatrix} x_2(t) \\ \min(x_1(t) + 2, \max(-x_1(t), x_1(t) - 2)) \end{bmatrix}.$$

With the following commands, the differential equations and boundary conditions are described. Function bvp5c() can be tried to solve the differential equations. However, the solution found is $u(t) \equiv 0$, and the number of points found is the same as the value of N. No matter how may times the commands are tried, the solution is the same. It should be noted that the analytical solution listed in [19] is not complete. Indeed, $u(t) \equiv 0$ is also a solution of the original differential equations. With the function call here, no other solutions can be found, which implies that for this particular problem, the general-purpose solver fails.

```
>> f=@(t,x)[x(2); min(x(1)+2,max(-x(1),x(1)-2))];
   g=@(xa,xb)[xa(1)-xb(1); xa(2)-xb(2)];
   ff=odeset; ff.RelTol=100*eps; ff.AbsTol=100*eps;
   N=5; S=bvpinit(linspace(0,2*pi,N),rand(2,1));
   sol=bvp5c(f,g,S,ff); t=sol.x; y=sol.y(1,:); size(t)
```

In Section 10.4, the same problem will be revisited again, and we will indicate that differential equations might have infinitely many solutions. The solver bvp5c() fails to find them.

10.3.7 Boundary value problems of integro-differential equations

In Section 4.5, some integro-differential equations were studied. Some of the equations could be converted into ordinary differential equations so that MATLAB could be used. In this section, an example is used to show the transformation and solution method for some boundary value problems for integro-differential equations.

Example 10.14. Consider the boundary value problem of the following integro-differential equation:[18]

$$x'(t) = x(t) - \frac{1}{1500}x^2(t) + \frac{1}{3000}\int_{-1}^{1} e^{2(t-s)}x^2(s)ds \tag{10.3.1}$$

where the boundary values are $x(-1) = 1$ and $x(1) = e^2$. It is known that the exact solution is $x(t) = e^{1+t}$. Convert the equation into ordinary differential equation, and find its numerical solution. Assess the precision of the numerical solutions.

Solutions. Although the integrals are different from those discussed in Section 4.5, a similar idea from that section can still be used to carry out manipulation. Moving the function of t out of the integrand, it is found from (10.3.1) that the explicit term of the integral can be written as

$$\int_{-1}^{1} \mathrm{e}^{-2s} x^2(s)\mathrm{d}s = \frac{3\,000}{\mathrm{e}^{2t}} \left[x'(t) - x(t) + \frac{1}{1\,500} x^2(t)\right].$$

Taking a derivative of (10.3.1) with respect to t, and substituting the integral expression back to the result

```
>> syms t s x(t)
   F=x-1/1500*x^2+1/3000*exp(2*t)*int(exp(-2*s)*x(s)^2,s,-1,1);
   F1=simplify(subs(diff(F,t),int(exp(-2*s)*x(s)^2,s,-1,1),...
         3000/exp(2*t)*(diff(x,t)-x+1/1500*x^2)))
```

the following second-order differential equation can be derived:

$$x''(t) = \frac{1}{750} x^2(t) - \frac{1}{750} x(t)x'(t) - 2x(t) + 3x'(t).$$

With the differential equation and boundary values, the following statements can be used to solve the boundary value problem directly. Compared with the exact solution, it is found that the norm of the error function is 4.0039×10^{-14}, and the elapsed time is 0.46 seconds. It can be seen that this method can be tried for some integro-differential equations.

```
>> f=@(t,x)[x(2); x(1)^2/750-x(1)*x(2)/750-2*x(1)+3*x(2)];
   g=@(xa,xb)[xa(1)-1; xb(1)-exp(2)];
   ff=odeset; ff.RelTol=100*eps; ff.AbsTol=100*eps;
   N=5; S=bvpinit(linspace(-1,1,N),rand(2,1));
   tic, sol=bvp5c(f,g,S,ff); toc        % solve boundary value problem
   t=sol.x; norm(sol.y(1,:)-exp(t+1)) % compute errors
```

10.4 Optimization-based boundary value problem solutions

In Volume IV of the series, systematic presentation on optimization techniques was given. Generally speaking, in an optimization problem, a meaningful objective function is defined. Then we adjust the values of the decision variables, such that the value of the objective function is minimized.

How to combine the benefits of optimization techniques and the differential equations solution processes together? In the previous chapters, initial value problems were studied, and the algorithms and solvers are mature. If the initial values are given, the differential equations can be solved directly. In boundary value problems, some initial values are known, while some are unknown. The unknown initial values can be selected as the decision variables for the optimization problem. Combined with the idea of a shooting method in Section 10.2, the final value obtained in simulation and the given final value may have a difference. The error between them can be used as the objective function. Therefore the boundary value problem can be converted into an optimization problem. If there is more than one final value, the objective function can be selected as the sum of absolute values of the errors. Through optimization process, the consistent initial values are found, from which the initial value problem can be solved again, yielding the solution of the original boundary value problem.

In this section, a simple boundary value problem is demonstrated first. Then the differential equations which are not suitable to be solved with the solver `bvp5c()` will be explored. These include implicit differential equations and delay differential equations. Simulation based boundary value problems and the boundary value problems of fractional-order differential equations are studied.

10.4.1 Illustrative example to simple boundary value problems

To demonstrate the conversion from boundary value problems into optimization problems, a simple boundary value problem is introduced first. For this simple boundary value problem, the solver `bvp5c()` may be more effective. There is no need to use the conversion process for this particular example.

Example 10.15. Solve the problem in Example 10.8 with the optimization method.

Solutions. It can be seen from the boundary values that the initial values of $y_1(t)$ and $y_3(t)$ are known, whereas the initial values of $y_2(t)$ and $y_4(t)$ are unknown. Therefore the decision variables can be selected as $x_1 = y_2(1)$ and $x_2 = y_4(1)$. Then the differential equations can be solved from the initial values. Of course, the solution obtained in this way may not be the one we are expecting, since the terminal conditions are not necessarily satisfied. There might be errors between the expected terminal values and those obtained by simulation methods. The errors can be used as the objective function. Therefore the MATLAB function given next can be used to compute this objective function. Note that to demonstrate solutions, the expected value "0" is also written into the objective function. In real applications, the term "−0" can be omitted.

```
function y=c10mbvp(x,f,ff)
y0=[0; x(1); 0; x(2)];                    % construct the initial values
[t,ym]=ode45(f,[1,2],y0,ff);              % solve differential equations
y=abs(ym(end,1)-0)+abs(ym(end,3)-0);      % compute the objective function
```

The following commands can be used to solve the boundary value problem. The consistent initial values are $y_2(1) = 0.017132$ and $y_4(1) = -0.5$. The elapsed time is 1.68 seconds, larger than that in Example 10.8. The norm of the error is 4.3324×10^{-15}. It can be seen that the method here is effective. The curves obtained are exactly the same as in Figure 10.4.

```
>> f=@(x,y)[y(2:4); -6*y(4)/x-6*y(3)/x^2+1/x^3];
   ff=odeset; ff.RelTol=100*eps; ff.AbsTol=100*eps;
   F=@(x)c10mbvp(x,f,ff); opt=optimset; opt.TolX=eps;
   tic, x=fminsearch(F,rand(2,1),opt)
   [t,y]=ode45(f,[1,2],[0;x(1);0;x(2)],ff); toc
   y0=(10*log(2)-3)*(1-t)/4+(1./t+(3+t).*log(t)-t)/2;
   norm(y(:,1)-y0), plot(t,y(:,[1 3])) % compute the errors
```

Since there are two terminal conditions, the sum of the absolute errors is used to define the objective function. In classical studies of optimization, many researcher are even interested in using the sum of the squared errors as the objective function, since it is "differentiable". In fact, the selection like this is not quite appropriate, since under the double precision framework, the maximum error tolerance can be set to 10^{-16}. If the squared sum is used as the objective function, the solution process may terminate prematurely at 10^{-8}. Therefore the precision of the solutions may not be as high. In modern optimization solvers, having "differentiable" objective functions is not thus important. If there is no special meaning, this consideration can be neglected. More meaningful objective functions should be selected, so as to better solve practical problems.

It can be seen from this example that the boundary value problem of differential equations can be converted into an optimization problem. Although for this simple example the efficiency of the optimization based solution method is slightly lower than that of the `bvp5c()` solver, `bvp5c()` solver has limitations. For instance, implicit differential equations and delay differential equations with boundary values cannot be solved at all with the `bvp5c()` solver. Optimization methods can be used to solve these differential equations. Several examples are used in this section to explore the boundary value problems for differential equations which cannot be solved with the `bvp5c()` solver.

10.4.2 Boundary value problems for implicit differential equations

Boundary value problems of implicit differential equations cannot be solved with the solvers such as `bvp5c()`. Under the current version of MATLAB, there is no such solver for implicit differential equations.

It can be seen from the previous introduction that, no matter what the boundary conditions, the conversion method discussed above can be used to convert the original problem into an optimization problem. More specifically, the unknown initial values can be selected as decision variables to solve the differential equations, while the solution at the other end can be compared with the given one, and the error can be used as the objective function. Therefore through an optimization technique, the consistent initial values can be found, from which the numerical solution of the differential equations can finally be established. In this section, an example is used to demonstrate the solution of boundary value problems of implicit differential equations.

Example 10.16. Consider the implicit differential equations in Example 5.11. If the given conditions are $x_1(0) = 0$ and $x_2(10) = 0$, solve the differential equations.

Solutions. The objective function of the problem is created as

```
function [y,xd0]=c10mimp1(x,f,ff)
x0=[0; x(1)]; x0F=ones(2,1);
[x0,xd0,f0]=decic_new(f,0,x0,x0F); %compute consistent initial values
[t,x]=ode15i(f,[0,10],x0,xd0,ff); y=abs(x(end,2)-0);
```

Compared with the earlier examples, there are two different points to consider: (1) the function handle f is an implicit differential equation; (2) the consistent initial values of the first-order derivatives are also required, but this is internal. An extra returned argument is left for the vector; (3) the internal call to the solver is different. No matter what the differences, the idea is the same, that is, computing the error between the final values and the expected terminal conditions.

With the following statements, the undetermined initial value $x_2(0) = 0.06266$ can be found. The consistent initial values of the first-order derivatives are $\boldsymbol{x}'(0) = [0.06279, 1.00197]^{\mathrm{T}}$. The total elapsed time is 24.7711 seconds. The curves obtained are very close to those shown in Figure 4.14, indicating that the solution process is successful.

```
>> f=@(t,x,xd)[xd(1)*sin(x(1))+xd(2)*cos(x(2))+x(1)-1;
              -xd(1)*cos(x(2))+xd(2)*sin(x(1))+x(2)];
   ff=odeset; ff.AbsTol=100*eps; ff.RelTol=100*eps;
   F=@(x)c10mimp1(x,f,ff); opt=optimset; opt.TolX=eps;
   tic, x=fminsearch(F,rand(1,1),opt); %optimization
   [~,xd0]=c10mimp1(x,f,ff)              %consistent initial values
   [t,y]=ode15i(f,[0,10],[0; x(1)],xd0,ff); toc
   plot(t,y)                             %solve implicit equations
```

It should be noted that since the final value $x_2(10) = 0$ is different from that in the original example, it is only an approximation. Therefore the value of $x_2(0)$ obtained is not 0, but a small number. The solutions are reasonable. Besides, in each step of the objective function evaluation, an implicit differential equation is solved once. Therefore the solution process is rather time consuming. It is a lucky coincidence that the solution of a boundary value problem for implicit differential equations was found in an acceptable time.

Example 10.17. Now consider the differential equations with multiple solutions in Example 4.9:

$$(y''(t))^2 = 4(ty'(t) - y(t)) + 2y'(t) + 1.$$

If the given conditions are $y(0) = 0$ and $y(1) = 1$, find the analytical and numerical solution and assess the errors in the numerical solution, and also the efficiency in the solution process.

Solutions. Since the original differential equation has analytical solutions, the following commands can be used to solve this differential equation directly:

```
>> syms t y(t); y1d=diff(y);
   yA=dsolve((diff(y,2))^2==4*(t*y1d-y)+2*y1d+1,...
         y(0)==0, y(1)==1);    % analytical solutions
```

The analytical solutions of the differential equations can be found, where there are two branches:

$$y(t) = \frac{t}{2}\left(\frac{1}{4}\left(\sqrt{11} \pm 1\right)^2 - 1\right) \mp \frac{t^2}{4}\left(\sqrt{11} \pm 1\right) + \frac{t^3}{6} + \frac{t^4}{12}.$$

As indicated earlier, it is rather complicated to convert it into explicit forms, since two branches are considered and the numerical solution accuracy is very low. An implicit format is used here. Introducing the state variables $x_1(t) = y(t)$ and $x_2(t) = y'(t)$, the standard form of the implicit differential equation can be written as follows:

$$\begin{bmatrix} x_1'(t) - x_2(t) \\ (x_2'(t))^2 - 4(tx_2(t) - x_1(t)) - 2x_2(t) - 1 \end{bmatrix} = \mathbf{0}$$

with $x_1(0) = 0$ and $x_1(1) = 1$. This problem cannot be solved with the solvers such as `bvp5c()`. An optimization technique should be introduced to find the solutions. The objective function in Example 5.13 can be embedded into the following MATLAB function:

```
function [y,xd0]=c10mimp2(x,f,ff)
x0=[0; x(1)]; x0F=ones(2,1);
[x0,xd0,f0]=decic_new(f,0,x0,x0F); % consistent initial values
[t,x]=ode15i(f,[0,1],x0,xd0,ff); y=abs(x(end,1)-1);
```

With the following statements, the optimization problem can be solved, and the consistent initial value is found, $x_2(0) = 1.8292$. The elapsed time of the solution process is about 3.36 seconds. Compared with the analytical solutions, the norm of the error is 2.4960×10^{-11}. It can be seen that the solution process is feasible.

```
>> f=@(t,x,xd)[xd(1)-x(2);
          xd(2)^2-4*(t*x(2)-x(1))-2*x(2)-1];
   ff=odeset; ff.AbsTol=100*eps; ff.RelTol=100*eps;
   F=@(x)c10mimp2(x,f,ff); opt=optimset; opt.TolX=eps;
   tic, x=fminsearch(F,rand(1,1),opt) % consistent initial values
   [~,xd0]=c10mimp2(x,f,ff)              % derivative initial values
   [t,y]=ode15i(f,[0,1],[0; x(1)],xd0,ff); plot(t,y), toc
   y0=t*((sqrt(11)+1)^2/4-1)/2-t.^2/4*(sqrt(11)+1)+t.^3/6+t.^4/12;
   norm(y(:,1)-y0)
```

Executing the above statements repeatedly, another solution can also be found. The consistent initial value is $x_2(0) = 0.1708$. Of course, sometimes the above commands may also yield error messages like "Exiting: Maximum number of function evaluations has been exceeded – increase `MaxFunEvals` option. Current function value: 0.202097". In fact, the expected objective function is 0, otherwise the solution process fails. The result x can be substituted into the objective function to see whether the value of the function is 0 or not, so as to judge whether the solution process is successful. If this error message appears, the statements should be executed again. It is not necessary to increase the value of `MaxFunEvals` as prompted.

10.4.3 Boundary value problems for delay differential equations

In Example 6.1, attempts were made to convert a delay differential equation into differential equations with no delays. This method is rather complicated, and does not have any universality, since most of the delay differential equations cannot be converted like this. Here, an optimization technique can be used to solve boundary value problems of delay differential equations. The basic idea is still the same. Consistent initial values can be found such that the errors in the final values are minimized. Since there are almost no examples on boundary value problems in delay differential equations in the literature, the examples here are converted from those in Chapter 6. In this section, examples are used to demonstrate the numerical solutions of boundary value problems of delay differential equations.

Example 10.18. Consider the delay differential equations in Example 6.13, with fixed delays. If the boundary values are $x_1(0) = 0$, $x_2(10) = 0.333$, and $x_1(10) = x_3(10) + 0.17$, and the history function are all constants, solve the boundary value problem of the given delay differential equations.

Solutions. It can be seen from the boundary conditions that $x_1(0)$ is given, therefore $x_2(0)$ and $x_3(0)$ are to be solved for with optimization techniques. They can be selected as decision variables. Since the boundary conditions are $x_2(10) = 0.333$ and $x_1(10) = x_3(10) + 0.17$, the following MATLAB function can be written, to compute the objective function, which is defined as the sum of the absolute errors in the final values.

```
function y=c10mdde1(x,f,tau,ff)
x0=[0; x(1); x(2)]; sol=dde23(f,tau,x0,[0,10],ff);
x=sol.y; y=abs(x(2,end)-0.333)+abs(x(1,end)-x(3,end)-0.17);
```

Therefore the following commands can be used to solve the boundary value problem for delay differential equations. The elapsed time is 29.35 seconds, and the consistent initial values are $x_2(0) = -2.7206$ and $x_3(10) = 2.0130$. The state curves are as shown in Figure 10.8.

```
>> f=@(t,x,Z)[1-3*x(1)-Z(2,1)-0.2*Z(1,2)^3-Z(1,2);
      x(3); 4*x(1)-2*x(2)-3*x(3)];
   ff=odeset; ff.RelTol=1e-8; ff.AbsTol=1e-8;
   tau=[1 0.5]; %set the two delay constants
   F=@(x)c10mdde1(x,f,tau,ff); opt=optimset; opt.TolX=eps;
   tic, x=fminsearch(F,rand(2,1),opt)
   sol=dde23(f,tau,[0; x(:)],[0,10],ff); plot(sol.x,sol.y), toc
```

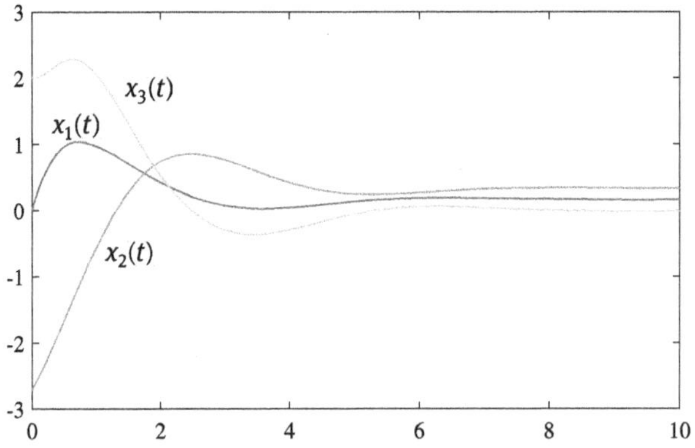

Figure 10.8: Solution of the delay differential equations.

Note that, as indicated in Chapter 6, the error tolerance in solving delay differential equations cannot be set to small quantities. Otherwise the solution process cannot be completed successfully. The error tolerance here is set to 10^{-8}.

Example 10.19. Consider the delay differential equations in Example 6.13, which is modified as follows:

$$\begin{cases} x_1'(t) = -2x_2(t) - 3x_1(t - 0.2|\sin t|), \\ x_2'(t) = -0.05x_1(t)x_3(t) - 2x_2(\alpha t) + 2, \\ x_3'(t) = 0.3x_1(t)x_2(t)x_3(t) + \cos(x_1(t)x_2(t)) + 2\sin 0.1t^2 \end{cases}$$

where $\alpha = 0.77$. If the boundary values are $x_1(0) = x_3(0) = 0$ and $x_3(10) = 5.5$, and the history functions of the differential equations are all constant, solve the boundary value problem.

Solutions. The current solvers in MATLAB cannot be used in solving boundary value problems of delay differential equations. The optimization method should be tried again for solving this kind of problems. In the specific problem, $x_1(0)$ and $x_3(0)$ are given, and $x_2(0)$ is unknown, which can be selected as a decision variable $x_2(0)$. The quantity $|x_3(10) - 5.5|$ is used as an objective function. The following MATLAB function can be written to compute the objective function:

```
function y=c10mdde2(x,f,tau,ff)
x0=[0; x(1); 0]; sol=ddesd(f,tau,x0,[0,10],ff);
x=sol.y; y=abs(x(3,end)-5.5);
```

To solve this optimization problem, the consistent initial value obtained is $x_2(0) = 0.0543$, and the elapsed time is 8.96 seconds. The curves obtained are very close to those in Figure 6.10.

```
>> tau=@(t,x)[t-0.2*abs(sin(t)); 0.77*t]; %describing delays
   f=@(t,x,Z)[-2*x(2)-3*Z(1,1);
          -0.05*x(1)*x(3)-2*Z(2,2)+2;         %delay differential equations
          0.3*x(1)*x(2)*x(3)+cos(x(1)*x(2))+2*sin(0.1*t^2)];
   ff=odeset; ff.RelTol=1e-8; ff.AbsTol=1e-8;
   F=@(x)c10mdde2(x,f,tau,ff); opt=optimset; opt.TolX=eps;
   tic, x=fminsearch(F,rand(1,1),opt)       %optimization
   sol=ddesd(f,tau,[0;x;0],[0,10],ff); plot(sol.x,sol.y), toc
```

10.4.4 Multipoint boundary value problems

Function `bvp5c()` discussed earlier can only be used to solve two-point boundary value problems, while in real applications, more than two points for the differential equations may be given. Therefore function `bvp5c()` cannot be used. An optimization technique can be introduced to solve multipoint boundary value problems. In this section, examples are used to demonstrate multipoint boundary value problems.

Example 10.20. It is seen from Example 2.30 that, in the differential equations, the information at $t = 0, \pi, 2\pi$ is known. Such a problem cannot be solved with two-point solvers. Use an optimization technique to find the solution of the differential equations.

Solutions. Before solving this problem, the mathematical form of the right-hand side of the equation for the given input $u(t)$ should be derived. This can be done with the symbolic commands

```
>> syms t; u(t)=exp(-5*t)*cos(2*t+1)+5;
   u1=5*diff(u,t,2)+4*diff(u,t)+2*u; simplify(u1)
```

It can be seen that the right-hand side of the equation is

$$u_1(t) = 87\mathrm{e}^{-5t}\cos(2t+1) + 92\mathrm{e}^{-5t}\sin(2t+1) + 10.$$

Selecting the state variables $x_1(t) = y(t)$, $x_2(t) = y'(t)$, $x_3(t) = y''(t)$, and $x_4(t) = y'''(t)$, the first-order explicit differential equations can be constructed as

$$\mathbf{x}'(t) = \begin{bmatrix} x_2(t) \\ x_3(t) \\ x_4(t) \\ u_1(t) - 10x_4(t) - 35x_3(t) - 50x_2(t) - 24x_1(t) \end{bmatrix}.$$

With the following anonymous function, this equation can be described:

```
>> u1=@(t)87*exp(-5*t)*cos(2*t+1)+92*exp(-5*t)*sin(2*t+1)+10;
   f=@(t,x)[x(2); x(3); x(4);
            u1(t)-10*x(4)-35*x(3)-50*x(2)-24*x(1)];
```

It can be seen from the initial values that only $x_1(t)$ is known, the other three states can be used to construct the decision variables. Besides, the terminal values $y'(\pi) = y''(2\pi) = y'(2\pi) = 0$ can also be written as $x_2(\pi) = x_3(2\pi) = x_2(2\pi) = 0$.

To implement this multipoint problem, the solution interval can be divided into two parts, $[0, \pi]$ and $[\pi, 2\pi]$. The result of the first interval is regarded as the initial value of the second one. Therefore the following objective function can be written:

```
function y=c10mmult(x,f,ff)
x0=[1/2; x(:)];
[t1,x1]=ode45(f,x0,[0,pi],ff);
[t2,x2]=ode45(f,x1(end,:).',[pi,2*pi],ff); %piecewise solution
y=abs(x1(end,2)-0)+abs(x2(end,3)-0)+abs(x2(end,2)-0);
```

With repeated trials, or with the global optimization solver `fminunc_global()` from Volume IV, the consistent initial values are $x_2(0) = 0.7985$, $x_3(0) = -4.0258$, and $x_3(0) = -9.0637$. The value of the objective function is $F(x) = 9.1898 \times 10^{-17}$. From the consistent initial values, the numerical solution of the original boundary value problem can be found.

```
>> ff=odeset; ff.RelTol=100*eps; ff.AbsTol=100*eps;
   F=@(x)c10mmult(x,f,ff); opt=optimset; opt.TolX=eps;
   tic, x=fminsearch(F,-10+20*rand(3,1),opt), F(x)
   [t0,y0]=ode45(f,[0,2*pi],[1/2; x(:)],ff); toc
```

Note that the above commands should be executed repeatedly, until consistent initial values are found, such that the value of the objective function is very close to zero.

The numerical solutions of the boundary value problem can alternatively be found by solving initial value problems. Compared with the analytical solutions, the norm of the error is 6.6661×10^{-9}. It can be seen that the results are more satisfactory.

```
>> syms t y(t); u(t)=exp(-5*t)*cos(2*t+1)+5;
   d1y=diff(y); d2y=diff(y,2); d3y=diff(y,3); d4y=diff(y,4);
   y(t)=dsolve(d4y+10*d3y+35*d2y+50*d1y+24*y==...
                  5*diff(u,t,2)+4*diff(u,t)+2*u,...
             y(0)==1/2,d1y(pi)==0,d2y(2*pi)==0,d1y(2*pi)==0);
   y1=double(y(t0)), norm(y0(:,1)-y1)  %compute the error
```

10.4.5 Floating boundary value problems revisited

In Section 10.3, an example was used to show that the general purpose solver `bvp5c()` failed. The characteristics of the example are that the values of the unknowns at the boundaries are not exactly known. Only the relationship of the boundary values is known. The solution of these differential equations may not be unique. The solver cannot be used to find any other solutions satisfying the floating boundary conditions. Here the optimization based technique is used to solve boundary value problems.

Example 10.21. Solve again the boundary value problem in Example 10.13.

Solutions. A function handle f for the first-order explicit differential equations in Example 10.13 is used. With such a handle and the given boundary conditions, the following objective function can be written, where the sum of absolute values of the errors is defined as the objective function:

```
function y=c10mnon(x,f,ff)
x0=x(:); [t,x]=ode45(f,[0,2*pi],x0,ff);
y=abs(x(1,1)-x(end,1))+abs(x(1,2)-x(end,2)); %objective function
```

With such an objective function, the following commands can be tried directly to solve the differential equations. Since the problem has infinitely many solutions, each time the following statements are called, a set of consistent initial values are found. For instance, the consistent initial value vector is $\boldsymbol{x}_0 = [0.3224, -0.9271]^{\mathrm{T}}$. The solutions are shown in Figure 10.9, superimposed by two broken lines, indicating that the boundary conditions are also satisfied.

```
>> f=@(t,x)[x(2); min(x(1)+2,max(-x(1),x(1)-2))];
   ff=odeset; ff.RelTol=100*eps; ff.AbsTol=100*eps;
   F=@(x)c10mnon(x,f,ff); opt=optimset; opt.TolX=eps;
   tic, x0=fminsearch(F,-2+4*rand(2,1),opt) %optimization
   [t0,y0]=ode45(f,[0,2*pi],x0,ff); toc      %solve differential equations
   plot(t0,y0,[0 2*pi],kron(x0(:),[1 1])','--')
```

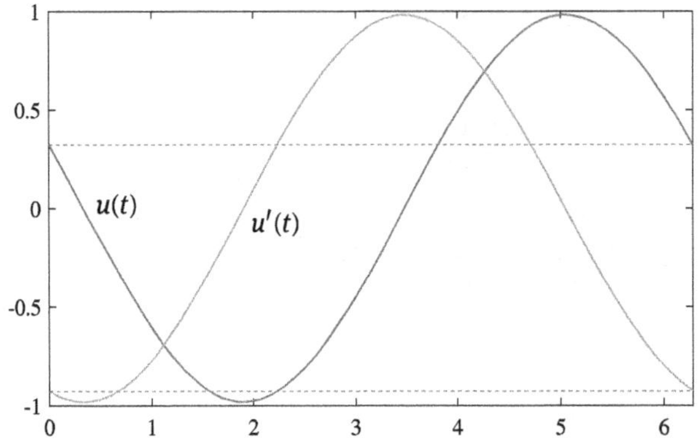

Figure 10.9: A solution of the problem with floating bounds.

If the above code is executed again, since the initial values are randomly chosen, it is quite probable that another solution is found. Further execution may yield even more solutions. It can be seen that the solutions are sine and cosine curves in one cycle, which can be regarded as the left–right translation of the curves in Figure 10.9, since the boundary conditions are satisfied. It can be seen that the analytical solution in [19], $u(t) = \pm\sin t$, is incomplete, since $u(t) = A\sin(t+\theta)$ ($|A| \leqslant 1$) also satisfies the original boundary value problem. Besides, $u(t) \equiv 0, \pm 2$ are also solutions of the original problem.

10.4.6 Boundary value problems for block diagrams

If complicate differential equations are not suitable to be solved with command-driven methods, Simulink can be used to draw the block diagram model. Then simulation methods can be used in solving the boundary value problems. Here only an example is given to demonstrate the solution method.

Example 10.22. Use Simulink and the optimization technique to solve again the semi-infinite interval boundary value problem in Example 10.12. The differential equation is a linear time-varying one.

Solutions. From the given differential equations, it is not difficult to build up the Simulink model, as shown in Figure 10.10. The functions $p(t)$, $q(t)$, and $f(t)$ are described by Fcn blocks, whose contents are respectively:

```
-2*(cos(u+1)+2)^2/(sin(u+1)+2)^2
-16*sin(u+1)/(sin(u+1)+2)^2
4*exp(-2*u)*(4*cos(u+1)+9)/(sin(u+1)+2)^2
```

and the initial value of the left integrator is set to variable a. This variable can be selected as the decision variable. Set the stoping time to $L = 10$.

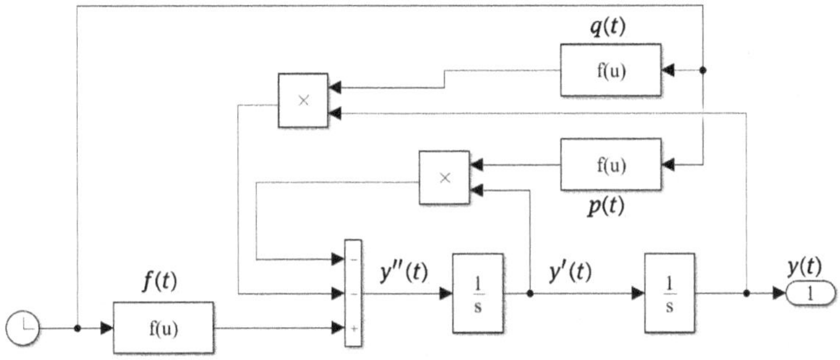

Figure 10.10: Simulink model (file: `c10mbvp1.slx`).

With the Simulink model, the following MATLAB function can be written, to describe the objective function. The physical meaning is that the value $|y(L)|$ is made as small as possible (as close as possible to zero, such that the boundary condition is satisfied).

```
function y=c10mbvp1a(x)
assignin('base','a',x)
[t,~,y0]=sim('c10mbvp1'); y=abs(y0(end)); %objective function
```

The consistent initial value found is −1.999999999999975, very close to the exact value of −2. Computing from the initial values, the solution of the original differential equations is then found.

```
>> a=1; opt=optimset; opt.TolX=eps; format long
   tic, a=fminsearch(@c10mbvp1a,a,opt)                  %optimization
   [t,~,y]=sim('c10mbvp1'); norm(y-exp(-2*t)), toc %solution and errors
```

Since these differential equations are very sensitive to the initial values, a slight error in the initial value may yield huge errors. For this particular example, the norm of the error is 1.2225×10^{-4}, with the elapsed time of 7.96 seconds. Although the efficiency is significantly higher than that in Example 10.12, the method presented here may have broader application fields, since differential equations of any complexity can be tackled. For instance, we can solve the boundary value problems of fractional-order differential equations to be discussed next.

10.4.7 Boundary value problems for fractional-order differential equations

To the author's knowledge, there are no methods capable of solving boundary value problems for fractional-order differential equations. In this section, a feasible method is proposed to solve such problems.

An effective block diagram based method is proposed in [74], which can be used for solving initial value problems of nonlinear fractional-order differential equations. Therefore, the method can be combined with optimization techniques, such that the boundary value problems for fractional-order differential equations can be solved.

Example 10.23. Consider the nonlinear fractional-order differential equation in Example 8.11. If the boundary values are $y(0) = 1$ and $y(1) = e^{-1}$, solve the boundary value problem. If the analytical solution is $y(t) = e^{-t}$, assess the precision of the numerical solution.

Solutions. In Example 9.22, a Simulink model `c9mfode3.slx` was created for the fractional-order differential equation. If the initial value of the $y'(t)$ integrator, the left one in Figure 9.35, is changed from −1 to variable a, a new model `c10mfode.slx` can be constructed. This model can be used to solve initial value problems of this fractional-order differential equation.

In normal cases, if a certain a, that is, $x_2(0)$, makes the problem singular, the solution process may abnormally abort, and error message is given. This is not what we are expecting. A trial structure can be used, if the solution process yields errors, and the objective function is artificially set to a large value. The singular points are avoided such that the optimization process can be continued. Based on the idea, the following objective function can be written:

```
function y=c10mfode1(x)
assignin('base','a',x)
try
   [t,~,y0]=sim('c10mfode',1); y=abs(y0(end)-exp(-t(end)));
catch, y=10; end
```

With the objective function, suitable parameters of Oustaloup filter are selected. The following commands can be used to solve the boundary value problem. The elapsed time of the solution process is 69.3 seconds. The consistent initial value is $x_2(0)$, that is, $a = -0.999998506$, which is very close to the theoretical value −1. Starting from the point, the numerical solution can be found, and the norm of the error is 1.3728×10^{-6}. It can be seen that the solution process here is successful.

```
>> a=rand(1); opt=optimset; opt.TolX=1e-13;
   ww=[1e-5,1e5]; N=25; format long                  % filter parameters
   tic, a=fminsearch(@c10mfode1,a,opt)               % optimization
   [t,~,y]=sim('c10mfode'); norm(y-exp(-t)), toc % solve equation
```

10.5 Exercises

10.1 Solve the following boundary value problem:[61]

$$y''(t) = -0.05y'(t) - 0.02y^2(t)\sin t + 0.00005\sin t\cos^2 t - 0.05\cos t - 0.0025\sin t$$

where $y(0) = y(2\pi)$, $y'(0) = y'(2\pi)$, with known analytical solution of $y(t) = 0.05\cos t$.

10.2 Solve the following boundary value problem:

$$y^{(4)}(t) = 6\mathrm{e}^{-4y(t)} - \frac{12}{(1+t)^4}$$

where $y(0) = 0$, $y(1) = \ln 2$, $y''(0) = -1$, and $y''(1) = -0.25$. The analytical solution is $y(t) = \ln(1+t)$.

10.3 Solve the following boundary value problem:[61]

$$y''(x) - 400y(x) = 400\cos^2\pi x + 2\pi^2\cos 2\pi x$$

where $y(0) = y(1) = 0$. The analytical solution is

$$y(x) = \frac{\mathrm{e}^{-20}}{1+\mathrm{e}^{-20}}\mathrm{e}^{20x} + \frac{1}{1+\mathrm{e}^{-20}}\mathrm{e}^{-20x} - \cos^2\pi x.$$

10.4 Magnetic monopoles mapped to the $(0,1)$ interval are described by differential equations[2]

$$\begin{cases} y_1''(x) = \dfrac{2y_1'(x)}{1-x} + \dfrac{y_1(x)}{x^2(1-x)^2}\left[y_1^2(x) - 1 + \dfrac{y_3^2(x) - y_2^2(x)}{(1-x)^2}\right], \\ y_2''(x) = \dfrac{2y_2(x)y_1^2(x)}{x^2(1-x)^2}, \\ y_3''(x) = \dfrac{y_3(x)}{x^2(1-x)^2}\left[y_1^2(x) + \dfrac{\beta(y_3^2(x) - x^2)}{(1-x)^2}\right] \end{cases}$$

where $y_1(0) = 1$, $y_2(0) = y_3(0) = 0$, $y_1(1) = 0$, $y_2(1) = \eta$, and $y_3(1) = 1$, with $0 \leqslant \beta \leqslant 20$ and $0 < \eta < 1$ both being constants. Letting $\eta = 0.5$, solve the boundary value problem for $x \in (0,1)$.

10.5 Solve the following boundary value problem:[61]

$$\begin{cases} y_1'(t) = ay_1(t)(y_3(t) - y_1(t))/y_2(t), \\ y_2'(t) = -a(y_3(t) - y_1(t)), \\ y_3'(t) = [b - c(y_3(t) - y_5(t)) - ay_3(t)(y_3(t) - y_1(t))]/y_4(t), \\ y_4'(t) = a(y_3(t) - y_1(t)), \\ y_5'(t) = -c(y_5(t) - y_3(t))/d \end{cases}$$

where $a = 100$, $b = 0.9$, $c = 1\,000$, and $d = 10$. The boundary values are $y_1(0) = y_2(0) = y_3(0) = 1$, $y_4(0) = -10$, and $y_3(1) = y_5(1)$.

10.6 Solve the following boundary value problem:[63]

$$\begin{cases} y_1'(t) = y_2(t), \\ y_2'(t) = y_3(t), \\ y_3'(t) = -(3-n)y_1(t)y_3(t)/2 - ny_2^2(t) + 1 - y_4^2(t) + sy_2(t), \\ y_4'(t) = y_5(t), \\ y_5'(t) = -(3-n)y_1(t)y_3(t)/2 - (n-1)y_2(t)y_4(t) + s(y_4(t) - 1) \end{cases}$$

where $n = -0.1$ and $s = 0.2$. The boundary values are $y_1(0) = y_2(0) = y_4(0) = y_2(b) = 0$, and $y_4(b) = 1$, $b = 11.3$.

10.7 Solve the following semi-infinite interval boundary value problem:[27]

$$\begin{cases} 3y(t)y''(t) = 2(y'(t) - z(t)), \\ z'''(t) = -y(t)z'(t), \end{cases}$$

with boundary conditions $y(0) = z(0) = 1$ and $y'(\infty) = z'(\infty) = 0$.

10.8 Consider the following boundary value problem:[2]

$$\begin{cases} f'''(t) - R[(f'(t))^2 - f(t)f''(t)] + AR = 0, \\ h''(t) + Rf(t)h'(t) + 1 = 0, \\ \theta''(t) + Pf(t)\theta'(t) = 0 \end{cases}$$

where $R = 10$ and $P = 0.7R$; A is an undetermined coefficient. If the boundary values are $f(0) = f'(0) = 0$, $f(1) = 0$, $f'(1) = 0$, $h(0) = h(1) = 0$, $\theta(0) = 0$, and $\theta(1) = 1$, solve the differential equations. If $R = 10\,000$, solve them again.

10.9 In [2], a kidney model is described by a boundary value problem of differential equations. Since the model is too complicated, it is not given here. The interested readers may refer to the cited reference, and find the numerical solution of the differential equations.

11 Partial differential equations

Up to now, the analytical and numerical methods were studied for ordinary differential equations. Ordinary differential equations are used to describe the relationship between unknown functions and their derivatives with respect to time. Ordinary differential equations are used in the dynamic modeling and simulation of lumped parameter systems.

Partial differential equations (PDEs) are different from ordinary differential equations. In partial differential equations, the derivatives of the unknowns with respect to time may be contained, and also the derivatives of the unknown with respect to spatial variables such as x, y, and z are involved. If only x is involved, such equations are referred to as one-dimensional partial differential equations; if x and y are involved, they are called two-dimensional partial differential equations. Of course, three- or even high-dimensional partial differential equations can be defined. Partial differential equations can be used in the modeling of distributed parameter dynamical systems.

In this chapter, we concentrate on studying partial differential equations. Generally speaking, partial differential equations do not have analytical solutions. Numerical solutions are mainly discussed in this chapter. Of course, some particular examples with analytical solutions are used for comparative studies of the algorithms, so that useful information may be provided for the readers in algorithm selection and parameter setting.

In Section 11.1, simple diffusion equations are studied. Numerical algorithms and MATLAB implementations are presented. A fast algorithm based on sparse matrices is introduced. Unfortunately, the algorithm is not extended to multidimensional cases. The Partial Differential Equation Toolbox provided in MATLAB can be used to solve other problems directly. The subsequent problems in the chapter will be solved with this toolbox.

In Section 11.2, classifications of partial differential equations and boundary conditions are proposed. The mathematical forms of elliptic, hyperbolic, parabolic, and eigenvalue partial differential equations are presented. Also the Dirichlet and Neumann boundary conditions are addressed. In Section 11.3, a partial differential equation solution interface is introduced, and based on it, the solution process of typical two-dimensional partial differential equations is demonstrated. The concepts such as geometric region design, equation and condition specification, and partial differential equation solution process are all demonstrated. The limitation of the method is that it applies only to two-dimensional partial differential equations. In Section 11.4, a general partial differential equation solution pattern is introduced. The modeling of `PDEModel` object is demonstrated through examples.

https://doi.org/10.1515/9783110675252-011

11.1 Numerical solutions of diffusion equations

Diffusion equations is a class of simple partial differential equations. In this section, the fundamental concepts of one-dimensional diffusion equations are introduced, followed by the numerical solution of discretized diffusion equations and their MATLAB implementations. Multidimensional diffusion equation are also formulated.

11.1.1 Mathematical form and analytical solutions of one-dimensional diffusion equations

Diffusion equations are also known as heat equations. Just consider one case. Assume that there is an extremely thin heat-conducting rod, whose thickness can be neglected. Both sides of the rod are connected to heat sources, with initial temperature. At the initial time, the temperature at each point on the rod is also known. The temperature at position x and time t can be described by a partial differential equation, known as homogeneous diffusion equation.

Definition 11.1. The mathematical form of a homogeneous diffusion equation is

$$\frac{\partial u(t,x)}{\partial t} = K\frac{\partial^2 u(t,x)}{\partial x^2} \tag{11.1.1}$$

where K is referred to as the diffusion coefficient.

Observing the differential equation itself, it is immediately seen that $u(t,x) = Kt + x^2/2 + C$ satisfies the homogeneous diffusion equation, where C can be any constant. In fact, the analytical solution of the equation may have different forms. An analytical solution will be shown next with MATLAB.

Example 11.1. Show that for any real numbers C and γ, function $u(t,x) = \mathrm{e}^{-\gamma^2 Kt}\sin\gamma x + C$ satisfies the homogeneous diffusion equation.

Solutions. If MATLAB is used to show whether an equality holds, the simplest method is to express the difference between the two sides of the equality using symbolic computation, and then simplify the result. If the result is zero, then the equality is proven. If it is not zero, the equality does not hold. For this specific problem, the following commands can be used to directly verify the analytical solution, because the simplified error is zero.

```
>> syms K t x; syms gam C real
   u(t,x)=exp(-gam^2*K*t)*sin(gam*x)+C; % construct the solution
   simplify(diff(u,t)-K*diff(u,x,x))     % check the analytical solution
```

Similar to the case when solving ordinary differential equations, for partial differential equations, the boundary as well as initial conditions are needed. For the unknown function $u(t, x)$, the boundary conditions can be regarded as functions when x is fixed at x_0 or x_1, that is, at both ends of the rod. Boundary conditions are usually functions of t:

$$u(t, x_0) = g_0(t), \quad u(t, x_1) = g_1(t). \tag{11.1.2}$$

Initial conditions are similar to the case of ordinary differential equations. They are the values of the unknown function at the initial time $t = t_0$. The difference is that the initial conditions are usually a function of x,

$$u(t_0, x) = \eta(x). \tag{11.1.3}$$

11.1.2 Discretizing diffusion equations

How can we discretize a diffusion equation? An apparent consideration is follows: selecting step-sizes k and h for t and x axes, respectively, that is, taking $h = \Delta x$ and $k = \Delta t$, equally-spaced points are created in the x and t axes such that $x_i = x_0 + (i-1)h$, $t_j = t_0 + (j-1)k$, where $i = 1, 2, \ldots, m$, $j = 1, 2, \ldots, n$. Denoting $U_i^j = u(t_j, x_i)$, a discrete form of the diffusion equation can be established as

$$\frac{U_i^{j+1} - U_i^j}{k} = \frac{K}{2}(\mathrm{D}^2 U_i^j + \mathrm{D}^2 U_i^{j+1}) = \frac{K}{2h^2}(U_{i-1}^j - 2U_i^j + U_{i+1}^j + U_{i-1}^{j+1} - 2U_i^{j+1} + U_{i+1}^{j+1}). \tag{11.1.4}$$

The following Crank–Nicolson formula can be derived:[44]

$$-rU_{i-1}^{j+1} + (1+2r)U_i^{j+1} - rU_{i+1}^{j+1} = rU_{i-1}^j + (1-2r)U_i^j + rU_{i+1}^j \tag{11.1.5}$$

where constant $r = kK/(2h^2)$ is defined.

For different j, the following linear algebraic equation can be established:

$$\boldsymbol{A}\boldsymbol{U}_{j+1} = \boldsymbol{B}_j \tag{11.1.6}$$

where the coefficient matrix $\boldsymbol{A}$ is a tri-diagonal one

$$\boldsymbol{A} = \begin{bmatrix} (1+2r) & -r & & & & \\ -r & (1+2r) & -r & & & \\ & -r & (1+2r) & -r & & \\ & & \ddots & \ddots & \ddots & \\ & & & -r & (1+2r) & -r \\ & & & & -r & (1+2r) \end{bmatrix}. \tag{11.1.7}$$

Besides, the known vector $\boldsymbol{B}_j$ and unknown vector $\boldsymbol{U}_j$ are defined as

$$\boldsymbol{B}_j=\begin{bmatrix} r(g_0(t_j)+g_0(t_{j+1}))+(1-2r)U_1^j+rU_2^j \\ rU_1^j+(1-2r)U_2^j+rU_3^j \\ rU_2^j+(1-2r)U_3^j+rU_4^j \\ \vdots \\ rU_{m-2}^j+(1-2r)U_{m-1}^j+rU_m^j \\ rU_{m-1}^j+(1-2r)U_m^j+r(g_1(t_j)+g_1(t_{j+1})) \end{bmatrix},\quad \boldsymbol{U}_{j+1}=\begin{bmatrix} U_1^{j+1} \\ U_2^{j+1} \\ U_3^{j+1} \\ \vdots \\ U_{m-1}^{j+1} \\ U_m^{j+1} \end{bmatrix} \tag{11.1.8}$$

where $j=1,2,\dots,n-1$. At time $j=0$, $\boldsymbol{U}_{j+1}=\boldsymbol{U}_1=u(t_0,x)=\eta(x)$ are the initial conditions. It can be seen from the equations that, since at $j=1$, $\boldsymbol{U}_1$ is determined by the initial conditions, it can be substituted into (11.1.8), and, by solving a linear algebraic equation, $\boldsymbol{U}_2$ can be found. Using a loop structure, for each j, vector $\boldsymbol{U}_{j+1}$ can be found, which spans the numerical solution of the partial differential equation.

Based on the above algorithm, it is not hard to write the following MATLAB function to solve the diffusion equation. The user needs to specify the initial condition function handle `eta`, the boundary condition function handles `g0` and `g1`, the step-sizes, number of computing points, and the diffusion coefficients to call the function and find the solution matrix $\boldsymbol{U}$. Since $\boldsymbol{A}$ is a constant matrix, its inverse can be evaluated first. In the following solution process, the inverse can be multiplied by $\boldsymbol{B}_j$, such that the computational efficiency may be boosted.

```
function [t,x,U]=diffusion_sol1(h,k,t0,x0,n,m,eta,g0,g1,K)
r=k*K/2/h^2; x=x0+(0:m-1)*h; t=t0+(0:n-1)*k;
U1=eta(x); U1=U1(:); U=U1; v=-r*ones(1,m-1);
A=(1+2*r)*eye(m)+diag(v,1)+diag(v,-1); iA=inv(A);
for j=1:n-1
   B=r*(g0(t(j))+g0(t(j+1)))+(1-2*r)*U1(1)+r*U1(2);
   for i=1:m-2, B(i+1,1)=[r,1-2*r,r]*U1(i:i+2); end
   B(m,1)=r*U1(m-1)+(1-2*r)*U1(m)+r*(g1(t(j))+g1(t(j+1)));
   U1=iA*B; U=[U U1];
end
```

Example 11.2. Solve the following one-dimensional diffusion equation:

$$\frac{\partial u(t,x)}{\partial t}-\frac{1}{\pi^2}\frac{\partial^2 u(t,x)}{\partial x^2}=0$$

where the boundary conditions are $u(t,0)=u(t,1)=0$, and initial condition is $u(0,x)=\sin\pi x$. If the analytical solution of the equation is $u(t,x)=\mathrm{e}^{-t}\sin\pi x$. Selecting $x\in[0,1]$, $t\in(0,T)$, with $T=1$, solve the partial differential equation and assess the accuracy.

Solutions. It can be seen from the diffusion equation that $K = 1/\pi^2$. Selecting the step-sizes $h = k = 0.01$, the following MATLAB commands can be written to describe the boundary and initial conditions, and then solve the diffusion equation. The solution surface of $u(t,x)$ is obtained, as shown in Figure 11.1.

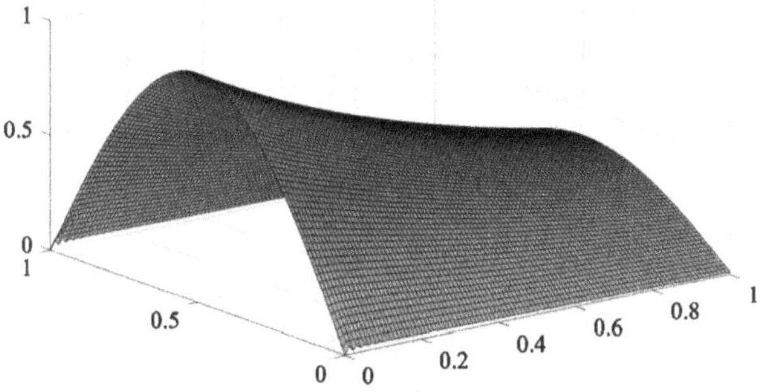

Figure 11.1: Solution surface of function $u(t,x)$.

```
>> g0=@(t)zeros(size(t)); g1=g0; eta=@(x)sin(pi*x);
   n=101; m=101; h=0.01; k=0.01; t0=0; x0=0; K=1/pi^2;
   [t,x,U]=diffusion_sol1(h,k,t0,x0,n,m,eta,g0,g1,K);
   surf(t,x,U)  % draw the solution surface
```

Since the analytical solution is known, and mesh grids are made in vectors $\boldsymbol{t}$ and $\boldsymbol{x}$, the maximum error with respect to the exact values is 0.0402. The error surface can also be obtained, as shown in Figure 11.2. It can be seen that large errors happen at both boundaries of x, and also when t is large.

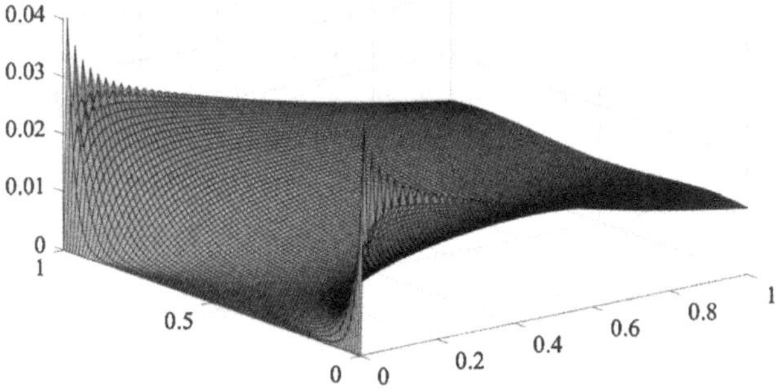

Figure 11.2: Error surface of the solutions.

```
>> [tx,xx]=meshgrid(t,x); U0=exp(-tx).*sin(pi*xx);
   norm(U(:)-U0(:),inf), contour(tx,xx,U-U0,200)
```

Example 11.3. Consider again Example 11.2. The accuracy in the previous example is relatively low. A natural way is to reduce the step-size so as to increase the accuracy. Assess the impact of step-size on the accuracy and also efficiency.

Solutions. Selecting different step-sizes k and letting $h = k$, the elapsed time, accuracy and sizes of the solution matrices are measured, as shown in Table 11.1 (see Algorithm 1 entries). It can be seen that if the step-size is reduced, the accuracy is increased, while the number of points and the elapsed time increase significantly. Therefore it is not suitable to select too small step-sizes to increase the accuracy.

```
>> g0=@(t)zeros(size(t)); g1=g0; eta=@(x)sin(pi*x);
   t0=0; x0=0; K=1/pi^2;
   for h=[0.01,0.005,0.0025,0.00125,0.000625,3.125e-4,1.5625e-4]
      k=h; n=1/k+1; m=n; tic %compare the errors
      [t,x,U]=diffusion_sol1(h,k,t0,x0,n,m,eta,g0,g1,K); toc
      N=prod(size(U)), [tx,xx]=meshgrid(t,x);
      U0=exp(-tx).*sin(pi*xx); norm(U(:)-U0(:),inf)
   end
```

Table 11.1: Step-size and elapsed time.

step-sizes $h = k$	0.01	0.005	0.0025	0.00125	0.000625	3.125×10^{-4}	1.5625×10^{-4}
Algorithm 1 (s)	0.0358	0.0958	0.3203	0.971	5.484	35.895	236.62
Algorithm 2 (s)	0.0450	0.0816	0.2419	0.575	2.049	8.977	34.956
maximum error	0.0402	0.0229	0.0126	0.0067	0.0035	0.0018	9.2865×10^{-4}
number of points, N	10 201	40 401	160 801	641 601	2 563 201	10 246 401	40 972 801

Let us now examine function `diffusion_sol1()`. Since a regular matrix $\boldsymbol{A}$ is used, it is rather time consuming and waste of matrix storage. It is not possible to handle large matrices. Since matrix $\boldsymbol{A}$ is sparse, the utilities of storing and handling sparse matrices are employed to write a new solver function `diffusion_sol2()`. Note that the inverse $\boldsymbol{A}^{-1}$ of a tri-diagonal matrix $\boldsymbol{A}$ is no longer sparse, and it should not be evaluated first as in the case of the previous example. Still the solver is much faster than the previous one, since a sparse matrix is used.

```
function [t,x,U]=diffusion_sol2(h,k,t0,x0,n,m,eta,g0,g1,K)
r=k*K/2/h^2; x=x0+(0:m-1)*h; t=t0+(0:n-1)*k;
U1=eta(x); U1=U1(:); U=U1;
v=-r*ones(1,m-1); i=1:m-1; v1=(1+2*r)*ones(1,m);
```

```
A=sparse([i i+1 i m],[i+1 i i m],[v,v,v1]); %create sparse matrix
for j=1:n-1
   B=r*(g0(t(j))+g0(t(j+1)))+(1-2*r)*U1(1)+r*U1(2);
   for i=1:m-2, B(i+1,1)=[r,1-2*r,r]*U1(i:i+2); end
   B(m,1)=r*U1(m-1)+(1-2*r)*U1(m)+r*(g1(t(j))+g1(t(j+1)));
   U1=A\B; U=[U U1]; %backslash operation is used to avoid computinginverse
end
```

Example 11.4. Solve the problem in Example 11.3 again with the sparse matrix based solver.

Solutions. With almost the same commands as before, the elapsed time and other information can be found as given in Table 11.1 (see Algorithm 2). It can be seen that, with a sparse matrix, the accuracy and number of points are exactly the same, but the elapsed time is significantly reduced. Therefore it is recommended to use such a solver for diffusion equations.

```
>> g0=@(t)zeros(size(t)); g1=g0; eta=@(x)sin(pi*x);
   t0=0; x0=0; K=1/pi^2;
   for h=[0.01,0.005,0.0025,0.00125,0.000625,3.125e-4,1.5625e-4]
      k=h; n=1/k+1; m=n; tic
      [t,x,U]=diffusion_sol2(h,k,t0,x0,n,m,eta,g0,g1,K); toc
      N=prod(size(U)), [tx,xx]=meshgrid(t,x);
      U0=exp(-tx).*sin(pi*xx); norm(U(:)-U0(:),inf)
   end
```

Even though sparse matrices are introduced, the step-size should not be selected too small. If possible, high-precision and effective computation methods and tools should be used.

11.1.3 Inhomogeneous diffusion equations

Slightly extending the diffusion equation in Definition 11.1, a more general form of a diffusion equation can be defined. Here inhomogeneous diffusion equation are introduced, and numerical solution methods for inhomogeneous diffusion equations are presented.

Definition 11.2. The mathematical form of an inhomogeneous diffusion equation is

$$\frac{\partial u(t,x)}{\partial t} = K\frac{\partial^2 u(t,x)}{\partial x^2} + f(u(t,x)) + \psi(t,x) \tag{11.1.9}$$

where K is the diffusion coefficient, $f(u(t,x))$ is any nonlinear function containing the unknown function $u(t,x)$, and $\psi(t,x)$ is any given function.

From the discretized model in (11.1.4), it is not hard to write down a discretized version of the inhomogeneous diffusion equation

$$\begin{aligned}\frac{U_i^{j+1}-U_i^j}{k} &= \psi(t_j,x_i)+f(U_i^j)\\ &\quad+\frac{K}{2h^2}(U_{i-1}^j-2U_i^j+U_{i+1}^j+U_{i-1}^{j+1}-2U_i^{j+1}+U_{i+1}^{j+1}).\end{aligned} \tag{11.1.10}$$

Therefore the equation obtained also satisfies the linear algebraic equation in (11.1.6), where matrix $\boldsymbol{A}$ and vector $\boldsymbol{U}_j$ are exactly the same. Vector $\boldsymbol{B}_j$ is modified as

$$\boldsymbol{B}_j=\begin{bmatrix} r(g_0(t_j)+g_0(t_{j+1}))+(1-2r)U_1^j+rU_2^j+k\Psi_1^j+kf(U_1^j)\\ rU_1^j+(1-2r)U_2^j+rU_3^j+k\Psi_2^j+kf(U_2^j)\\ rU_2^j+(1-2r)U_3^j+rU_4^j+k\Psi_3^j+kf(U_3^j)\\ \vdots\\ rU_{m-2}^j+(1-2r)U_{m-1}^j+rU_m^j+k\Psi_{m-1}^j+kf(U_{m-1}^j)\\ rU_{m-1}^j+(1-2r)U_m^j+r(g_1(t_j)+g_1(t_{j+1}))+k\Psi_m^j+kf(U_m^j)\end{bmatrix} \tag{11.1.11}$$

where $\Psi(t_j,x_i)$ is simply denoted as Ψ_i^j. Similar to the solver `duffusion_sol2()`, the following MATLAB code is written for inhomogeneous diffusion equations:

```
function [t,x,U]=diffusion_sol(h,k,t0,x0,n,m,eta,g0,g1,K,f,Psi)
r=k*K/2/h^2; x=x0+(0:m-1)*h; t=t0+(0:n-1)*k;
U1=eta(x); U1=U1(:); U=U1;
v=-r*ones(1,m-1); i=1:m-1; v1=(1+2*r)*ones(1,m);
A=sparse([i i+1 i m],[i+1 i i m],[v,v,v1]); %create sparse matrix
for j=1:n-1
   B=r*(g0(t(j))+g0(t(j+1)))+(1-2*r)*U1(1)+r*U1(2);
   for i=1:m-2, B(i+1,1)=[r,1-2*r,r]*U1(i:i+2); end
   B(m,1)=r*U1(m-1)+(1-2*r)*U1(m)+r*(g1(t(j))+g1(t(j+1)));
   for i=1:m, B(i,1)=B(i,1)+k*Psi(t(j),x(i))+k*f(U1(i)); end
   U1=A\B; U=[U U1];  %backslash operation and record the results
end
```

Example 11.5. Consider an inhomogeneous diffusion equation, for which the boundary conditions are $u(t,0)=u(t,1)=0$, the initial condition is given by $u(0,x)=\sin\pi x$, and also the given functions are $\Psi(t,x)=x(1-x)\cos t$ and $f(u)=-2u\cos u$. Solve this inhomogeneous diffusion equation.

Solutions. The boundary and initiation conditions, as well as the $\Psi(t,x)$ function, are described with anonymous functions. With the following commands, the inhomogeneous diffusion equation can be solved, and the solution surface with contours can be obtained, as shown in Figure 11.3.

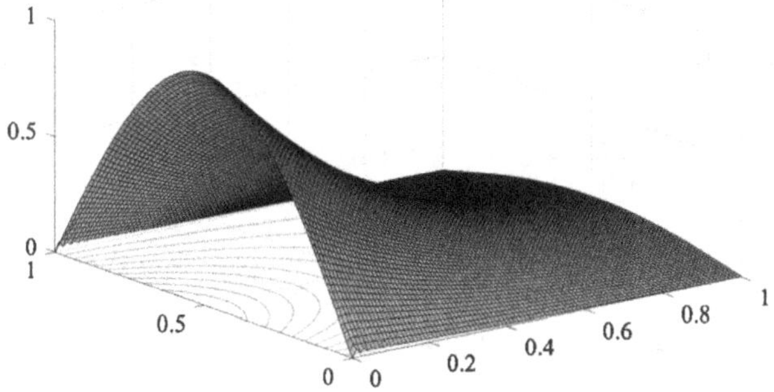

Figure 11.3: Surface of the solution $u(t,x)$.

```
>> g0=@(t)zeros(size(t)); g1=g0; K=1/pi^2; f=@(u)-2*u*cos(u);
   eta=@(x)sin(pi*x); Psi=@(t,x)x*(1-x)*cos(t);
   n=101; m=101; h=0.01; k=0.01; t0=0; x0=0;
   [t,x,U]=diffusion_sol(h,k,t0,x0,n,m,eta,g0,g1,K,f,Psi);
   surfc(t,x,U) % draw the solution surface
```

11.1.4 Multidimensional diffusion equations

The diffusion equations studied so far were one-dimensional diffusion equations. Their background was the heat transfer in a long thin rod. If the radius of the rod is not negligible, or if it is a three-dimensional solid body, two- or three-dimensional diffusion equations can be established. In mathematics, the diffusion equation can be extended to even higher dimensions. Here only the mathematical form of such diffusion equations is given, and in the following sections, two-dimensional diffusion equations will be demonstrated.

Definition 11.3. The mathematical form of an n-dimensional diffusion equation is

$$\frac{u(t,\boldsymbol{x})}{\partial t} = \psi(t,\boldsymbol{x},u) + K\left(\frac{\partial^2 u(t,\boldsymbol{x})}{\partial x_1^2} + \frac{\partial^2 u(t,\boldsymbol{x})}{\partial x_2^2} + \cdots + \frac{\partial^2 u(t,\boldsymbol{x})}{\partial x_n^2}\right). \tag{11.1.12}$$

11.2 Several special forms of partial differential equations

Apart from the partial differential equations discussed so far, MATLAB has its own Partial Differential Equation Toolbox. It can be used to describe and solve two-dimensional partial differential equations. Here, some special forms of two-dimensional partial differential equations are briefly described.

11.2.1 Classification of partial differential equations

The mathematical form of second-order partial differential equations is

$$a\frac{\partial^2 u(x,y)}{\partial x^2}+b\frac{\partial^2 u(x,y)}{\partial x\partial y}+c\frac{\partial^2 u(x,y)}{\partial y^2}+\cdots=f. \tag{11.2.1}$$

Based on the value of $\Delta = b^2 - 4ac$, such partial differential equations can be classified as follows:
(1) If $\Delta < 0$, the partial differential equation is elliptic;
(2) If $\Delta = 0$, the equation is parabolic;
(3) If $\Delta > 0$, the equation is hyperbolic.

Different types of partial differential equations will be further described later.

Definition 11.4. The mathematical form of an elliptic partial differential equation is

$$-\operatorname{div}(c\nabla u)+au=f(t,\boldsymbol{x}) \tag{11.2.2}$$

where $u = u(t, x_1, x_2, \dots, x_n) = u(t,\boldsymbol{x})$, and ∇u is the gradient of u, defined as

$$\nabla u=\left[\frac{\partial}{\partial x_1},\frac{\partial}{\partial x_2},\dots,\frac{\partial}{\partial x_n}\right]u. \tag{11.2.3}$$

Definition 11.5. Divergence $\operatorname{div}(v)$ of function v is defined as

$$\operatorname{div}(v)=\left(\frac{\partial}{\partial x_1}+\frac{\partial}{\partial x_2}+\cdots+\frac{\partial}{\partial x_n}\right)v. \tag{11.2.4}$$

The divergence $\operatorname{div}(c\nabla u)$ can further be expressed as

$$\operatorname{div}(c\nabla u)=\left[\frac{\partial}{\partial x_1}\left(c\frac{\partial u}{\partial x_1}\right)+\frac{\partial}{\partial x_2}\left(c\frac{\partial u}{\partial x_2}\right)+\cdots+\frac{\partial}{\partial x_n}\left(c\frac{\partial u}{\partial x_n}\right)\right] \tag{11.2.5}$$

where $\operatorname{div}(c\nabla u)$ is also denoted as $\nabla\cdot(c\nabla u)$.

If c is a constant, the term can further be simplified as

$$\operatorname{div}(c\nabla u)=c\left(\frac{\partial^2}{\partial x_1^2}+\frac{\partial^2}{\partial x_2^2}+\cdots+\frac{\partial^2}{\partial x_n^2}\right)u=c\Delta u \tag{11.2.6}$$

where Δ is the Laplace operator. Therefore, elliptic partial differential equation can simply be written as

$$-c\left(\frac{\partial^2}{\partial x_1^2}+\frac{\partial^2}{\partial x_2^2}+\cdots+\frac{\partial^2}{\partial x_n^2}\right)u+au=f(t,\boldsymbol{x}). \tag{11.2.7}$$

Definition 11.6. The general form of a parabolic partial differential equation is

$$d\frac{\partial u}{\partial t} - \operatorname{div}(c\nabla u) + au = f(t, \boldsymbol{x}). \tag{11.2.8}$$

Based on the above descriptions, if c is a constant, the equation can be simply written as

$$d\frac{\partial u}{\partial t} - c\left(\frac{\partial^2 u}{\partial x_1^2} + \frac{\partial^2 u}{\partial x_2^2} + \cdots + \frac{\partial^2 u}{\partial x_n^2}\right) + au = f(t, \boldsymbol{x}). \tag{11.2.9}$$

Definition 11.7. The general form of a hyperbolic partial differential equation is

$$m\frac{\partial^2 u}{\partial t^2} - \operatorname{div}(c\nabla u) + au = f(t, \boldsymbol{x}). \tag{11.2.10}$$

If c is a constant, the equation can be simplified as

$$m\frac{\partial^2 u}{\partial t^2} - c\left(\frac{\partial^2 u}{\partial x_1^2} + \frac{\partial^2 u}{\partial x_2^2} + \cdots + \frac{\partial^2 u}{\partial x_n^2}\right) + au = f(t, \boldsymbol{x}). \tag{11.2.11}$$

It can be seen from the above three types of equations that the difference lies in the order of the derivative of the function u with respect to t. If there is no derivative, it can be regarded as a constant, such that the equation is like an elliptic algebraic equation $ax_1^2 + bx_2^2 = c$. Therefore, it is named an elliptic partial differential equation; if the first-order derivative of u is involved, it is like a parabolic equation $y = ax_1^2 + b_2^2$, therefore it is called a parabolic partial differential equation; if the second-order of the derivative of u is involved, it is like $y^2 = ax_1^2 + bx_2^2$, and the equation is referred to as a hyperbolic partial differential equation.

The finite element method is used in MATLAB Partial Differential Equation Toolbox, for handling such equations. In elliptic partial differential equations, the coefficients c, a, d, and f can be described as any functions, while in other forms, they must be constants.

11.2.2 Eigenvalue-type partial differential equations

Definition 11.8. The eigenvalue-type partial differential equation is defined as

$$-\operatorname{div}(c\nabla u) + au = \lambda du. \tag{11.2.12}$$

For constant c, the equation can further be simplified as

$$-c\left(\frac{\partial^2 u}{\partial x_1^2} + \frac{\partial^2 u}{\partial x_2^2} + \cdots + \frac{\partial^2 u}{\partial x_n^2}\right) + au = \lambda du. \tag{11.2.13}$$

Comparing (11.2.13) and (11.2.2), it is found that if the term λdu is moved to the left-hand side, the equation can be converted into an elliptic partial differential equation. Therefore, such equation is a special case of an elliptic partial differential equation.

11.2.3 Classification of boundary conditions

Generally speaking, there are three types of boundary conditions:

(1) If the values of function $u(t, \boldsymbol{x})$ on the boundaries are known, the condition is referred to as Dirichlet boundary conditions, which is named after a German mathematician Peter Gustav Lejeune Dirichlet (1805–1859).

Definition 11.9. The mathematical form of Dirichlet boundary conditions is

$$h\left(\boldsymbol{x}, t, u, \frac{\partial u}{\partial \boldsymbol{x}}\right) u\big|_{\partial\Omega} = r\left(\boldsymbol{x}, t, u, \frac{\partial u}{\partial \boldsymbol{x}}\right) \tag{11.2.14}$$

where $\partial\Omega$ represents the boundary of the geometric region of interest. Assuming that the boundary conditions are satisfied, one should specify r and h. They can be constants, or functions of t, $\boldsymbol{x}$ and even u and $\partial u/\partial \boldsymbol{x}$. For simplicity, let $h = 1$.

(2) If the values of $\partial u(t, \boldsymbol{x})/\partial \boldsymbol{x}$ at the boundaries are known, the boundary conditions are referred to as Neumann boundary conditions, named after a German mathematician Carl Gottfried Neumann (1832–1925).

Definition 11.10. The general form of Neumann boundary conditions is

$$\left[\frac{\partial}{\partial \boldsymbol{n}}(c\nabla u) + qu\right]\bigg|_{\partial\Omega} = g \tag{11.2.15}$$

where $\partial u/\partial \boldsymbol{n}$ is the partial derivative of $\boldsymbol{x}$ in the normal direction.

(3) There is another form, which is a certain combination of $u(t, \boldsymbol{x})$ and $\partial u(t, \boldsymbol{x})/\partial \boldsymbol{x}$, the so-called Robin boundary condition, named after a French mathematician Victor Gustave Robin (1855–1897).

11.3 User interface of typical two-dimensional partial differential equations

11.3.1 An introduction of the user interface

In the Partial Differential Equation Toolbox provided in MATLAB, a graphical user interface is given to solve a two-dimensional partial differential equation for a function $u(x_1, x_2)$. The interface supports the geometric region description by basic elements, such as circles and elliptic, rectangular, and polygon shapes, in graphical methods. Set operations such as union, intersection, and difference are supported to draw complicated geometric regions. After defining the solution region, Delaunay triangulation of the geometric region is allowed, and a mesh grid can be generated automatically.

In MATLAB environment, typing `pdetool` at the MATLAB prompt, the interface can be launched, as shown in Figure 11.4.

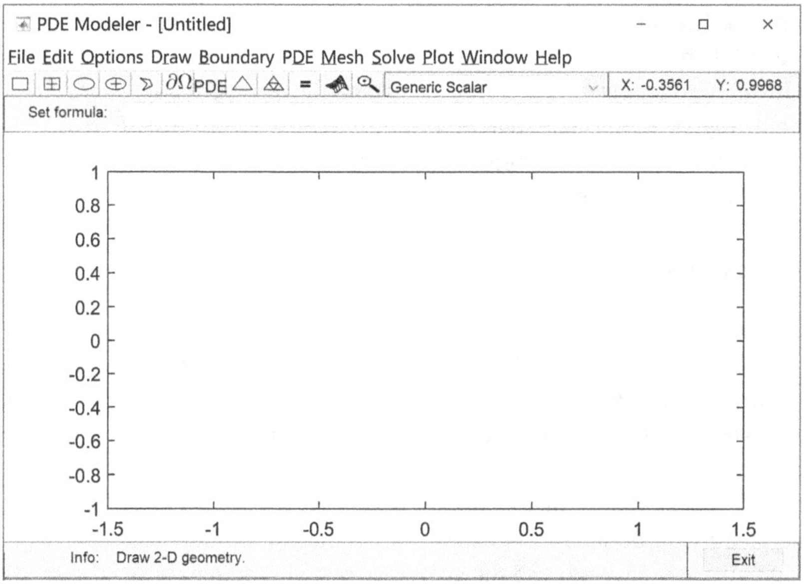

Figure 11.4: Graphical user interface.

There are several parts in the user interface:

(1) Menu system. A very comprehensive menu system is provided in the interface. Most of the menu items can also be implemented also from the buttons of the toolbar.

(2) Toolbar. The buttons in the toolbar are explained in Figure 11.5. It can be seen that the geometric region definition, equation and condition parameter descriptions, as well as partial differential equation solutions and result display can all be set with the interface. In the listbox to the right, the common partial differential equation types solvable by the interface can be selected.

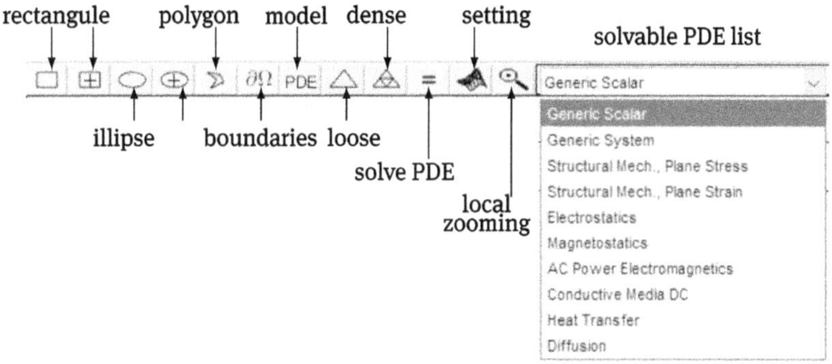

Figure 11.5: Toolbar of the interface.

(3) **Set formula.** The geometric region can be regarded as the sets composed of some fundamental shapes. Set computations such as union, intersection, and difference are allowed to precisely describe the geometric regions.
(4) **Geometric region.** In the major part of the interface, the users are allowed to draw the geometric region with different shapes. Three-dimensional displays are supported in MATLAB, but new graphics windows will be needed.

11.3.2 Geometric region design

In this section, an example is used to show how to use the interface to define the geometric region. Some ellipses and rectangles can be drawn with the tools as shown in Figure 11.6. Each shape can be regarded as a set. In **Set formula** edit box, the formula can be expressed as (R1+R2+E1)−E2, meaning the union of sets R1, R2, and E1, and the removal of set E2.

Clicking the $\partial\Omega$ button in the toolbar, the geometric region can be obtained. Selecting the menu item Boundary →Remove All Subdomain Borders, the borders in adjacent domains are removed, and the geometric region is automatically drawn, as shown in Figure 11.7.

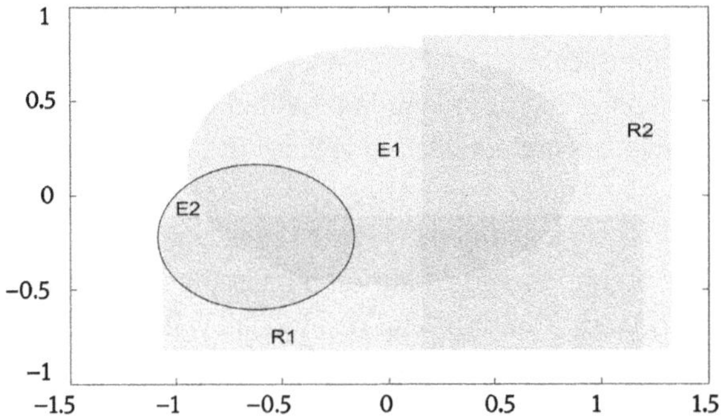

Figure 11.6: Geometric region setting.

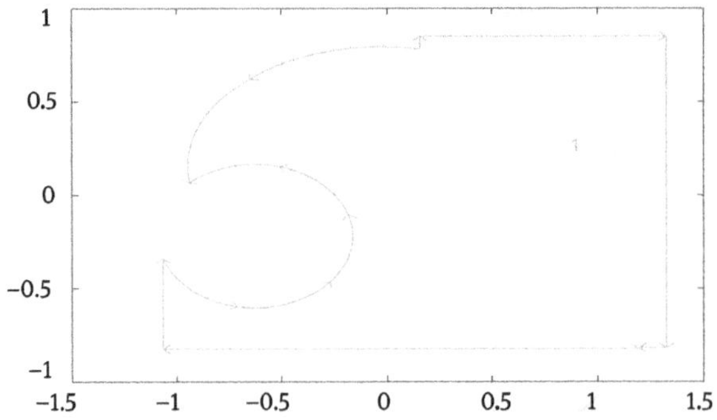

Figure 11.7: The actual geometric region.

With the geometric region, the Δ button in the toolbar can be clicked such that Delaunay triangulation in the geometric region is made, as shown in Figure 11.8. If a dense triangulation is expected, the appropriate button can be clicked, and the new mesh grids are shown in Figure 11.9. It should be noted that the denser the grids, the more accurate the solutions, while the longer the time.

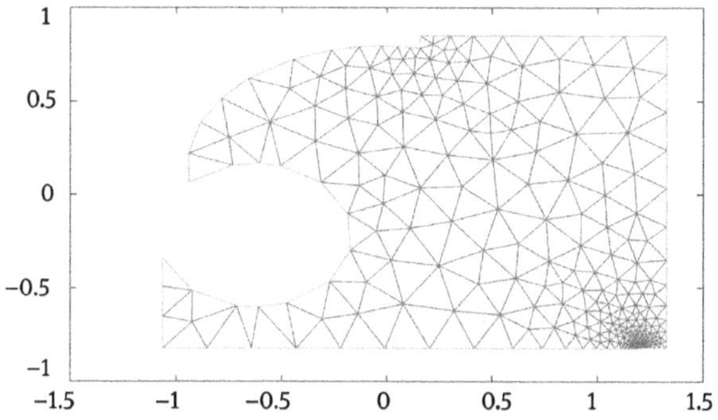

Figure 11.8: Triangulation in the geometric region.

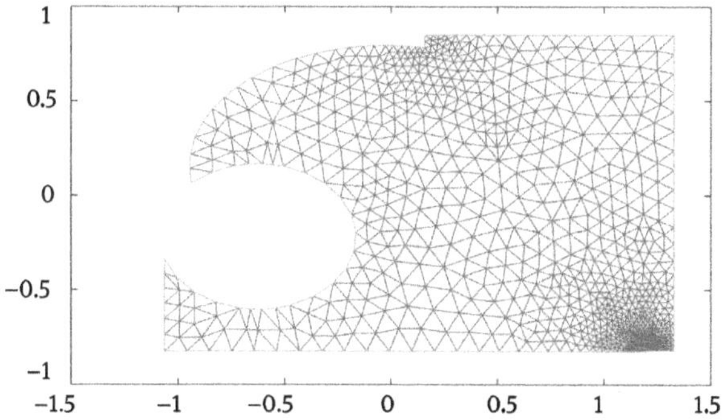

Figure 11.9: Dense grids.

11.3.3 Boundary condition description

Clicking the ∂Ω button in the toolbar, the boundary conditions can be specified. The Dirichlet and Neumann conditions are supported in the interface.

Selecting the menu item Boundary→Specify Boundary Conditions, a dialog box show in Figure 11.10 is opened. The boundary conditions can be specified in the dialog box. If the conditions on all the bounders are expected to be 0, then the r edit box can be filled with 0.

Boundary Condition — □ ×

Boundary condition equation:	h*u=r		

Condition type:	Coefficient	Value	Description
○ Neumann	g	0	
◉ Dirichlet	q	0	
	h	1	
	r	0	

OK   Cancel

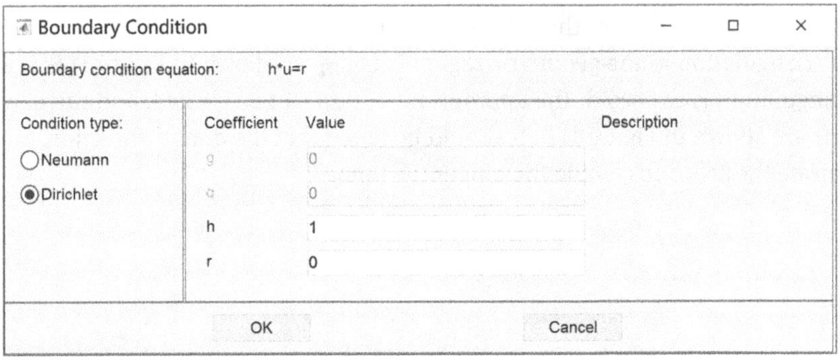

Figure 11.10: Boundary condition dialog box.

11.3.4 Examples of partial differential equation solutions

When the geometric region and boundary conditions are specified, and the partial differential equation is described, the = button in the toolbar can be clicked to solve the partial differential equation directly. An example is given next to demonstrate the solution process.

Example 11.6. Solve the hyperbolic partial differential equation

$$\frac{d^2u}{dt^2} - \frac{\partial^2 u}{\partial x^2} - \frac{\partial^2 u}{\partial y^2} + 2u = 10.$$

Solutions. It can be seen from the given equation that $c = 1$, $a = 2$, $f = 10$, and $d = 1$. Clicking the PDE label in the toolbar, a dialog box shown in Figure 11.11 is opened. Various partial differential equation types are listed on the left. The Hyperbolic item can be selected, and the parameters in the dialog box can be specified.

PDE Specification — □ ×

Equation:	d*u'-div(c*grad(u))+a*u=f	

Type of PDE:	Coefficient	Value
○ Elliptic	c	1.0
◉ Parabolic	a	2
○ Hyperbolic	f	10.0
○ Eigenmodes	d	1

OK   Cancel

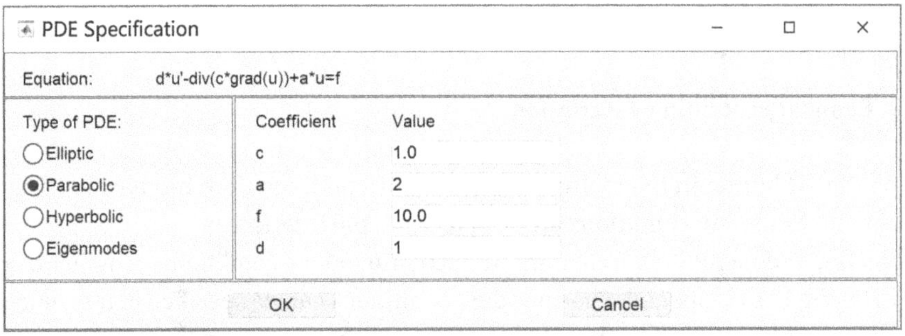

Figure 11.11: Partial differential equation parameter dialog box.

To solve this partial differential equation, the equal button in the toolbar can be clicked, and the solution of the equation can be found, as shown graphically in Figure 11.12. The pseudocolor in the figure reflects the solution $u(x, y)$. Note that if time is involved, the $u(x, y)$ values at time $t = 0$ are displayed. The display at other time t will be illustrated later.

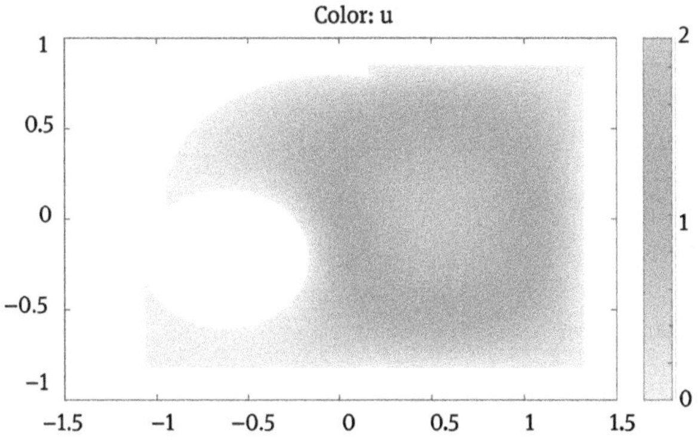

Figure 11.12: Equation solutions.

The boundary conditions can be modified. For instance, in the dialog box shown in Figure 11.10, the Dirichlet condition is still used, with the values of u on the boundaries set to 10. This can be set by filling the r edit box by 10. The partial differential equation can be solved again, and a solution similar to that in Figure 11.12 can be found. The difference is that the scale on the right has changed to 10~12.

11.3.5 Other solution display methods

In the previous example it can be seen that a visual method can be used to input the partial differential equations and boundary conditions in the `pdetool` interface, and such equations can then be solved. The solutions can be displayed in contours, in pseudocolor. Apart from the pseudocolor display, other display methods are supported. For instance, the surface or even animation methods are allowed. The display formats are demonstrated next as an example.

Example 11.7. Consider the partial differential equation and geometric region in Example 11.6. The boundary conditions are the same as in that example. Display the solution with contours.

Solutions. The original problem in the user interface was saved in file `c11mpde1.m`. Load the file directly into the user interface. Clicking the button in the toolbar, a dialog box shown in Figure 11.13 appears. If Contour and Arrows options are clicked, the contours with arrows will be shown as in Figure 11.14.

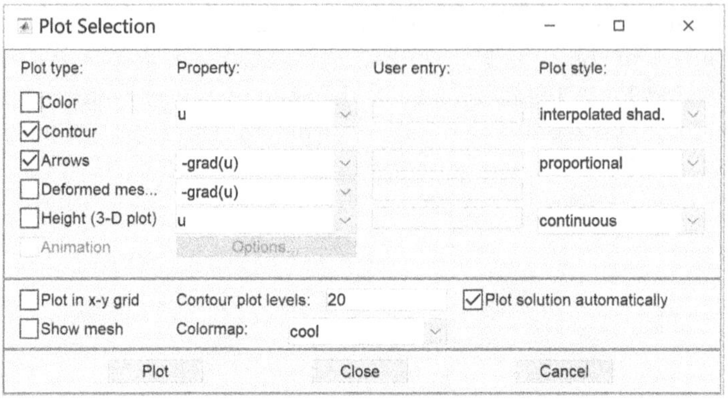

Figure 11.13: Display setting dialog box.

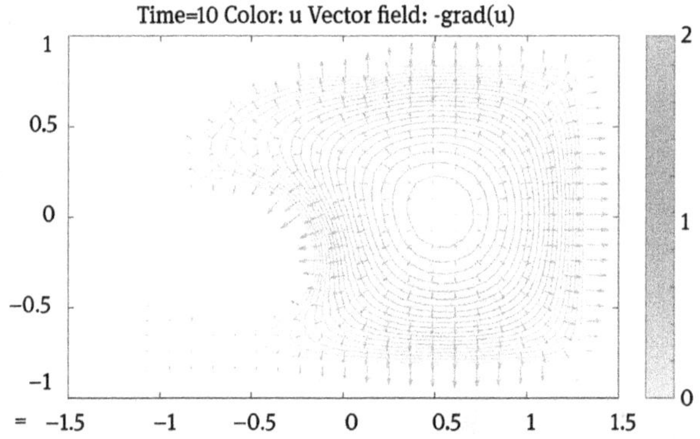

Figure 11.14: Solution representation by contours and arrows.

Note also that, in the dialog box in Figure 11.13, the Property items all have listboxes. The first one is the default u, indicating the display is for the solution $u(x, y)$. If the solutions of other functions are expected, they can be assigned from the listboxes.

If the option Height (3d-plot) is clicked, another figure window appears, and a three-dimensional surface plot in mesh grid form can be obtained as shown in Figure 11.15.

Example 11.8. Consider again the partial differential equation and geometric region in Example 11.6. Display the equation solutions in three-dimensional animation format.

Solutions. The default time vector for the user interface is $\boldsymbol{t}$=0:10. The solution at the final time $t = 10$ in shown in Figure 11.12. It can be seen for a hyperbolic partial differential equation that the solution should also be a function of time. Therefore the animation format can be used to illustrate solutions in a dynamical way. The hyperbolic partial differential equation in Example 11.6 is still used. We illustrate how to display the dynamic results of time-varying equations.

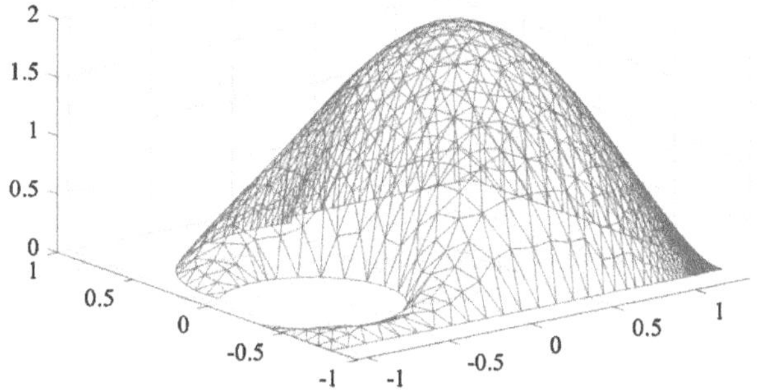

Figure 11.15: Three-dimensional mesh grid surface.

The menu item Solve →Parameters can be selected, and then a dialog box in Figure 11.13 appears. In this dialog box, the time interval can be assigned to `0:0.1:4`. Also if Animation is clicked, then clicking the Options button, a further dialog box appears, which allows the user to assign video play speed, with the default of 6 fps, i. e., 6 frames per second. The solution of the partial differential equation can be displayed in animation. The user may use the menu item Plot →Export Movie to export the video into MATLAB workspace, for instance, by saving as variable `M`. Function `movie(M)` can be used to play videos in MATLAB Graphics window. Command `movie2avi(M,'myavi.avi')` can be used to save the animation into a video file `myavi.avi`, for later viewing.

11.3.6 Partial differential equations with functional coefficients

In the discussion earlier, the coefficients c, a, d, and f were assumed to be constants. In real partial differential equations, sometimes functional coefficients are needed. In the Partial Differential Equation Toolbox, nonlinear functional coefficients can be handled for elliptic partial differential equations. In the strings for the coefficients, the characters `x`, `y` are allowed to describe x_1, x_2 or x, y in the equation; characters `ux` and `uy` are used for describing $\partial u/\partial x$ and $\partial u/\partial y$. They can be of any nonlinear function. Examples will be used to show the solution of such partial differential equations.

Example 11.9. Assume that a partial differential equation is given by

$$-\mathrm{div}\left(\frac{1}{1+|\nabla u|^2}\nabla u\right)+(x^2+y^2)u=\mathrm{e}^{-x^2-y^2}$$

where the boundaries are with zero conditions. Solve this partial differential equation.

Solutions. Observing the equation, it can be seen that this is an elliptic partial differential equation with

$$c = \frac{1}{\sqrt{(1 + (\partial u/\partial x)^2 + (\partial u/\partial y)^2}}, \quad a = x^2 + y^2, \quad f = \mathrm{e}^{-x^2-y^2},$$

and on the boundary u is 0. Using the Partial Differential Equation Toolbox, the dialog box shown in Figure 11.11 appears. One may select **Elliptic** item to show the elliptic partial differential equation. In the equation model, fill `1./sqrt(1+ux.^2+uy.^2)` in item c. In a and f items, one should fill in `x.^2+y.^2` and `exp(-x.^2-y.^2)`. With **Solve → Parameters** menu, a dialog box is displayed, from which the item **Use nonlinear solver** should be ticked. The functional coefficients apply only to elliptic partial differential equations. Clicking the equality sign in the toolbar, the solutions can be found, and the results are shown in Figure 11.16.

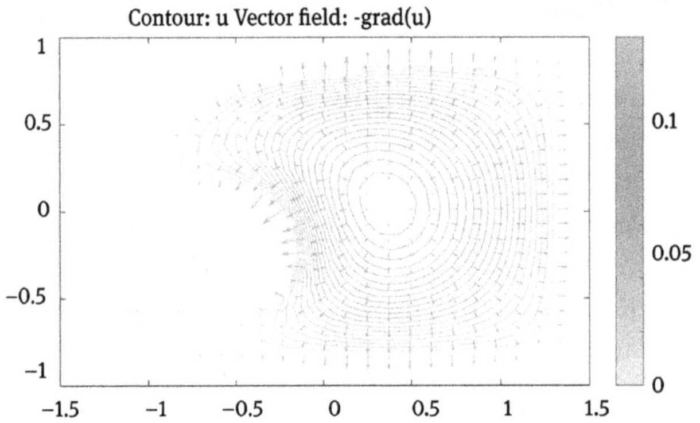

Figure 11.16: Contours with arrows.

11.4 Solutions of partial differential equations

Apart from the user interface based solution technique, command driven methods are supported also in MATLAB Partial Differential Equation Toolbox. Since MATLAB R2016a, a `PDEModel` object can used to represent partial differential equation models, and it can also be used in partial differential equation solutions. In this section, describing of geometric regions, partial differential equation and boundary condition descriptions, as well solutions are all presented. The command driven technique can also be used in solving partial differential equations which may be difficult to handle in other methods.

11.4.1 Creating a blank partial differential equation object

Before discussing the modeling and solution of partial differential equations, the command `createpde()` can be used to create a blank partial differential equation object, with the syntax M=`createpde`. The name of the object is `PDEModel`, whose commonly used members are provided in Table 11.2.

Table 11.2: Members of the `PDEModel` object.

member names	member explanation
`PDESystemSize`	numbers of PDEs N, the default one is 1. Function can also be called with `createpde(`N`)`
`IsTimeDependent`	whether the PDE contains explicitly time t; 0 for no, and 1 for yes. The member can be set by a command, or set automatically when specifying coefficients
`Geometry`	geometric region, the default one is an empty matrix. It can also be set with functions such as `geometryFromEdges()`
`EquationCoefficients`	coefficients of PDEs, which can be set by the functions such as `specifyCoefficients()`
`BoundaryConditions`	boundary conditions set by `applyBoundaryCondition()`
`InitialConditions`	initial conditions set by `setInitialConditions()`
`Mesh`	mesh grid format, with default empty matrix. It can be set to assigned individually with function `generateMesh()`
`SolverOptions`	control parameters, similar to the case of ODEs, the member can be set to `RelativeTolerance` and `AbsoluteTolerance`

11.4.2 Statement description of geometric regions

Two-dimensional geometric region can be created graphically with `pdetool` user interface, and it can be stored in a file. Unfortunately, the geometric domain thus generated cannot be loaded directly into the `PDEModel` object. Relevant MATLAB statements should be issued to define the region again. In this section, the design method is illustrated.

Example 11.10. In Example 11.6, a MATLAB file `c11mpde0.m` was created. If the file is opened, it can be seen that inside the function a paragraph

```
pderect([-1.064151 1.1949686 -0.1031447 -0.8226415],'R1');
pderect([0.1584906 1.3257862 0.85031447 -0.8176101],'R2');
pdeellip(-0.023899,0.1534591,0.92201258,0.6415094,0,'E1');
pdeellip(-0.625157,-0.218868,0.46163522,0.384905660,0,'E2');
set(findobj(get(pde_fig,'Children'),...
                    'Tag','PDEEval'),'String','(R1+R2+E1)-E2')
```

is present. Do not attempt to execute the code, since pde_fig is an internal function, which cannot be executed in MATLAB directly. It can be seen that two rectangles and two ellipses are defined. Set operations are also described. With the internal commands, the geometric region can be described.

Four basic shapes are provided in the Partial Differential Equation Toolbox. Details of the shapes are described in Table 11.3. Each geometric shape is described by a column vector, with different lengths. If set operations are expected, the end of the shorter ones should be appended by 0's, such that their lengths are unified.

Table 11.3: Basic shapes and parameters.

shape names	sign x_1	x_2	x_3	x_4	x_5	x_6	x_7	x_8	x_9	x_{10}
circle	1	x_0	y_0	r	0	0	0	0	0	0
polygon	2	n	start of each segment x_i				end of the segments y_i			
rectangle	3	4	x_m	x_M	x_M	x_m	y_m	y_m	y_M	y_M
ellipses	4	x_0	y_0	r_x	r_y	θ	0	0	0	0

When the basic shapes are described, all the shapes can be spanned into a matrix. Each shape is assigned to a string for its name, and set operations are defined in a certain format. Function decsg() can then be called to compute the bounds matrix. The subbounds are still retained in the matrix. If one wants to remove the interior bounds, function csgdel() can be used so that only the outer bounds are extracted. Function pdegplot() can be called to draw the boundaries. The syntaxes of these functions are fixed, and can be better understood after learning the demonstrative commands in the examples.

Example 11.11. Use low-level statements to declare the boundaries in Figure 11.17. The geometric region is defined by the union of circle C_1 and rectangle R, and subtracting circle C_2.

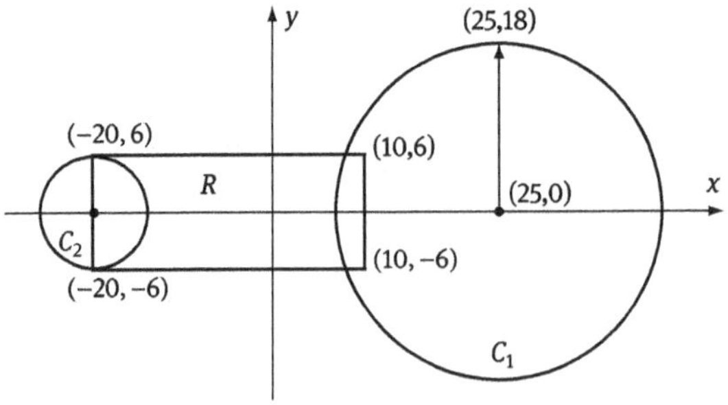

Figure 11.17: Illustration of the expected geometric region.

Solutions. The three basic shapes should be created first, with the parameters specified in the illustration. With the basic shapes, set operations are defined, and functions can be called sequentially to extract the bounds. The finalized bounds are as shown in Figure 11.18. The syntaxes of these functions are fixed, and one may read and compare with the display. Detailed information may be seen in [51].

```
>> C1=[1; 25; 0; 18; zeros(6,1)];  %the larger circle
   C2=[1; -20; 0; 6; zeros(6,1)];  %the smaller circle
   R=[3; 4; -20; 10; 10; -20; -6; -6; 6; 6]; %draw the rectangle
   smat=[R,C1,C2];      %form the matrix from all these shapes
   ns=char('R','C1','C2')';           %set names, not transpose is made
   shape='(C1+R)-C2';                 %set operation expression
   [g,g0]=decsg(smat,shape,ns);   %carry out set operation
   [g1,g10]=csgdel(g,g0);           %remove the interior bounds in the set
   pdegplot(g1,'EdgeLabels','on','SubdomainLabels','on') %draw bounds
```

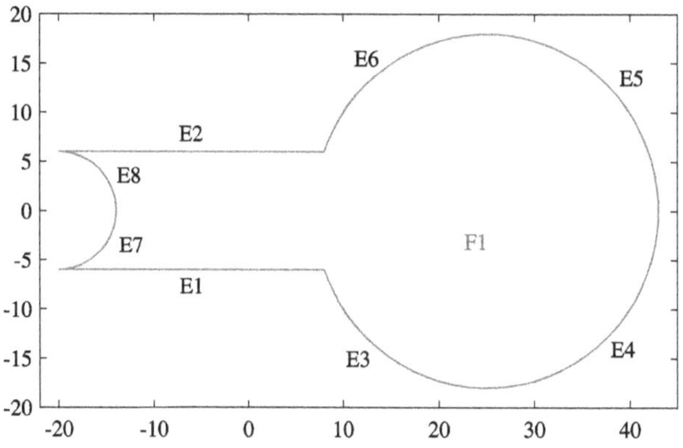

Figure 11.18: The bounds of the geometric region.

It can be seen that with the above commands, the boundaries are extracted successfully. Also each edge is described and assigned a name. Normally, a circle can be described by 4 edges. The numbered edges can be used in subsequent boundary condition specifications.

Example 11.12. It seems that the bounds created with the `pdetool` user interface cannot be extracted directly. Use the information extracted from the model file in Example 11.10, and redefine the boundaries with basic shapes.

Solutions. The information in the file of Example 11.10 is listed below:

```
pderect([-1.064151 1.1949686 -0.1031447 -0.8226415],'R1');
pderect([0.1584906 1.3257862 0.85031447 -0.8176101],'R2');
```

```
pdeellip(-0.023899,0.1534591,0.92201258,0.6415094,0,'E1');
pdeellip(-0.625157,-0.218868,0.46163522,0.384905660,0,'E2');
```

It can be seen that R_1 is a rectangle, whose x range is $(-1.064, 1.195)$ and y range is $(-0.103, -0.822)$. It can be described directly by the following statements. Other shapes can also be redefined with commands, so that the boundaries can be recreated as shown in Figure 11.19.

```
>> R1=[3;4;-1.064;1.195;1.195;-1.064;-0.103;-0.103;-0.822;-0.822];
   R2=[3;4;0.158;1.326;1.326;0.158;-0.818;-0.818;0.850;0.850];
   E1=[4;-0.0239;0.153;0.922;0.642;0;0;0;0;0];
   E2=[4;-0.625;-0.21;0.462;0.385;0;0;0;0;0];
   smat=[R1,R2,E1,E2]; ns=char('R1','R2','E1','E2')';
   shape='(R1+R2+E1)-E2'; [g,g0]=decsg(smat,shape,ns);
   [g1,g10]=csgdel(g,g0); %remove the interior bounds
   pdegplot(g1,'EdgeLabels','on','SubdomainLabels','on') %draw bounds
   axis([-1.5,1.5,-1,1]), save c11exbnd g1 %save the bounds to a file
```

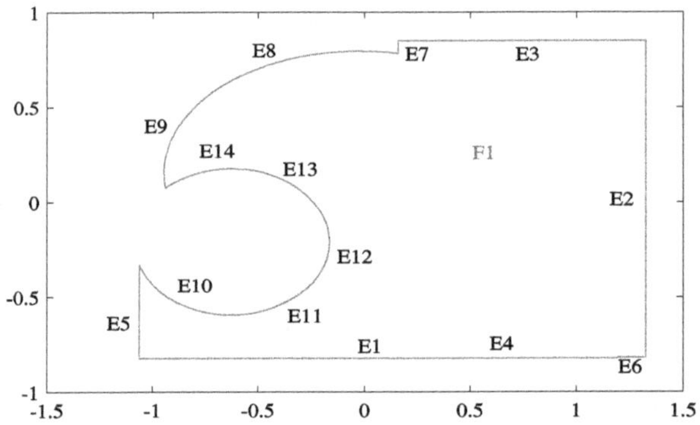

Figure 11.19: The geometric region boundary redefined.

Compared with the two-dimensional boundaries created earlier, the descriptions of three-dimensional geometric regions are even more complicated. There are no suitable low-level commands to describe such geometric regions. There is a user interface provided in the Partial Differential Equation Toolbox, with other CAD softwares such as SolidWorks. If one has already established an `*.stl` file in SolidWorks, function `importGeometry()` can be used to load the geometric information into MATLAB workspace.[51] Then, function `pdegplot()` can be used to draw three-dimensional geometric regions directly.

If the geometric region model has been created, `geometryFromEdges(`M`,g1)` function can be used to assign the information to the blank `PDEModel` object M.

11.4.3 Boundary and initial condition descriptions

From the previously established boundaries of the geometric region, each edge of the boundary can be assigned individual boundary conditions. From the given `PDEModel` object M, the command `applyBoundaryCondition()` can be called to assign a boundary condition, with the syntaxes

```
applyBoundaryCondition(M,'edge',k,'u',v)  % Dirichlet boundary
applyBoundaryCondition(M,'edge',k,'g',v)  % Neumann boundary
```

where $\boldsymbol{k}$ is the serial number of the edge, and it can be a vector, so that the boundary conditions for several edges can be set simultaneously. Vector $\boldsymbol{v}$ is composed of boundary values or dot operations of functions. If a certain boundary edge was not assigned a boundary condition, default zero Dirichlet condition is assigned automatically. For instance, if edges 3 and 7 in the M object are expected to be constant 2, the command

```
applyBoundaryCondition(M,'edge',[3,7],'u',2)
```

can be used. Each set of boundary conditions should be set individually by calling `applyBoundaryCondition()` function once.

Similar to boundary conditions, function `setInitialConditions(`M`, `u_0`)` can be used to set initial conditions, where u_0 can be a constant or a function handle for the initial conditions. If $m \neq 0$, the initial conditions $u'(t)$ should also be assigned, with `setInitialConditions(`M`,`u_0`,`u_0'`)`. Initial value specification will be illustrated later through examples.

11.4.4 Partial differential equations descriptions

Under the `PDEModel` object, a standard form of partial differential equations is defined. In this section, this standard form is presented. Then, the method of solving partial differential equations with MATLAB is illustrated.

Definition 11.11. The standard partial differential equation expressed for `PDEModel` object is given by

$$m\frac{\partial^2 u}{\partial t^2} + d\frac{\partial u}{\partial t} - \operatorname{div}(c\nabla u) + au = f \tag{11.4.1}$$

where m, d, c, a, and f are all referred to as coefficients. They can be constants, zeros, or known functions of t and u.

It can be seen from the standard form that the elliptic ($m = d = 0$), parabolic ($m = 0$), and hyperbolic ($d = 0$) partial differential equations are just special cases. Besides, diffusion equation ($m = a = 0$) and Poisson equation ($m = d = a = 0$) are also special cases.

Function `specifyCoefficients()` can be used to specify coefficients of the given partial differential equations with the syntax

`specifyCoefficients(`M`,name1,value2,name2,value2, ⋯)`

where "name" can be selected as `'m'`, `'d'`, `'c'`, `'a'`, and `'f'`, while the "value" can be set to values and function handles. All the five items must be assigned, otherwise an error message will be displayed.

Example 11.13. Use `PDEModel` object to express the hyperbolic partial differential equation in Example 11.6.

Solutions. Comparing with the standard form in (11.4.1), it is immediately recognized that $m = c = 1$, $a = 2$, $f = 10$, and $d = 0$. All of these parameters are constants. With these preparations, the following commands can be used to describe the hyperbolic partial differential equation:

```
>> M=createpde; load c11exbnd; geometryFromEdges(M,g1)
   specifyCoefficients(M,'m',1,'c',1,'a',2,'f',10,'d',0);
```

For the coefficients which are not constants, anonymous or MATLAB functions can be used to describe them. The input arguments of the functions are `region` and `state`. They are both structured variables. In the structure `region`, the members are x, y, and z, for spatial information, while in `state`, the members are u, `ux` (for $\partial u/\partial x$), `uy`, `uz`, and `time` (for t). Therefore the structured variables with member names should be used to describe the coefficients. Dot operations should be used in describing the coefficients. An example will be given next to show how to describe variable coefficients in anonymous functions.

Example 11.14. The elliptic partial differential equation in Example 11.9 is with variable coefficients. Create such a model.

Solutions. For convenience, the variable coefficients in Example 11.9 are rewritten below:

$$c = \frac{1}{\sqrt{(1 + (\partial u/\partial x)^2 + (\partial u/\partial y)^2}}, \quad a = x^2 + y^2, \quad f = \mathrm{e}^{-x^2-y^2}.$$

From the mathematical expressions, the anonymous functions for the coefficients a, c, and f can be established, from them then handles can be set to the partial differential equation object:

```
>> f=@(region,state)exp(-region.x.^2-region.y.^2);
   a=@(region,state)region.x.^2+region.y.^2;
   c=@(region,state)1./sqrt(1+state.ux.^2+state.uy.^2);
   M=createpde; load c11exbnd; geometryFromEdges(M,g1);
   specifyCoefficients(M,'m',0,'d',0,'c',c,'a',a,'f',f);
```

11.4.5 Numerical solutions of partial differential equations

Similar to the case when solving partial differential equation, Delaunay mesh grid should be generated first before the solution process, so that the partial differential equations solutions on the mesh grid points can be found. More about Delaunay mesh grid can be found in Volume I. Function `generateMesh()` can be used to generate the mesh grid, with the syntaxes

```
m=generateMesh(M)
m=generateMesh(M,name1,value,name2,value2, ...)
```

The commonly used property is `'Hmax'`, which determines the sizes of the mesh grid. For accurate solutions, it should be set to very small numbers, while this will in turn reduce the speed in the solution process. A tradeoff should be made in the selection. Normally, it should be set to 0.02 or 0.01. Of course, other properties are also allowed, see [51]. With the generated mesh grid, the information in member `Mesh` of the object M will be updated automatically.

When the mesh grids are generated, function `s=solvepe(M)` can be called to solve the partial differential equations numerically. A structured variable `s` can be returned, with the members `NodalSolution`, `XGradients`, and others. The member `NodalSolution` contains the information of the solution on the mesh grids, while `XGradients` contains the information of the partial derivatives. If variable t is explicitly contained in the partial differential equations, a time vector $\boldsymbol{t}$ should be generated first, by using the command `s=solvepe(M,t)` to solve the partial differential equations.

It should be pointed out that the eigenvalue-type partial differential equations cannot be solved with `solvepde()` function. They should be solved with function `solvepdeeig()` instead. Such partial differential equations are not covered in this book.

The solution u obtained is a collection of the values on the mesh grid points. It cannot be converted to those in other mesh grids by default. The dedicated function `pdeplot()` can be used to draw the solution. It can be extracted with the command `u=s.NodalSolution` first. There are several ways to call function `pdeplot()`:

```
pdeplot(M,'XYData',u)                   % draw the surface
pdeplot(M,'XYData',u,'contour',on)      % draw contours
pdeplot(M,'XYData',u,'ZData',u)         % draw 3D surface
pdeplot(M,'XYData',u,'FlowData',[s.XGradients,s.YGradients])
```

where in the last one, the arrows are added to the results. If the partial differential equations are three-dimensional, function `pdeplot3D()` can be called to draw the results.

The following examples can be used to demonstrate the partial differential equation solution process.

Example 11.15. Solve again the hyperbolic partial differential equation studied in Example 11.6 with MATLAB commands.

Solutions. In fact, the model in Example 11.13 has been demonstrated earlier. With this model, mesh grids can be generated and the partial differential equation can be solved. Unfortunately, the result obtained here is different from that obtained by the user interface.

```
>> M=createpde; load c11exbnd; geometryFromEdges(M,g1)
   specifyCoefficients(M,'m',1,'c',1,'a',2,'f',10,'d',0);
   u0=0; ud0=0; setInitialConditions(M,u0,ud0); %initial condition
   generateMesh(M,'Hmax',0.05); t=0:0.1:10; %generate mesh grid
   sol=solvepde(M,t); u=sol.NodalSolution;  %extract the solutions
   pdeplot(M,'XYData',u,'ZData',u(:,end))   %draw the solutions
```

Example 11.16. Solve the elliptic partial differential equation with variable coefficients studied in Example 11.9.

Solutions. The model for the partial differential equation has been created in Example 11.14. From it, the partial differential equation can be solved directly with the following commands. Again the solution obtained is different from that found using the user interface.

```
>> f=@(region,state)exp(-region.x.^2-region.y.^2);
   a=@(region,state)region.x.^2+region.y.^2;
   c=@(region,state)1./sqrt(1+state.ux.^2+state.uy.^2);
   M=createpde; load c11exbnd; geometryFromEdges(M,g1);
   specifyCoefficients(M,'m',0,'d',0,'c',c,'a',a,'f',f);
   generateMesh(M,'Hmax',0.1); sol=solvepde(M); %generate mesh grids
   pdeplot(M,'XYData',sol.NodalSolution) %draw the solutions
```

Example 11.17. Consider the partial differential equation given below[45]

$$\frac{\partial u}{\partial t}=\frac{\partial^2 u}{\partial x^2}+\frac{\partial^2 u}{\partial y^2}+(16\pi^2-1)\mathrm{e}^{-t}(\sin 4\pi x+\sin 4\pi y)$$

where the initial conditions at $t = 0$ are given by $u(0,x,y) = \sin 4\pi x + \sin 4\pi y$. The solution region is $0 \leqslant x,y \leqslant 1$. At the boundaries $x = 0$ and $x = 1$, the boundary conditions are $\mathrm{e}^{-t}\sin 4\pi y$, while for $y = 0$ and $y = 1$, the boundary conditions are $\mathrm{e}^{-t}\sin 4\pi x$. Solve the partial differential equation. If the solution is given by $u(t,x,y) = \mathrm{e}^{-t}(\sin 4\pi x + \sin 4\pi y)$, assess the efficiency and accuracy.

Solutions. Compared with the standard form of `PDEModel` object, it can be seen from the given partial differential equation that

$$m=a=0,\quad d=c=1,\quad f=(16\pi^2-1)\mathrm{e}^{-t}(\sin 4\pi x+\sin 4\pi y).$$

Therefore, the following statements can be used to describe the square geometric region, and describe the coefficients of the partial differential equation. It can also be seen from the boundary plot that edges 1 and 3 describe the boundaries at $y = 0$ and $y = 1$, respectively, while the edges 2 and 4 describe the boundaries at $x = 0$ and $x = 1$, respectively. The boundary conditions can be described by appropriate commands. The initial condition can also be expressed. With this information, the partial differential equation can be solved, and the solution surface is as shown in Figure 11.20. The result looks the same as for the analytical solution.

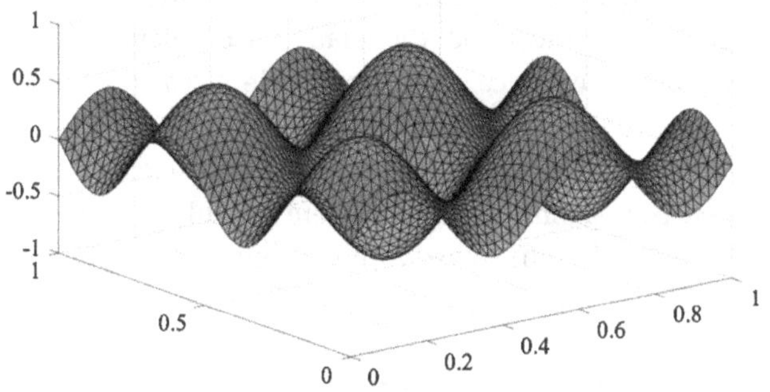

Figure 11.20: The solution surface at $t = 1$.

```
>> R=[3;4;0;1;1;0;0;0;1;1]; [g,h]=decsg(R,'R','R');
   M=createpde(1); geometryFromEdges(M,g);    %geometric region
   f=@(R,S)(16*pi^2-1)*exp(-S.time).*(sin(4*pi*R.x)+sin(4*pi*R.y));
   specifyCoefficients(M,'m',0,'d',1,'c',1,'a',0,'f',f);
   m13bnd=@(R,S)exp(-S.time).*sin(4*pi*R.x); %edges 1 and 3
   m24bnd=@(R,S)exp(-S.time).*sin(4*pi*R.y); %edges 2 and 4
   applyBoundaryCondition(M,'edge',[1,3],'u',m13bnd);
   applyBoundaryCondition(M,'edge',[2,4],'u',m24bnd);
   z=@(S)sin(4*pi*S.x)+sin(4*pi*S.y); setInitialConditions(M,z);
   tic, generateMesh(M,'Hmax',0.02);       %generate mesh grid Hmax
   ee=1e-6; M.SolverOptions.RelativeTolerance=ee;
   M.SolverOptions.AbsoluteTolerance=ee; %set error tolerances
   t=[0 1]; sol=solvepde(M,t); toc, u=sol.NodalSolution(:,end);
   pdeplot(M,'XYData',u,'ZData',u,'colorbar','off')
   shading faceted, grid %note Delaunay mesh grids are used here
```

Delaunay triangulation mesh grids are generated automatically for the geometric regions by the Partial Differential Equation Toolbox in MATLAB. The coordinates of the vertexes are stored in matrix $\boldsymbol{p}$, whose two row vectors are, in fact, the $\boldsymbol{x}$ and $\boldsymbol{y}$ vectors of the vertices. Matrix $\boldsymbol{p}$ can be extracted with `meshToPet()` function. Since the analytical solution is known, the exact values at the vertices can be found, and compared with the numerical solution obtained, so as to find the maximum error. Under the current setting, the maximum error is 2.3960×10^{-5}.

```
>> u0=@(t,x,y)exp(-t).*(sin(4*pi*x)+sin(4*pi*y));  % exact values
   [p,t,e]=meshToPet(M.Mesh); size(p)  % find vertexes and number
   u1=u0(1,p(1,:),p(2,:)).'; norm(u(:,end)-u1,inf) % maximum error
```

In the following code, different values of `'Hmax'` and error tolerance `ee` are probed, and the comparative results obtained are given in Table 11.4. In this table, the combinations (`'Hmax'`, `ee`) are used. It can be seen that the relative error tolerance does not contribute much to the accuracy. The choice of 10^{-6} is sufficient. The parameter `'Hmax'` is a very important factor to both the accuracy and elapsed time. The smaller the value of `'Hmax'`, the smaller the error, and the heavier the cost (elapsed time and number of vertices are increased significantly). Therefore in real applications, these parameters should be properly chosen, so as to find the numerical solutions effectively.

```
>> tic, generateMesh(M,'Hmax',0.1);          % generate mesh grids for Hmax
   ee=1e-6; M.SolverOptions.RelativeTolerance=ee;
   M.SolverOptions.AbsoluteTolerance=ee; % error tolerance
   t=[0 1]; sol=solvepde(M,t); toc, u=sol.NodalSolution(:,end);
   [p,t,e]=meshToPet(M.Mesh); size(p)      % generate mesh grids
   u1=u0(1,p(1,:),p(2,:)).'; norm(u(:,end)-u1,inf) % compute error
```

Table 11.4: Comparative studies of algorithm parameters.

parameters	$(0.1, 10^{-6})$	$(0.05, 10^{-6})$	$(0.02, 10^{-6})$	$(0.01, 10^{-6})$	$(0.005, 10^{-6})$	$(0.005, 10^{-7})$
elapsed time	2.186	4.156	11.293	34.134	141.946	245.042
numbers	497	1 893	11 637	46 337	182 409	182 409
maximum errors	0.0050	5.8×10^{-4}	2.4×10^{-5}	3.2×10^{-6}	4.9×10^{-7}	4.9×10^{-7}

Example 11.18. Consider the two-dimensional inhomogeneous diffusion in the $0 \leqslant x, y \leqslant 1$ region:

$$\frac{\partial y}{\partial t} = \frac{\partial^2 u}{\partial x^2} + \frac{\partial^2 u}{\partial y^2} + (1 - x)xt.$$

If on the boundaries, zero Dirichlet conditions are assumed, and the initial condition is $u(0, x, y) = \sin \pi x/2$, solve the diffusion equation and make the animation into a video file.

Solutions. Compared with the standard form in (11.4.1), it can be seen that $c = K = 1$, $d = 1$, $m = a = 0$, and $f = (1 - x)xt$. Therefore, the following commands can be used to describe the partial differential equation, and the geometric region and conditions. The equation can then be solved numerically. Animation can then be used to show the solution and be saved into a video file `c11mdiff.avi`.

```
>> R=[3;4;0;1;1;0;0;0;1;1]; [g,h]=decsg(R,'R','R');
   M=createpde(1); geometryFromEdges(M,g); % geometric region
   f=@(R,S)(1-R.x).*R.x.*cos(S.time);        % anonymous function
   specifyCoefficients(M,'m',0,'d',1,'c',1,'a',0,'f',f);
   applyBoundaryCondition(M,'edge',1:4,'u',0); % Dirichlet condition
   u0=@(S)sin(pi*S.x/2); setInitialConditions(M,0); % initial condition
   tic, generateMesh(M,'Hmax',0.02);                % mesh grids
   t=0:0.01:1; sol=solvepde(M,t); toc, u=sol.NodalSolution;
   vid=VideoWriter('c11mdiff.avi'); open(vid); % open blank video file
   for k=1:length(t),k % make solution into animation
      pdeplot(M,'XYData',u,'ZData',u(:,k),'colorbar','off')
      shading faceted; grid, zlim([0 0.02]), drawnow
      hVid=getframe; writeVideo(vid,hVid); % write a frame
   end, close(vid)                         % when completed, close video file
```

Example 11.19. Consider the following partial differential equation:[32]

$$\begin{cases} \dfrac{\partial u}{\partial t} = 1 + u^2 v - 4.4u + \alpha\left(\dfrac{\partial^2 u}{\partial x^2} + \dfrac{\partial^2 u}{\partial y^2}\right), \\ \dfrac{\partial v}{\partial t} = 3.4u - u^2 v + \alpha\left(\dfrac{\partial^2 v}{\partial x^2} + \dfrac{\partial^2 v}{\partial y^2}\right) \end{cases}$$

where $\alpha = 2 \times 10^{-3}$. It is known that the initial conditions are $u(0,x,y) = 0.5 + y$, $v(0,x,y) = 1 + 5x$, and the Neumann boundary conditions are zero. Solve the partial differential equation over the square region $0 \leqslant x, y \leqslant 1$, for $0 \leqslant t \leqslant 11.5$.

Solutions. It can be seen from the given equation that there are two partial differential equations. Letting $u_1 = u$ and $u_2 = v$, compared with the standard form, it is seen that $m = a = 0$, $d = 1$, and $c = 2 \times 10^{-3}$. The coefficient $\boldsymbol{f} = [1 + u_1^2 u_2 - 4.4u_1, 3.4u_1 - u_1^2 u_2]^{\mathrm{T}}$, and the initial condition $\boldsymbol{u}_0 = [0.5 + y, 1 + 5x]^{\mathrm{T}}$.

Selecting a step-size of 0.5 for time t, the following statements can be used to construct the square geometric region, describe the partial differential equations, set the conditions, and generate mesh grids, then finally find the numerical solution. Since $t = 11.5$ is the last sample in vector $\boldsymbol{t}$, the surfaces of the solutions can be found in Figures 11.21, (a) and (b).

```
>> R=[3;4;0;1;1;0;0;0;1;1]; [g,h]=decsg(R,'R','R'); % square region
   f=@(R,S)[1+S.u(1,:).^2.*S.u(2,:)-4.4*S.u(1,:);
```

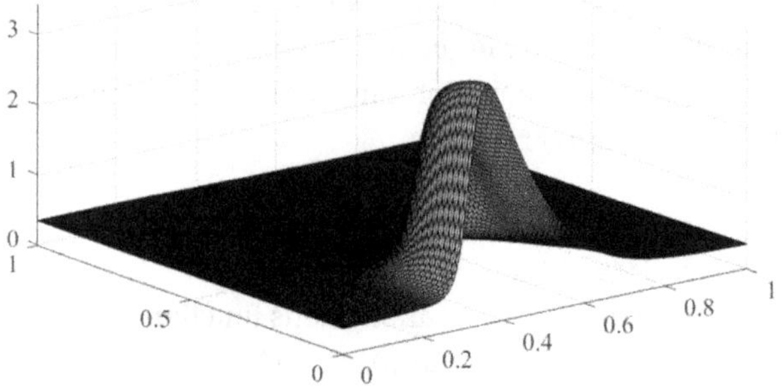

(a) surface of function u

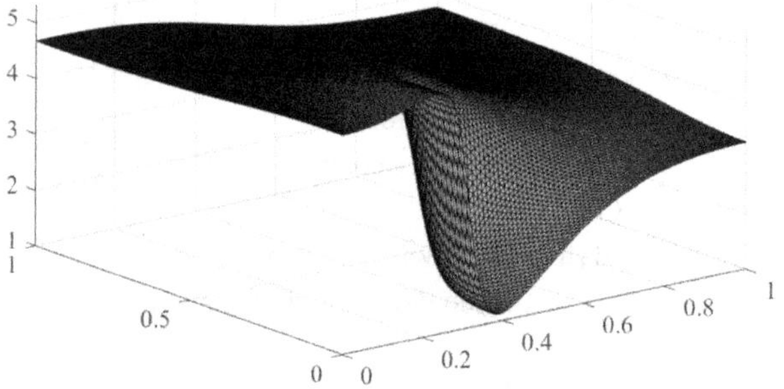

(b) surface of function v

Figure 11.21: Solutions at $t = 11.5$.

```
          3.4*S.u(1,:)-S.u(1,:).^2.*S.u(2,:)];
M=createpde(2); geometryFromEdges(M,g);       % geometric region
specifyCoefficients(M,'m',0,'d',1,'c',2e-3,'a',0,'f',f);
applyBoundaryCondition(M,'edge',1:4,'g',0); % Neumann condition
u0=@(R)[0.5+R.y; 1+5*R.x]; setInitialConditions(M,u0); % initial
tic, generateMesh(M,'Hmax',0.01);      % generate mesh grids with Hmax
t=0:0.5:11.5; sol=solvepde(M,t); toc % solve PDE
u1=sol.NodalSolution(:,1,:); u2=sol.NodalSolution(:,2,:);
pdeplot(M,'XYData',u1,'ZData',u1(:,:,end),'colorbar','off')
shading faceted, grid, figure  % draw next surface
pdeplot(M,'XYData',u1,'ZData',u2(:,:,end),'colorbar','off')
shading faceted, grid               % set mesh grid format
```

If Hmax is selected relatively small, the surface obtained is much smoother than that in [32], and more accurate. The cost is the increase in elapsed time, for this problem, which is about 366.9 seconds. If the precision in the reference is acceptable, `Hmax` can be set to 0.05, and then the elapsed time is 12.79 seconds.

11.5 Exercises

11.1 Show that for any real constants C and γ, $u(t,x) = \mathrm{e}^{-\gamma^2 Kt}\cos\gamma x + C$ satisfies the homogeneous diffusion equation.

11.2 High precision $o(h^p)$ numerical differentiation algorithms were presented in Volume II. In Section 11.1, the algorithm with $p = 1$ is used, and tri-diagonal coefficient matrix can be established. If larger values of p are taken, derive high precision algorithms for diffusion problems, and solve again for the diffusion equations in the examples.

11.3 Solve the extended diffusion equation in (11.1.1), where $K = 1$ and the boundary conditions are $u(t,0) = 0$ and $u(t,1) = 0$, also the initial condition is $u(0,x) = \eta(x) = \sin\pi x$. Find numerical solutions of the diffusion equations for $t \in (0,1)$, and measure the elapsed time for different step-sizes.

11.4 Solve the following diffusion equation, where u_t is the short-hand notation of $\partial u/\partial t$, while u_{xx} is $\partial^2 u/\partial x^2$:[17]

(a) $u_t = u_{xx} + x$, $u(0,x) = \sin 2x$, $u(t,0) = u(t,\pi) = 0$;

(b) $u_t = u_{xx} + 10$, $u(0,x) = 3\sin x - 4\sin 2x + 5\sin 3x$, $u(t,0) = u(t,\pi) = 0$.

11.5 Solve the diffusion equation $u_t = 3u_{xx}$, with the boundary conditions $u(t,0) = u(t,\pi) = 0$, and the initial condition $u(0,x) = x(\pi - x)$. The solution is needed for $t \in (0,3)$. It is known that the analytical solution of the original equation can be expressed as an infinite series.[17] Assess the accuracy and efficiency if the analytical solution is

$$u(t,x) = \frac{8}{\pi}\sum_{k=1}^{\infty}\frac{1}{(2k-1)^3}\mathrm{e}^{-3(2k-1)^2 t}\sin(2k-1)x.$$

11.6 For the homogeneous diffusion equation $u_t = u_{xx}$, with $x_l \leqslant x \leqslant x_r$, $0 \leqslant t \leqslant t_f$, $x_l = 0$, $x_r = 1$, the initial condition is $u(0,x) = \sin\pi x + \sin 2\pi x$, and boundary conditions are

$$u(t,x_l) = \sin\pi x_l \mathrm{e}^{-\pi^2 t} + \sin 2\pi x_l \mathrm{e}^{-4\pi^2 t}, \quad u(t,x_r) = \sin\pi x_r \mathrm{e}^{-\pi^2 t} + \sin 2\pi x_r \mathrm{e}^{-4\pi^2 t}.$$

If the analytical solution is $u(t,x) = \sin\pi x \mathrm{e}^{-\pi^2 t} + \sin 2\pi x \mathrm{e}^{-4\pi^2 t}$,[45] find the numerical solution and check its accuracy and efficiency.

11.7 Show that for any real constants γ, C_1, and C, the function

$$u(t,x) = C + C_1 x + \mathrm{e}^{-K\gamma^2 t}\sin\gamma x - \frac{1}{K\beta^2}\sin\beta x,$$

where $\beta \neq 0$, satisfies the following inhomogeneous diffusion equation:

$$u_t = u_{xx} + \sin\beta x.$$

11.8 Solve the following partial differential equation:

$$\begin{cases} \dfrac{\partial u_1}{\partial t} = 0.024\dfrac{\partial^2 u_1}{\partial x^2} - F(u_1 - u_2), \\ \dfrac{\partial u_2}{\partial t} = 0.17\dfrac{\partial^2 u_2}{\partial x^2} + F(u_1 - u_2) \end{cases}$$

where $F(x) = \mathrm{e}^{5.73x} - \mathrm{e}^{-11.46x}$. The initial conditions are $u_1(x,0) = 1$, $u_2(x,1) = 0$, while the boundary conditions are

$$\frac{\partial}{\partial x}u_1(0,t) = 0, \quad u_2(0,t) = 0, \quad u_1(1,t) = 1, \quad \frac{\partial}{\partial x}u_2(1,t) = 0.$$

Bibliography

[1] Abramowita M, Stegun I A. Handbook of Mathematical Functions with Formulas, Graphs and Mathematical Tables. 9th edition. Washengton D.C.: United States Department of Commerce, National Bureau of Standards, 1970.

[2] Ascher U M, Mattheij R M M, Russel R D. Numerical Solution of Boundary Value Problems for Ordinary Differential Equations. Philadelphia: SIAM Press, 1995.

[3] Ascher U M, Petzold L R. Computer Methods for Ordinary Differential Equations and Differential–Algebraic Equations. Philadelphia: SIAM Press, 1998.

[4] Åström K J. Introduction to Stochastic Control Theory. London: Academic Press, 1970.

[5] Bashford F, Adams J C. An Attempt to Test the Theories of Capillary Action by Comparing the Theoretical and Measured Forms of Drops of Fluid, with an Explanation of the Method of Integration Employed in Constructing the Tables which Give the Theoretical Forms of Such Drops. Cambridge: Cambridge University Press, 1883.

[6] Bellen A, Zennaro M. Numerical Methods for Delay Differential Equations. Oxford: Oxford University Press, 2003.

[7] Bilotta E, Pantano P. A Gallery of Chua Attractors. Singapore: World Scientific, 2008.

[8] Blanchard P, Devaney R L, Hall G R. Differential Equations. 4th edition. Boston: Brooks/Cole, 2012.

[9] Bogdanov A. Optimal control of a double inverted pendulum on a cart. Technical Report CSE-04-006, Department of Computer Science & Electrical Engineering, OGI School of Science & Engineering, OHSU, 2004.

[10] Brown C. Differential Equations — A Modeling Approach. Los Angeles: SAGE Publications, 2007.

[11] Burger M, Gerdts M. A survey on numerical methods for the simulation of initial value problems with sDAEs. In: Ilchmann A, Reis T, eds., Surveys in Differential–Algebraic Equations IV. Switzerland: Springer, 2017.

[12] Butcher J C. Coefficients for the study of Runge–Kutta integration processes. Journal of the Australian Mathematical Society, 1963, 3: 185–201.

[13] Butcher J C. Numerical Methods for Ordinary Differential Equations. 3nd edition. Chichester: Wiley, 2016.

[14] Caputo M, Mainardi F. A new dissipation model based on memory mechanism. Pure and Applied Geophysics, 1971, 91(8): 134–147.

[15] Chamati H, Tonchev N S. Generalized Mittag-Leffler functions in the theory of finite-size scaling for systems with strong anisotropy and/or long-range interaction. Journal of Physics A: Mathematical and General, 2006, 39: 469–478.

[16] Chicone C. An Invitation to Applied Mathematics: Differential Equations, Modeling and Computation. London: Elsevier, 2018.

[17] Coleman M P. An Introduction to Partial Differential Equations with MATLAB. 2nd edition. Boca Raton: CRC Press, 2013.

[18] Cryer C W. Numerical methods for functional differential equations. In: Schmitt, K, ed., Delay and Functional Differential Equations and Their Applications, New York: Academic Press, 1972.

[19] De Coster C. The lower and upper solutions method for boundary value problems. In: Cañada A, Drábek P and Fonda A, eds., Handbook of Differential Equations — Ordinary Differential Equations, Volume 1. Amsterdam: Elsevier, 2004.

[20] Diethelm K. The Analysis of Fractional Differential Equations: An Application-Oriented Exposition Using Differential Operators of Caputo Type. New York: Springer, 2010.

[21] Doshi H. Numerical techniques for Volterra equations. MathWorks File Exchange #23623, 2015.

https://doi.org/10.1515/9783110675252-012

[22] Enns R H, McGuire G C. Nonlinear Physics with MAPLE for Scientists and Engineers. 2nd edition. Boston: Birkhäuser, 2000.
[23] Enright W H. Optimal second derivative methods for stiff systems. In: Willoughby R A, ed., Stiff Differential Systems. New York: Plenum Press, 1974.
[24] Felhberg E. Classical fifth-, sixth, seventh- and eighth-order Runge–Kutta formulas with stepsize control. Technical report, Washengton D C: NASA Technical Report TR R-287, 1968.
[25] Felhberg E. Low-order classical Runge–Kutta formulas with stepsize control and their applications to some heat transfer problems. Technical report, Washengton D C: NASA Technical Report TR R-315, 1969.
[26] Gear C W. Automatic detection and treatment of oscillatory and/or stiff ordinary differential equations. In: Hinze J, ed., Numerical Integration of Differential Equations and Large Linear Systems, Berlin: Springer-Verlag, 1982, 190–206.
[27] Gladwell I. The development of the boundary-value codes in the ordinary differential equations — Chapter of the NAG library. In: Childs B, eds., Codes for Boundary-Value Problems in Ordinary Differential Equations. Berlin: Springer-Verlag, 1979.
[28] Gleick J. Chaos — Making a New Science. New York: Penguin Books, 1987.
[29] Govorukhin V. Ode87 integrator, MATLAB Central File ID: #3616, 2003.
[30] Hairer E. A Runge–Kutta method of order 10. Journal of the Institute of Mathematics and Its Applications, 1978, 21: 47–59.
[31] Hairer E, Lubich C, Roche M. The Numerical Solution of Differential–Algebraic Systems by Runge–Kutta Methods, Lecture Notes in Mathematics. Berlin: Springer-Verlag, 1980.
[32] Hairer E, Nørsett S P, Wanner G. Solving Ordinary Differential Equations I: Nonstiff Problems. 2nd edition. Berlin: Springer-Verlag, 1993.
[33] Hairer E, Wanner G. Solving Ordinary Differential Equations II: Stiff and Differential–Algebraic Problems. 2nd edition. Berlin: Springer-Verlag, 1996.
[34] Hartung F, Krisztin T, Walther H-O, Wu J. Functional differential equations with state-dependent delays: Theory and applications. In: Cañada A, Drábek P and Fonda A, eds., Handbook of Differential Equations — Ordinary Differential Equations, Volume 3. Amsterdam: Elsevier, 2006.
[35] Hermann M, Saravi M. Nonlinear Ordinary Differential Equations — Analytical Approximation and Numerical Methods. India: Springer, 2016.
[36] Hethcote H W, Stech H W, van den Driessohe P. Periodicity and stability in epidemic models: A survey. In: Busenberg S N and Cooke K L, eds., Differential Equations and Applications in Ecology, Epidemics, and Population Problems. New York: Academic Press, 1981.
[37] Hilfer R. Applications of Fractional Calculus in Physics. Singapore: World Scientific, 2000.
[38] Kalbaugh D V. Differential Equations for Engineers — The Essentials. Boca Raton: CRC Press, 2018.
[39] Kermack W O, McKendrick A G. A contribution to the mathematical theory of epidemics. Proceedings of the Royal Society A, 1927, 115(772): 700–721.
[40] Keskin A Ü. Ordinary Differential Equations for Engineers — Problems with MATLAB Solutions. Switzerland: Springer, 2019.
[41] Kundel P, Mehrmann V. Differential–Algebraic Equations — Analysis and Numerical Solution. Zürich: European Mathematical Society, 2006.
[42] Lapidus L, Aiken R C, Liu Y A. The occurrence and numerical solution of physical and chemical systems having widely varying time constants. In: Willoughby R A, ed., Stiff Differential Systems. New York: Plenum Press, 1974.
[43] Laub A J. Matrix Analysis for Scientists and Engineers. Philadelphia: SIAM Press, 2005.
[44] LeVeque R J. Finite Difference Methods for Ordinary and Partial Differential Equations — Steady-State and Time-Dependent Problems. Philadelphia: SIAM Press, 2007.

[45] Li J C, Chen Y T. Computational Partial Differential Equations using MATLAB. Boca Raton: CRC Press, 2008.
[46] Li Z G, Soh Y C, Wen C Y. Switched and Impulsive Systems — Analysis, Design, and Applications. Berlin: Springer, 2005.
[47] Liao X X, Wang L Q, Yu P. Stability of Dynamic Systems. Oxford: Elsevier, 2007.
[48] Liberzon D, Morse A S. Basic problems in stability and design of switched systems. IEEE Control Systems Magazine, 1999, 19(5): 59–70.
[49] Lorenz E N. Deterministic nonperiodic flow. Journal of the Atmospheric Sciences, 1963, 20(2): 130–141.
[50] Lorenz E N. The Essense of Chaos. Seattle: University of Washington Press, 1993.
[51] MathWorks. Partial differential equation toolbox — User's guide, 2016.
[52] Mazzia F, Iavernaro F. Test set for initial value problem solvers [R/OL]. Technical Report 40, Department of Mathematics, University of Bari and INdAM. http://pitagora.dm.uniba.it/~testset/, 2003.
[53] Molor C B. Numerical Computing with MATLAB. MathWorks Inc, 2004.
[54] Oustaloup A, Levron F, Nanot F, Mathieu B. Frequency band complex non integer differentiator: characterization and synthesis. IEEE Transactions on Circuits and Systems I: Fundamental Theory and Applications, 2000, 47(1): 25–40.
[55] Petráš I, Podlubny I, O'Leary P. Analogue Realization of Fractional Order Controllers. TU Košice: Fakulta BERG, 2002.
[56] Podlubny I. Fractional Differential Equations. San Diego: Academic Press, 1999.
[57] Podlubny I. Matrix approach to discrete fractional calculus. Fractional Calculus and Applied Analysis, 2000, 3(4): 359–386.
[58] Polyanin A D, Zaitsev V F. Handbook of Ordinary Differential Equations — Exact Solutions, Methods and Problems. Boca Raton: CRC Press, 2018.
[59] Richardson L F. Arms and Insecurity. Chicago: Quadrangle Books, 1960.
[60] Routh E J. A Treatise on the Stability of a Given State of Motions. London: Cambridge University Press, 1877.
[61] Scott M R, Watts H A. A systematized collection of codes for solving two-point boundary-value problems. In: Lapidus L and Schiesser W E, eds., Numerical Methods for Differential Systems — Recent Developments in Algorithms, Software, and Applications. New York: Academic Press, 1976.
[62] Shampine L F, Allen Jr R C, Pruess S. Fundamentals of Numerical Computing. New York: John Wiley & Sons, 1997.
[63] Shampine L F, Gladwell I, Thompson S. Solving ODEs with MATLAB. Cambridge: Cambridge University Press, 2003.
[64] Shukla A K, Prajapati J C. On a generalization of Mittag-Leffler function and its properties. Journal of Mathematical Analysis and Applications, 2007, 336: 797–811.
[65] Stroud A H. Numerical Quadrature and Solution of Ordinary Differential Equations. New York: Springer-Verlag, 1974.
[66] Sun Z D, Ge S S. Stability Theory of Switched Dynamical Systems. London: Springer, 2011.
[67] Sun Z Q, Yuan Z R. Control System Computer Aided Design. Beijing: Tsinghua University Press, 1988 (in Chinese).
[68] The MathWorks Inc. Simulink user's manual, 2007.
[69] Van der Pol B, Van der Mark J. Frequency demultiplication. Nature, 1927, 120: 363–364.
[70] Vinagre B M, Chen Y Q. Fractional calculus applications in automatic control and robotics, 41st IEEE CDC, Tutorial workshop 2, Las Vegas, 2002.
[71] Waldschmidt M. Transcendence of periods: the state of the art. Pure and Applied Mathematics Quarterly, 2006, 2(2): 435–463.

[72] Willoughby R A. Stiff Differential Systems. New York: Plenum Press, 1974.
[73] Wuhan University, Shandong University. Computing Methods. Beijing: People's Education Press, 1979 (in Chinese).
[74] Xue D Y. Fractional-Order Control Systems — Fundamentals and Numerical Implementations. Berlin: de Gruyter, 2017.
[75] Xue D Y. FOTF Toolbox. MATLAB Central File ID: #60874, 2017.
[76] Xue D Y. Fractional Calculus and Fractional-Order Control. Beijing: Science Press, 2018 (in Chinese).
[77] Xue D Y, Chen Y Q. System Simulation Techniques with MATLAB and Simulink. London: Wiley, 2013.
[78] Yang H, Jiang B, Cocquempot V. Stabilization of Switched Nonlinear Systems with Unstable Modes. Switzerland: Springer, 2014.
[79] Zhang H G, Wang Z L, Huang W. Control Theory of Chaotic Systems. Shenyang: Northeastern University Press, 2003 (in Chinese).
[80] Zhao C N, Xue D Y. Closed-form solutions to fractional-order linear differential equations. Frantiers of Electrical and Electronic Engineering in China, 2008, 3(2): 214–217.
[81] Zhao X D, Kao Y G, Niu B, Wu T T. Control Synthesis of Switched Systems. Switzerland: Springer, 2017.

MATLAB function index

Bold page numbers indicate where to find the syntax explanation of the function. The function or model name marked by * are the ones developed by the authors. The items marked with ‡ are those downloadable freely from internet.

Index

www.ingramcontent.com/pod-product-compliance
Lightning Source LLC
Chambersburg PA
CBHW080133220326
41598CB00032B/5055
9783110675245